METHODS IN MOLECULAR

Series Editor
John M. Walker
School of Life Sciences
University of Hertfordshire
Hatfield, Hertfordshire, AL10 9AB, UK

For other titles published in this series, go to
www.springer.com/series/7651

Viral Vectors for Gene Therapy

Methods and Protocols

Edited by

Otto-Wilhelm Merten

Généthon, Evry, France

Mohamed Al-Rubeai

School of Chemical and Bioprocess Engineering, and Conway Institute for Biomolecuar and Biomedical Research, University College Dublin, Belfield, Dublin, Ireland

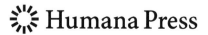

Humana Press

Editors
Otto-Wilhelm Merten
Généthon, Evry
France
omerten@genethon.fr

Mohamed Al-Rubeai
School of Chemical & Bioprocess Engineering
and Conway Institute for Biomolecular and
Biomedical Research, University College Dublin
Belfield, Dublin, Ireland
m.al-rubeai@ucd.ie

ISSN 1064-3745 e-ISSN 1940-6029
ISBN 978-1-4939-5828-3 e-ISBN 978-1-61779-095-9
DOI 10.1007/978-1-61779-095-9
Springer New York Dordrecht Heidelberg London

Humana Press is part of Springer Science+Business Media (www.springer.com)

Preface

The huge potential for gene therapy to cure a wide range of diseases has led to high expectations and a great increase in research efforts in this area. The first human gene therapy protocol was conducted in 1990 by W. French Anderson and showed promising results. Over the following years, more than 1,500 gene therapy protocols were approved for clinical trials, illustrating the rapid growth of this field. Furthermore, with the sequencing of the human genome and the development of advanced technologies for the identification of genes and their function, the number of candidate diseases for gene therapy has continued to increase. However, the efficient transfer of a therapeutic gene into human cells depends upon the technology used for gene therapy. A number of delivery systems are in use, which either involve physical delivery of naked DNA or the use of viral vectors. The protocols of the latter system are the subject of this book.

There is a large and rapidly growing body of literature on methods for gene delivery involving the use of viral vectors. This is because genes are delivered more efficiently by viral vectors, compared to DNA transfection. Vectors derived from retroviruses and adenoviruses are used in the majority of gene therapy clinical trials to date. However, vectors derived from adeno-associated viruses, poxviruses, herpes simplex viruses, and baculoviruses are receiving increasingly more attention in the field of gene therapy. The properties of each of these vectors are described in Chapter 1, while Chapter 2 gives answers based on examples of clinical trials to the question of why gene therapy has not yet become an effective treatment for genetic disease.

Methods in Molecular Biology: Viral Vectors for Gene Therapy brings together the knowledge and experience of those who are employing methodology of virus production, transferring protocols, and evaluating the efficacy of gene product. This is a comprehensive methods book that provides basic principles for the development of gene therapy viral products that are safe and effective. Chapters presenting protocols in readily reproducible, step-by-step fashion, opening with an introductory overview, a list of the materials and reagents needed to complete the experiment, and followed by a detailed procedure that is supported with a helpful notes section offering tips and tricks of the trade as well as troubleshooting advice. There are chapters on production, purification, and characterization of the most popular viral vector systems of adenovirus, retrovirus, and adeno-associated virus (Chapters 5–11). The methodologies are in most cases simple, tested, and robust processes. The protocols for the less common viral vector systems of baculovirus, herpes virus, and measles virus are presented in Chapters 12–14. The growing interest in these vectors has created a strong demand for large-scale manufacturing and purification procedures.

In view of the interest of many laboratories and practitioners in the preclinical and clinical application of gene therapy vectors, it seems appropriate to include chapters to describe protocols on the in vivo gene delivery into CD34 and mesenchymal cells as non-exhaustive examples for in vivo gene transfer (Chapters 15 and 16). In this context, we have also included Chapter 11 on characterization and quality control testing of in vivo gene delivery of AAV viral vectors for the treatment of muscular and eye diseases to present an example on a subject which is still today very much en vogue for most scientists.

Chapter 3 presents basic considerations concerning the characterization of cell banks for the production of viral vectors. It describes the advantages and disadvantages of the most widely used cell lines, HEK293. The importance of viral purification in manufacturing is now widely recognized, and information is presented here (Chapter 4) on the most commonly used purification methods and chromatographic options available for large-scale processes.

Gene therapy raises many unique ethical concerns. Although germ line gene therapy is controversial, somatic gene therapy is morally acceptable for treating diseases since all effects of therapy end with the life of the patient, at the very latest. Chapter 17 explores some of the ethical issues surrounding human gene therapy. The final chapter (Chapter 18) presents examples of clinical trials and examines the processes of good clinical practice, good manufacturing practice, and regulations for conducting gene therapy trials.

Protocols in gene therapy are not well understood by many scientists who will find this book to be of interest. The material is addressed primarily to those interested in viral gene therapy, but topics will also be of interest to scientists in virology, biomedicine, molecular biology, cell culture, preclinical and clinical trials, cell banking, manufacturing, quality control as well as medical practitioners. It will provide an invaluable resource for students and researchers involved in the development of expression systems, gene delivery systems, and therapeutic products. The editors come from industrial gene therapy (O.-W. Merten) and academic bioprocessing (M. Al-Rubeai) backgrounds and are therefore well placed to ensure that the contents are addressed to and understandable by a wide range of readers. We are enthusiastic for the cause of gene therapy – we hope that our readers find inspiration to explore further its potential themselves and that this work helps their rapid progress.

Finally, we thank all the contributors, the series editor John Walker, and Humana Press for their efforts which made this volume possible.

Otto-Wilhelm Merten
Mohamed Al-Rubeai

Contents

Contributors

KARI J. AIRENNE • *Department of Molecular Medicine, A.I. Virtanen Institute, University of Eastern Finland, Kuopio, Finland*

RAMON ALEMANY • *Laboratori de Recerca Traslacional, Institut Català d'Oncologia – IDIBELL, L'Hospitalet de Llobregat, Barcelona, Spain*

MOHAMED AL-RUBEAI • *School of Chemical & Bioprocess Engineering and Conway Institute for Biomolecular and Biomedical Research, University College Dublin, Belfield, Dublin, Ireland*

PAULA M. ALVES • *IBET/ITQB-UNL, Oeiras, Portugal*

ANA I. AMARAL • *IBET/ITQB-UNL, Oeiras, Portugal*

KATHLEEN B. BAMFORD • *Department of Microbiology, Imperial College Healthcare NHS Trust, London, UK; Department of Infectious Diseases and Immunity, Imperial College London, London, UK*

MARLENE CARMO • *ICH, UCL, London, UK*

HAIFENG CHEN • *Virovek, Inc., San Francisco, CA, USA*

MIGUEL CHILLON • *Biochemistry and Molecular Biology Department, Laboratory of Gene Therapy for Autoimmune Diseases, CBATEG, Universitat Autònoma Barcelona, Barcelona, Spain*

ANA S. COROADINHA • *Animal Cell Technology Laboratory, IBET/ITQB-UNL, Oeiras, Portugal*

FRANÇOIS-LOÏC COSSET • *Human Virology Department, INSERM U758, Ecole Normale Supérieure de Lyon, and Université de Lyon 1, Lyon, France*

PEDRO E. CRUZ • *Animal Cell Technology Laboratory, IBET/ITQB-UNL, and ECBIO, Oeiras, Portugal*

CLAIRE DAIGRE • *Department of Agricultural and Biological Engineering, Mississippi State University, Starkville, MS, USA*

PATRICE P. DENÈFLE • *Translational Sciences, IPSEN, and Biotherapies, ParisTech Institute, Paris-Descartes University, Paris, France*

EDWIGE DORMOND • *Baxter Bioscience Manufacturing SARL, Neuchâtel, Switzerland*

ALBERTO L. EPSTEIN • *Université Lyon 1, Lyon, France, and Centre de Génétique et Physiologie Moléculaire et Cellulaire, CNRS, UMR5534, Villeurbanne, France*

MARK J. FEDERSPIEL • *Department of Molecular Medicine, Gene and Virus Therapy Shared Resource, Viral Vector Production Laboratory, Mayo Clinic Comprehensive Cancer Center, Mayo Clinic, Rochester, MN, USA*

CORNEL FRAEFEL • *University of Zurich, Institute of Virology, Zurich, Switzerland*

CECILIA FRECHA • *Human Virology Department, INSERM U758, Ecole Normale Supérieure de Lyon, and Université de Lyon 1, Lyon, France*

MICHAEL FUCHS • *Institut für Wissenschaft und Ethik, University Bonn, Bonn, Germany*
FLORIANE FUSIL • *Human Virology Department, INSERM U758, Ecole Normale Supérieure de Lyon, and Université de Lyon 1, Lyon, France*
ALAIN GARNIER • *Department of Chemical Engineering, Centre de Recherche PROTEO, Université Laval, Laval, QC, Canada*
KOLJA GELSE • *Department of Pediatrics, Nikolaus Fiebiger Center of Molecular Medicine, University of Erlangen-Nürnberg, Erlangen, Germany*
MARC GIRY-LATERRIÈRE • *Faculty of Medicine, Department of Neurosciences, CMU, Geneva, Switzerland*
LINDA C. GREGORY • *Department of Molecular Medicine, Gene and Virus Therapy Shared Resource, Viral Vector Production Laboratory, Mayo Clinic Comprehensive Cancer Center, Mayo Clinic, Rochester, MN, USA*
AMINE A. KAMEN • *Biotechnology Research Institute, NRC, Montreal, QC, Canada*
KIRSTEN K. LANGFIELD • *Department of Molecular Medicine, Gene and Virus Therapy Shared Resource, Viral Vector Production Laboratory, Mayo Clinic Comprehensive Cancer Center, Mayo Clinic, Rochester, MN, USA*
ANSSI J. MÄHÖNEN • *Department of Molecular Medicine, A.I. Virtanen Institute, University of Eastern Finland, and Ark Therapeutics Oy, Kuopio, Finland*
KAISA-EMILIA MAKKONEN • *Department of Molecular Medicine, A.I. Virtanen Institute, University of Eastern Finland, Kuopio, Finland*
PEGGY MARCONI • *Department of Experimental and Diagnostic Medicine, Section of Microbiology, University of Ferrara, Ferrara, Italy*
OTTO-WILHELM MERTEN • *Généthon, Evry, France*
MAURO MEZZINA • *European Association for Scientific Career Orientation (CNRS/EASCO), Paris, France*
TERESA RODRIGUES • *Oxford Biomedica, Oxford, UK*
PATRICK SALMON • *Faculty of Medicine, Department of Neurosciences, CMU, Geneva, Switzerland*
HOLM SCHNEIDER • *Department of Pediatrics, Nikolaus Fiebiger Center of Molecular Medicine, University of Erlangen-Nürnberg, Erlangen, Germany*
MARÍA MERCEDES SEGURA • *Department of Biochemistry and Molecular Biology, Center of Animal Biotechnology and Gene Therapy (CBATEG), Universitat Autònoma de Barcelona, Barcelona, Spain*
GLYN N. STACEY • *National Institute for Biological Standards and Control (An Operating Centre of the Health Protection Agency), South Mimms, UK*
ELS VERHOEYEN • *Human Virology Department, INSERM U758, Ecole Normale Supérieure de Lyon, and Université de Lyon 1, Lyon, France*
HENRY J. WALKER • *Department of Molecular Medicine, Gene and Virus Therapy Shared Resource, Viral Vector Production Laboratory, Mayo Clinic Comprehensive Cancer Center, Mayo Clinic, Rochester, MN, USA*
JAMES N. WARNOCK • *Department of Agricultural and Biological Engineering, Mississippi State University, Starkville, MS, USA*
DAGMAR WIRTH • *Helmholtz Centre for Infection Research, Braunschweig, Germany*

J. FRASER WRIGHT • *Clinical Vector Core, Center for Cellular and Molecular Therapeutics, The Childrens Hospital of Philadelphia and University of Pennsylvania School of Medicine, Philadelphia, PA, USA*
SEPPO YLÄ-HERTTUALA • *Department of Molecular Medicine and Department of Medicine and Gene Therapy Unit, A.I. Virtanen Institute, University of Eastern Finland, Kuopio, Finland*
OLGA ZELENAIA • *Clinical Vector Core, Center for Cellular and Molecular Therapeutics, The Childrens Hospital of Philadelphia, Philadelphia, PA, USA*

Chapter 1

Introduction to Viral Vectors

James N. Warnock, Claire Daigre, and Mohamed Al-Rubeai

Abstract

Viral vector is the most effective means of gene transfer to modify specific cell type or tissue and can be manipulated to express therapeutic genes. Several virus types are currently being investigated for use to deliver genes to cells to provide either transient or permanent transgene expression. These include adenoviruses (Ads), retroviruses (γ-retroviruses and lentiviruses), poxviruses, adeno-associated viruses, baculoviruses, and herpes simplex viruses. The choice of virus for routine clinical use will depend on the efficiency of transgene expression, ease of production, safety, toxicity, and stability. This chapter provides an introductory overview of the general characteristics of viral vectors commonly used in gene transfer and their advantages and disadvantages for gene therapy use.

Key words: Adenovirus, Adeno-associated virus, Lentivirus, Retrovirus, Baculovirus, Poxvirus, Herpes virus, Virus infection, Virus structure

1. Introduction

The success of gene therapy relies on the ability to safely and effectively deliver genetic information to target cells, through either an ex vivo or an in vivo route. The former requires target cells to be extracted from the patient, transfected with the therapeutic gene, and returned to the patient once the gene transfer is complete. The in vivo route requires the vector to be introduced into the host, where it transduces target cells within the whole organism. Gene transfer has traditionally been achieved by the use of either viral or nonviral vectors. While nonviral methods are generally considered to be safer than viral transduction (1, 2), the production yield for plasmid DNA needs to be increased, and costs need to be decreased to make this a commercially viable gene-delivery method (3, 4); in addition, the gene transfer efficiency has to be improved. Consequently, only 17.9% of gene

Otto-Wilhelm Merten and Mohamed Al-Rubeai (eds.), *Viral Vectors for Gene Therapy: Methods and Protocols*,
Methods in Molecular Biology, vol. 737, DOI 10.1007/978-1-61779-095-9_1,

therapy clinical trials employ naked or plasmid DNA, whereas 45% of trials use either retroviral or adenoviral vectors (http://www.wiley.co.uk/genmed/clinical/).

Viruses have complex and precise structural features, which have adjusted through natural evolution for efficient transfection of specific host cells or tissues (5). A number of virus types are currently being investigated for use as gene-delivery vectors. These include adenoviruses (Ads), retroviruses (γ-retroviruses and lentiviruses), poxviruses, adeno-associated viruses (AAV), and herpes simplex viruses (HSV) (6). It is unlikely that any one of these vectors will emerge as a suitable vector for all applications. Instead, a range of vectors will be necessary to fulfill the objectives of each treatment (7).

2. Adenoviruses

Adenovirus (Ad) was first discovered in 1953 in human adipose tissue (8). This virus has since been classified into six species (A–F) that infect humans, and these species are subdivided into over 50 infective serotypes (9). From the variety of known Ads, researchers have concluded that viruses Ad2 and Ad5 of species C are the most effective for creating viral vectors for use in gene therapy (10). Ad vectors, now one of the most widely studied vector forms, are prominently used in worldwide clinical trials. As of March 2011, 402 of the total 1,703 gene-therapy clinical trials included studies with Ad vectors (http://www.wiley.co.uk/genetherapy/clinical).

2.1. Structure

2.1.1. The Capsid

The Ad capsid is a nonenveloped, icosahedral protein shell (70–100 nm in diameter) that surrounds the inner DNA-containing core. The capsid comprises 12 identical copies of the trimeric hexon protein (9). A pentameric penton base protein is located at each vertex of the capsid, and from it extends a trimeric fiber protein that terminates in a globular knob domain, as seen in Fig. 1 (11).

2.1.2. The Genome

The genome of the Ad is a linear, double-stranded DNA (dsDNA) ranging from 26 to 40 kb in length (12). This linear form is organized into a compact, nucleosome-like structure within the viral capsid and is known to have inverted terminal repeat (ITR) sequences (103 base pairs in length) on each end of the strand (11). The viral genome comprises two major transcription regions, termed the early region and the late region (13, 14). The early region of the genome contains four important transcription units (E1, E2, E3, and E4). Table 1 outlines the functions of each unit of the early region.

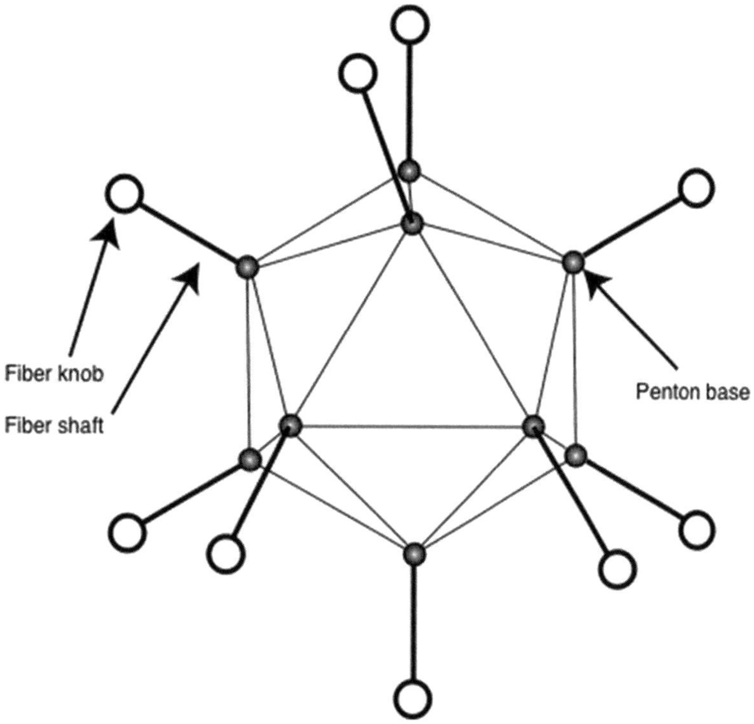

Fig. 1. Schematic diagram of an Ad capsid. The major structural protein of the capsid is the hexon. Penton capsomers, formed by association of the penton base and fiber, are localized at each of the 12 vertices of the Ad capsid.

Table 1
Early transcription units and their functions (Ad virus)

Transcription unit	Function
E1A	Activates early-phase transcription and induces the S phase of the host cell
E1B	Codes for E1B 19K and E1B 55K, which inhibit apoptosis and allow for viral replication
E2	Codes for DNA polymerase (pol), preterminal protein (pTP), and DNA-binding protein (DBP)
E3	Codes for proteins that block natural cellular responses to viral infection
E4	Codes for a variety of proteins that perform in DNA replication, mRNA transport, and splicing

2.2. Life Cycle

The early phase of adenoviral DNA invasion begins when the virus comes in contact with a host cell and ends at the onset of DNA replication. The globular knob domain of the viral capsid has a high affinity for the coxsackievirus and adenovirus receptor (CAR), which can be found on a variety of cells throughout the human body (15, 16). When the virus locates a host cell, the process of binding and internalization begins. The virus–host cell affinity between the fibrous knob and the CAR is heightened by the interaction of the penton base protein with secondary cellular receptors. The virus then travels through the cell membrane via receptor-mediated endocytosis, the virion is released, and the genome escapes the protein capsid and makes its way into the host cell nucleus, as depicted in Fig. 2.

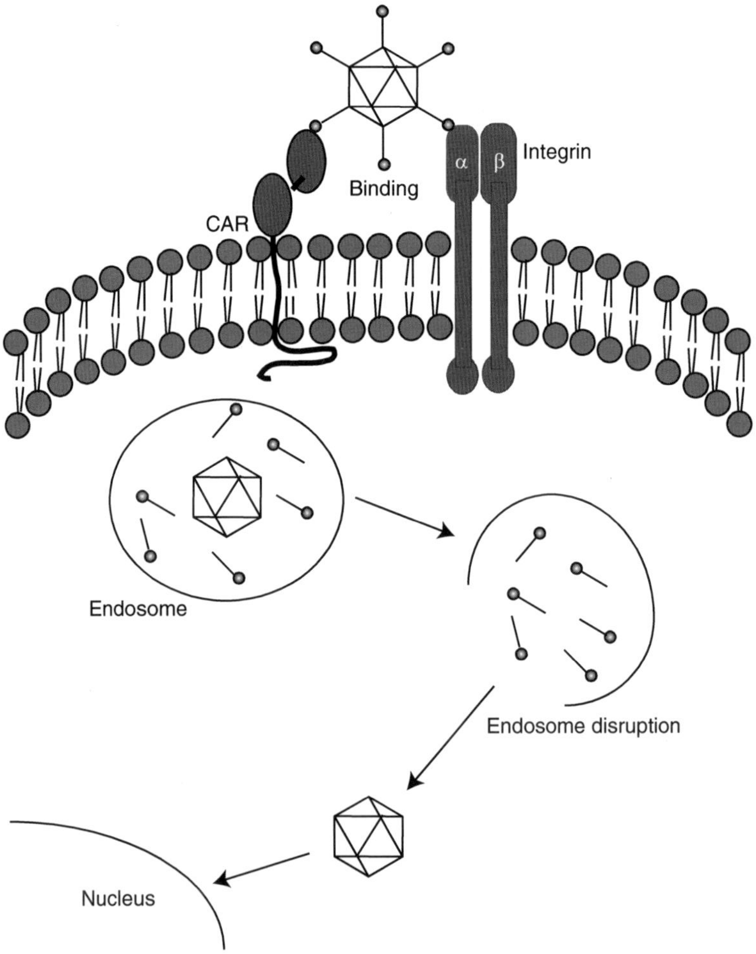

Fig. 2. Schematic representation of adenoviral infection. The Ad virion attaches to the host cell surface by CAR and integrin receptors. The virus enters the cell through clathrin-mediated endocytosis before viral DNA replication and transcription occur in the host nucleus.

Transcription of viral DNA begins when the genome enters the host cell nucleus. At this time, the E1A transcription unit of the early phase is transcribed, followed quickly by the E1B unit. Together, these two units help to prepare the genome for further transcription, shift the host cell into the S phase of replication, and inhibit apoptosis of the host cell. The E2 unit, the next to transcribe, encodes for DNA polymerase, a preterminal protein, and a DNA-binding protein, all of which are necessary for DNA replication. This process is followed by the transcription of the E3 unit, which inhibits the host cell from responding to the viral invasion. Finally, the E4 unit is transcribed to produce a variety of proteins required for DNA replication and movement into the late phase.

The late phase begins at the onset of viral DNA replication. This process begins at the origins of replication in the ITRs on either end of the viral genome, and the terminal protein at each end of the chromosome acts as the DNA primer. The products of late-phase transcription are expressed after a 20-kb section of the major late promoter has been transcribed. This section then undergoes multiple splicing cycles to return five encoding proteins of the late mRNA. These proteins are later used either to form the viral capsid or to assist in assembling the viral progeny. The host cell finally disintegrates and the virus is released.

The first-generation vectors are the most commonly used viral vectors in gene-therapy trials (11). These vectors, based on Ads 2 and 5 of species C, have the E1 region of the genome deleted to allow more genomic space for foreign DNA (10, 17). The E3 region may also be deleted for viral DNA to be replicated in culture. These eliminations allow the insertion of approximately 7.5 kb of DNA into the vector. Another vector form used in gene therapy is known as the "gutted" vector, in which all adenoviral DNA is excised except for the ITRs and packing signals. These vectors allow up to 36 kb of foreign DNA to be accommodated within the viral vector.

2.3. Preclinical Gene Transfer and Clinical Trials

Adenoviral vectors have many benefits that account for their growing popularity in gene-therapy trials; however, they also have some limitations that must be overcome before they can be used for a wide range of treatment options. Some of these advantages and disadvantages are listed in Table 2.

The Ad vector is most commonly associated with studies of cancer treatment. In one study, these vectors successfully delivered tumor suppressor genes p53 and p16 to tumor growths. The Ad vector responsible for the delivery of the p53 gene was the first to be approved for gene-therapy treatment (18). Suicide gene therapy, or prodrug therapy, has also been studied as a cancer treatment option. Suicide therapy uses viral proteins to metabolize

Table 2
Advantages and disadvantages of adenoviral vectors

Advantages	Disadvantages
Ability to infect both dividing and quiescent cells	Long-term correction not allowed
Stability of recombinant vectors	Humoral and cellular immune response from high vector doses
Large insert capacity	
Nononcogenic	
Can be produced at high titers	

nontoxic drugs into a toxic form, resulting in cell death. Recently, a phase I/II suicide-gene-therapy clinical trial has been completed in prostate-cancer patients, using an E1/E3-deleted replication-deficient Ad (CTL102) encoding the bacterial nitroreductase enzyme in combination with prodrug CB1954 (19). A total of 19 patients received virus plus prodrug, and 14 of these had a repeat treatment. Minimal toxicity was observed in patients, including those that received repeated dosages. The greatest reduction in prostate-specific antigen (PSA) was 72%; however, less than 40% of patients showed a PSA reduction greater than 10% (20). Furthermore, an increased frequency of T cells recognizing PSA was detected in 3 out of 11 patients following therapy, suggesting that this direct cytotoxic strategy can also stimulate tumor-specific immunity (19).

Gene therapy using adenoviral vectors has also been employed in the study of various liver diseases because of the vector's ability to affect nondividing cells and its high concentration in the liver after administration (21, 22). A recent study has assessed the therapeutic effect of an Ad vector carrying PAI-1 small interfering RNA (siRNA) on hepatic fibrosis. Histological and immunohistochemical analysis showed a significant reduction of liver fibrosis in rats that received the vector. The vector was able to correct the levels of matrix metalloproteinases and their inhibitors and to stimulate hepatocyte proliferation while concurrently inhibiting apoptosis (23).

Other popular research done with Ad vectors includes studies of stem cell differentiation (24), AIDS (25), cardiovascular disease (26), and pulmonary tuberculosis (27, 28). Adenoviral vectors have been widely studied and are likely to be prominent in the future of gene therapy.

3. Adeno-Associated Virus

AAV originates from the *Dependovirus* genus of the *Parvovirus* family and was first discovered in 1965 as a coinfecting agent of the Ad (29). This small virus is naturally replication-defective and requires the assistance of either a helper virus, such as the Ad or the herpes virus, or some form of genotoxic stress to replicate within a host cell nucleus (30).

3.1. Structure

3.1.1. The Capsid

The AAV capsid is a nonenveloped, icosahedral protein shell, 22 nm in diameter (30). Each serotype of this virus has its own characteristic capsid with a special affinity for certain host cell receptors, allowing it to be used to target a variety of tissue types (31–33).

3.1.2. The Genome

The genome of AAV is composed of a linear, single-stranded DNA with two open reading frames flanked on each end by a 145-bp ITR sequence (30–32). The 5′ open reading frame contains nucleotides that code for four important replication proteins, Rep 78, Rep 68, Rep 52, and Rep 40. The 3′ open reading frame codes for three capsid proteins, VP1, VP2, and VP3. Table 3 outlines the functions of each of these proteins.

3.2. Life Cycle

AAV serotype 2 is the most commonly used AAV vector in gene-therapy clinical trials. The life cycle of this viral serotype begins with the binding of the viral capsid to the host cell via negatively charged heparan sulfate proteoglycans (HSPGs); this attachment is enhanced by coreceptor integrins and various growth factor receptors (29) that help to bind the viral vector to the host cell surface. The vector is taken up by the cell through clathrin-mediated endocytosis (30, 31). Internalization is quickly followed

Table 3
Functions of Rep and Cap proteins (AAV)

Protein	Function
Rep 40 Rep 52	Participate in the generation and accumulation of single-stranded viral genome from the double-stranded replicative intermediates
Rep 68 Rep 78	Interact with Rep-binding elements and ITR terminal resolution sites to assist in the DNA replication process
Cap (vp1, vp2, vp3)	All share the same V3 regions but have different N-termini – used to form the capsid structure in a ratio of 1:1:10

by acidification of the endosome and release of the viral genome. It is not fully understood how the viral genome is able to integrate with the host cell nucleus; however, researchers have found that a helper virus is required to penetrate the host nuclear membrane before the AAV genome can begin replication (29–31). Once inside the nucleus, the AAV DNA integrates with the S1 site of chromosome 19 (33) and replication commences, producing the four Rep proteins and the three Cap proteins outlined in Table 3.

The process for creating vectors from AAVs begins with the deletion of genes coding for the Rep and Cap proteins. This deletion provides approximately 5 kb of packing space for foreign DNA. The new DNA is inserted into the "gutted" virus that contains only the ITRs. The ITRs contain all *cis*-acting elements necessary for replication and packaging in the presence of a helper virus. The Rep and Cap proteins and all necessary adenoviral helper genes are expressed on either one or two plasmids. The expression of Ad genes from a plasmid eliminates the need for coinfection with wild-type adenovirus. Production of AAV vectors requires cotransfection of human embryonic kidney cells (HEK293) with the gutless AAV and one or two helper plasmids (29, 34, 35).

3.3. Preclinical Gene Transfer and Clinical Trials

Adeno-associated viral vectors are most widely used in tissue engineering studies. For such applications, these vectors possess a wide range of advantages; however, some obstacles must still be overcome for these vectors to become commercially approved and be available for treatment. A list of the benefits and limitations of the AAV vectors may be found in Table 4.

Table 4
Advantages and disadvantages of AAV vectors (36)

Advantages	Disadvantages
Nonpathogenic	Smaller size limits the amount of foreign genes that can be inserted
Broad host and cell type tropism range	Slow onset of gene expression[a]
Transduce both dividing and nondividing cells	
Maintain high levels of gene expression over a long period of time (years) in vivo	

[a] *Note*: In the case that single-stranded AAV vectors are used; using self-complementary AAV vectors (double-stranded AAV vectors), the gene expression is more rapid, as the transduction is independent of DNA synthesis (37)

In various animal studies, AAV vectors have been used to treat skin burns (38), excision wounds (39), and incision wounds (40) and have shown great promise for the future. Researchers have also found AAV vectors to be stable in various tissues, including the brain (41), as well as in many different cell types, including muscle (42) and retina cells (43).

The wide range of tissues that can be affected by AAV vectors is due in large part to the unique capsid of each AAV serotype. For example, AAV2 (the most commonly used AAV serotype) has a high affinity for HSPGs (44) – receptors found in a variety of cell types – whereas AAV5 will bind to the platelet-derived growth factor receptor (PDGFR), commonly found on the cells of the brain, lung, and retina (45, 46). Other serotypes whose receptors have not been determined still show an obvious affinity for specific cell types. For example, AAV1, AAV6, AAV7, and AAV8 are attracted to muscle, lung, muscle and liver, and liver cells, respectively (47–49). Further studies have been done with the so-called "mosaic" serotypes, where researchers combined two different AAV vectors and discovered that these mosaics often maintained affinity for the receptors associated with both serotypes (50, 51). Once inside the host cell, rAAV vectors stay mostly episomal (in both human and nonhuman cells) (52). However, stable expression of the vector is possible for extended time periods, often in excess of 1 year, for several cell types including brain (41), muscle (42), and eye (43).

Unlike vectors used in other gene-therapy trials, the main focus of AAV trials has been on monogenetic diseases (53%), followed by cancer (23%) (36). Cystic fibrosis is the most frequently targeted disease. Repeated administration of aerosolized AAV vector containing the cystic fibrosis transmembrane regulator (AAV-CFTR) is well tolerated and safe (53, 54); however, in a phase 2B clinical trial, no statistically significant improvement was seen in patients receiving AAV-CFTR compared to placebo (55). AAV vectors have also been used to treat hemophilia B with some success. In a phase 1/2 dose-escalation clinical study, rAAV-2 vector expressing human F.IX was infused through the hepatic artery into seven subjects. There was no acute or long-lasting toxicity observed at the highest dose, which was able to produce a therapeutic effect. However, in contrast to previous work performed in dogs (56), the expression of therapeutic levels of F.IX only lasted 8 weeks as a result of immunogenic destruction of hepatocytes expressing the AAV antigen (57). Other diseases that have been treated with rAAV vectors are Canavan disease (58), infantile neuronal ceroid lipofuscinosis (59), Parkinson's disease (60), and α1-anti-trypsin deficiency (61).

4. Retroviruses

Retroviruses are known for their ability to reverse the transcription of their single-stranded RNA genome, thus creating dsDNA to replicate after infecting host cells. These viruses are most generally categorized as either simple (oncogenic retroviruses) or complex (lentiviruses and spumaviruses) (62). This section discusses the simple oncogenic retroviruses – most commonly the murine leukemia virus – before discussing the complex retroviruses in the lentivirus section. The oncogenic retroviruses are limited by their inability to infect non-dividing cells; however, they are considered extremely useful for tissue engineering studies, particularly those concerning bone repair.

4.1. Structure

4.1.1. The Capsid

The retroviral capsid is an enveloped protein shell that is 80–100 nm in diameter and contains the viral genome (52). The envelope structure surrounding the capsid is actually a lipid bilayer that originates from the host cell and contains both virus-encoded surface glycoproteins and transmembrane glycoproteins (63). The basic retroviral structure is similar to lentiviruses (HIV-1 – shown in Fig. 3).

4.1.2. The Genome

The genome of the retrovirus is a linear, nonsegmented, single-stranded RNA, 7–12 kb in length (62). The simple class of retroviruses contains three major coding segments and one small coding domain. The major segments contain three genes – gag, pol, and env – which code for proteins important in viral integration,

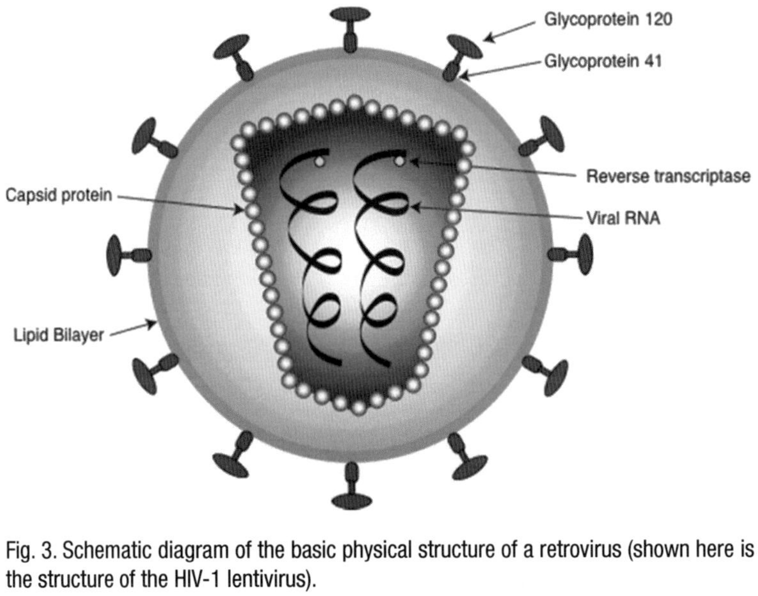

Fig. 3. Schematic diagram of the basic physical structure of a retrovirus (shown here is the structure of the HIV-1 lentivirus).

Table 5
Functions of retroviral genes

Protein	Function
gag	Codes for the viral core
pol	Codes for reverse transcriptase and integrase
env	Codes for surface and transmembrane components of the viral envelope proteins
pro (small coding domain)	Encodes a viral protease

replication, and encapsulation (52). The small coding domain contains the pro gene, which encodes for viral protease (63). A more detailed description of the coding segments and their protein products may be found in Table 5.

4.2. Life Cycle

The life cycle of the retrovirus begins when the glycoproteins of the viral envelope attach to specific host cell receptors. The viral envelope then fuses with the cellular membrane of the host, and the viral core is released into the host cell cytoplasm. Proteins coded for by the pol gene are then used to begin viral transformation. Viral reverse transcriptase is used to create a dsDNA genome from the original single-stranded RNA of the virus. As the viral dsDNA cannot pass through the nuclear membrane in nondividing cells, the integration of the virus is only possible when the cell is cycling, after the breakdown of the nuclear membrane. Subsequently, viral integrase helps the newly formed dsDNA integrate into the host cell genome, where it will remain a permanent part of the host cell, now known as a provirus. With respect to virus replication, RNA polymerase II transcribes the provirus to mRNA, which codes for viral proteins. After the virus reassembles in the cytoplasm, it escapes the cell by budding out from the cellular membrane, where the capsid receives its envelope (64).

Retroviral vectors must be replication defective. To achieve this, all of the *trans*-acting elements of the genome (gag, pol, and env genes) are removed, leaving only the attachment sites, the long terminal repeat, the packaging signals, and the sites important for viral gene expression. Removal of gag, pol, and env genes provide space for the gene of interest to be inserted into the viral genome. Vector replication can only occur in packaging cell lines. Packaging cells are transfected with plasmids containing the gag, pol, and env genes, which they consistently express allowing for retroviral proliferation (65, 66). To increase safety, the gag and pol genes are contained in one plasmid, while the env gene is contained in another. The vector gene is contained in a third plasmid (split-genome packaging plasmid). This allows multiple regenerations of the vector to be produced without any risk of a replication-competent retrovirus being formed (66).

4.3. Preclinical Gene Transfer and Clinical Trials

Retroviral vectors are widely used in studies of tissue repair and engineering. Because these vectors can be used to infect dividing cells without producing any immunogenic viral proteins while also becoming a permanent part of the host cell genome, they have proven to be an extremely useful tool in gene-therapy research. These vectors are limited only by their relatively small carrying capacity and their inability to infect nondividing cells; however, these disadvantages have not kept them from being the most widely used vectors in the research of gene and cell therapy (67).

One of the most common types of study done with these viral vectors involves bone repair. Current methods of bone grafting are limited by the availability of source grafting material and the dangers of disease transfer. However, retroviral vectors have recently been used to deliver various growth factors and differentiation factors to both mature bone cells and stem cells that have been used in tissue scaffolding, and various animal studies have yielded promising results (68, 69). Retroviral vectors have also been used in the repair of damaged cartilage (70, 71) and in the formation of tissue-engineered blood vessels for the treatment of cardiovascular disease (72).

In addition, retroviral gene therapy has also been used in clinical trials, among others, to treat X-linked severe combined immunodeficiency (X-SCID) in infants and preadolescents (73, 74). Autologous CD34+ hematopoietic cells were transduced ex vivo with retroviral vectors containing the open reading frame of human *IL2RG* cDNA. Significant improvements in T-cell function have been observed, although one study reported leukemias in four patients secondary to retroviral insertional mutagenesis (75).

5. Lentiviruses

Lentiviruses, a subcategory of the retrovirus family, are known as complex retroviruses based on the details of the viral genome. The most common example of a lentivirus is the human immunodeficiency virus type 1 (HIV-1).

5.1. Structure

The lentiviral capsid is the same as that of the simple retroviruses described in Subheading 4. The lentiviral genome, like that of other retroviruses, contains a single-stranded RNA, 7–12 kb in length (62). However, while the genome contains the same genes

Table 6
Genes expressed in HIV-1 lentivirus in addition to the simple retrovirus genes described in Table 1

Protein	Function
rev	An RNA-binding protein that acts to induce the transition from the early to the late phase of HIV gene expression
tat	An RNA-binding protein that enhances transcription 1,000-fold
nef	Disturbs T-cell activation and stimulates HIV infectivity
vpr	Mediates HIV to infect nondividing cells
vpu	Enhances the release of HIV-1 from the cell surface to the cytoplasm
Vif	A polypeptide necessary for the replication of HIV-1

These genes are nonessential and absent in lentiviral vectors. The *rev* gene along with the simple genes *gag*, *pol*, and *env* are expressed on plasmids that are present in packaging cells

as the simple retroviruses (gag, pol, and env, see Table 5), it also comprises six other genes – two regulatory genes and four accessory genes – that code for proteins important for viral replication, binding, infection, and release. Table 6 outlines each of these six genes and the functions of their expressed proteins. The most common lentiviral vector is made from HIV-1. In these vectors, the original genes present in the simple virus, all four of the additional accessory genes, and one of the regulatory genes are deleted to create space for the insertion of foreign genes (76, 77). In contrast to the simple retroviruses, LV vectors are generally produced by transfection of HEK293 or 293T cells. The first of two necessary helper plasmids contains the gag, pol, and rev genes; the other plasmid contains the env gene (78). A further plasmid brings in the recombinant LV vector sequence.

5.2. Life Cycle

The life cycle of the lentivirus is representative of the retrovirus family, in that the glycoproteins of the viral envelope are attracted to specific cellular receptors; the envelope then fuses with the host cell membrane, and the core is released into the cytoplasm. Soon after this internalization, the single-stranded RNA is transcribed in reverse with the help of viral proteins to form a double-stranded genome that is incorporated into the host genome. However, some important differences do take place in the life cycle of the lentivirus (64). First of all, gene expression occurs in two separate phases, known as the early and late phases, which are separated by the binding of the rev protein (79). Second, the lentivirus is capable of infecting nondividing cells via proteins expressed from the vpr gene (80).

Finally, the tat gene, found only in complex retroviruses, is essential for the replication of HIV-1 (81).

The self-inactivating expression vector (SIN) is another vector form of the lentivirus, in which the U3 promoter is deleted, causing transcriptional inactivation of the provirus. This vector form limits both genome mobility and possibilities of recombination in the host cell (78, 82).

Oftentimes, the vectors used in gene-therapy trials are given an envelope surrounding the capsid structure that is composed of very specific glycoproteins, namely, the vesicular stomatitis virus glycoprotein (VSV-G), which allows the vector a high tropism and the ability to infect a wide variety of cell types (83).

5.3. Clinical Trials

Lentiviral vectors possess many advantages over other simple retroviral vectors. For example, lentiviral vectors can infect mouse and rat embryos to generate transgenic animals with high tissue-specific expression of transgene (84). Since these vectors also have a relatively large carrying capacity for foreign genomic material (52), it is suggested that they can be used to produce other transgenic animal species.

Lentiviral vectors have traditionally been used in studies dealing with nondividing host cells, such as those of the nervous and cardiac systems (85). The first clinical trial using a LV vector was approved in 2002. Since then, eight other protocols have received approval, and 11 others have been submitted or are under review (86). The first trial to be approved was for VRX496™ (lexgenleucel-T) anti-HIV RNA therapy (87). The vector has been shown to be safe and offers short-term efficacy and is currently in phase I/II trials. Other diseases to be targeted with LV vectors are adrenoleukodystrophy (ALD) (88), a progressive neurodegenerative disease that causes diffuse demyelination and primarily affects the CNS, Parkinson's disease (89), sickle cell anemia and β-thalassemia (90), HIV (91), and cancer immunotherapy (92). A comprehensive review of forthcoming clinical trials can be found in D'Costa et al. (86).

6. Baculoviruses

The most commonly studied baculovirus is known as the *Autographa californica multiple nucleopolyhedrovirus* (AcMNPV). It was originally thought that this virus was incapable of infecting mammalian cells; however, in 1983 several studies showed that baculovirus could be internalized by mammalian cells (93, 94), and they were used for the expression of human interferon β (95). Subsequently, AcMNPV has been successfully internalized in a number of human cells, with some of the viral genome reaching the host cell nucleus (96, 97). Though the baculovirus is not the most widely studied virus in gene therapy, it is nonpathogenic to human cell lines and is unable to replicate in mammalian cells. These are considerable safety advantages and may be a distinct advantage over other viral vectors.

6.1. Structure

The baculoviral capsid is a rod-shaped protein shell (40–50 nm in diameter and 200–400 nm in length) that is naturally protected by a polyhedron coat. While this coat does provide viral protection, the virus does not need it to exist. Some of the most studied of these viruses are actually in the "budded" form, which consists of an envelope that surrounds the capsid and that contains glycoproteins essential for viral binding to host cells (98). The genome of the baculovirus is a complex circular, dsDNA containing the genes necessary for viral infection of host cells. An in-depth description of AcMNPV genes can be found in Cohen et al. (99).

6.2. Life Cycle

The life cycle of the baculovirus is still not fully understood; however, researchers have come to some conclusions about its binding, internalization, and nuclear uptake. Though the method of virus–host cell interaction is not clear, researchers do agree that the glycoprotein Gp64 is necessary for this interaction to occur (98). Researchers also agree that the virus is then taken into the cell via some form of endocytosis, possibly clathrin-mediated endocytosis, though some other internalization methods may coexist (97, 100–102). Once inside the cytoplasm, the virus frees itself through acidification of the endosome (96, 103). Scientists agree that the transfer of the nucleocapsid is somehow blocked in the cytoplasm of the host cell, and some say that this is due to the various microtubules throughout the cytoplasm. This conclusion is supported by the fact that transport time decreases when these microtubules have been chemically disintegrated (104). Once the viral genome finally reaches the host cell nucleus, it is ready to be taken up into the cell; however, it appears that this uptake process may vary depending on the type of host cell. Some cells take up the genome directly through nuclear pores, while others seem to transport the viral genome to different subcellular compartments. Still others appear to degrade the viral genome prior to nuclear uptake (105). Though researchers have not yet completely understood the life cycle of the baculovirus, this virus and its capabilities are being continually studied.

6.3. Preclinical Gene Transfer and Clinical Applications

Though baculoviruses are not yet a widely studied vector form, they do possess a number of benefits that have awakened the curiosity of many researchers. First, they do not replicate inside mammalian host cells and are not toxic (106). Second, baculoviral DNA has been known to automatically degrade inside host cells over time (107, 108). Also, because the baculovirus only infects insects and invertebrates, humans do not appear to have preexisting antibodies or T-cells specifically against baculovirus (109). Finally, these viruses may be constructed into vectors with a DNA carrying capacity of up to 38 kb, allowing the delivery of a large amount of foreign genomic material to the host cells (110). The main drawback associated with baculovirus is a rapid, complement-mediated inactivation. To overcome this, researchers have successfully coated virus particles with polyethylenimine, protecting them against complement inactivation (111, 112).

Not only does this virus have a variety of promising characteristics, but it has also been proven to be practical in a number of gene-therapy trials. Baculoviruses have been used in animal studies to deliver genes to a wide range of cell types, including carotid artery (113), liver (114, 115), brain (116, 117), and skeletal muscle (118). This relatively new vector form has already caught the interest of many scientists and will likely play a large role in the future of gene therapy.

7. Herpes Simplex Virus

Actually, many different varieties of the HSV have been discovered. The most common of these, known as HSV-1, is well known by the average person as the viral cause for cold sores. One of the most intriguing aspects of this virus is its ability to infect a host and then remain latent for a period before reappearing again (119). Research on this virus continues in hopes that its unique characteristics will lead to a breakthrough in gene therapy.

7.1. Structure

7.1.1. The Capsid

The HSV has an icosohedral protein shell that is covered by a viral envelope. Embedded within the envelope are a variety of glycoproteins important for the viral attachment to host cellular receptors. Tegument is a layer of proteins and enzymes coded for by the viral genome that lies between the capsid core and the viral envelope (119). Figure 4 illustrates this capsid structure.

7.1.2. The Genome

The HSV genome consists of a dsDNA (152 kb in length) that codes for up to 90 different proteins important for viral attachment and replication (120). This genome is further organized into what are known as unique long and unique short segments, and these segments are capped on each end by inverted repeat sequences (52).

7.2. Life Cycle

One of the most interesting characteristics of the HSV is its ability to remain latent in host cells after the initial infection and then to reappear spontaneously (119). The life cycle of this virus begins

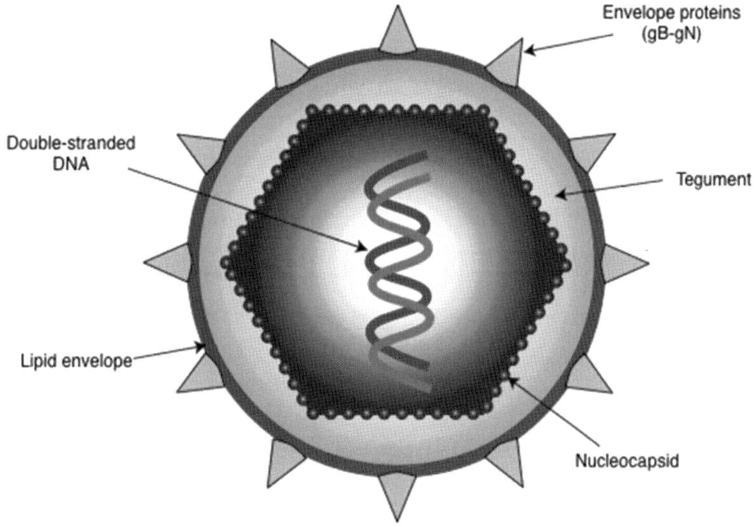

Fig. 4. Schematic diagram showing the structure of the herpes simplex virus.

Table 7
Functions of α, β, and γ genes of HSV-1

Protein class	Function
α	Major transcriptional regulatory proteins – necessary for the synthesis of β and γ proteins
β	Include DNA polymerase and transcriptional factors involved in viral replication
γ	Primarily serve as structural proteins

when it binds to host cell surface receptors via glycoproteins of the viral envelope. The virus is then taken into the cell, where it is delivered to the nucleus. Once the virus reaches the nucleus, the viral capsid binds to the nuclear membrane and releases the viral DNA into the host cell nucleus. Transcription of the viral DNA is a complex process involving multiple steps with the help of a variety of proteins. These proteins are classified into three groups, termed α, β, and γ proteins. The α proteins are also known as the immediate early proteins, and the β proteins are called the early proteins. The γ proteins are referred to as the late proteins, and DNA replication begins following their transcription (63). Table 7 provides a more detailed explanation of these gene products and their functions.

The HSV-1 has been used to develop two different types of viral vectors. The replication-defective vectors work in much the same way as the adenoviral and retroviral vectors. The α genes involved in viral DNA replication are deleted, and the foreign gene of interest is inserted into the viral genome. Helper cells are then used to take the place of the deleted genes, and the vector is ready for infection. The other type of HSV-1 vector is known as the amplicon vector. In this vector form, plasmids that contain the HSV-1 origin of replication, all necessary packaging signals, and the gene of interest are cotransfected with a helper virus and inserted into a cell line that supports the growth of the helper virus. Replication of this vector is prevented by either deletion of the α genes or temperature influence.

7.3. Preclinical Gene Transfer and Clinical Applications

The main advantage of the HSV is its ability to remain latent within host cells after infection. This distinctive feature, along with the fact that the virus is naturally neurotropic, allows it to infect neural cells and, therefore, to assist in treating neural diseases. Most of the animal studies performed with HSV have involved either the treatment of brain tumors or Parkinson's disease. In both cases, gene therapy using the herpes simplex viral vector has shown promising results (121).

8. Poxviruses

The poxvirus is most widely known for its use as a vaccine against smallpox. Recombinant gene expression of this virus was first performed in 1982 (122, 123), and it was one of the first animal viruses to be used as a gene-transfer vector (124). The most commonly studied strains of this virus include the MVA and NYVAC viruses, as they are naturally replication-defective in most human tissues and they lack the ability to produce infectious virus in human host cells (125).

8.1. Structure

The poxvirus capsid is acquired in the host cell cytoplasm after DNA replication. Some of the protein products of viral genome replication form the capsid during viral reassembly. These mature virions are wrapped by an envelope structure that originates from the *trans*-Golgi to form intracellular enveloped viruses. These viruses later fuse with the inner cell membrane, are released from the Golgi envelope, and are reenveloped by the host cell membrane upon escape (126).

The MVA and NYVAC poxviruses have unusually large dsDNA genomes (178 and 192 kb, respectively) (127). Approximately 100 genes are specifically conserved in poxviruses, while the existence of other genes helps to define the different viral strands. All of the genes present in the viral genome are valuable, and few introns, if any, exist in many of the viral strands (128).

8.2. Life Cycle

The life cycle of the poxvirus begins when the glycoproteins of the viral envelope attach to host cell surface receptors. Once the virus enters the cytoplasm, it is thought to form a type of replication center enclosed by the rough endoplasmic reticulum. Unique to the poxvirus, replication of the viral genome actually takes place in the host cell cytoplasm, as opposed to the host cell nucleus (129). Viral gene expression is an extremely complex cascade mechanism that leads to the production of early, intermediate, and late transcription factors, along with structural proteins and various enzymes (130). Immature virions are formed followed by the production of mature virions that are enveloped in a double membrane structure by the *trans*-Golgi and that are released at the host cell membrane and reenveloped by the lipid bilayer that makes up the host cell membrane (126).

8.3. Preclinical Gene Transfer and Clinical Applications

The most widely used poxviral vectors originate from the MVA and NYVAC viral strains. These strains are often chosen because they are naturally replication-defective and unable to produce infective viruses in human tissues (125).

These viruses have long been used as vaccines against diseases such as smallpox but are now being studied as viral vectors to be used against other viral, parasitic, and bacterial diseases, including HIV, West Nile fever, and tuberculosis (131, 132). These viral vectors are used to elicit an immune response against foreign diseases that have become resistant to the drugs once used to kill them (133). However, they are not dangerous because they are naturally replication-defective in human cells.

Poxviral vectors are also being used in immunomodulation gene therapy, in which they can safely deliver tumor-associated antigens to tumor cells, causing an immune response against those tumor cells (124). While not the most commonly studied vector forms, poxviral vectors have proven applicable to the treatment of many diseases for the treatment of various forms of cancer, and researchers are highly interested in their impact on treatment options in the future.

9. Summary

The ability of viruses to deliver foreign DNA to cells for therapeutic purposes has been exploited in numerous different contexts. The diverse nature of different vectors and the variability of different diseases mean that there will almost certainly be no "one size fits all" vector. Clinical trials have shown that certain vectors have great potential for specific diseases. For example, retroviral vectors have had great success in treating X-SCID, whereas lentiviral vectors have been used to target various neurological diseases, including

Parkinson's and ALD, and other clinical trials have employed AAV vectors to treat monogenic disorders, such as Duchenne muscular dystrophy and hemophilia B. Although no viral vector has yet received clinical approval in Europe or the USA, the encouraging results from clinical trials, coupled with continual improvements in vector design and safety, shows that this technology has immense potential.

References

1. Schalk, J. A., Mooi, F. R., Berbers, G. A., van Aerts, L. A., Ovelgonne, H., and Kimman, T. G. (2006) Preclinical and clinical safety studies on DNA vaccines, *Hum. Vaccin* **2**, 45–53.
2. Thomas, C. E., Ehrhardt, A., and Kay, M. A. (2003) Progress and problems with the use of viral vectors for gene therapy, *Nat. Rev. Genet.* **4**, 346–358.
3. Mairhofer, J., and Grabherr, R. (2008) Rational vector design for efficient non-viral gene delivery: challenges facing the use of plasmid DNA, *Mol. Biotechnol* **39**, 97–104.
4. Bower, D. M., and Prather, K. L. (2009) Engineering of bacterial strains and vectors for the production of plasmid DNA, *Appl. Microbiol. Biotechnol.* **82**, 805–813.
5. Navarro, J., Oudrhiri, N., Fabrega, S., and Lehn, P. (1998) Gene delivery systems: Bridging the gap between recombinant viruses and artificial vectors, *Adv. Drug Deliv. Rev.* **30**, 5–11.
6. Bouard, D., Alazard-Dany, D., and Cosset, F. L. (2009) Viral vectors: from virology to transgene expression, *Br. J. Pharmacol.* **157**, 153–165.
7. McTaggart, S., and Al-Rubeai, M. (2002) Retroviral vectors for human gene delivery, *Biotechnol. Adv.* **20**, 1–31.
8. Rowe, W. P., Huebner, R. J., Gilmore, L. K., Parrott, R. H., and Ward, T. G. (1953) Isolation of a cytopathogenic agent from human adenoids undergoing spontaneous degeneration in tissue culture, *Proc. Soc. Exp. Biol. Med.* **84**, 570–573.
9. Shenk, T. (1996) Adenoviridae: The viruses and their replication in *Fields Virology* (Fields, B. N., Ed.), pp 2111-2148, Lippincott-Raven Publishers, Philadelphia, PA.
10. Kovesdi, I., Brough, D. E., Bruder, J. T., and Wickham, T. J. (1997) Adenoviral vectors for gene transfer, *Curr. Opin. Biotechnol.* **8**, 583–589.
11. Douglas, J. T. (2007) Adenoviral vectors for gene therapy, *Mol. Biotechnol.* **36**, 71–80.
12. Campos, S. K., and Barry, M. A. (2007) Current advances and future challenges in Adenoviral vector biology and targeting, *Curr. Gene Ther.* 7, 189–204.
13. Boyer, J., and Ketner, G. (1999) Adenovirus Late Gene Expression, in *Aednovirues: Basic Biology to Gene Therapy* (Seth, P., Ed.), pp 69–77, R. G. Landes Bioscience, Austin, TX.
14. Rux, J. J., and Burnett, R. (1999) in *Adenoviruses: Basic Biology to Gene Therapy* (Seth, P., Ed.), pp 5-15, Medical Intelligence 15. R. G. Landes Bioscience, Austin, TX.
15. Bergelson, J. M., Cunningham, J. A., Droguett, G., Kurt-Jones, E. A., Krithivas, A., Hong, J. S., et al. (1997) Isolation of a common receptor for Coxsackie B viruses and adenoviruses 2 and 5, *Science* **275**, 1320–1323.
16. Tomko, R. P., Xu, R., and Philipson, L. (1997) HCAR and MCAR: the human and mouse cellular receptors for subgroup C adenoviruses and group B coxsackieviruses, *Proc. Natl. Acad. Sci. USA* **94**, 3352–3356.
17. Berkner, K. L. (1988) Development of adenovirus vectors for the expression of heterologous genes, *Biotechniques* **6**, 616–629.
18. Roth, J. A. (2006) Adenovirus p53 gene therapy, *Expert Opin. Biol. Ther.* **6**, 55–61.
19. Onion, D., Patel, P., Pineda, R. G., James, N. D., and Mautner, V. (2009) Anti-Vector and Tumor Immune Responses following Adenovirus Directed Enzyme Pro-drug Therapy for the Treatment of Prostate Cancer, *Hum. Gene Ther.* **20**, 1249–1258.
20. Patel, P., Young, J. G., Mautner, V., Ashdown, D., Bonney, S., Pineda, R. G., et al. (2009) A Phase I/II Clinical Trial in Localized Prostate Cancer of an Adenovirus Expressing Nitroreductase with CB1984, *Mol. Ther.* **17**, 1292–1299.
21. Pastore, L., Morral, N., Zhou, H., Garcia, R., Parks, R. J., Kochanek, S., et al. (1999) Use of a liver-specific promoter reduces immune response to the transgene in adenoviral vectors, *Hum. Gene Ther.* **10**, 1773–1781.
22. Ghosh, S. S., Gopinath, P., and Ramesh, A. (2006) Adenoviral vectors: a promising tool

for gene therapy, *Appl Biochem Biotechnol* **133**, 9–29.
23. Hu, P.-F., Chen, H., Zhong, W., Lin, Y., Zhang, X., Chen, Y.-X., et al. (2009) Adenovirus-mediated transfer of siRNA against PAI-1 mRNA ameliorates hepatic fibrosis in rats, *J. Hepatol* **51**, 102–113.
24. Okuda, T., Tagawa, K., Qi, M. L., Hoshio, M., Ueda, H., Kawano, H., et al. (2004) Oct-3/4 repression accelerates differentiation of neural progenitor cells in vitro and in vivo, *Brain Res. Mol. Brain Res.* **132**, 18–30.
25. Shiver, J. W., Fu, T. M., Chen, L., Casimiro, D. R., Davies, M. E., Evans, R. K., et al. (2002) Replication-incompetent adenoviral vaccine vector elicits effective anti-immunodeficiency-virus immunity, *Nature* **415**, 331–335.
26. Suckau, L., Fechner, H., Chemaly, E., Krohn, S., Hadri, L., Kockskamper, J., et al. (2009) Long-Term Cardiac-Targeted RNA Interference for the Treatment of Heart Failure Restores Cardiac Function and Reduces Pathological Hypertrophy, *Circulation* **119**, 1241–1252.
27. Mata-Espinosa, D. A., Mendoza-Rodriguez, V., Aguilar-Leon, D., Rosales, R., Lopez-Casillas, F., and Hernandez-Pando, R. (2008) Therapeutic Effect of Recombinant Adenovirus Encoding Interferon-[gamma] in a Murine Model of Progressive Pulmonary Tuberculosis, *Mol. Ther.* **16**, 1065–1072.
28. Happel, K. I., Lockhart, E. A., Mason, C. M., Porretta, E., Keoshkerian, E., Odden, et al. (2005) Pulmonary Interleukin-23 Gene Delivery Increases Local T-Cell Immunity and Controls Growth of Mycobacterium tuberculosis in the Lungs, *Infect. Immun.* **73**, 5782–5788.
29. Goncalves, M. A. (2005) Adeno-associated virus: from defective virus to effective vector, *Virol. J.* **2**, 43.
30. Büning, H., Braun-Falco, M., and Hallek, M. (2004) Progress in the use of adeno-associated viral vectors for gene therapy, *Cells Tissues Organs* **177**, 139–150.
31. Ding, W., Zhang, L., Yan, Z., and Engelhardt, J. F. (2005) Intracellular trafficking of adeno-associated viral vectors, *Gene Ther.* **12**, 873–880.
32. McCarty, D. M., Fu, H., Monahan, P. E., Toulson, C. E., Naik, P., and Samulski, R. J. (2003) Adeno-associated virus terminal repeat (TR) mutant generates self-complementary vectors to overcome the rate-limiting step to transduction in vivo, *Gene Ther.* **10**, 2112–2118.
33. Kotin, R. M., Menninger, J. C., Ward, D. C., and Berns, K. I. (1991) Mapping and direct visualization of a region-specific viral DNA integration site on chromosome 19q13-qter, *Genomics* **10**, 831–834.
34. Matsushita, T., Elliger, S., Elliger, C., Podsakoff, G., Villarreal, L., Kurtzman, G. J., et al. (1998) Adeno-associated virus vectors can be efficiently produced without helper virus, *Gene Ther.* **5**, 938–945.
35. Russell, D. W., and Kay, M. A. (1999) Adeno-associated virus vectors and hematology, *Blood* **94**, 864–874.
36. Coura Rdos, S., and Nardi, N. B. (2007) The state of the art of adeno-associated virus-based vectors in gene therapy, *Virol. J.* **4**, 99.
37. McCarty, D. M., Monahan, P. E., and Samulski; R. J. (2001) Self-complementary recombinant adeno-associated virus (scAAV) vectors promote efficient transduction independently of DNA synthesis. *Gene Ther.* **8**, 1248–1254.
38. Galeano, M., Deodato, B., Altavilla, D., Squadrito, G., Seminara, P., Marini, H., et al. (2003) Effect of recombinant adeno-associated virus vector-mediated vascular endothelial growth factor gene transfer on wound healing after burn injury, *Crit. Care Med.* **31**, 1017–1025.
39. Deodato, B., Arsic, N., Zentilin, L., Galeano, M., Santoro, D., Torre, V., et al. (2002) Recombinant AAV vector encoding human VEGF165 enhances wound healing, *Gene Ther.* **9**, 777–785.
40. Galeano, M., Deodato, B., Altavilla, D., Cucinotta, D., Arsic, N., Marini, H., et al. (2003) Adeno-associated viral vector-mediated human vascular endothelial growth factor gene transfer stimulates angiogenesis and wound healing in the genetically diabetic mouse, *Diabetologia* **46**, 546–555.
41. Paterna, J. C., and Bueler, H. (2002) Recombinant adeno-associated virus vector design and gene expression in the mammalian brain, *Methods* **28**, 208–218.
42. Wang, Z., Zhu, T., Qiao, C., Zhou, L., Wang, B., Zhang, J., et al. (2005) Adeno-associated virus serotype 8 efficiently delivers genes to muscle and heart, *Nat. Biotechnol.* **23**, 321–328.
43. Martin, K. R., Klein, R. L., and Quigley, H. A. (2002) Gene delivery to the eye using adeno-associated viral vectors, *Methods* **28**, 267–275.
44. Summerford, C., and Samulski, R. J. (1998) Membrane-associated heparan sulfate proteoglycan is a receptor for adeno-associated virus type 2 virions, *J. Virol.* **72**, 1438–1445.

45. Di Pasquale, G., Davidson, B. L., Stein, C. S., Martins, I., Scudiero, D., Monks, A., et al. (2003) Identification of PDGFR as a receptor for AAV-5 transduction, *Nat. Med.* **9**, 1306–1312.
46. Rabinowitz, J. E., Rolling, F., Li, C., Conrath, H., Xiao, W., Xiao, X., et al. (2002) Cross-packaging of a single adeno-associated virus (AAV) type 2 vector genome into multiple AAV serotypes enables transduction with broad specificity, *J. Virol.* **76**, 791–801.
47. Chao, H., Liu, Y., Rabinowitz, J., Li, C., Samulski, R. J., and Walsh, C. E. (2000) Several log increase in therapeutic transgene delivery by distinct adeno-associated viral serotype vectors, *Mol. Ther.* **2**, 619–623.
48. Gao, G. P., Alvira, M. R., Wang, L., Calcedo, R., Johnston, J., and Wilson, J. M. (2002) Novel adeno-associated viruses from rhesus monkeys as vectors for human gene therapy, *Proc. Natl. Acad. Sci. USA* **99**, 11854–11859.
49. Halbert, C. L., Allen, J. M., and Miller, A. D. (2001) Adeno-associated virus type 6 (AAV6) vectors mediate efficient transduction of airway epithelial cells in mouse lungs compared to that of AAV2 vectors, *J. Virol.* **75**, 6615–6624.
50. Hauck, B., Chen, L., and Xiao, W. (2003) Generation and characterization of chimeric recombinant AAV vectors, *Mol. Ther.* 7, 419–425.
51. Stachler, M. D., and Bartlett, J. S. (2006) Mosaic vectors comprised of modified AAV1 capsid proteins for efficient vector purification and targeting to vascular endothelial cells, *Gene Ther.* **13**, 926–931.
52. Osten, P., Grinevich, V., and Cetin, A. (2007) Viral vectors: a wide range of choices and high levels of service, *Handb. Exp. Pharmacol.*, 177–202.
53. Moss, R. B., Rodman, D., Spencer, L. T., Aitken, M. L., Zeitlin, P. L., Waltz, D., et al. (2004) Repeated adeno-associated virus serotype 2 aerosol-mediated cystic fibrosis transmembrane regulator gene transfer to the lungs of patients with cystic fibrosis: a multicenter, double-blind, placebo-controlled trial, *Chest* **125**, 509–521.
54. Wagner, J. A., Messner, A. H., Moran, M. L., Daifuku, R., Kouyama, K., Desch, J. K., et al. (1999) Safety and biological efficacy of an adeno-associated virus vector-cystic fibrosis transmembrane regulator (AAV-CFTR) in the cystic fibrosis maxillary sinus, *Laryngoscope* **109**, 266–274.
55. Moss, R. B., Milla, C., Colombo, J., Accurso, F., Zeitlin, P. L., Clancy, J. P., et al. (2007) Repeated aerosolized AAV-CFTR for treatment of cystic fibrosis: a randomized placebo-controlled phase 2B trial, *Hum. Gene Ther.* **18**, 726–732.
56. Mount, J. D., Herzog, R. W., Tillson, D. M., Goodman, S. A., Robinson, N., McCleland, M. L., et al. (2002) Sustained phenotypic correction of hemophilia B dogs with a factor IX null mutation by liver-directed gene therapy, *Blood* **99**, 2670–2676.
57. Manno, C. S., Pierce, G. F., Arruda, V. R., Glader, B., Ragni, M., Rasko, J. J., et al. (2006) Successful transduction of liver in hemophilia by AAV-Factor IX and limitations imposed by the host immune response, *Nat. Med.* **12**, 342–347.
58. McPhee, S. W., Janson, C. G., Li, C., Samulski, R. J., Camp, A. S., Francis, J., et al. (2006) Immune responses to AAV in a phase I study for Canavan disease, *J. Gene Med.* **8**, 577–588.
59. Crystal, R. G., Sondhi, D., Hackett, N. R., Kaminsky, S. M., Worgall, S., Stieg, P., et al. (2004) Clinical protocol. Administration of a replication-deficient adeno-associated virus gene transfer vector expressing the human CLN2 cDNA to the brain of children with late infantile neuronal ceroid lipofuscinosis, *Hum. Gene Ther.* **15**, 1131–1154.
60. During, M. J., Kaplitt, M. G., Stern, M. B., and Eidelberg, D. (2001) Subthalamic GAD gene transfer in Parkinson disease patients who are candidates for deep brain stimulation, *Hum. Gene Ther.* **12**, 1589–1591.
61. Flotte, T. R., Brantly, M. L., Spencer, L. T., Byrne, B. J., Spencer, C. T., Baker, D. J., et al. (2004) Phase I trial of intramuscular injection of a recombinant adeno-associated virus alpha 1-antitrypsin (rAAV2-CB-hAAT) gene vector to AAT-deficient adults, *Hum. Gene Ther.* **15**, 93–128.
62. Coffin, J. M., Hughes, S. H., and Varmus, H. E. (1997) *Retroviruses*, Cold Spring Harbor Laboratory Press.
63. Zhang, X., and Godbey, W. T. (2006) Viral vectors for gene delivery in tissue engineering, *Adv. Drug Deliv. Rev.* **58**, 515–534.
64. Buchschacher, G. L. (2004) Safety considerations associated with development and clinical application of lentiviral vector systems for gene transfer, *Curr. Genom.* **5**, 19–35.
65. Buchschacher, G. L., Jr. (2001) Introduction to retroviruses and retroviral vectors, *Somat. Cell Mol. Genet.* **26**, 1–11.
66. Hu, W. S., and Pathak, V. K. (2000) Design of retroviral vectors and helper cells for gene therapy, *Pharmacol. Rev.* **52**, 493–511.

67. Robbins, P. D., Tahara, H., and Ghivizzani, S. C. (1998) Viral vectors for gene therapy, *Trends Biotechnol.* **16**, 35–40.
68. Charles, H. R., Donna, D. S., Shin-Tai, C., Thomas, A. L., Matilda, H. C. S., Jon, E. W., et al. (2008) Retroviral-based gene therapy with cyclooxygenase-2 promotes the union of bony callus tissues and accelerates fracture healing in the rat, *J. Gene Med.* **10**, 229–241.
69. Phillips, J. E., and García, A. J. (2008) Retroviral-mediated gene therapy for the differentiation of primary cells into a mineralizing osteoblastic phenotype, *Methods Mol. Biol.* **433**, 333–354.
70. Li, Y., Tew, S. R., Russell, A. M., Gonzalez, K. R., Hardingham, T. E., and Hawkins, R. E. (2004) Transduction of passaged human articular chondrocytes with adenoviral, retroviral, and lentiviral vectors and the effects of enhanced expression of SOX9, *Tissue Eng.* **10**, 575–584.
71. Tew, S. R., Li, Y., Pothacharoen, P., Tweats, L. M., Hawkins, R. E., and Hardingham, T. E. (2005) Retroviral transduction with SOX9 enhances re-expression of the chondrocyte phenotype in passaged osteoarthritic human articular chondrocytes, *Osteoarthritis Cartilage* **13**, 80–89.
72. Yu, H., Eton, D., Wang, Y., Kumar, S. R., Tang, L., Terramani, T. T., et al. (1999) High efficiency in vitro gene transfer into vascular tissues using a pseudotyped retroviral vector without pseudotransduction, *Gene Ther.* **6**, 1876–1883.
73. Chinen, J., Davis, J., De Ravin, S. S., Hay, B. N., Hsu, A. P., Linton, G. F., et al. (2007) Gene therapy improves immune function in preadolescents with X-linked severe combined immunodeficiency, *Blood* **110**, 67–73.
74. Gaspar, H. B., Parsley, K. L., Howe, S., King, D., Gilmour, K. C., Sinclair, J., et al. (2004) Gene therapy of X-linked severe combined immunodeficiency by use of a pseudotyped gammaretroviral vector, *Lancet* **364**, 2181–2187.
75. Cavazzana-Calvo, M., and Fischer, A. (2007) Gene therapy for severe combined immunodeficiency: are we there yet? *J. Clin. Invest.* **117**, 1456–1465.
76. Naldini, L., Blomer, U., Gallay, P., Ory, D., Mulligan, R., Gage, F. H., et al. (1996) In vivo gene delivery and stable transduction of nondividing cells by a lentiviral vector, *Science* **272**, 263–267.
77. Dull, T., Zufferey, R., Kelly, M., Mandel, R. J., Nguyen, M., Trono, D., et al. (1998) A third-generation lentivirus vector with a conditional packaging system, *J. Virol.* **72**, 8463–8471.
78. Zufferey, R., Dull, T., Mandel, R. J., Bukovsky, A., Quiroz, D., Naldini, L., et al. (1998) Self-inactivating lentivirus vector for safe and efficient in vivo gene delivery, *J. Virol.* **72**, 9873–9880.
79. Malim, M. H., Hauber, J., Le, S. Y., Maizel, J. V., and Cullen, B. R. (1989) The HIV-1 rev trans-activator acts through a structured target sequence to activate nuclear export of unspliced viral mRNA, *Nature* **338**, 254–257.
80. Heinzinger, N. K., Bukinsky, M. I., Haggerty, S. A., Ragland, A. M., Kewalramani, V., Lee, M. A., et al. (1994) The Vpr protein of human immunodeficiency virus type 1 influences nuclear localization of viral nucleic acids in nondividing host cells, *Proc. Natl. Acad. Sci. USA* **91**, 7311–7315.
81. Feng, S., and Holland, E. C. (1988) HIV-1 tat trans-activation requires the loop sequence within tar, *Nature* **334**, 165–167.
82. Miyoshi, H., Blomer, U., Takahashi, M., Gage, F. H., and Verma, I. M. (1998) Development of a self-inactivating lentivirus vector, *J. Virol.* **72**, 8150–8157.
83. Wiznerowicz, M., and Trono, D. (2005) Harnessing HIV for therapy, basic research and biotechnology, *Trends Biotechnol.* **23**, 42–47.
84. Lois, C., Hong, E. J., Pease, S., Brown, E. J., and Baltimore, D. (2002) Germline transmission and tissue-specific expression of transgenes delivered by lentiviral vectors, *Science* **295**, 868–872.
85. Warnock, J. N., Merten, O.-W., and Al-Rubeai, M. (2006) Cell culture processes for the production of viral vectors for gene therapy purposes, *Cytotechnology* **50**, 141–162.
86. D'Costa, J., Mansfield, S. G., and Humeau, L. M. (2009) Lentiviral vectors in clinical trials: Current status, *Curr. Opin. Mol. Ther.* **11**, 554–564.
87. Levine, B. L., Humeau, L. M., Boyer, J., MacGregor, R. R., Rebello, T., Lu, X., et al. (2006) Gene transfer in humans using a conditionally replicating lentiviral vector, *Proc. Natl. Acad. Sci. USA* **103**, 17372–17377.
88. Cartier, N., Hacein-Bey-Abina, S., Bartholomae, C. C., Veres, G., Schmidt, M., Kutschera, I., et al. (2009) Hematopoietic stem cell gene therapy with a lentiviral vector in X-linked adrenoleukodystrophy, *Science* **326**, 818–823.
89. Williams, D. A. (2009) ESCGT 2008: progress in clinical gene therapy, *Mol Ther* **17**, 1–2.
90. Bank, A., Dorazio, R., and Leboulch, P. (2005) A phase I/II clinical trial of beta-globin

gene therapy for beta-thalassemia, *Ann. N Y Acad. Sci.* **1054**, 308–316.

91. Li, M. J., Kim, J., Li, S., Zaia, J., Yee, J. K., Anderson, J., et al. (2005) Long-term inhibition of HIV-1 infection in primary hematopoietic cells by lentiviral vector delivery of a triple combination of anti-HIV shRNA, anti-CCR5 ribozyme, and a nucleolar-localizing TAR decoy, *Mol. Ther.* **12**, 900–909.
92. Chhabra, A., Yang, L., Wang, P., Comin-Anduix, B., Das, R., Chakraborty, N. G., et al. (2008) CD4+CD25- T cells transduced to express MHC class I-restricted epitope-specific TCR synthesize Th1 cytokines and exhibit MHC class I-restricted cytolytic effector function in a human melanoma model, *J. Immunol.* **181**, 1063–1070.
93. Tjia, S. T., zu Altenschildesche, G. M., and Doerfler, W. (1983) Autographa californica nuclear polyhedrosis virus (AcNPV) DNA does not persist in mass cultures of mammalian cells, *Virology* **125**, 107–117.
94. Volkman, L. E., and Goldsmith, P. A. (1983) In Vitro Survey of Autographa californica Nuclear Polyhedrosis Virus Interaction with Nontarget Vertebrate Host Cells, *Appl. Environ. Microbiol.* **45**, 1085–1093.
95. Smith, G. E., Summers, M. D., and Fraser, M. J. (1983) Production of human beta interferon in insect cells infected with a baculovirus expression vector, *Mol. Cell Biol.* **3**, 2156–2165.
96. Boyce, F. M., and Bucher, N. L. (1996) Baculovirus-mediated gene transfer into mammalian cells, *Proc. Natl. Acad. Sci. USA* **93**, 2348–2352.
97. Condreay, J. P., Witherspoon, S. M., Clay, W. C., and Kost, T. A. (1999) Transient and stable gene expression in mammalian cells transduced with a recombinant baculovirus vector, *Proc. Natl. Acad. Sci. USA* **96**, 127–132.
98. Hofmann, C., Lehnet, W., and Strauss, M. (1998) The baculovirus system for gene delivery into hepatocytes, *Gene Ther. Mol. Biol.* **1**, 231–239.
99. Cohen, D. P. A., Marek, M., Davies, B. G., Vlak, J. M., and van Oers, M. M. (2009) Encyclopedia of Autographa californica nucleopolyhedrovirus genes, *Virologica Sinica* **24**, 359–414.
100. Long, G., Pan, X., Kormelink, R., and Vlak, J. M. (2006) Functional entry of baculovirus into insect and mammalian cells is dependent on clathrin-mediated endocytosis, *J. Virol.* **80**, 8830–8833.
101. Matilainen, H., Rinne, J., Gilbert, L., Marjomaki, V., Reunanen, H., and Oker-Blom, C. (2005) Baculovirus entry into human hepatoma cells, *J. Virol.* **79**, 15452–15459.
102. van Loo, N. D., Fortunati, E., Ehlert, E., Rabelink, M., Grosveld, F., and Scholte, B. J. (2001) Baculovirus infection of nondividing mammalian cells: mechanisms of entry and nuclear transport of capsids, *J. Virol.* **75**, 961–970.
103. Hofmann, C., Sandig, V., Jennings, G., Rudolph, M., Schlag, P., and Strauss, M. (1995) Efficient gene transfer into human hepatocytes by baculovirus vectors, *Proc. Natl. Acad. Sci. USA* **92**, 10099–10103.
104. Salminen, M., Airenne, K. J., Rinnankoski, R., Reimari, J., Valilehto, O., Rinne, J., et al. (2005) Improvement in nuclear entry and transgene expression of baculoviruses by disintegration of microtubules in human hepatocytes, *J. Virol.* **79**, 2720–2728.
105. Abe, T., Hemmi, H., Miyamoto, H., Moriishi, K., Tamura, S., Takaku, H., et al. (2005) Involvement of the Toll-like receptor 9 signaling pathway in the induction of innate immunity by baculovirus, *J. Virol.* 79, 2847–2858.
106. Kost, T. A., Condreay, J. P., and Jarvis, D. L. (2005) Baculovirus as versatile vectors for protein expression in insect and mammalian cells, *Nat. Biotechnol.* **23**, 567–575.
107. Ho, Y. C., Chung, Y. C., Hwang, S. M., Wang, K. C., and Hu, Y. C. (2005) Transgene expression and differentiation of baculovirus-transduced human mesenchymal stem cells, *J. Gene Med.* **7**, 860–868.
108. Sung, L. Y., Lo, W. H., Chiu, H. Y., Chen, H. C., Chung, C. K., Lee, H. P., et al. (2007) Modulation of chondrocyte phenotype via baculovirus-mediated growth factor expression, *Biomaterials* **28**, 3437–3447.
109. Strauss, R., Huser, A., Ni, S., Tuve, S., Kiviat, N., Sow, P. S., et al. (2007) Baculovirus-based vaccination vectors allow for efficient induction of immune responses against plasmodium falciparum circumsporozoite protein, *Mol. Ther.* 15, 193–202.
110. Cheshenko, N., Krougliak, N., Eisensmith, R. C., and Krougliak, V. A. (2001) A novel system for the production of fully deleted adenovirus vectors that does not require helper adenovirus, *Gene Ther.* **8**, 846–854.
111. Kim, Y.-K., Choi, J. Y., Jiang, H.-L., Arote, R., Jere, D., Cho, M.-H., et al. (2009) Hybrid of baculovirus and galactosylated PEI for efficient gene carrier, *Virology* **387**, 89–97.
112. Yang, Y., Lo, S.-L., Yang, J., Yang, J., Goh, S. S. L., Wu, C., et al. (2009) Polyethylenimine coating to produce serum-resistant baculoviral

vectors for in vivo gene delivery, *Biomaterials* **30**, 5767–5774.

113. Airenne, K. J., Hiltunen, M. O., Turunen, M. P., Turunen, A. M., Laitinen, O. H., Kulomaa, M. S., et al. (2000) Baculovirus-mediated periadventitial gene transfer to rabbit carotid artery, *Gene Ther.* 7, 1499–1504.
114. Hoare, J., Waddington, S., Thomas, H. C., Coutelle, C., and McGarvey, M. J. (2005) Complement inhibition rescued mice allowing observation of transgene expression following intraportal delivery of baculovirus in mice, *J. Gene Med.* 7, 325–333.
115. Huser, A., Rudolph, M., and Hofmann, C. (2001) Incorporation of decay-accelerating factor into the baculovirus envelope generates complement-resistant gene transfer vectors, *Nat. Biotechnol.* **19**, 451–455.
116. Lehtolainen, P., Tyynela, K., Kannasto, J., Airenne, K. J., and Yla-Herttuala, S. (2002) Baculoviruses exhibit restricted cell type specificity in rat brain: a comparison of baculovirus- and adenovirus-mediated intracerebral gene transfer in vivo, *Gene Ther.* 9, 1693–1699.
117. Wang, C. Y., and Wang, S. (2005) Adeno-associated virus inverted terminal repeats improve neuronal transgene expression mediated by baculoviral vectors in rat brain, *Hum. Gene Ther.* **16**, 1219–1226.
118. Pieroni, L., Maione, D., and La Monica, N. (2001) In vivo gene transfer in mouse skeletal muscle mediated by baculovirus vectors, *Hum. Gene Ther.* 12, 871–881.
119. Baron, S. (1996) *Medical Microbiology*, 4th ed., The University of Texas Medical Branch at Galveston.
120. Kufe, D. W., Frei III, E., Holland, J. F., Weichselbaum, R. R., Pollock, R. E., Bast, R. C., et al. (2006) *Cancer Medicine*, 7th ed., BC Decker, Columbia.
121. Casper, D., Engstrom, S. J., Mirchandani, G. R., Pidel, A., Palencia, D., Cho, P. H., et al. (2002) Enhanced vascularization and survival of neural transplants with ex vivo angiogenic gene transfer, *Cell Transplant.* **11**, 331–349.
122. Mackett, M., Smith, G. L., and Moss, B. (1982) Vaccinia virus: a selectable eukaryotic cloning and expression vector, *Proc. Natl. Acad. Sci. USA* 79, 7415–7419.
123. Panicali, D., and Paoletti, E. (1982) Construction of poxviruses as cloning vectors: insertion of the thymidine kinase gene from herpes simplex virus into the DNA of infectious vaccinia virus, *Proc. Natl. Acad. Sci. USA* **79**, 4927–4931.
124. Gomez, C. E., Najera, J. L., Krupa, M., and Esteban, M. (2008) The poxvirus vectors MVA and NYVAC as gene delivery systems for vaccination against infectious diseases and cancer, *Curr. Gene Ther.* **8**, 97–120.
125. Tartaglia, J., Perkus, M. E., Taylor, J., Norton, E. K., Audonnet, J. C., Cox, W. I., et al. (1992) NYVAC: a highly attenuated strain of vaccinia virus, *Virology* **188**, 217–232.
126. Moss, B. (2006) Poxvirus entry and membrane fusion, *Virology* **344**, 48–54.
127. Antoine, G., Scheiflinger, F., Dorner, F., and Falkner, F. G. (1998) The complete genomic sequence of the modified vaccinia Ankara strain: comparison with other orthopoxviruses, *Virology* **244**, 365–396.
128. Upton, C., Slack, S., Hunter, A. L., Ehlers, A., and Roper, R. L. (2003) Poxvirus orthologous clusters: toward defining the minimum essential poxvirus genome, *J. Virol.* 77, 7590–7600.
129. Tolonen, N., Doglio, L., Schleich, S., and Krijnse Locker, J. (2001) Vaccinia virus DNA replication occurs in endoplasmic reticulum-enclosed cytoplasmic mini-nuclei, *Mol. Biol. Cell.* **12**, 2031–2046.
130. Moss, B. (2001) Poxviridae: the viruses and their replication, in *Fields Virology* 4th ed., pp 2849-2883, Lippincott/The Williams & Wilkins Co, Philadelphia.
131. Fauci, A. S. (2001) Infectious diseases: considerations for the 21st century, *Clin. Infect. Dis.* **32**, 675–685.
132. Morens, D. M., Folkers, G. K., and Fauci, A. S. (2004) The challenge of emerging and re-emerging infectious diseases, *Nature* **430**, 242–249.
133. Sutter, G., and Staib, C. (2003) Vaccinia vectors as candidate vaccines: the development of modified vaccinia virus Ankara for antigen delivery, *Curr. Drug Targets Infect. Disord.* **3**, 263–271.

Chapter 2

Introduction to Gene Therapy: A Clinical Aftermath

Patrice P. Denèfle

Abstract

Despite three decades of huge progress in molecular genetics, in cloning of disease causative gene as well as technology breakthroughs in viral biotechnology, out of thousands of gene therapy clinical trials that have been initiated, only very few are now reaching regulatory approval. We shall review some of the major hurdles, and based on the current either positive or negative examples, we try to initiate drawing a learning curve from experience and possibly identify the major drivers for future successful achievement of human gene therapy trials.

Key words: Gene therapy, Clinical trials, Viral and nonviral approaches, Systemic delivery, Local delivery, Ex vivo gene therapy

1. Three Decades of Human Clinical Gene Therapy

The invention of recombinant DNA technology (1) consequently led to the immediate inception of engineered gene transfer into human cells, aiming at reversing a cellular dysfunction or creating new cellular function. The concept of direct therapeutic benefit based on a gene defect correction in human cells or on gene therapy was born.

Exactly 30 years ago, Martin Cline made a first early and certainly premature human gene therapy attempt in 1979 at treating severe thalassemia patients through an ex vivo β-globing gene transfer protocol in the bone marrow of two patients in Italy and Israel (2). As the protocol had not received any otherwise mandatory approval by regulatory bodies, the study was promptly terminated and Cline was forced to resign his department chairmanship at UCLA (University of California, Los Angeles) and lost several research grants. Subsequently, the Recombinant

Otto-Wilhelm Merten and Mohamed Al-Rubeai (eds.), *Viral Vectors for Gene Therapy: Methods and Protocols*, Methods in Molecular Biology, vol. 737, DOI 10.1007/978-1-61779-095-9_2,

DNA Advisory Committee (RAC) at the National Institute of Health (NIH) was urged in 1980 to expand its regulatory function beyond recombinant DNA experiments so as to include human gene therapy studies.

In 1982, a seminar was held at the Branbury Conference Center of Cold Spring Harbors Labs. A group of scientists, led by Ted Friedmann and Paul Berg, came together to build the foundations of gene therapy and to draw what its future might be. As an outcome, the first book on gene therapy (3) was and is still a landmark reference to this field.

In 1989, Rosenberg et al. initiated the first RAC-approved gene therapy clinical trial, which was actually a "gene-labeling" study targeting a neomycin-resistance gene transfer into tumor-infiltrating lymphocytes using a retroviral construct, for the treatment of metastatic melanoma with Interleukin-2 (4).

Effectively, a therapeutic gene clinical trial took place in 1990 to treat severe combined immunodeficiency (SCID) by transferring the adenosine deaminase (ADA) gene into T-cells using a retroviral vector. No significant clinical benefit was observed, albeit the protocol appeared to be safe for the patients (5, 6).

These pioneer clinical studies, as well as some others, landmarked the inception in the 1990s of a major burst of academic, clinical, biotechnological, and sustained financial efforts lasting for more than two decades (7). Even today, there are thousand clinical trials registered as ongoing. Among which, 65 trials that are declared in late stage (i.e., phase II–phase III) have proven to be safe and would be in the clinical benefit evaluation phase (Table 1).

Factually, one can also notice a sustained input of about a hundred new clinical trials per year since 1999 (7). This seems in

Table 1
Number of gene therapy trials worldwide (7)

	Gene therapy clinical trials	
Phase	**Number**	**%**
Phase I	928	60.4
Phase I/II	288	18.7
Phase II	254	16.5
Phase II/III	13	0.8
Phase III	52	3.4
Single subject	2	0.1
Total	1,537	

clear contrast with the commonly held opinion that gene therapy would be no longer active because of disengagement, especially from certain large pharmaceutical industries, after a "1990s golden age."

Despite this constant entry flow into clinical trials, the quasi-absence of a registered drug after 20 years is quite compelling and worth revisiting from a pure clinical development strategy perspective.

Most of the initial failures were most probably due to very naive "science-driven" approach to clinical practice, but even today, many projects are simply blocked because of fundamental absence of translational research practice and still a strong under-estimation of some key technical challenges. The rest of this book addresses the fundamentals to be considered at the molecular biology and the bioengineering level, but one should also pay attention to the most standard clinical development parameters, which sometimes are simply lacking in the project development plans.

In the late 1990s, a news & views section in a major journal was entitled: "Gene therapy has been keeping for long pretending to be 5 years from the clinics." With more than a thousand clinical trials launched, the goal is no longer to enter man study for the sake of a nice publication. The goal is set to complete successfully human clinical trials and get to product registration, which we are closer now than ever.

2. Gene Therapy: Definition and Basic Prerequisites

As a source of major hope for many incurable human diseases, the concept of human gene therapy was immediately perceived as the highest promise for curative treatment: a therapy acting at the root of the genetic dysfunction.

The concept of gene therapy is relying on gene intervention. From a pure pharmacokinetic point of view, nucleic acid has a poor cell penetration capacity. For the past 30 years, an incredible armada of viral and nonviral vectors has been engineered to formulate the nucleic-acid-based "active principle." Therefore, virus-derived gene delivery vectors were thought from the beginning to be optimized biomimetic vehicles. However, since they have also evolved under a very high selection environment of infectious agents, humans are also naturally equipped with very sophisticated defense systems. These defense systems, which are often specific to higher primates, cannot be ignored in the context of a gene therapy clinical development plan, especially when it comes to use of a natural human-derived virus. Other hurdles are the active virus loads and the amount of virus particles to be used to achieve therapeutic effects, which combined with the administration route are very difficult to predict in terms of clinical pharmacology and drug safety, imposing extremely careful clinical development protocols.

As foreign DNA cannot stay freely in a dividing cell, it does not get associated with the host DNA replication machinery. On one hand, one has engineered integrative vectors enabling the "therapeutic gene" to be integrated into the host DNA, thereby enabling long-term expression potential (e.g. use of oncoretroviral or lentiviral vectors). A major drawback is the random insertion into the host genome that can lead to serious adverse effect (SAE) (8). On the other hand, one has tailored "nonintegrative vectors," which are mainly used to transfer DNA into quiescent cells but which will be lost after a few replication cycles in dividing cells (e.g. adenoviral or adeno-associated viral vectors).

The nature of target tissue/cell and the length of desired therapeutic effect have, therefore, to be taken into consideration in the gene therapy project charter.

In addition, the routes of administration of a therapeutic principle can have major consequences both in terms of efficacy and safety. Routinely, one classifies gene therapy protocols into three main categories: ex vivo, local in vivo, and systemic in vivo administrations (see Table 2).

In other words, the field has been facing major challenges, from novelty to translational research, which have often been complicated by specific ethical concerns (9) led by the subjective perception of gene therapy practice as a "Sorcerer's apprentice."

For the sake of clarity, we now focus on specific sets of examples, including dead ends, mixed successes to the most promising, clinical studies that are intended to contribute to the frame into which the field should continue to contribute to the improvement of human health.

3. Current Status: Clinical Trials and Case Studies

3.1. Systemic Delivery Has Not Been Delivered

After several years of clinical attempts, lack of clinical efficacy, major SAEs, and often unsurmounted industrial bioproduction issues, one should ask the question of clinical plausibility of systemic gene therapy protocols. The treatment of human diseases often requires systemic administration procedures, and most often oral or intraparenteral routes. Using viral or nonviral approaches via the oral route, no protocol has yet been able to achieve satisfactory results in preclinical studies; therefore, most studies have focused on parental routes. Given the classical multiplicity of infection (MOI) in the range of 10–10,000, authors are considering a routine dose ranging from 10^8 to 10^{15} viral particles per kg of body weight. This effective dose definition immediately triggers

Table 2
Routes of administration used in gene therapy protocols

Routes of administration	Ex vivo	Local in vivo	Systemic
Definition	Gene transfer is performed out of the living organism; the therapeutic agent is the "reinfused cells"	Gene transfer product is injected into a local, and possibly isolated body compartment (intramuscular (IM), intratumoral, locoregional, stereotactic)	Product is administered through oral or intraparenteral route so that it can reach all parts of the body
Examples	SCID-ADA protocol	IM: NV1FGF Locoregional: Duchenne's dystrophy with plasmids or AAV Stereotactic: Parkinson's disease with AAV	AAV-FIX
Comments	Autologous cells are handled in a dedicated processing center The efficacy of treatment is related to the ability of transduced cells to perform sustainable effects	Local administration is preferred if therapeutic benefit can be achieved Local immune reaction can be specific Product leakage has to be documented	Dose-limiting rate and major reaction to large viral load are commonly encountered, generally limiting the practical translational approaches, from mouse-based experiments to human clinical development

several major technical, pharmacological, and immunological hurdles to consider. We can schematically classify them as follows:

- Mastering an industrial bioprocess that is scalable to the Good Manufacturing Practice (GMP)-compliant production of clinical and eventually commercial batches
- Defining a purification process and a formulation that is on line with the vector physicochemical properties and the desired volume to be injected
- Documenting the pharmacokinetics and ADMET (adsorption, desorption, metabolism, elimination, and toxicity) properties of vectors in human at such high doses
- Documenting, in terms of long-term potential side effects, the immunoreactivity against the vector itself or the therapeutic cells, and the fate of the product if it needs to be readministered

Below are two examples of gene therapy concepts that have emerged more than 20 years ago, for which clinical realization is desperately kept on being delayed, i.e., in cystic fibrosis (10) and Duchenne's muscular dystrophy (DMD) (11).

3.1.1. Cystic Fibrosis

Although predominantly used in the pioneering days of CF gene therapy, adenovirus-based vector usage has dropped in the last decade due to poor transduction efficiency in human airway epithelial cells and the inability for readministration. In addition, a study by Tosi et al. raised concerns that antiadenovirus immune responses, in particular cytotoxic T-lymphocyte-mediated (CTL) responses and major histocompatibility complex class I antigen (MHC-I) presentation, may be further enhanced if the host has a preexisting *Pseudomonas* infection (12). These data highlighted potential problems for adenovirus-based vectors in CF gene therapy and definitely confined the use of adenovirus-based vectors for CF gene transfer to upstream research studies.

As a potential alternative to adenovirus, adeno-associated virus (AAV) (13) was assessed for lung transduction in clinical cystic fibrosis gene therapy trials. However, the feasibility of repeated AAV administration is still unresolved, and the limited capacity of AAV to carry the full-length cystic fibrosis transmembrane conductance regulator (CFTR) gene and a suitably strong promoter remains a significant problem. However, Lai et al. (14) have recently shown that the efficiency of AAV *trans*-splicing can be greatly improved through rational vector design and may, therefore, allow the CFTR cDNA to be split between two viral vectors.

So far, two human gene therapy phase I/II protocols have been undertaken with incremental and repeat doses of AAV, up to 2×10^{12} and 2×10^{13} DNase-resistant particles, respectively (13, 15).

In both studies, viral shedding and increases in neutralizing antibodies were observed, but no serious adverse event (8) was associated to the virus administration. Importantly, a significant reduction in sputum IL-8 and some improvement in lung function were noted after the first administration, but not after the second or third administration.

On the basis of these studies, Targeted Genetics Corporation initiated a large repeat-administration multicentric phase IIb study (100 subjects), sufficiently powered to detect significant changes in lung function. Eligible subjects were randomized to two aerosolized doses of either AAV-CF or placebo 30 days apart. The subjects underwent pulmonary function testing every 2 weeks during the active portion of the study (3 months) and were followed for safety for a total of 7 months. No publication is available 4 years after the study was completed, but the company announced that the trial had not met its primary end point and, therefore, the CF program has been discontinued (16).

There may be several reasons for these new disappointing outcomes: (1) As for adenoviral vector, AAV-2 was still too inefficient in reaching airway epithelial cells via the apical membrane, (2) the inverted terminal repeat (ITR) promoter used to drive CFTR expression was not strong enough, and (3) repeat administration of AAV-2 to the lung was actually not possible owing to the mounting of an antiviral immune response. Finally, on the back of previously published AAV-2 aerosolization studies, Croteau et al. (17) evaluated the effects of exposure of healthy volunteers to AAV2. Based on airborne vector particle calculations, the authors estimated exposure to 0.0006% of the administered dose. At such an infradose, no deleterious health effects were detectable, but this underlies the strong requirement in improving the general ADMET properties of the vector system and the necessity to perform these studies even before going into phase I.

Studies are currently underway to assess the feasibility of repeated administration of lentivirus-based vectors into airways by several groups (18, 19), and further data will be needed before the relevance of such viruses for CF gene therapy can be decided. In addition, the safety profile of virus insertion into the genome of airway epithelial cells will have to be carefully monitored.

With the concept that bone marrow-derived hematopoietic or mesenchymal stem cells may have the capacity to differentiate into airway epithelial cells (20), some groups have entered this very challenging and controversial approach for the treatment of CF (21, 22).

On the nonviral side, parallel work had been made regarding the formulation of vectors (23), and the United Kingdom (UK) CF Gene Therapy Consortium clinical trial program has been carefully comparing these agents and is now assessing whether the most efficient currently available nonviral gene transfer agent is able to alter CF lung disease. As the extension of gene transfer achieved is still too small and transient to drive any clear therapeutic benefit, most research for CF gene therapy has returned to the laboratory. In UK, there are no more trials ongoing at present, but it remains the goal of the UK Consortium to work together to meet the challenges and enhance progress to a phase III (large-scale) study this year.

Finally, electroporation and some emerging physical delivery methods such as ultrasound and magnetofection have shown encouraging results in vitro and in rodent models, and again, translational research into larger animal models, such as sheep, and hopefully in the clinic is challenging (24, 25).

In perspective as of today, one can expect the promise for a curative therapy for CFTR may not rely on gene therapy, but on "protein-decay" therapy, with the phase II clinical development of a small molecule, miglustat, by Actelion, which has been shown to slow down the mutated protein degradation and enables it to be exported to the membrane (26).

3.1.2. Duchenne's Muscular Dystrophy

DMD is an X-linked inherited disorder that leads to major systemic muscle weakness and degeneration. Muscle fiber necrosis is related to the dystrophin gene deficiency itself (27). Becker muscular dystrophy (BMD) has clinical picture similar to that of DMD but is generally milder than DMD, and the onset of symptoms usually occurs later. The clinical distinction between the two conditions is relatively easy because (1) less severe muscle weakness is observed in patients with BMD and (2) affected maternal uncles with BMD continue to be ambulatory after age 15–20 years. The cloning of the dystrophin gene opened the door for gene therapy (27–30). However, as in systemic disorders, there are major roadblocks including (1) the large amount of skeletal muscle (basically half the body weight of a healthy human being), (2) the involvement of cardiac and the peritoneal muscles in the disease, and (3) the extremely large size of the dystrophin protein, 427 kDa, encoded by a 79 exons gene (28, 31, 32).

In one study, nine DMD/BMD patients were injected with a naked dystrophin gene-carrying plasmid into the radialis muscle. Patients were divided into three cohorts, each injected with one of following three doses: 200 μg once, 600 μg once, or 600 μg twice (2 weeks apart). Biopsies were then retrieved 3 weeks postinjection, and amplicon DNA could be detected only in 6/9 patients. Patients from the first cohort and one patient from the second cohort exhibited 0.8–8% of weak, complete sarcolemma labeling (29), while 3–26% of muscle fibers showed incomplete/partial labeling. The third group showed 2–5% complete sarcolemma labeling and 6–7% showed partial labeling. There were no observed adverse effects to the treatment. The study concluded that the expression of dystrophin was low (29), and thus, the study was not pursued. One may question why the study was initiated despite the product obviously failed to meet basic efficacy requirements to reach future clinical application and even worse was facing major industrial bioproduction pitfalls given the clinical doses that could be inferred from preclinical studies.

For several years, several preclinical studies have been initiated, and finally several concurrent clinical trials were initiated using various pseudotyped adeno-associated viruses (33) as a vehicle to deliver either truncated versions of the gene (mini or microdystrophin) or an exon-skipping RNA structure, all thought to achieve truncated albeit functional dystrophin protein expression (28, 34, 35). The AAV vector, whatever the serotype, provides superior transduction efficiency to the skeletal muscle but is also a source for potential immune response that remains to be carefully understood (36–38). No conclusive result has been drawn yet from the current clinical studies. However, the intramuscular high-dose pharmacokinetic profile in relevant preclinical models and eventually in humans is yet to be thoroughly documented prior to launching any efficacy clinical gene therapy.

However, the last 5–7 years, reviewed elsewhere (11), have seen unrivaled progress in efficient systemic delivery of synthetic and chemically modified oligonucleotides again used to enforce mutated exon splicing (39). This progress has led to several more clinical trials, which are labeled as "small molecule" trials, i.e., out of the boundaries of gene therapy. The most advanced clinical trial, led by a company called Prosensa in Holland, is completing a phase IIb and has led to finalize a collaborative agreement with GSK in October 2009, marking the return of large pharmaceutical companies in the plain field.

3.2. Gene Therapy Potential Promise to Disease Treatments

The above examples clearly illustrate how gene therapy has progressively moved from "systemic" administration routes toward more pragmatic local administration regimen or to alternative small molecule innovative therapeutics. We now review the most promising local gene therapy clinical protocols.

3.2.1. Parkinson's Disease

Parkinson's disease is primarily due to the local degeneration of nigrostriatal neurons projecting into striatum, and a subsequent shortage of dopamine in this target region. Predisposing and risk factors are numerous but disease mechanism remains unclear. More than a million patients are affected both in Europe and the USA. So far, the main treatment has been oral administration of L-DOPA, a dopamine precursor, but patients generally encounter motor complications after 5 years of treatment. Deep stimulation surgery, therefore, becomes the second phase of disease management for 0.5% of patients in France each year.

The therapeutic challenge is then to trigger continuous release of dopamine into striatum neurons. Gene therapy is a plausible approach, as far as cellular therapies could be. In addition to be continuous, dopamine release should remain local, to avoid dyskinesia effects observed in systemic administration of the precursor in the pharmacologic treatment.

Several clinical trials have been undertaken (40, 41). In California, Avigen, later taken over by Genzyme, initiated a trial with an AAV-vector to express the L-DOPA converting enzyme, and another biotechnology company, Ceregene, conducted a phase I open label study with 12 patients, then a phase II trial with an AAV-based vector expressing neurturin (CERE-120), a neuron survival factor (42). Very recently, Ceregene has reported additional clinical data from a double-blinded, controlled phase II trial of CERE-120 in 58 patients with advanced Parkinson's disease. The company, however, announced that the phase II trial did not meet its primary end point of improvement in the Unified Parkinson's Disease Rating Scale (UPDRS) motor off score at 12 months of follow-up, although several secondary end points suggested a modest clinical benefit. An additional, protocol-prescribed analysis reported focused on further analysis of the data from the 30 subjects who continued to be evaluated under double-blinded conditions for up to 18 months, which indicate increasing effects of CERE-120 over time. A clinically modest but statistically significant treatment effect in the primary efficacy measure (UPDRS motor off; $p = 0.025$), as well as similar effects on several more secondary motor measures ($p < 0.05$), was seen at the 18 months end point. Not a single measure similarly favored sham surgery at either the 12 or 18 months time points. Additionally, CERE-120 appears safe when administered to advanced Parkinson's disease patients, with no significant concerns related to the neurosurgical procedure, the gene therapy vector, or the expression of neurturin in the Parkinson's disease brain. Long-term safety was also performed in a primate model and was satisfactory (43). The company also reported the results of an analysis of neurturin gene expression in the brains from two CERE-120 treated subjects who died of causes unrelated to treatment. These analyses revealed that CERE-120 produced a clear evidence of neurturin expression in the targeted putamen but no evidence for transport of this protein to the cell bodies of the degenerating neurons, located in the substantia nigra. In addition to the known cell loss in Parkinson's disease, and in agreement with the perspectives defined elsewhere (44), these findings suggest that deficient axonal transport in degenerating nigrostriatal neurons in advanced Parkinson's disease impaired transport of CERE-120 and/or neurturin from putaminal terminals to nigral cell bodies, reducing the therapeutic effect of CERE-120.

In parallel to this study, Oxford Biomedica, in collaboration with a group in Hospital H. Mondor in France, has built an equine lentivirus-based vector to express three genes involved in dopamine synthesis. The product (ProSavin) is administered locally to the region of the brain called the striatum, converting cells into a replacement dopamine factory within the brain, thus replacing the patient's own lost source of the neurotransmitter. A phase I/II study was initiated in December 2007 in France with patients with mid- to late-stage Parkinson's disease who are failing on current treatment with L-DOPA but have not progressed to experiencing drug-induced movement disorders called dyskinesia. After a first cohort of three patients who showed no side effect or an antibody response (42), the dose-escalation stage of the study has progressed to the second dose level. The 6-month data from the first dose level suggest ProSavin is safe and well tolerated and showed encouraging evidence of efficacy (42).

3.2.2. Severe Combined Immunologic Disorders

Another successful albeit often controversial is the case of ex vivo gene therapy. This is the case of severe combined immunologic disorders (SCID) treatment. Soon after the first US trial led by Blaese and colleagues (5), a network of European groups led by A. Fischer in France, A. Trascher in the UK, and M. Roncarolo in Italy initiated similar protocols for the treatment of SCID. The successful treatment of the first patients was greeted with a lot of enthusiasm when it was first reported in 2000 and 2002 (45–47). However, this euphoria turned to a serious alert at the end of 2002 when two of the first ten children treated in France developed SAE, described as leukemia-like conditions (48). As demonstrated later, the insertion of the therapeutic DNA into the patient cells had occurred next to one specific locus LMO2 (the proto-oncogene LIP domain only two locus) (49–51). With the news of this devastating event, most SCID-X1 gene therapy trials were placed on hold worldwide. However, in view of patient overall and lack of alternative treatment, some ADA and SCID-X1 trials were pursued, with extremely careful monitoring and better vector types designed so as to reduce the odds of such adverse effect. Work is now focusing on correcting the gene without triggering an insertional oncogenic event.

Between 1999 and 2007, gene therapy has restored the immune systems of at least 26 children with two forms [ADA-SCID (nine children) and SCID-X1 (ten children)) of the disorder, and four of the ten SCID-X1 patients had developed leukemia-related SAE (52). As of today, 20 children have been treated, four of them have developed leukemia-like adverse effects and one patient has unfortunately died from leukemia. From a clinical point of view, patients, who have been able to lead a normal life for periods up to 3 years, should be considered cured by this pioneering gene therapy treatment. Otherwise, 10 years later, none of these 20 children would be alive today without gene therapy.

Based on this clinical success, several important protocols are now entering the clinical stage. A major example is that of the

Wiskott–Aldrich syndrome (WAS), which is a complex primary immunodeficiency disorder associated with microthrombocytopenia, autoimmunity, and susceptibility to malignant lymphoma. At the molecular level, WAS is caused by mutations in the gene encoding the Wiskott–Aldrich syndrome protein (WASP). WASP is a cytosolic adaptor protein mediating the rearrangement of the actin cytoskeleton upon surface receptor signaling, which in turn is instrumental for cognate and innate immunity, cell motility, and protection against autoimmune disease (53). WASP confers selective advantage for specific hematopoietic cell populations and serves a unique role in marginal zone B-cell homeostasis and function (54).

The success of such blood stem cell transplantation is related to the patient's age, the conditioning regimen precell infusion, and the extent of reconstitution postcell reinfusion. Since WASP is expressed exclusively in hematopoietic stem cells, and because WASP exerts a strong selective pressure, gene therapy is expected to cure the disease (55). Cumulative preclinical data obtained from WASP-deficient murine models and human cells indicate a marked improvement of the impaired cellular and immunological phenotypes associated with WASP deficiency. A first clinical trial is currently being conducted with a retroviral construct (55, 56) with a careful monitoring of insertional events (57). However, capitalizing on experience with SCID-ADA and establishing a solid European network, A. Galy and colleagues have engineered, validated, and GMP-produced a very potent lentiviral product (58) and a three-site clinical study is due to start in 2010 (59).

3.3. Two Clear-Cut Examples of Products Successfully Reaching Registration

As stated above, the most promising gene therapy clinical results are obtained with local delivery procedures. In addition to the above examples, two key examples of successful development of candidate drugs up to the phase III are in the field of vascular/metabolic disorders.

3.3.1. Lipoprotein Lipase

The first example is that of lipoprotein lipase gene for the treatment of familial lipoprotein lipase deficiency. The product initially cloned into adenovirus and retroviruses by us in the 1990s (60–62) is now carried onto an AAV vector (63). Very encouraging data have been obtained through a direct multiple intramuscular (IM) injection in the inner limb with corrective expression obtained for several weeks postinjection (64), and the product registration has been started by European Medical Agency (EMA) in January 2010 as a centralized procedure, which is the standard route for all advanced therapies.

3.3.2. Peripheral Vascular Disease

The second example is that of peripheral vascular disease (PVD), which is predominantly affecting the lower extremities. PVD has a relatively low mortality but results in considerable morbidity and disability.

Even though angioplasty and reconstructive surgery are somewhat effective treatment options for many patients with peripheral arterial insufficiency, these procedures are associated with considerable risks, notably restenosis after peripheral angioplasty. In addition, the severity and progressive nature of this disease often limit these treatment options, resulting in persistent, disabling symptoms or limb loss. PVD, therefore, represents an attractive target for a gene therapy approach to restoration of effective limb perfusion in selected patients (65).

Dr. Jeffrey Isner and his colleagues have taken a novel approach (66) to the problem of peripheral artery insufficiency with encouraging results. This group has been at the forefront of angiogenic gene therapy for peripheral artery insufficiency, publishing several studies over the past 15 years that have set the ground (65–69) for the clinical study by Sanofi-Aventis.

Fibroblast growth factor 1, FGF1, is a proangiogenic factor acting on various cellular subtypes, and more particularly involved in preexisting microvessels sprouting, microcapillary network genesis, and arteriolic maturation. Pharmacodynamic studies of an FGF-encoding plasmid (70, 71) in two animal models confirmed the therapeutic potential of such an vector (70, 71). Several preclinical toxicity studies were also performed to document vector lack of integration as well as lack of neither oncogenic nor retinopathic potential of the product.

Two human clinical trials (phase I–IIa) were performed and have documented good tolerance to NV1 FGF as well as local angiogenesis effects limited to the injection point, confirming product safety (72, 73). Consequently, a first phase II double-blinded clinical study was performed with 125 patients, to document product efficacy and has achieved a remarkable twofold reduction of amputation in the treated group vs. placebo (74).

As of today, a large-scale pivotal phase III trial, called TAMARIS, is ongoing (75) since November 2007 (490 patients, 130 clinical centers) to document reduction of amputation and increase of life span. The study is aimed to be completed by July 2010 (76). These results, if proven positive, will most probably result in a long-awaited milestone, i.e., the registration of the first gene therapy product for a large clinical indication.

4. Future Developments and Prospects

Several lines of observations can be drawn from these past 20 years of clinical trials.

First, yet the primordial concept was meant to tackle inheritable genetic disorders, seen as *low-hanging fruits* for a fast clinical proof of concept, most of the clinical protocols have been addressing acquired complex disorders, e.g. cancer, cardiovascular, neurodegenerative diseases.

Second, even though the science was sort of intuitively genuine, clinical gene therapy is now understood as a "difficult" clinical development field, and there is still a trend from private investors to stay away from this area, although major clinical successes are now emerging, such as for the SCID and now peripheral artery diseases (PAD).

Third, the driving force has remained often too long in the hands of academic research, and thus, clinical development has been failing repeatedly because of translational research issues, such as good laboratory practice (GLP) preclinical, clinical development, and GMP lack of expertise.

Fourth, although viral vector are considered as best in class to achieve efficacy in men, major adverse effects have been encountered such as vector-related oncogenesis in some trials and complex immunologic responses to the virus in most of systemic and local administration protocols.

However, watching the drug pipeline from the market approval end, several investigational new drugs are by now registered or close to approval, namely, RTV-ADA treated cells from the treatment of SCID-ADA in Italy (52), the AAV-LPL product in Europe (64), and NV1FGF for the treatment of PAD (76, 77).

In the new perspective of true clinical realization and positive learning experience, the mastering and practical application of the right set of tools such as vector design and scale-up production will become true strategic advantages for future gene therapy projects.

Acknowledgement

In memory of Prof. J. M. ISNER, who has pioneered gene therapy and has opened the avenue to the first ever gene therapy product registration for cardiovascular diseases.

References

1. Cohen, S., Chang, A., Boyer, H. and Helling, R. (1973) Construction of biologically functional bacterial plasmids in vitro. *Proc. Natl. Acad. Sci. USA* **70**, 3240–3244.
2. Friedmann, T. (ed.) (1999) *The development of Human gene therapy*, Cold Spring Harbor Press, Cold Spring Harbor, N.Y.
3. Friedmann, T. (1983) *Gene Therapy Facts and fictions in Biology's New Approaches to disease.*, pp. 91-96, (ed. Friedmann, T.), Cold Spring Harbor Laboratory Press, Cold Spring Harbor, N.Y.
4. Rosenberg, S., Aebersold, P., Cornetta, K., Kasid, A., Morgan, R., and Moen, R. (1990) Gene transfer into humans--immunotherapy of patients with advanced melanoma, using tumor-infiltrating lymphocytes modified by retroviral gene transduction. *N. Engl. J. Med.* **323**, 570–578.
5. Blaese, R.M., Culver, K.W., Miller, A.D., Carter, C.S., Fleisher, T., Clerici, M., et al. (1995) T lymphocyte-directed gene therapy for ADA- SCID: initial trial results after 4 years. *Science* **270**, 475–480.

6. Trent, R.J. and Alexander, I.E. (2004) Gene therapy: applications and progress towards the clinic. *Intern. Med. J.* **34**, 621–625.
7. Gene Therapy Clinbical Trials Worlwide. in *The Journal of Gene Medicine* (Eds Wiley et al, Interscience, 2009).
8. Tsurumi, C., Shimizu, Y., Saeki, M., Kato, S., Demartino, G.N., Slaughter, C.A., et al. (1996) cDNA cloning and functional analysis of the p97 subunit of the 26S proteasome, a polypeptide identical to the type-1 tumor-necrosis-factor-receptor-associated protein-2/55.11. *Eur. J. Biochem.* **239**, 912–921.
9. King, N.M. and Cohen-Haguenauer, O. (2008) En route to ethical recommendations for gene transfer clinical trials. *Mol. Ther.* **16**, 432–438.
10. Nash, P., Whitty, A., Handwerker, J., Macen, J., and McFadden, G. (1998) Inhibitory specificity of the anti-inflammatory myxoma virus serpin, SERP-1. *J. Biol. Chem.* **273**, 20982–20991.
11. Foster, K., Foster, H., and Dickson, J.G. (2006) Gene therapy progress and prospects: Duchenne muscular dystrophy. *Gene Ther.* **13**, 1677–1685.
12. Tosi, M.F., van Heeckeren, A., Ferkol, T.W., Askew, D., Harding, C.V., and Kaplan, J.M. (2004) Effect of Pseudomonas-induced chronic lung inflammation on specific cytotoxic T-cell responses to adenoviral vectors in mice. *Gene Ther.* **11**, 1427–1433.
13. Moss, R.B., Rodman, D., Spencer, L.T., Aitken, M.L., Zeitlin, P.L., Waltz, D., et al. (2004) Repeated adeno-associated virus serotype 2 aerosol-mediated cystic fibrosis transmembrane regulator gene transfer to the lungs of patients with cystic fibrosis: a multicenter, double-blind, placebo-controlled trial. *Chest* **125**, 509–521.
14. Lai, Y., Yue, Y., Liu, M., Ghosh, A., Engelhardt, J.F., Chamberlain, J.S., et al. (2005) Efficient in vivo gene expression by trans-splicing adeno-associated viral vectors. *Nat. Biotechnol.* **23**, 1435–1439.
15. Flotte, T.R., Zeitlin, P.L., Reynolds, T.C., Heald, A.E., Pedersen, P., Beck, S., et al. (2003) Phase I trial of intranasal and endobronchial administration of a recombinant adeno-associated virus serotype 2 (rAAV2)-CFTR vector in adult cystic fibrosis patients: a two-part clinical study. *Hum. Gene Ther.* **14**, 1079–1088.
16. http://clinicaltrials.gov/ct2/show/study/NCT00073463. Safety and Efficacy of Recombinant Adeno-Associated Virus Containing the CFTR Gene in the Treatment of Cystic Fibrosis. This study has been terminated. In *ClinicalTrials.gov* (2010).
17. Croteau, G.A., Martin, D.B., Camp, J., Yost, M., Conrad, C., Zeitlin, P.L., et al. (2004) Evaluation of exposure and health care worker response to nebulized administration of tgAAVCF to patients with cystic fibrosis. *Ann. Occup. Hyg.* **48**, 673–681.
18. Medina, M.F., Kobinger, G.P., Rux, J., Gasmi, M., Looney, D.J., Bates, P., et al. (2003) Lentiviral vectors pseudotyped with minimal filovirus envelopes increased gene transfer in murine lung. *Mol. Ther.* **8**, 777–789.
19. Sinn, P.L., Shah, A.J., Donovan, M.D., and McCray, P.B., Jr. (2005) Viscoelastic gel formulations enhance airway epithelial gene transfer with viral vectors. *Am. J. Respir. Cell Mol. Biol.* **32**, 404–410.
20. Krause, D.S., Theise, N.D., Collector, M.I., Henegariu, O., Hwang, S., Gardner, R., et al. (2001) Multi-organ, multi-lineage engraftment by a single bone marrow-derived stem cell. *Cell* **105**, 369–377.
21. Kotton, D.N., Fabian, A.J., and Mulligan, R.C. (2005) Failure of bone marrow to reconstitute lung epithelium. *Am. J. Respir. Cell Mol. Biol.* **33**, 328–334.
22. Macpherson, H., Keir, P., Webb, S., Samuel, K., Boyle, S., Bickmore, W., et al. (2005) Bone marrow-derived SP cells can contribute to the respiratory tract of mice in vivo. *J. Cell Sci.* **118**, 2441–2450.
23. Desigaux, L., Gourden, C., Bello-Roufai, M., Richard, P., Oudrhiri, N., Lehn, P., et al. (2005) Nonionic amphiphilic block copolymers promote gene transfer to the lung. *Hum. Gene Ther.* **16**, 821–829.
24. Pringle I., Davies L., McLachlan G., Collie D., Gill D. et al. (2004) Duration of reporter gene expression from naked pDNA in the mouse lung following direct electroporation and development of wire electrodes for sheep lung electroporation studies. *Mol. Ther.* **9**, S56.
25. Xenariou S., Liang H., Griesenbach U., Farley R., Somerton L. et al. (2005) Sonoporation increases non-viral gene transfer to the murine lung. *Mol. Ther.* **11**, S139.
26. Norez, C., Antigny, F., Noel, S., Vandebrouck, C., and Becq, F. (2009) A cystic fibrosis respiratory epithelial cell chronically treated by miglustat acquires a non-cystic fibrosis-like phenotype. *Am. J. Respir. Cell Mol. Biol.* **41**, 217–225.
27. Monaco, A.P., Neve, R.L., Colletti-Feener, C., Bertelson, C.J., Kurnit, D.M., and Kunkel, L.M. (1986) Isolation of candidate cDNAs for portions of the Duchenne muscular dystrophy gene. *Nature* **323**, 646–650.
28. Alexander, B.L., Ali, R.R., Alton, E.W., Bainbridge, J.W., Braun, S., Cheng, S.H.,

et al. (2007) Progress and prospects: gene therapy clinical trials (part 1). *Gene Ther.* **14**, 1439–1447.

29. Romero, N.B., Braun, S., Benveniste, O., Leturcq, F., Hogrel, J.Y., Morris, G.E., et al. (2004) Phase I study of dystrophin plasmid-based gene therapy in Duchenne/Becker muscular dystrophy. *Hum. Gene Ther.* **15**, 1065–1076.
30. Takeshima, Y., Yagi, M., Wada, H., Ishibashi, K., Nishiyama, A., Kakumoto, M., et al. (2006) Intravenous infusion of an antisense oligonucleotide results in exon skipping in muscle dystrophin mRNA of Duchenne muscular dystrophy. *Pediatr. Res.* **59**, 690–694.
31. Chamberlain, J.S. (2002) Gene therapy of muscular dystrophy. *Hum. Mol. Genet.* **11**, 2355–2362.
32. Rafael, J.A. and Brown, S.C. (2000) Dystrophin and utrophin: genetic analyses of their role in skeletal muscle. *Microsc. Res. Tech.* **48**, 155–166.
33. Le Bourdelles, B., Horellou, P., Le Caer, J.P., Denefle, P., Latta, M., Haavik, J., et al. (1991) Phosphorylation of human recombinant tyrosine hydroxylase isoforms 1 and 2: an additional phosphorylated residue in isoform 2, generated through alternative splicing. *J. Biol. Chem.* **266**, 17124–17130.
34. Li, S., Kimura, E., Ng, R., Fall, B.M., Meuse, L., Reyes, M., et al. (2006) A highly functional mini-dystrophin/GFP fusion gene for cell and gene therapy studies of Duchenne muscular dystrophy. *Hum. Mol. Genet.* **15**, 1610–1622.
35. Rodino-Klapac, L.R., Janssen, P.M., Montgomery, C.L., Coley, B.D., Chicoine, L.G., Clark, K.R., et al. (2007) A translational approach for limb vascular delivery of the micro-dystrophin gene without high volume or high pressure for treatment of Duchenne muscular dystrophy. *J. Transl. Med.* **5**, 45.
36. Mingozzi, F. and High, K.A. (2007) Immune responses to AAV in clinical trials. *Curr. Gene Ther.* 7, 316–324.
37. Wang, Z., Storb, R., Lee, D., Kushmerick, M.J., Chu, B., Berger, C., et al. (2009) Immune Responses to AAV in Canine Muscle Monitored by Cellular Assays and Noninvasive Imaging. *Mol. Ther.* **18**, 617–624.
38. Cooper, M., Nayak, S., Hoffman, B.E., Terhorst, C., Cao, O., and Herzog, R.W. (2009) Improved induction of immune tolerance to factor IX by hepatic AAV-8 gene transfer. *Hum. Gene Ther.* **20**, 767–776.
39. Hoen, P.A., de Meijer, E.J., Boer, J.M., Vossen, R.H., Turk, R., Maatman, R.G., et al. (2008) Generation and characterization of transgenic mice with the full-length human DMD gene. *J. Biol. Chem.* **283**, 5899–5907.
40. Evans, J.R. and Barker, R.A. (2008) Neurotrophic factors as a therapeutic target for Parkinson's disease. *Expert Opin. Ther. Targets* **12**, 437–447.
41. Laguna Goya, R., Tyers, P., and Barker, R.A. (2008) The search for a curative cell therapy in Parkinson's disease. *J. Neurol. Sci.* **265**, 32–42.
42. Palfi, S. (2008) Towards gene therapy for Parkinson's disease. *Lancet Neurol.* 7, 375–376.
43. Herzog, C.D., Brown, L., Gammon, D., Kruegel, B., Lin, R., Wilson, A., et al. (2009) Expression, bioactivity, and safety 1 year after adeno-associated viral vector type 2-mediated delivery of neurturin to the monkey nigrostriatal system support cere-120 for Parkinson's disease. *Neurosurgery* **64**, 602–12; discussion 612–613.
44. Coelho, M., Ferreira, J., Rosa, M., and Sampaio, C. (2008) Treatment options for non-motor symptoms in late-stage Parkinson's disease. *Expert Opin. Pharmacother.* **9**, 523–535.
45. Cavazzana-Calvo, M., Hacein-Bey S., de Saint Basile G., Gross F., Yvon, E., Nusbaum P., et al. (2000) Gene therapy of human severe combined immunodeficiency (SCID)-X1 disease. *Science* **288**, 669–672.
46. Aiuti, A., Slavin, S., Aker, M., Ficara, F., Deola, S., Mortellaro, A., et al. (2002) Correction of ADA-SCID by stem cell gene therapy combined with nonmyeloablative conditioning. *Science* **296**, 2410–2413.
47. Hacein-Bey-Abina, S. Von Kalle C., Schmidt M., McCormack M.P., Wulffraat N., Leboulch P., et al. (2003) LMO2-associated clonal T cell proliferation in two patients after gene therapy for SCID-X1. *Science* **302**, 415–419.
48. (ESGT), E.S.G.T. (Press release January, 2003).
49. Rabbitts, T.H., Axelson, H., Forster, A., Grutz, G., Lavenir, I., Larson, R., et al. (1997) Chromosomal translocations and leukaemia: a role for LMO2 in T cell acute leukaemia, in transcription and in erythropoiesis. *Leukemia* **11 Suppl 3**, 271–272.
50. Nam, C.H. and Rabbitts, T.H. (2006) The role of LMO2 in development and in T cell leukemia after chromosomal translocation or retroviral insertion. *Mol. Ther.* **13**, 15–25.
51. Pike-Overzet, K., de Ridder, D., Weerkamp, F., Baert, M., Verstegen, M., Brugmann, M., et al. (2006) Gene therapy: is IL2RG oncogenic in T-cell development? *Nature* **443**, E5.

52. Cavazzana-Calvo, M. and Fischer, A. (2007) Gene therapy for severe combined immunodeficiency: are we there yet? *J. Clin. Invest.* **117**, 1456–1465.
53. Notarangelo, L., Miao, C., and Ochs, H. (2008) Wiskott-Aldrich syndrome. *Curr. Opin. Hematol.* **15**, 30–36.
54. Westerberg, L.S., de la Fuente, M.A., Wermeling, F., Ochs, H.D., Karlsson, M.C., Snapper, S.B., et al. (2008) WASP confers selective advantage for specific hematopoietic cell populations and serves a unique role in marginal zone B-cell homeostasis and function. *Blood* **112**, 4139–4147.
55. Marangoni, F., Bosticardo, M., Charrier, S., Draghici, E., Locci, M., Scaramuzza, S., et al. (2009) Evidence for long-term efficacy and safety of gene therapy for Wiskott-Aldrich syndrome in preclinical models. *Mol. Ther.* **17**, 1073–1082.
56. Dewey, R.A., Avedillo Diez, I., Ballmaier, M., Filipovich, A., Greil, J., Gungor, T., et al. (2006) Retroviral WASP gene transfer into human hematopoietic stem cells reconstitutes the actin cytoskeleton in myeloid progeny cells differentiated in vitro. *Exp. Hematol.* **34**, 1161–1169.
57. Mantovani, J., Charrier, S., Eckenberg, R., Saurin, W., Danos, O., Perea, J., et al. (2009) Diverse genomic integration of a lentiviral vector developed for the treatment of Wiskott-Aldrich syndrome. *J. Gene Med.* **11**, 645–654.
58. Charrier, S., Stockholm, D., Seye, K., Opolon, P., Taveau, M., Gross, D.A., et al. (2005) A lentiviral vector encoding the human Wiskott-Aldrich syndrome protein corrects immune and cytoskeletal defects in WASP knockout mice. *Gene Ther.* **12**, 597–606.
59. Galy, A., Roncarolo, M.G., and Thrasher, A.J. (2008) Development of lentiviral gene therapy for Wiskott Aldrich syndrome. *Expert Opin. Biol. Ther.* **8**, 181–190.
60. Excoffon, K.J., Liu, G., Miao, L., Wilson, J.E., McManus, B.M., Semenkovich, C.F., et al. (1997) Correction of hypertriglyceridemia and impaired fat tolerance in lipoprotein lipase-deficient mice by adenovirus-mediated expression of human lipoprotein lipase. *Arterioscler. Thromb. Vasc. Biol.* **17**, 2532–2539.
61. Liu, G., Excoffon, K.J., Benoit, P., Ginzinger, D.G., Miao, L., Ehrenborg, E., et al. (1997) Efficient adenovirus-mediated ectopic gene expression of human lipoprotein lipase in human hepatic (HepG2) cells. *Hum. Gene Ther.* **8**, 205–214.
62. Liu, G., Excoffon, K.J., Wilson, J.E., McManus, B.M., Miao, L., Benoit, P., et al. (1998) Enhanced lipolysis in normal mice expressing liver-derived human lipoprotein lipase after adenoviral gene transfer. *Clin. Invest. Med.* **21**, 172–185.
63. Rip, J., van Dijk, K.W., Sierts, J.A., Kastelein, J.J., Twisk, J., and Kuivenhoven, J.A. (2006) AAV1-LPL(S447X) gene therapy reduces hypertriglyceridemia in apoE2 knock in mice. *Biochim. Biophys. Acta* **1761**, 1163–1168.
64. Vaessen, S.F., Twisk, J., Kastelein, J.J., and Kuivenhoven, J.A. (2007) Gene therapy in disorders of lipoprotein metabolism. *Curr. Gene Ther.* 7, 35–47.
65. Fox, J.C. and Swain, J.L. (1996) Angiogenic gene therapy. A leg to stand on? *Circulation* **94**, 3065–3066.
66. Takeshita, S., Zheng, L.P., Asahara, T., Reissen, R., Brogi, E., Ferrara, N., et al. (1993) In vivo evidence of enhanced angiogenesis following direct arterial transfer of the plasmid encoding vascular endothelial growth factor. *Circulation* **88 (suppl I)**, I-476, abstract.
67. Takeshita, S., Zheng, L.P., Brogi, E., Kearney, M., Pu, L.Q., Bunting, S., et al. (1994) Therapeutic angiogenesis: a single intraarterial bolus of vascular endothelial growth factor augments revascularization in a rabbit ischemic hind limb model. *J. Clin. Invest.* **93**, 662–670.
68. Feldman, L.J., Tahlil, O., and Steg, P.G. (1996) Adenovirus-mediated arterial gene therapy for restenosis: problems and perspectives. *Semin. Interv. Cardiol.* **1**, 203–208.
69. Tsurumi, Y., Takeshita, S., Chen, D., Kearney, M., Rossow, S.T., Passeri, J., et al. (1996) Direct intramuscular gene transfer of naked DNA encoding vascular endothelial growth factor augments collateral development and tissue perfusion. *Circulation* **94**, 3281–3290.
70. Caron, A., Michelet, S., Sordello, S., Ivanov, M.A., Delaere, P., Branellec, D., et al. (2004) Human FGF-1 gene transfer promotes the formation of collateral vessels and arterioles in ischemic muscles of hypercholesterolemic hamsters. *J. Gene Med.* **6**, 1033–1045.
71. Witzenbichler, B., Mahfoudi, A., Soubrier, F., Le Roux, A., Branellec, D., Schultheiss, H.P., et al. (2006) Intramuscular gene transfer of fibroblast growth factor-1 using improved pCOR plasmid design stimulates collateral formation in a rabbit ischemic hindlimb model. *J. Mol. Med.* **84**, 491–502.
72. Comerota, A.J., Throm, R.C., Miller, K.A., Henry, T., Chronos, N., Laird, J., et al. (2002) Naked plasmid DNA encoding fibroblast growth factor type 1 for the treatment of end-stage unreconstructible lower extremity

ischemia: preliminary results of a phase I trial. *J. Vasc. Surg.* **35**, 930–936.

73. Baumgartner, I., Pieczek, A., Manor, O., Blair, R., Kearney, M., Walsh, K., et al. (1998) Constitutive expression of phVEGF165 after intramuscular gene transfer promotes collateral vessel development in patients with critical limb ischemia. *Circulation* **97**, 1114–1123.
74. Nikol, S., Baumgartner, I., Van Belle, E., Diehm, C., Visona, A., Capogrossi, M.C., et al. (2008) Therapeutic angiogenesis with intramuscular NV1FGF improves amputation-free survival in patients with critical limb ischemia. *Mol. Ther.* **16**, 972–978.
75. Baumgartner, I., Chronos, N., Comerota, A., Henry, T., Pasquet, J.P., Finiels, F., et al. (2009) Local gene transfer and expression following intramuscular administration of FGF-1 plasmid DNA in patients with critical limb ischemia. *Mol. Ther.* **17**, 914–921.
76. Keo, H., Hirsch, A., Baumgartner, I., Nikol, S., and Henry, T. (2009) Gene Therapy in Critical Limb Ischemia. *Vascular Disease Management* **6**, 118–124.
77. Ruck, A. and Sylven, C. (2008) Therapeutic angiogenesis gains a leg to stand on. *Mol. Ther.* **16**, 808–810.

Chapter 3

Host Cells and Cell Banking

Glyn N. Stacey and Otto-Wilhelm Merten

Abstract

Gene therapy based on the use of viral vectors is entirely dependent on the use of animal cell lines, mainly of mammalian origin, but also of insect origin. As for any biotechnology product for clinical use, viral vectors have to be produced with cells derived from an extensively characterized cell bank to maintain the appropriate standard for assuring the lowest risk for the patients to be treated. Although many different cell types and lines have been used for the production of viral vectors, HEK293 cells or their derivatives have been extensively used for production of different vector types: adenovirus, oncorectrovirus, lentivirus, and AAV vectors, because of their easy handling and the possibility to grow them adherently in serum-containing medium as well as in suspension in serum-free culture medium. Despite this, these cells are not necessarily the best for the production of a given viral vector, and there are many other cell lines with significant advantages including superior growth and/or production characteristics, which have been tested and also used for the production of clinical vector batches. This chapter presents basic considerations concerning the characterization of cell banks, in the first part, and, in the second part, practically all cell lines (at least when public information was available) established and developed for the production of the most important viral vectors (adenoviral, oncoretroviral, lentiviral, AAV, baculovirus).

Key words: Cell bank, Cell-line characterization, Host cell lines, Oncoretroviral vector, Lentiviral vector, Adeno-associated viral vector, Adenoviral vector, Baculovirus

1. Introduction

During the development of gene therapy, a wide range of packaging cell lines have been used to generate the recombinant viral vector to be used as the therapeutic product. While these cell lines are not used directly in the patients themselves, regulatory requirements for application of the final therapeutic vector will include a full technical history for the packaging cells used to identify any risk factors and establish their suitability for manufacturing a

Otto-Wilhelm Merten and Mohamed Al-Rubeai (eds.), *Viral Vectors for Gene Therapy: Methods and Protocols*, Methods in Molecular Biology, vol. 737, DOI 10.1007/978-1-61779-095-9_3, © Springer Science+Business Media, LLC 2011

clinical product. This dossier will include information on the origin, characterization, culture history, and cell banks of the host cell line. It is an important principle that for any single product all of these various stages in the development of the final therapy, including process development, preclinical testing, and manufacturing of the licensed product, should be performed on cells of the same origin. Accordingly, an important early step is establishment of a cell bank for use as the sole source of packaging cells throughout the development and manufacturing of the final product. This chapter outlines the cell banking process and reviews the various cell lines used as host cells for packaging of gene therapy vectors and the generic requirements for testing cell banks for manufacturing purposes.

2. Cell Banking Procedures

In the biotechnology industry, where microorganisms and cell cultures have been used for manufacturing purposes for many decades, the establishment of a well-characterized cryopreserved master cell bank is a key step to assure the provision of a source of reliable cells for all future work. Vials from the master cell bank are recovered and expanded to prepare working cell banks from which individual vials are used to initiate cultures for each production run or period of experimentation. This master-working bank system is crucial to assure long-term provision of reproducible cells for consistent product quality.

A desirable addition to the already described cell banking process is to analyze cells beyond the anticipated limit of use to check the stability of their characteristics. Regulatory requirements state that cells should be analyzed in a production run, at or beyond the point of harvesting product or at an equivalent passage level equivalent (1–4).

3. QC and Safety Testing

There are a range of key issues for quality control of all cell cultures that are important for the quality and safety of products derived from cell cultures. Central aspects are as follows:

- Viability
- Identity (i.e., the cells are what they are purported to be)
- Purity (i.e., freedom from microbiological contamination)
- Stability on growth or passage in vitro

The following sections provide an overview of the key quality-control procedures required for cell substrates used in manufacturing processes of therapeutic products.

3.1. Viability

The ability of the large majority of cells to recover from the cryopreserved state and regenerate a suitably growing culture is key to reliable culture processes for production. It is often determined using a dye exclusion test such as trypan blue (5, 6), which, typically for mammalian cell lines, will yield viability values in the range of 80–100%. While this means of measuring membrane integrity is a useful indicator of the viability of cells, it is important to remember that any one technique will only give a specific and incomplete perspective on the overall status of a cell culture regarding its ability to grow and replicate. In addition, viability measurements at a single time point may not predict the ongoing fate of the culture, for example, "post-thaw" cells observed to be "viable" by dye exclusion test could in fact be in the early stages of programmed cell death. For certain cultures, it may be helpful to use additional parameters of viability such as early markers of apoptosis, for example, annexin IV expression.

A more generally meaningful measure of viability is the ability to grow and replicate at an acceptable rate. A number of approaches are used, often in combination, to assess this capacity of a cell bank. Examples include "cloning efficency," "plating efficiency," and "population doubling rate" (7). It is also important that the culture recovered from a working cell bank is representative of the original master cell bank, and this will require characterization, although this may not be as extensive as for the master bank.

A further important use of cell viability and growth tests is to evaluate consistency of cultures recovered from different vials from within each cell bank (homogeneity testing) (8, 9).

3.2. Identity and Authenticity of Cell Lines

Many mammalian cell lines used for basic research may have been passed from one laboratory to another and in the process may become permanently altered due to extensive passaging, variation in local culture procedures and reagents, and microbial contamination events. In addition, accidental cross-contamination, switching or mislabeling of mammalian cell lines used in research laboratories has been reported many times in the literature (10–13) and clearly will have led to the publication of invalid data. Such events may go unrecognized where there is similarity in the morphology of the original and replacement cell line.

Clearly, before committing the significant resource involved in making GMP banks of cells for manufacturing, it will be vital to confirm that a selected source of production cells is authentic. This can be achieved first by confirming the cell line provenance through a well-documented history and traceability, ideally to the laboratory in which it was derived. Second, the candidate source of production cells can be characterized to confirm cell line authenticity directly. Short Tandem Repeat (STR) DNA profiling methods provide the capacity to make highly specific identification for human cell lines (14, 15). However, specific identification of nonhuman cells remains challenging, and equivalent STR profiling methods are not developed for many species, although there are useful methods for identity in a range of species including mouse, dogs (16), and some primate species (17, 18). For other species, improvements in identification of the species of origin can be achieved using conserved intron analysis (e.g., (19)) and, more recently, Cox 1 gene sequencing (20, 21). However, further work is required to qualify methods for specific identification of the range of cell lines used as gene therapy host cells.

It is important to recognize that the value of any direct analysis of cell line identity is dependent on having material from the original animal/donor or a consensus on the cell line identity profile from multiple sources.

3.3. Microbial Contamination

Bacterial and fungal contamination from the environment will destroy cell cultures, and if such contamination is from spore-forming organisms, which can survive readily in the environment and cell culture conditions, the problem can reemerge periodically. Standard pharmacopoeial sterility testing methods can be used to give assurance that aseptic processing and other controls are excluding contaminants from the environment (22, 23). It is important to note that such methods are not intended to detect breaches in aseptic processing and are not capable of detecting all potential bacterial and fungal contaminants. In particular, contamination by *Mycoplasma* and *Acholeplasma* spp. can go unnoticed as these organisms require special isolation media and growth conditions. In addition, they can persist without necessarily affecting the growth of the cells and may fail to show obvious signs of contamination such as medium turbidity and appearence of microbial colonies. Standard methodologies for detecting these organisms have been established and can be obtained from the US and European Pharmacopoeia (24, 25). Mycobacteria have also been isolated from animal cell substrates (26), and specific isolation methods may be recommended for cell line testing for such contamination.

Virological testing is usually based on risks associated with the original cells used to derive the cell line (e.g., original donor or animal colony used), other aspects of derivation (such as genetic constructs), and exposure to materials of biological origin during derivation, culture, and processing of the cells. For each aspect, risk evaluation is important to identify those most likely potential contaminants for which the candidate cell lines should be tested. The risk of microbial contamination from the original cells can be evaluated based on factors including the species and tissue of cells, their geographical origin, and level of isolation of donor animals from the environment (i.e., husbandry and health controls) (27). Typically long established cell lines will have been exposed to traceable lots of fetal calf serum and porcine trypsin, in which case guidance is available on typical agents for which testing should be perfomed and similar lists exist for exposure to materials of murine and human origin (1, 2, 4). However, it is important to recognize that these are generic recommendations that may need to be adjusted based on changes in the viruses most prevalent among source herds, and testing requirements will need to be reviewed in the light of the specific exposure history of each cell line under consideration. Risk evaluation of a cell line should ideally be supported by detailed information on the culture history so that potential virological contaminants can be identified and an appropriate testing regime can be applied.

For nonmammalian cell types such as insect cell lines, the adventitious agent aspects and other safety issues may differ considerably, and detection of organisms that grow at lower temperature optima (e.g., spiroplasma) and persistent reverse transcriptase activity may require experimental evaluation to demonstrate that it cannot be transferred to mammalian cells. Other cell types such as embryonic stem cell lines and those from avian origins may also have specific additional adventitious agent issues that will need to be evaluated (see revised WHO (4) guidance in preparation referred to in Knezevic et al. (3)). However, despite the development of technologies that could herald the ability to identify all viral contaminants including unknown viruses (e.g., (28)) and microarray detection systems that may enable broad ranging screening of cell substrates, it is currently not feasible to test cell banks to cover all potential viral contaminants.

3.4. Characterization, Replication-Competent Virus Testing, and Stability Testing

Characterization of each bank will depend on the cell type and key phenotypic and genotypic markers for the stem cell line. Typically, the master cell bank will receive the most detailed characterization, and more focussed quality control is performed on working banks starting with viability, sterility, mycoplasma, homogeneity, and identity. However, regulators may leave open the option of carrying out most safety testing on each working cell bank where this can be justified by the manufacturer. A typical scheme for a testing regime for cell banks is shown in Table 1, but the testing should be established based on a specific risk assessment for each cell line. Additional characterization of the working cell bank is performed by passaging the cells to the normal anticipated cell generation level (end of production cells) required if instability is suspected in certain characteristics that may affect the final product. This should be documented as part of the cell banking process.

Table 1
Key elements of a typical testing regime for cell banks of production cells

Test specification	Examples of test methods
Bacteria/fungi	Inoculation of microbiological culture media to detect growth of bacteria and fungi
Mycoplasma	Direct culture in broth and agar and indirect test using indicator culture/DNA stain
Postthaw recovery on a proportion of vials (homogeneity testing)	Trypan blue dye exclusion Markers of apoptosis Doubling time
Identity and stability	
Cell characteristics	Short Tandem Repeat (STR) DNA profile Karyotype by Giemsa-band analysis of metaphase spreads Fluorescent In Situ Hybridization Human Leukocyte Antigen (HLA) genotype Single Nucleotide Polymorphisms. Comparative Genome Hybridization by DNA microarray methodology.
Genetic contructs and construct stability (beyond production)	Intergration site(s) analysis FISH Vector and helper sequences: mRNA analsyis (Northern blot) cellular DNA analysis (Southern blot, DNA sequencing, Restriction endonuclease fragment mapping) Functionality Product titer (e.g., viral titer) Product characterization Absence of RCR (replication competent retrovirus) appearance in the case of retroviral vector producer cell lines
Adventitous agents	
Specific	PCR/RT PCR for specified viral sequences (based on risk assessment) Antibody production tests in rodents inoculated with test cells, e.g., MAP, HAP, RAP
Nonspecific	Cell culture inoculation for detection of CPE and Hemagglutinin Animal inoculation and observation for pathology Electron microscopy (SEM of ultracentrifuged supernatants and TEM of sections of multiple cells) Reverse transcriptase assays
Tumorigenicity/oncogenicity[a]	Inoculation of immunocompromised mice with viable test cells (tumorigenicity) or disrupted cells (oncogenicity)

[a] It has been considered that testing for tumorigenicity/oncogenicity should be part of early evaluation of a cell substrate and not a routine test for cell banks and is not necessary or useful where the cell line is known to be tumorigenic (3)

In addition, the copy number of the inserted expression cassettes as well as the identity of the chromosomally inserted sequences has to be determined using, for instance, Fluorescence In Situ Hybridization (FISH), partial or complete DNA sequencing of the integrated coding sequences, and analysis of mRNA transcripts encoding the gene product such as Northern blotting. Further methods used include restriction endonuclease fragment mapping (examination for insertion/mutation/integration sites/rearrangements) and Southern blotting.

Concerning retroviral vector producer cell lines, specific tests on the potential appearance of RCR (replication-competent retroviruses) have to be performed on the banked producer cells (MCB) as well as at the end of the production cycle on 1% of pooled producer cells or 10^8 cells, whichever is fewer. Cocultures with permissive cell lines (e.g., *Mus dunni* cells for amphotropic and ecotropic retroviruses, HEK293 or HCT116 cells for amphotropic, VSV-G, and GaLV enveloped RCRs (29)), including several blind passages. Supernatants from the coculture should be tested by PG4S + L- or an alternative assay (30). Similar tests have to be performed for lentiviral vector producer cell lines. With respect to vector preparations, it is evident that they have to be tested for the absence of replication-competent viruses (RCR (retrovirus) (31–37), RCL (lentivirus) (38), RCA (adenovirus) (39, 40), RCAAV (41, 42)). As an example, in the case of retroviral vector containing supernatant, it is recommended to test 5% of clinical-grade supernatant by amplification on a permissive cell line (e.g., *Mus dunni*) including several blind passages, followed by the PG4S + L- or an alternative assay (30).

4. Good Cell Culture Practice

In order to avoid the various hazards outlined above that could disqualify a cell substrate from use in manufacturing of therapeutic products, there are a number of fundamental principles for good cell culture practice, which were captured in a consensus guidance document from an ECVAM task force on Good Cell Culture Practice (43). This guidance described seven core principles of GCCP, now being incorporated into some regulatory guidance (e.g., revision of WHO (4), see Knezevic et al. (3)) and some of these are discussed below.

4.1. Understanding the Cells and the Culture System

The variations that occur in the in vitro cell culture environment, particularly in terms of the composition of the culture medium, will clearly influence the cell biology and responses of the cells. Accordingly, basal media and additives such as serum, growth factors, amino acids, and other growth-promoting compounds should, ideally all be specified and documented according to their chemical composition, purity, and, where relevant, biological activity. The use of serum or other poorly defined reagents may not be avoidable but, in addition to raising safety concerns, will reduce the degree of definition and standardization that can be obtained in a cell culture manufacturing process. However, there may be compromise to be stuck between the benefits of a closely defined growth medium and the potential shortcomings of a completely defined system that may not meet the full biological needs of cells. Where complex biological reagents (e.g., FBS) continue to be required, they should be carefully controlled by preuse selection of batches. Given that cell–extracellular contacts often have a fundamental influence on the survival, growth, and function of cells, it follows that similarly careful specification and selection should also apply to cell culture surfaces, i.e., using specified culture vessels and surface coatings where relevant.

Variations in the general physical and chemical environment (e.g., pH, temperature, gas atmosphere) can clearly have a significant influence on viability, growth, and function of cells and should be quantified with acceptable tolerance limits. However, it is important to be aware that, using standard laboratory equipment such as 5% CO_2/air incubators and T flasks, the physical environmental characteristics typically undergo significant and regular changes when cultures are removed from the incubator and passaged exposing them to the laboratory environment.

Physical stresses on cells due to manipulation during production can also have a significant influence on the quality of the cells and the final product. Care should be taken to minimize the impact of manipulation of cells by prompt processing and return of cells to standard culture conditions. The process of passaging cells includes some of the more disruptive events such as cell detachment, washing, and centrifugation, but there are alternative culture systems such as "spinner" flasks and other bioreactor systems where shear forces on cells may be significant and in this respect even pipetting cell suspensions too vigorously can be damaging. Cell harvesting and passaging procedures should be specified to ensure consistency of cell output and exposure to adverse effects. A further methodology prone to causing adverse effects on cells is cryopreservation. As for passaging, standardised methodology is important (typically using slow rate cooling at −1 °C/min, following addition of a cryoprotectant, typically 5–10% v/v DMSO) to ensure reproducible removal of intracellular water to prevent ice damage (for a review, see (44)). In addition, it is important to select healthy cultures and check the viability of each preserved batch immediately after preservation (43).

4.2. Contamination

The various sources of microbiological contamination have already been discussed above. In general cell culture work it is clearly critical to adopt rigorous aseptic technique and provide appropriate environmental controls and air quality for cell culture processing and preparation of growth media. The presence of any antimicrobial in a biological process or product could mask contaminants which have some degree of resistance (e.g., mycoplasma, Achromobacter (26, 45)) and even in the case of some commonly used antibiotics may affect cellular function (e.g., (46)). In addition, penicillin and other beta-lactam antibiotics are recommended to be specifically excluded from production cell cultures (4) and the new draft revised recommendations on cell substrates, see Knezevic et al. (3).

4.3. Other Aspects of GCCP

The GCCP guidance (43) also addresses the core needs for training, laboratory safety, and recording and reporting on cell culture-based work. In addition, following any cloning procedure, the process of selecting a suitable recombinant cell line from multiple cell clones is of critical importance in delivering and efficient and economic production process. In new guidance under development by WHO, best practice in cloning and selection of cell lines has also been considered (3).

5. Host Cell Lines Utilized in the Development of Gene Therapy

Viral vectors for use in gene therapy applications have been produced with many different human and nonhuman animal cell lines (Table 2). Although various cell lines have been evaluated and some of them have been developed up to the GMP production of viral vectors, one cell line should be mentioned here in particular, HEK293 and its derivatives, because it is the cell line that has been most frequently used for the production of a wide range of viral vectors. In this chapter, most of the existing host cell lines are described, however, with a particular emphasis on HEK293 cells and their derivatives.

5.1. HEK293 Cells

HEK 293 cells were generated by transformation of cultures of normal human embryonic kidney cells with sheared adenovirus 5 DNA in the laboratory of Alex van der Eb in Leiden, Holland in the early 1970s. They were obtained from a healthy aborted fetus and originally cultured by van der Eb himself, and the transformation by adenovirus was performed by Frank Graham who published his findings in 1977 after he left Leiden for McMaster University in Canada (47). They are called HEK to reflect their origin in human embryonic kidney, while the number 293 comes from Graham's habit of numbering his experiments, with the

Table 2
Sources of host cell lines, mentioned/described in the present book

Cell line	ATCC	DSMZ	ECACC	ICLC	JCRB	Others
HEK293	CRL1573	ACC 305	85120602	HTL03003	JCRB9068	
293T	CRL-11268	ACC 635		HTL04001		
293T/17	CRL-11268					
293FT						Invitrogen R700-07
293E						Invitrogen
293-6E						NRC-BRI
ANJOU 65	CRL-11269					
Bing	CRL-11554					
ProPak A	CRL-12006					
ProPak X	CRL-12007					
NIH 3T3	CRL-1658	ACC 173 (Swiss albino)	93061524 (Swiss NIH embryo)	ATL95002 (Swiss albino embryo)	JCRB0615	
GPE + 86	CRL-9642			HTL06007		
GP + envAM12	CRL-9641					
PG13	CRL-10686					

HT-1080	CCL-121	ACC 315		HTL98016	JCRB9113, IFO50354	
RD	CCL-136			HTL97021 (TE671 Subline No. 2)	JCRB9072	
CEM	CCL-119				JCRB0033, JCRB9023, IFO50412	
8E5	CRL-8993					
HeLa	CCL-2	ACC 57		HTL95023	JCRB0160	
HeLa-S3	CCL-2.2	ACC 161		HTL95020	JCRB0713, JCRB9010	
HeRC32	CRL-2972					
NCI-H1299	CRL-5803					
A549	CCL-185			HTL03001	JCRB0076, IFO50153	
BHK-21 C13	CCL-10				JCRB9020	
Vero	CCL-81		84113001		JCRB0111, IFO50471, JCRB9013	
Sf9	CRL-1711	ACC 125	89070101	ATL95005		Invitrogen 12659017

Note: Only those cell lines that can be obtained from one of the mentioned repositories are indicated in this table. All other cell lines can only be obtained from the authors of the papers describing the specific cell line original HEK293 cell clone arising from the product of his 293rd experiment (Wikipedia).

Subsequent analysis has shown that the transformation was brought about by an insert consisting of ~4.5 kilobases from the left arm of the viral genome genome (17% of the left-hand region → 4,344 bp of the left-hand of Ad5 viral DNA, El region, pIX gene), which became incorporated into human chromosome 19 (19q13.2) (48).

HEK293 cells and their derivatives are known to be tumorigenic (49). For many years, it was assumed that HEK293 cells were generated by transformation of either a fibroblastic or an endothelial or an epithelial cell, all of which are abundant in kidney. However, the HEK cell cultures may contain small numbers of almost all cell types of the body. In fact, Graham and coworkers more recently have provided evidence that HEK293 cells and several other human cell lines generated by adenovirus transformation of human embryonic kidney cells have many properties of immature neurons, suggesting that the adenovirus was taken up and transformed a neuronal lineage cell in the original kidney culture (50). Further confirmation on this fact was provided by van der Eb who speculated that these cells may have originated from a rare neuronal cell in the kidney cell cultures at an FDA meeting entitled "Vaccines and related biological products advisory committee," which took place in May 2001 (http://www.fda.gov/ohrms/dockets/ac/01/transcripts/3750t1_01.pdf, see page 85 of this report for comment on the potential neuronal origin of 293 cells). The obtained data have been put together into a database on HEK293 cells available at http://www.mbi.ufl.edu/~shaw/293.html.

5.1.1. Traceability

The traceability of HEK293 is not excellent. Although they have been established in 1977, the passages during the first years after the establishment have not really been traced. Only more recently established subclones have a certain traceability that is often sufficient for clinical studies.

As an example, the traceability of 293FT cells is presented here. The real traceability starts with 1988 when Life Technologies got the HEK293 cells from R. Horlick via R. Swanson (both from Pharmacopeia in the USA). Today, it is practically impossible to trace back the way how the cells came from Graham's lab in Canada to Pharmacopeia.

In 1998, Life Technologies selected the 293F cells ("fast-growing" clone of HEK293), and 1 year later Life Technologies generated the 293FT cells after having stably transfected the 293F cells with pCMVSPORT6Tag.neo for overexpressing the SV40 T antigen (the expression of the SV40 T antigen is controlled by the human CMV promoter (→ high level, constitutive expression). The gene encoding the SV40 T antigen permits the episomal replication of plasmids containing the SV40 early promoter and origin. Today, the cells are available from Invitrogen and are traceable back to 1988.

Since 1988, these cells are traceable for the medium and serum (USDA approved) used; no trypsin was used since 1988.

5.1.2. Stability

HEK293 cells are continuous cells and as such show a certain tendency to change over many subpassages. Although this is not very extensively documented, a study by Park et al. (51) presented this issue. Whereas suspension adapted HEK293 cells (293S) cultured in a specific serum-free medium (293SFMII) maintained specific cell growth, cell size, and adenovirus production over 40 passages after thawing, the adherent cell clone (293M) did not show constant culture parameters. The cells had been received from ATCC at a passage number 31. The specific growth rate increased from 0.29/day (at passage 43) to 0.74 ± 0.01/day (at passages 66–86). In parallel, the cells became smaller in size at later passages.

Complementary to these results, a recent paper has presented results on the evolution of the tumorigenicity of HEK293 as function of the passage level. Whereas cells thawed from a cell bank at passage 21 (from the China Center for Type Culture Collection) and cultivated for further 31 passages did not induce tumors in nude mice when injected sc at cell numbers of $1–2 \times 10^6$ c/0.2 ml (during an observation period of 8 weeks) (thus confirming results from Graham et al. (47)), later passages (P65 and P71) led to the formation of solid tumors of about 0.5 cm in all injected nude mice within 2 weeks (52).

These data indicate very clearly, that first, a well-characterized cell stock (at a low passage number) has to be established, and second, thawed cells should only be used over a limited and in beforehand validated number of passages to maintain critical cellular characteristics such as cell growth and vector production and also to maintain a reduced tumorigenicity.

5.1.3. Use of HEK 293 Cells

Basic culture conditions: The original cell line is an adherently growing cell line when cultivated in standard medium (e.g., DMEM) with serum. However, in the absence of serum and in media with low calcium ion concentration (different formulations have been developed and can be purchased from different vendors), these cells have a high tendency to detach to suspension and they can then be subcultured in suspension in stirred tank bioreactors.

As with the vast majority of mammalian cells, HEK293 cells and their derivatives are cultured at 37°C (classical temperature), but for certain vector productions, lower temperatures have been shown to be optimal (e.g., Jardon and Garnier (53)) for r-adenovirus production using 293S cells, Kotani et al. (54) and

Kaptein et al. (55) for the production of MLV using PA317 cells, Le Doux et al. (56) for the production of MLV using ψCRIP cells), as a result of the balance between production and degradation. The atmosphere is 95% air, 5% CO_2; under reactor conditions, optimal pO_2 level as well as pCO_2 level depend on the set points chosen, which in general lead to improved culture conditions and, therefore, to improved growth and/or vector production (53).

HEK293 cells are used for the transient production of adenoviral vectors (by infection) or AAV, MLV, or LV vectors (via transfection). In order to improve certain functions, derivatives of these cells have been established by inserting either the T-Ag of SV40 or the EBV nuclear antigen (see Subheading 5.1.4). In addition, HEK293 cells have also been used for the establishment of stable producer cell lines for the production of MLV and LV vectors. More details on HEK293-based MLV producer cell lines can be found in Subheading 5.1.5.

As for the production of MLV producer cell lines, similar attemps have been performed to develop LV packaging/producer cell lines to facilitate and optimize the production of this vector of high interest. More details on HEK293-based LV producer cell lines can be found in Subheading 5.1.6.

5.1.4. HEK 293 Subclones

HEK293 cells have also been the base for the development of derivatives, such as 293T cells or 293E cells:

1. 293T cells and derived cell lines
   a. 293T cells (293tsA1609neo) (ATCC CRL-11268)

      The 293T cell line (293tsA1609neo) is a highly transfectable derivative of the HEK293 cell line into which the temperature-sensitive gene of SV40 T-antigen was inserted. It could be shown that these cells can be much more efficiently transfected and they show a higher specific growth rate than the HEK293 cells. In addition, certain vector production rates are higher than for HEK293 cells (57). A subclone is the 293T/17 cell line.

      This cell line has been used for the transient production of MLV and LV vectors using small-scale and large-scale transfection methods. The same cells have also been used for the establishment of stable producer cell lines of MLV vectors. They were the base for the establishment of the ANJOU65 (ATCC CRL-11269), which were the base for the BOSC23 (58) and Bing (ATCC CRL-11554) (59) cells, producing ecotropic and amphotropic MLV vectors, respectively. Another group (G. Nolan) developed the Phoenix helper-free retrovirus producer lines using 293T cells. As for the ANJOU65 cell system, ecotropic and amphotropic packaging and producer cell lines have been developed. With respect to the establishment of HIV-1 and FIV based lentiviral vectors, a similar approach (as for the Phoenix (http://www.stanford.edu/group/nolan/retroviral_systems/phx.html) system) had been employed, leading to the Helix (http://www.stanford.edu/group/nolan/retroviral_systems/helix.html) and Felix (http://www.stanford.edu/group/nolan/retroviral_systems/felix.html) retrovirus system.

Concerning the establishment of stable inducible lentiviral producer cell lines, 293T cells have been evaluated by different groups for the production of HIV-1-based LV vectors (60, 61) and EIAV-based LV vectors (62).

b. 293FT cells (Invitrogen R700-07):

Another subclone is the 293FT cells, which have been established via the transfection of HEK293 cells with pCMVSPORT6Tag.neo for overexpressing the SV40 T antigen. The advantage of this cell line is the fact that it is traceable back to 1988 (see above). In principle, their use is as for the 293T cells with similar growth and production levels; however, they have been less used.

2. 293E Cells (Invitrogen), 293-6E cells (NRC-BRI, Montreal, www.nrc-cnrc.gc.ca)

293EBNA-1 – either 293E or 293-6E (expressing the Epstein-Barr virus (EBV) nuclear antigen-1, EBNA-1) are cell lines that allow increasing expression levels by permitting plasmid replication or episomal persistence, respectively, in the transfected cells throughout the production phase. Essential for the plasmid replication or episomal maintenance is the presence of EVB oriP replication origins, respectively, in the plasmid backbones. The EBV oriP-EBNA1 system also serves as a strong cis transcriptional enhancer for many viral and nonviral promoters.

293E cells have been evaluated for the transient production of LV and AAV vectors in bioreactors via plasmid transfection (63, 64).

5.1.5. HEK293 Retroviral Vector Producer Cell Lines

From a historical point of view, packaging cell lines were based on the use of the mouse cell line NIH 3T3 (see Subheading 6.1). However, certain limitations of these packaging cell lines have initiated the search to improve them. Principally, mouse cell lines are associated with the following drawbacks: They produce relatively low titers. Furthermore, murine retroviral sequences that are present in murine packaging cells can be selectively packaged into retroviral particles (65), increasing the possibility of generating RCRs. In addition, human packaging cell lines generate vector particles that are less likely to be inactivated by human serum

(66, 67). Therefore, the use of human cell lines for the establishment of packaging cells is a step toward increased biological safety, as they lack endogenous murine retroviruses (67–69). In fact, viral supernatants or producer cells derived from human cells have never given a positive test result for RCRs in small- or large-scale assays (68, 70). Human MLV packaging cells are described in Subheadings 5.1.5.1. – 5.1.5.4. and 6.2.

Patience et al. (71) showed that retroviral vectors interact with human packaging cells (FLY cells that are based on HT1080 cells; see Table 4, Subheading 6.2) to produce retroviral particles that are far less contaminated by endogenous viral sequences or other types of extraneous particles than murine packaging cells (for instance, the murine GP+envAm12 packaging system). Endogenous C-type proviral genome sequences can be reactivated (74). Hatzoglou et al. (75) reported the efficient packaging of a specific VL30 retroelement by psi 2 cells, and Farson et al. (76) showed that in contrast to the mouse system, where the ratio of the transmission of recombinant retrovirus and of the murine VL30 was about 1:1, the HEK293-based production system transmitted HERV-H elements at a ratio of at least as low as $1{:}5 \times 10^5$. In addition, as packaging cell lines derived from human cells lack endogenous murine retroviral sequences, the likelihood of producing RCRs is minimized (77). Dog-based producer cell lines are also characterized by the absence of endogenous retroviral sequences homologous with MLV vector sequences (78).

One further argument for the use of nonmurine (mainly human) cell lines for the establishment of MLV vector packing cell lines are differences in the glycosylation of retroviral particles. It seems that the glycosylation of the retroviral particles (glycosylation of the env protein and of cellular proteins incorporated into the particles, lipid-associated carbohydrates) has an impact on the stability/retention of retroviral particles in human serum. It is known that retroviral particles when produced with mouse packaging cell lines are inactivated by the human complement within 20 min after injection. This is generally considered to be due to the presence of the galactosyl(α1-3)galactosyl carbohydrate moiety on the vectors produced by murine packaging cells, whereas such vectors produced in human or primate cells do not have this glycostructure and, therefore, are resistant to complement inactivation (66–68). However, it has been shown that even retroviral vectors produced from a galactosyl(α1-3)galactosyl carbohydrate positive ferret brain cell line (Mpf) are resistant to complement inactivation, signifying that it is not only the structure of the glycosylation of the env protein but also other epitopes, such as lipid-associated carbohydrates (galactosyl(α1-3) galactyosyl carbohydrate moiety being only one of them), present on the surface of the viral membrane where antibody binds and/or complement acts (79).

In the following section, retroviral vector (MLV) producer cell lines using HEK293 or 293T cells are presented:

5.1.5.1. ProPak A and ProPak X (ATCC CRL-12006, ATCC CRL-12007)

The ProPak packaging cell lines (69) produce either murine leukemia virus (MLV) xenotropic particles (ProPak-X cells) or amphotropic particles (ProPak-A cells). They were derived from the human embryonic kidney line HEK293 (see ATCC CRL-1573) (68).

To derive the amphotropic packaging cell line ProPakA.6 (characterized by separate gag-pol and env packaging functions expressed from a heterologous (non-MLV) promoter to maximally reduce homology between packaging and vector sequences), the pCMV*Ea plasmid (4070A – env plasmid) was introduced into HEK293 cells by cotransfection with the pHA58 plasmid conferring resistance to hygromycin B (250 μg/ml). Clones were subsequently transfected with gag-pol and vector plasmids. Next, the pCMV-gp construct (gag-pol construct) was stably transfected into the HEK293-Env clones by cotransfection with the plasmid pSV2pac. The cells are puromycin-resistant (1 μg/ml). They secrete defective (noninfectious) murine leukemia virus (MLV) particles composed of gag-pol and env proteins (69, 80). The ProPak A6 cell line was deposited at the ATCC (catalogue no. CRL-12006).

Stable producer cell lines have been established by transduction with a MLV vector. These cells showed stability for more than 3 months. Titers of up to 2×10^6 G418 CFU/ml were obtained (end-point titers on NIH 3T3 cells) (68).

To construct the ProPak-X (xentrope envelope) cell line, the pMoMLVgp plasmid (coding for the gag-pol function under control of an MoMLV-LTR) and the pHA58 plasmid (conferring resistance to hygromycin) were cotransfected into HEK293 cells using calcium phosphate coprecipitation. Clones were screened for the level of Gag secretion and one clone secreting high levels of Gag was selected (designated ProGag); this clone yielded high viral titers in transient transfection. The expression plasmid containing the murine xenotropic env gene, pCI-Ex, was cotransfected with pSV2pac into the ProGag cell line by calcium phosphate precipitation and puromycin-resistant cells were selected. The resulting cells were screened for Env expression and clones designated ProPak-X, expressing high levels of Env were screened for ability to produce transducing vector. Clone 36 designated ProPak-X.36 was deposited at ATCC (catalogue no. CRL-12007) (80). A similar development was performed to generate the ProPak-A-52 cell line producing retroviral vectors pseudotyped with the 10A1 envelope protein (80).

Certain vectors that consistently give rise to replication-competent retrovirus (RCR) in PA317 cells do not give rise to detectable RCR in ProPak-A-based producer cultures (68).

ProPak-based producer cells were demonstrated to be free of replication-competent retrovirus (RCR) by stringent testing. Consistently, higher transduction of target cells was achieved with ProPak-derived amphotropic vector than with PA317-packaged amphotropic vector (69).

The amphotropic and xenotropic vectors produced with ProPak cells have been shown to be resistant to human serum (complement resistance). With respect to transduction efficiency, ProPak vector preparations have a 2–3 times better transduction efficiency than PA317 cells (69), however, vector titers have not been communicated.

5.1.5.2. "Kat" Cells

CellGenesys developed HEK293-based 3rd generation producer cell lines (split genome approach with reduced homology between the packaging and the vector constructs) for the production of amphotropic (4070A→PUZIkat2, 10A1→STRAkat) and xenotropic ($NZB_{9\text{-}1}$→ALLIkat) retroviral vectors (76).

To derive the packaging cell line "TOMkat," which is the basis for the different packaging cell lines providing different env proteins, the pkat2gagpol plasmid was introduced into 293S cells (M. Mathew, Cold Spring Harbor Laboratories, Cold Spring Harbor, NY) by cotransfection with the neomycin-resistance plasmid. 100 gagpol transfectant clones were picked and evaluated for RT activity. The 12 clones displaying the highest RT activity were evaluated for vector production capacity in a transient transfection assay. The four best clones were evaluated for stability over 12 passages (6 weeks). One of these clones (after evaluation of stability and absence of RCR appearance in long-term passages) called "TOMkat" was selected for further stable transfection with envelope plasmids. For establishing the amphotropic packaging cell line ("PUZIkat"), "TOMkat" cells were cotransfected with pkat2amenv and a hygromycin-resistance marker plasmid. Clones were selected via FACS analysis and evaluated transiently for vector production in a second term for absence of RCR production. The best clone produced titers of $5\text{–}10 \times 10^5$ TU per ml (on NIH 3T3). Similarly, two other split-genome packaging cell lines have been developed: "ALLIkat2" and "STRAkat" by cotransfecting with pkat2xenoenv and a hygromycin-resistance marker plasmid or with pkat210A1env and a hygromycin-resistance marker plasmid, respectively. Characterization was performed as for the "PUZIkat" packaging cells. All packaging cells have been evaluated for stability and absence of RCR appearance (76).

Using PUZIkat-based vector producer cells (CC49ζ), large-scale clinical batches with a mean titer of 5.2×10^6 TU/ml (on NIH 3T3) have been produced.

5.1.5.3. 293GP-A2 Cells

These amphotrope packaging cells published by Ghani et al. (81) are based on a suspension and serum-free adapted clone of

HEK293 cells, the 293SF cells (adapted from 293S cells (from Dr. Michael Matthew (Cold Spring Harbor Laboratories))). Subconfluent 293SF cells plated in Petri dishes were transfected with pMD2.GPiZeo^r (plasmid containing the gag-pol gene of Moloney murine leukemia virus with downstream an IRES element and the Zeo-resistance gene–selection at 400 μg/ml for 2 weeks). The highest RT expression clone was then transfected with the pMD2.AIPuro^r plasmid (conferring the amphotrope env gene and downstream an IRES element followed by the puromycin resistance gene–selection at 0.2 μg/ml for 2 weeks). The selected packaging clone was able to produce up to 4×10^7 IVP/ml after generation of a producer clone (titered on HT1080, transgene: GFP).

The cells can be cultured under serum-free conditions in a stirred tank reactor using SFM from Invitrogen. Under these conditions, titers of $>10^7$ IVP/ml were obtained, and the cells were stable for more than 3 months. No RCR formation could be detected.

5.1.5.4. New Generation of Retroviral Producer Cell Lines Using Flp-Mediated Site-Specific Integration of Retroviral Vectors

Further HEK293-based cell lines have been developed, allowing the predictable and stable virus production through Flp-mediated site-specific integration of retroviral vectors (protocols for their use are described in Chapter 7). These cell lines, Flp293A (82) and 293 FLEX (83), produce retroviral vectors pseudotyped with amphotropic and GaLV envelopes, respectively. Their particularity is the possibility to exchange the MLV vector insert to the vector construct of choice via Flp sites.

The 293 FLEX cells have been established using the following strategy: HEK293 cells have been tagged by transduction with a retroviral vector (MSCV based – pIRESGALEO) produced with PA317 producer cells where *LacZ* is under the control of a LTR promoter and the fusion protein gene hygromycin B phosphotransferase/thymidine kinase (*hygtk*) (= positive/negative selection marker) is under the control of the encephalomyocarditis virus internal ribosome entry site (EMCV-IRES). This vector contains two FRT sites in the U3 of 3′LTR, a wild-type FRT site (*wt*) and a spacer mutant FRT site (F5) followed by an ATG-defective neomycin phosphotransferase gene (Fig. 1a). The tagged cells have been selected using 200 μg/ml of hygromycin B. All selected cells were tag-positive, and the integrated tagging cassette is shown in Fig. 1b. The choice of transduction instead of transfection with a plasmid is based on the fact that retroviral vectors integrate specifically into active chromosomal sites, which is not the case for plasmid transfection.

The packaging functions were introduced by lipofectamine-based cotransfection using the pCeb containing MoMLV gag-pol and the blasticidin-resistance gene (*bsr*), both driven by the MoMLV 5′LTR ((66); see establishment of FLY cells) and pGALV containing the gene of the Gibbon Ape Leukemia Virus

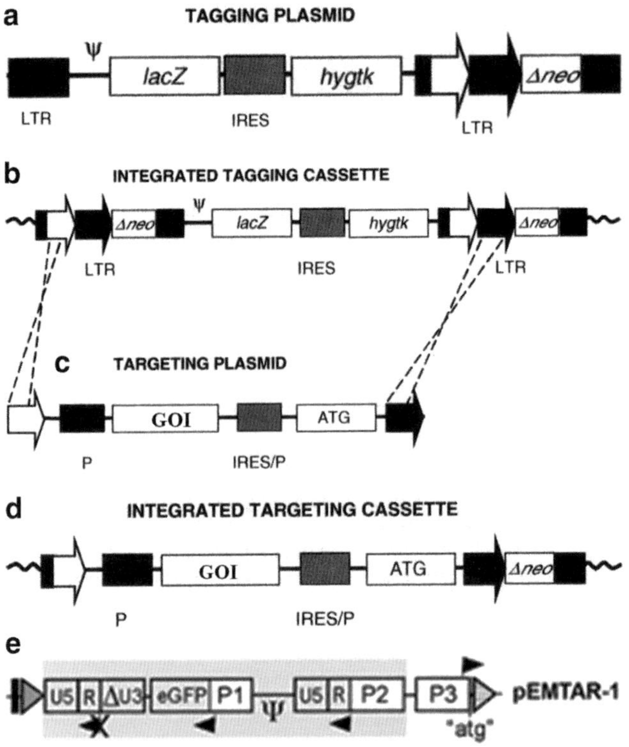

Fig. 1. Tag and target methodology for the cassette exchange by Flp mediated recombination. (**a**) Schematic representation of the tagging retroviral plasmid containing a *lacZ* gene, *hygtk* positive and negative selection markers and two FRT sites, a wild type represented by a *white arrow* and a mutant represented by a *black arrow*, in the 3′LTR, followed by a defective *neo* gene. (**b**) Tagging construct after proviral integration, resulting in a duplication and transfer of the two FRT sites on the 5′LTR. (**c**) Schematic representation of the targeting plasmid containing the two FRT sequences flanking the gene of interest and an ATG sequence that will restore the open reading frame of the *neo* gene in the tagged clone after Flp recombinase mediated exchange (**d**) (83). (**e**) Design of an optimized targeting construct (pEMTAR-1). In pEMTAR-1, the orientation of viral vector genome was inverted, and the restored open reading frame is under the control of a constitutive PGK promoter (P3). Promoter P2 (e.g., MPSV) is driving the expression of the viral genomic RNA. Promoter P1 (e.g., EFS, fes, CMV) controls the expression of the transgene. The transcriptional start sites are indicated by *arrows*. Abbreviations: *P* promoter, *GOI* gene of interest.

envelope protein and the zeocin (*Zeo*) resistance marker under the control of the CMV promoter (Cosset, pers. commun., see establishment of TEFLY GA cells). The cells have been selected with 10 μg/ml of blasticidin and 100 μg/ml zeocin.

For targeting (site-specific cassette exchange of the tag vector by the vector containing the transgene of choice), the established cells (293 FLEX) have been transfected with two plasmids, the targeting vector plasmid (e.g., pEmMFG – targeting vector containing a FRT *wt*, an MoMLV-based retroviral MFG-LTR that drives the eGFPgene (in this model, the transgene of choice) followed by an EMCV-IRES element next to an ATG and F5 FRT site (Fig. 1c)) and the FLP recombinase containing plasmid (pSVFLPe – obtained by cloning the FLP recombinase 1,300 bp fragment from pFLPe downstream of a SV40 promoter) using the calcium phosphate coprecipitation method. The cassette exchange leads to the appearance of a complete functional ORF of the *neo*-gene; thus, the clones are selected in the presence of 1000 μg/ml of G418 24 h posttransfection (Fig. 1d – integrated targeting cassette). The surviving clones are picked, amplified, and analyzed for vector production and cassette exchange (Southern blotting). In the case of MLV-eGFP producer clone (MFG-eGFP), the titers obtained ranged from 1.1 to 1.4×10^6 iu/ml.

In the case of the Flp293A cells, after tagging for introducing the tagging vector cassette via MLV-transduction (the MLV vector was produced by PG13 cells transfected with the pTAGeGFP plasmid using calcium phosphate transfection and the cells were selected for hygromycine resistance), the selected clone (1B2) was cotransfected (use of calcium phosphate method) with the gag-pol containing plasmid pCeB ((66), see establishment of the FLY cell lines) and pENVAhis (84) (containing the wt amphotropic *env* gene of 4070A and the histidinol-dehydrogenase gene for selection purpose) and selected for resistance to blasticidin and histidinol. Clone Flp293A was obtained. The targeting is performed as for the 293 FLEX cells (see above). In the case of MLV-eGFP producer clones (MSIReGFP and MLIReGFP), the average titers obtained were $8.1 \times 10^6 \pm 1.5 \times 10^6$ ip/10^6 cells $\times 24$ h and $2.5 \times 10^7 \pm 1.3 \times 10^7$ ip/10^6 cells $\times 24$ h, respectively.

Both cell lines have been characterized by a certain read-through, leading to a low-level contamination of the MLV vector batches by MLV vector with the resistance gene (*neo* gene) as transgene (85).

5.1.6. HEK293: 293T Lentiviral Vector Producer Cell Lines

For lentiviral vectors, small- and large-scale productions are essentially based on the use of transfection methods using either HEK293 cells or their derivatives. Owing to traceability uncertainties, some GMP large-scale productions have been performed with HEK293 cells (86), and other large-scale productions (57, 87–89) have been performed using 293T cells because of their superior growth, transfection, and vector production characteristics (http://www.fda.gov/ohrms/dockets/ac/01/briefing/3794b3.htm). Attempts to produce lentiviral vectors at a larger scale in suspension by maintaining a transfection method have successfully been performed using a serum-free suspension clone HEK293SF-3F6 cells (see above, (90)), and under optimized conditions infectious titers of about 10^8 TU/ml have been obtained. This result had been validated in a 3-L stirred tank reactor (91).

Although of high interest, stable LV producer cell lines have only been developed and evaluated in a research environment, but no further transition to a larger scale/industrial scale had been performed. Most studies made use of HEK293 cells (92–94) growing in an attached mode. Brousseau et al. (95) successfully developed a suspension producer cell line based on a suspension adapted HEK293 clone (HEK293SF-3F6) established by Côté et al. (90).

On the other side, Ikeda et al. (60), Throm et al. (61), and Stewart et al. (62) developed HIV-1- and EIAV-, respectively, based lentiviral vector producer cell lines using 293T cells.

In general, the development of stable cell lines has been hindered by the toxicity of one or more components, including the most commonly used env glycoprotein, the vesicular stomatitis virus G protein (VSV-g) (96), or the expression of HIV gag-pol (97, 98). Packaging cell lines with conditionally regulated VSV-g expression have been developed (tet-off approach as used by Kafri et al. (99), Klages et al. (92), Xu et al. (94), Farson et al. (38), Ni et al. (93), and Throm et al. (61)), but this approach lacks flexibility for changing envelopes and accessory genes. The cell lines require several days of induction (removal of doxycyclin, which suppresses the expression of VSV-g) prior to maximal vector production, and the titer is often lower than for the transient transfection process (92, 99, 100). In addition, producer cells have often been shown to be unstable (Delenda et al. unpublished observations) due to background expression of the VSV-g protein (as a result of leaky expression) (38, 93) which can lead to cell fusion and syncytia formation (101) and finally to cell death.

In addition, many published producer cell lines are first- or second-generation LV vector producer cell lines and should not be used for the production of clinical batches due to safety considerations because the constructs used for the establishment of these cells contained at least the tat function if not other accessory genes (38, 60, 61, 93, 94, 99).

Real advances toward a large-scale LV vector production using producer cell lines (293SF-PacLV and PC48.2, respectively) were achieved by Brousseau et al. (95) and Stewart et al. (62) because both groups used inducible systems that are based on the addition of an inducing compound and not on the withdrawal of a suppressor as used in the tet-off approach (see above). This is of high relevance in the case of large-scale productions performed in bioreactors. Whereas Brousseau et al. (95) used the 293SF-3F6 cell line, which is adapted to growth in suspension under serum-free conditions, Stewart et al. (62) used 293T cells. In both cases, third-generation LV constructs were used. On one side, Brousseau et al. (95) used a double induction system (tetracycline and cumate: cells expressing the repressor (CymR) of the cumate switch and the reverse transactivation (rtTA2S-M2) of the tetracycline (Tet) switch (selection via neomycin and hygromycin genes, respectively) contained the inducible genes (VSV-g (pTR5-CuO-VSV-g-IRES-GFP) and rev (pTR5-CuO-Rev)) and the constitutively expressed gag-pol (pMPG-CMV-gag/polRRE) genes;

the cells were established via cotransfection of these three plasmids), and on the other side, Stewart et al. (62) used a simple induction (tetracycline: cells constitutively express coTetR (use of pPuro. coTetR, containing the puromycin selectable marker) and contain pgag-pol and pVSVG under control of an hCMV promoter containing 2xTetO2 sequences; selectable markers were Hygromycin and Zeocin, respectively. Two of all selected clones (PC48, PC71) showed highest vector production). In both cases, the induction factor was between 2100 and 3700. After induction, the producer cell lines can provide vector titers beyond 10^7 TU/ml 2–6 days postinduction (95). Stewart et al. (62) indicated similar vector titers for the producer cells and for the transient vector production.

Chapter 8 of the present volume describes a standard production and purification method of LV vectors (HIV-1)) based on the tritransfection of 293T cells.

5.1.7. HEK293 Adenoviral Vector Producer Cell Lines

Only human adenoviral vectors are considered here. Since the development of HEK293 cells in 1977, many more different adenovirus complementing cell lines have been developed because this first complementation cell line was not optimal. The drawback with HEK293 cells is that they have the tendency to produce replication-competent adenoviruses (RCA) due to homologous recombination events between the adenoviral vector construct and the adenoviral sequences in the cellular genome (102, 103). In order to remedy this problem, new complementation cell lines have been developed with reduced adenoviral sequences to reduce or even avoid any risk of recombination events and thus the generation of RCAs. These complementation cell lines are all based on human cells and in almost all cases on primary neuronal or retinal cells. All more recent complementation cell lines have been immortalized with the E1 gene of adenovirus 5. They are briefly described in Subheading 7.

6. Other Cell Lines Used as MLV Packaging/Producer Cells

6.1. NIH 3T3 (ATCC CRL-1658) Derived MLV Packaging/Producer Cells

For historical reasons, NIH 3T3-based packaging/producer cell lines should be mentioned here, as the first packaging cell lines were based on these mouse cells due to the mouse origin of gamma-retroviral vectors. NIH 3T3 is an adherently growing fibroblast-like cell line isolated from mouse embryonic tissue (*Mus musculus*). It is highly sensitive to sarcoma virus focus formation and leukemia virus propagation and has proven to be very useful in DNA transfection studies (104, 105). They have been used as basal cell line for the establishment of various first-, second-, and third-generation cell lines for the production of MLV vectors. As for NIH 3T3 cells, all derived packaging/producer cell lines grow adherently. Third-generation packaging cell lines based on the use of NIH 3T3 cells are summarized in Table 3.

6.2. FLY-Packaging Cell Lines

A range of other cell lines of human origin have been used in the development of new packaging cell lines and are included in Table 4.

Table 3
Third-generation packaging cell lines derived using NIH 3T3

Cell line name, references	Generation of the cell lines	Comments
GPE + 86 (ATCC CRL-9642) (34, 106)	Established by electroporation of two plasmids into NIH 3T3 cells: The gag-pol plasmid (pGag-Pol/Gpt – gag and pol regions from Moloney murine leukemia virus (Mo-MuLV) contained the selectable marker gene *gpt*, and was deficient for ψ, env, and 3′LTR, and the env plasmid (pEnv – env region from Mo-MuLV→ecotropic envelope), deficient for ψ, gag-pol, and the 3′LTR) were originally derived from the 3P0 parent plasmid (Mo-MuLV). After electroporation of 10^7 cells with 5 μg and 10 μg DNA of pEnv and pGag-Pol/Gpt, respectively, the cells were resuspended in DMEM + serum and plated. 48–72 h later selective medium containing 15 μg hypoxanthine, 250 μg xanthine, and 25 μg mycophenolic acid was added and surviving cells were analyzed for RT activity. Clones were selected using the same medium and characterized for RT activity and vector production after electroporation with Δneo and N2 plasmid	This line is capable of packaging nucleic acids containing a psi packaging sequence into recombinant ecotropic retrovirus genomes. It can be used to produce retroviral vectors for delivery of foreign genes into susceptible eukaryotic cells. Stable producer clones generate titers of up to 4×10^6 colony forming units per ml (N2 vector titered on NIH 3T3 cells) (34)
GP + *env*AM12 (ATCC CRL-9641) (33, 106)	The line was established by electroporation into NIH 3T3 cells of two plasmids that separately encode the env region of a murine amphotropic MuLV and the gag, pol and other sequences needed for viral packaging (→ amphotropic envelope). The same establishment protocol was used as for the establishment of the GP + 86 cells. The difference was the pEnv plasmid: PenvAm (- plasmid contained the 5′LTR and 5′ donor splice site of the 3P0 plasmid (Mo-MLV) and 4070A env derived from the pL1 plasmid (amphotropic murine leukemia virus clone 4070A))	This line is capable of packaging nucleic acids containing a psi packaging sequence into recombinant amphotropic retrovirus genomes. It can be used to produce retroviral vectors for delivery of foreign genes into susceptible eukaryotic cells. Stable producer clones generate titers of >2×10^6 colony forming units per ml (N2 vector titered on NIH 3T3 cells) (33)

(continued)

Table 3 (continued)

Cell line name, references	Generation of the cell lines	Comments
ψCRE and ψCRIP (107)	Established by sequential Ca-phosphate transfection of 2 plasmids into NIH 3T3 cells. In the first round of transfection, the pCRIPenv – plasmid (pCRIPenv- (a LTR-, ψ-, gag-pol+, env-plasmid derived from the CRIP plasmid)) was cotransfected with the plasmid pSVHm confering resistance to hygromycin B. Stable transformants were selected using 200 µg/ml hygromycin B. 2/16 positive clones which showed highest release of RT activity (env-1, env-15) were further used: env-1 was chosen to derive the amphotropic ψCRIP packaging line and env-15 was selected as the parental clone for the ecotropic ψCRE line, since it showed a twofold higher level of Mo-MuLV-specific transcripts in an RNA gel blot analysis. In a second step, either the pCRIPAMgag- (a LTR-, ψ-, gag-, pol-, 4070A env+plasmid derived from pCRIP) or the pCRIPgag-2 plasmid (a LTR-, ψ-, gag-, pol-, env+(ecotropic env) plasmid derived from pCRIP) was introduced into the env- cells, along with the plasmid pSV2gpt (containing the bacterial xanthine-guanine phosphoribosyltransferase gene as a dominant selectable marker), conferring resistance to G418. Isolated resistant clones were tested for their ability to package the BAG retroviral vector into infectious particles. Each clone to be tested was infected by a helper-free stock of BAG virus and populations of 50-100 G418-resistant colonies were derived from each infection amplified. Two clones showing the highest packaging activity were selected: CRIP14 (→ ψCRIP, a retrovirus (amphotropic (4070A) pseudotyped) packaging cell line) and CRE25 (→ ψCRE, a retrovirus (ecotropic pseudotyped) packaging cell line)	Selection against loss of the DNAs conferring the packaging functions can be performed by growing the cells in medium containing fetal bovine serum and hygromycin B, mycophenolic acid, adenine, and xanthine. For the selection of vector producers, geneticin is used. Stable producer clones generate titers of 10^6 colony forming units per ml (HSGneo vector titered on NIH 3T3 cells) (107). Merten et al. (108) have reported titers of up to 6×10^6 FFU/ml (ψCRIP-LLZA: vector tittered on NIH 3T3 cells).

(continued)

Table 3 (continued)

Cell line name, references	Generation of the cell lines	Comments
PG13 (ATCC CRL-10686) (109)	Established by sequential Ca-phosphate transfection of 2 plasmids into NIH 3T3 cells. The MoMLV gag-pol expression construct pLGPS (consisting of a 5′-truncated MoMLV long terminal repeat (LTR) promoter without the adjacent retroviral packaging signal, the MoMLV gag-pol coding region, MoMLV sequences from 7676 (ClaI) to 7774 (end of the env gene), and the SV40 early polyadenylation signal, cloned into a modified version of the poison-sequence-minus pBR322 derivative pML, called pMLCN) was cotransfected using a plasmid containing the herpes simplex virus thymidine kinase gene as a selectable marker (ratio of selectable marker plasmid to the pLGPS plasmid was 1:20 or 1:100) into NIH 3T3 cells. After selection in HAT medium (30 μM hypoxathine, 1 μM amethopterin, 20 μM thymidine) and test for RT production, the clone (1/17 positive clones) with the highest RT production (4×>than any other clone) was selected for further development. The GaLV env expression plasmid (pMOV-GaLV Seato env consisting of a MoMLV LTR promoter without the adjacent ψ function, the GaLV env coding region and the SV40 late polyadenylation signal in a pBR322 plasmid backbone) was cotransfected with a plasmid containing the mutant methotrexate-resistant dihydrofolate reductase gene (dhfr*): pFR400 into the cells expressing MoMLV gag-pol (ratio of selectable marker plasmid to pMOV-GaLV Seato env plasmid: 1:20 or 1:100). Cell colonies containing the genes were selected in medium containing 100 nM methotrexate and dialyzed fetal bovine serum and were isolated. Clone PG13 (1 out of 20 clones) produced the highest-titer virus in a transient production test using retrovirus vector plasmid pLN	Selection against loss of the DNAs conferring the packaging functions can be performed by growing the cells in medium containing dialyzed fetal bovine serum and 100 nM amethopterin for 5 days, followed by cultivation in medium containing HAT and untreated fetal bovine serum for an additional 5 days. After selection, the cells should be maintained in medium containing HT for 2 days to avoid toxic effects due to residual amethopterin. Stable producer clones generate titers of 5×10^4 to 3×10^6 colony forming units per ml (LN vector titered on different nonmurine cells (rat, hamster, bovine, cat, dog, monkey, human)) (109)

(continued)

Table 3 (continued)

Cell line name, references	Generation of the cell lines	Comments
PG368 (85)	Established from PG13 cells, using a similar cassette exchange mechanism as described for the 293 FLEX and the Flp293A cells: PG13 cells have been tagged with the retroviral ragging vector TAGeGFP as used by Schucht et al. (82). In order to avoid the problem of read-through and thus the production of small quantities of MLV vector with the resistance gene as transgene (see Fig. 1d), the targeting construct had been newly designed which is essentially characterized by the inversion of the orientation of the vector. The most efficient targeting construct is shown in Fig. 1e): promoter 1 (P1): SF, P2: MPSV, P3: hPGK	Titers: the titers obtained were in the range of $5–8 \times 10^5$ ip/ml, which was about two to four times lower than when the vector orientation in the targeted producer cells was in the normal sense. In the case of therapeutic vectors SIN11(fes-GP91) and SIN11(EFS-γc) the titers ranged from 1 to 4.1×10^5 ip/ml and were thus 5–20-fold lower than the titers detected for SIN11-SF vectors oriented in the normal sense

7. Advanced Adenovirus Vector Packaging Cell Lines

The general problem of HEK293 cells in context of the production of adenoviral vectors is the generation of RCAs due to considerable overlap of the adenoviral sequences used to transform the HEK cells and the sequences of the adenoviral vector. Therefore, in the last 15 years or so, several authors have developed new, improved adenoviral vector producer cell lines with reduced transforming adenoviral sequences and in the case of the more recently developed cell lines, practically without any overlap. These cell lines are presented in Table 5, providing their key features.

Concerning adenoviral vector production, no real comparisons have been performed between the different complementation cell lines. Only Nadeau and Kamen (113) have published a literature-data-based comparison between HEK293 and Per.C6 cells, which is presented in Table 6, indicating that both cell lines show comparable specific vector production.

Chapter 5 presents methods to construct recombinant adenovirus vectors.

Table 4
Cell lines used for the establishment of third-generation "FLY" packaging cell lines

Cell line/origin	Packaging lines	Generation of cells	Notes/features/references
HT-1080 Human ras expressing fibrosarcoma (ATCC CCL-121)	FLY-packaging cell lines: FLY-A and FLY-RD	Established by sequential Ca-phosphate transfection of 2 plasmids into HT-1080 cells. They were transfected with the CeB plasmid containing the MoMLV (Moloney murine leukemia virus) gag-pol gene and 74 nucleotides (nts) downstream the *bsr* selectable marker (translation reinitiation mechanism) conferring resistance to blasticidin S (selection: 4 μg/ml), both driven by the MoMLV 5′LTR promoter. Clones have been selected for high RT activity. After selection of the best producer clone via vector transduction packaging cells have been established via transfection with an env-plasmid, containing the env-gene and 76 nts downstream the *phleo* selectable marker (AF plasmid – 4070A env protein-gene or the RDF plasmid – env protein from the cat endogenous virus RD114). The best clones (FLYA13 and FLYRD18) have been selected by evaluating the vector production capacities of subcultures	These cells were able to produce titers of $1–7 \times 10^6$ iu/ml and 1×10^6 to 3×10^7 iu/ml for the amphotrope and RD114 pseudotyped MLV vectors (LacZ vector tittered on TE671 cells), respectively. No RCR could be detected per 10^7 vector particles (66)
RD (Formerly also named TE671) Human rhabdomyosarcoma (ATCC CCL-136)	TE FLY packaging cell lines: TEFLY-A and TEFLY-GA	TEFLY A lines producing amphotrope MLV vectors, TEFLY GA lines producing GALV pseudotyped MLV vectors. For the establishment of the latter cells plasmid containing the GALV env gene and downstream the *phleo* selectable marker were used. The strategy for the establishment of these cells was the same as for FLY cells (72)	Viral titers obtained are similar to FLY lines
CEM Human T lymphoblast from an acute lymphoblastic leukemia (ATCC CCL-119)	CEM FLY packaging cell lines. (NB cells grow in suspension)	CEM FLY A cell line – an amphotrope pseudotyped (4070A) vector producer cell line, established using the same strategy as for the FLY cells (see above). The absence of the generation of RCRs was shown using a mobilisation assay	These cells are able to produce titers of up to 1×10^7 iu/ml (LacZ vector titered on TE671 cells). No RCR could be detected per 10^7 vector particles.The advantage of these cells is that they are absolute suspension cells and can be grown to cell densities beyond 2×10^7 c/ml in spinner flasks (73)

Table 5
Improved packaging cell lines for adenoviral vector production

Name/references	Origin	Transformation	Characteristics/comments
911 (110)	Human embryonic retinoblasts	Plasmid containing base pairs 79–5789 of the Ad5 genome	RCA appearance as for HEK293 cells Propagation: comparable with HEK293 cells. Highly transfectable. Excellent for plaque assays (appearance of plaques at days 3–4 instead of 4–10 for HEK293 cells). Up to three times higher virus titers than for HEK293 cells
NCL (102)	A549 (human lung cells)	Plasmid containing base pairs 505–2760 of the Ad5 genome – no sequence overlap with Ad vectors (E1A under the control of PGK promoter and E1B under control of rabbit β globulin gene)	At least two recombination events necessary for getting generation of RCAs – one of these recombinations has to be nonhomologous Propagation: no information available; reduced vector amplification in comparison to HEK293 cells due to the lack of the 1 EB 55 kDa gene
PER.C6 (103)	Human embryonic retinoblasts	Plasmid containing base pairs 459–3510 (under control of PGK promoter) of the Ad5 genom (1EA/1EB) – no sequence overlap with Ad vectors	No reported generation of RCAs Propagation: comparable with HEK293 cells. Equivalent production of E1-deleted Ad vectors as for HEK293 or 911 cells
N52.E6 (111)	Primary human amniocytes	Plasmid containing base pairs 505–3522 (under control of PGK promoter) of the Ad5 genome (1EA/1EB) – no sequence overlap with Ad vectors	No reported generation of RCAs Propagation: no information available; equivalent production of E1-deleted Ad vectors as for HEK293 or 911 cells
NC5T11 (ProBioGen, designer cell line)	Human fetal brain cells	E1 sequence of the Ad genome (defined sequence not reported)	Establishment, propagation: no information available Generation of RCA: no information available, but probably low risk as this is a "designer" cell line
UR (112)		Ad5 E1A/E1B sequence (nucleotides 459–3510)	No reported generation of RCAs Propagation: development of a serum-free suspension process planned. Equivalent production of E1-deleted Ad vectors as for HEK293

Table 6
Comparison of HEK293 and PerC6 for Ad vector production

Cell line, process mode	Titer	Specific production	References
HEK293, batch	$3.9–5.1 \times 10^{10}$ vp/ml $3.3–14.5 \times 10^{9}$ ip/ml 7.5×10^{9} ip/ml	$2.5–5.4 \times 10^{4}$ vp/c $1.9–6.4 \times 10^{3}$ pfu/c 5.6×10^{3} ip/c	Zhang et al. (114) Iyer et al. (115) Nadeau et al. (116)
Per.C6, batch	7.5×10^{10} vp/ml 5.75×10^{10} vp/ml	7.5×10^{4} vp/c 3.6×10^{4} vp/c	Irish et al. (117) Liu and Shoupeng (118)
HEK293, perfusion	2×10^{11} vp/ml $3.2–7.8 \times 10^{9}$ ip/ml 9×10^{9} ip/ml	1×10^{5} vp/c 1.3×10^{3} ip/c 1.5×10^{3} vp/c	Chaubard (119) Garnier et al. (120) Nadeau et al. (121)
Per.C6, perfusion	1.5×10^{11} vp/ml	2×10^{4} vp/c	Irish et al. (117)

Note: MOI used: 5–50

8. Complementing Cell Lines for Production of Gutless Adenovirus Vectors

These vectors are very attractive for gene therapy because the associated in vivo immuno response is highly reduced compared to first- and second-generation adenovirus vectors, while maintaining high transduction efficiency and tropism. However, since they are devoid of all viral regions, gutless vectors require viral proteins supplied in *trans* by helper virus. To remove contamination by a helper virus from the final preparation, different systems based on the excision of the helper-packaging signal have been generated. Among them, Cre-loxP system is mostly used. With this system, the helper adenovirus has a packaging signal flanked by two loxP sites and amplification is performed in Cre-recombinase-expressing cell lines, e.g., 293-Cre (122). When the helper adenovirus enters the cell, its packaging signal is excised, preventing the inclusion of its genome into the viral particle, but retaining all coding regions for the viral proteins needed to produce the gutless vectors. The downside is that contamination levels still are 0.1–1% too high to be used in clinical trials, needing specific purification methods.

In general, this production system is somehow problematic regarding RCAs because RCAs have a selective growth advantage over gutless vectors.

For more information, see Alba et al. (123).

Chapter 6 presents manufacturing methods for gutless adenoviral vectors.

8.1. C7-Cre

C7-Cre is a HEK293-based packaging line containing the Ad E2b genes encoding DNA polymerase and preterminal protein as well as the cre-recombinase gene (124). The developed helper adenovirus is deleted in the E1, E3 region and in the viral DNA polymerase gene region. As for all other helper adenoviruses in this context, the packaging signal is flanked by two loxP sites. The advantage with respect to 293-Cre cells is that only two passages were necessary for obtaining titers of 10^7 TU/ml, in comparison to 293-Cre cells, for which six to seven passages were necessary. Contamination with packaging-competent helper virus levels at about 3–4% at passage level 3; gradient centrifugation can reduce the contamination level to 0.2–1% (124).

8.2. PerC6-Cre

The PerC6-Cre packaging line is known to be contaminated with packaging-competent helper virus levels of approximately 0.63%.

9. Cell Lines for Production of Oncolytic Adenoviral Vectors

In view of the production of oncolytic adenoviral vectors, HeLa-S3 (clone of HeLa cells (ATCC CCL-2), which grows in suspension in serum-free medium (125, 126)), and H1299 cells (lung large cell carcinoma – suspension growth in serum-free medium) (ATCC CRL-5803) have been selected for production purposes due to their superior specific vector production (127). Such oncolytic adenoviral vectors are replication-selective that specifically target and destroy human cancer cells. This viral vector is engineered to replicate only in human tumor cells and not in normal cells, based on their abnormal retinoblastoma protein (pRB) tumor suppressor function. For replication, either helper cell lines providing E1 and E4 functions or cell lines that are defective in the pRB signalling pathway are required (128). The cell lines have been evaluated and used for the establishment of a large-scale manufacturing process for the production of oncolytic adenoviruses.

10. Cell Lines for Adeno-Associated Viral Vectors (129)

Classically, AAV vectors have been produced by bitransfection followed by adenovirus infection or by tritransfection of HEK293 cells (see above and Chapter 9). As transfection systems are rather limited in their capacity for scale-up, other production systems have been developed for the production of different AAV serotypes and these include production systems based on the use of HeLa (130) or A549 (131) with (recombinant) adenovirus or HEK293 or BHK-21 cells with recombinant Herpes virus (132, 133) or Sf9 with recombinant baculovirus (134) (see Chapter 10). When using the HeLa- or A549-based production system, these cells have been modified to contain the rep-cap functions of AAV (- packaging cells) (130, 131) and the recombinant AAV vector (- producer cells) (135). After infection with *wt* adenovirus or adenovirus defective in E2b and an rAAV/Ad-hybrid vector (the AAV cassette is cloned into the E1 region) (in the case of the packaging cells) (130) or only with Adenovirus (in the case of the producer cells), AAV vector contaminated by adenovirus is produced. The producer cell line approach had been scaled to 100 L. Concerning the use of the recombinant herpes simplex type 1 production of AAV, either suspension culture adapted HEK293 or BHK-21 cells are used (133). For more details, see Table 7.

11. Sf9 Cells for Baculoviral Vectors

As baculovirus can infect mammalian cells without being able to replicate, this virus represents a very safe alternative to transfer genes (141–147). In principle, different cell lines can be used, Sf9, Sf21, and High-Five; however, only the Sf9 cells are presented in the following. Caution should be taken with High-Five cells, since these cells have recently been shown to suffer from the latent infection by a novel nodavirus, Tn5 cell line (TNCL) virus (148):

Sf9 cells (ATCC CRL-1711) can be grown in attached mode in T-flasks using serum-containing media (e.g., Grace's Insect Medium with L-glutamine and 500 mg/L $CaCl_2$, 2.6 g/L KCl, 3.3 g/L lactalbumin hydrolysate, 3.3 g/L yeastolate, 10% FCS) or serum-free media as well as in suspension in agitated systems (spinner, stirred tank reactor system) in commercial serum-free media. The media are formulated for use without CO_2; however, the omission of yeastolate or lactalbumin hydrolysate will lead to poor performance by this line. In contrast to mammalian cells, the optimal culture temperature is 27–28°C.

For baculovirus production, the cells are infected with an MOI of 0.1 of a titered baculovirus stock (production of baculovirus as gene vector is presented in Chapter 12). This production system was also established for the production of AAV vectors, see above and Chapter 10.

Table 7
Large-scale AAV production systems based on the use of mammalian cells: packaging cell lines and the rHSV/BHK-21 based system

Name/references	Description	Culture/production conditions	Further information on selected clones
HeLa B50 (130)	HeLa cells were transfected with a rep-cap2 expressing vector carrying the neomycine gene (pP5-rep-cap-Neo plasmid) (since the rep gene products are cytotoxic, the expression of the rep protein was inducible using the endogenous p5 promoter which is induced by E1a of the adenovirus). After G418 selection, altogether 8/708 isolated clones were able to *trans*-complement *rep/cap* of which clone B50 yielded 100× more vector than the other clones. The induction of AAV production was done by infection with wt Ad5 or a temperature sensitive mutant in E2b of Ad5 followed 24 h later by infection with an Ad-AAV hybrid virus (constructs consist of the 5′ sequence of Ad (map units 0–1), a copy of the rAAV genome, and Ad sequence spanning map units 9–16.1) (MOI = 10 for both vectors)	Adherent growth in 100× 15 cm^2 plates, infection at a cell density of 10^7 cells per plate	Production Ad5 wt: 1.9×10^{10} total TU, 2.2×10^{13} total gc Sub100r: 7.8 (± 0.8) × 10^{10} TU, 5.9 (± 0.6) × 10^{14} gc, about 100× higher yields than for the HEK293 tritransfection system
K209 (131)	In order to eliminate the potential risk of the presence of human papilloma virus sequences a similar cell line was developed based on the use of A549 (ATCC CCL-185) cells. Establishment as for the HeLa B50 clone, after selection, 3 out of 800 clones showed high vector production; one of them was K209. The same production protocol as for HeLa B50 cells used	Adherent growth in F-12 K, 10% FCS	Production A tenfold lower MOI of Ad is sufficient for inducing vector production comparable to HeLa B50 cells

(continued)

Table 7 (continued)

Name/references	Description	Culture/production conditions	Further information on selected clones
HeRC32 (136)	A HeLa based packaging cell line was established by cotransfecting the rep/cap containing plasmid pspRC (the ITR deleted AAV genome (part 190–4484 bp) was excised as an Xba1 fragment from psub201 plasmid (137) and inserted into the Xba1 site of pSP72 (→pspRC plasmid)) together with the PGK-neo	DMEM, 10%FCS	One of the selected clones is the HeRC32 cell line. After transfection with AAVCMVnlsLacZ vector plasmid and infection with adenovirus (MOI = 50), the cells produced about 10^5 particles/cell which is about 1–2 orders of magnitude more than measured for HEK293 cells. For developing a stable producer cell line, this cell line was cotransfected with a rAAV plasmid (e.g., AAVCMVnlzLacZ) and a plasmid conferring resistance to a drug, such as hygromycine B (138). For production the cells are infected with adenovirus or replication defective adenovirus (MOI 35), the production rates were comparable to tritransfected HEK293 cells (139)
A549 clon 10 (135)	An A549 based packaging cell line was established by transfecting the rep/cap containing plasmid pUC-ACGrep/cap (the ATG start codon was replaced by ACG in order to reduce the expression of toxic rep proteins) with the neo^r selectable marker. 1/1296 clones was rep/cap positive clone 10	Adherent growth in medium, 5%FCS; for suspension growth, a proprietary serum-free medium was used	For establishing a producer cell line, clone 10 was transduced with a recombinant AAV vector. Clones were selected via PCR analysis and evaluated by induction of AAV production (infection with adenovirus, MOI = 5) Production levels were $>10^4$ particles/cell. The production system was evaluated at a 15 L scale. The cells were free of wtAAV and were stable for more than 40 passages

BHK-21 C13 (Baby Hamster Kidney; ATCC CCL-10) (132, 133)	Fibroblast like cell line established from *Mesocricetus auratus* (hamster, Syrian golden). This cell line was used for the production of rAAV using the recombinant herpes simplex type 1 production system (133). A two HSV infection system is used: one rHSV containing the rAAV-GOI (called rHSV/AAV-GOI; GOI for gene of interest) is based on the replication-deficient HSV d27.1 containing the rAAV cassette in the TK gene of the HSV backbone, and the second one containing the AAV rep and cap genes (called: rHSV-rep2/capx; x for the different capsid serotypes) Advantages vis-à-vis the use of HEK293 cells: 3 times lower MOI is needed for similar AAV production and BHK-21 cells grow faster than HEK293 cells	BHK-21 C13 cells grow adherently in serum-containing medium, but can be adapted to suspension growth in serum-free medium with a reduced Ca^{2+} concentration Adherent growth in EMEM or DMEM, 5%FCS; for suspension growth, different serum-free media can be used: MDSS2 ((140) and others)	Vector production Basic studies were done using HEK293 cells (132): production of AAV2-GFP: 6×10^3 TU/c; 1.5×10^5 VG/c, which is about 30 times higher than for the HEK293 based tritransfection method Improved method is based on the use of BHK-21 C13 cells (133): production of AAV1-AAT (use of rHSV-rep2/cap1 at MOI = 4, and rHSV-AAT vector at MOI = 2): 85400 DRP/c, whereas, the HEK293 cell system produced at most 74600 DRP/c when the MOI of the rHSV-rep2/cap1 was 12) AAV serotypes 1, 2, 5, and 8 have been produced with this system (133)

Glossary

Ad	Adenovirus
AAV	Adeno-associated virus
ATCC	American Type Culture Collection
BHK	Baby hamster kidney (cell line)
CBER	Centers for Biologics Evaluation and Research
CMV	Cytomegalovirus
Cox 1	Cytochrome oxidase 1
CuO	Cumate operator
DMEM	Dulbecco's Modified Eagle Medium
DMSO	Dimethylsulphoxide
DRP	DNAse-resistant particles
DSMZ	Deutsche Sammlung von Mikroorganismen und Zellkulturen: (German Resource Centre for Biological Material)
E1, E4	Early genes of adenovirus
EBV	Epstein-Barr virus
ECACC	European Collection of Cell Cultures
EMEA (EMA)	European Medicine Agency
FBS	Fetal bovine serum
FCS	Fetal calf serum
FDA	Food and Drug Administration
G418	Geneticin
GaLV	Gibbon Ape leukemia Virus
GCCP	Good cell culture practice
GFP	Green fluorescent protein
GOI	Gene of interest
GP	Gag-pol
HAT	Hypoxanthine–Aminopterin–Thymidine
HBV	Hepatitis B virus
HCV	Hepatitis C virus
HEK	Human embryonic kidney (cell line)
HIV	Human immunodeficiency virus
hGPK	Human phosphoglycerate kinase promoter
HSV	Herpes simplex virus
ICH	International Conference on Harmonisation
ICLC	Interlab Cell Line Collection
IP	Infectious particle
IRES	Internal ribosomal entry sites
ITR	Inverted terminal repeat
JCRB	Japanese Collection of Research Bioresources
LV	Lentivirus/lentiviral
LTR	Long terminal repeat
MCB	Master cell bank
MLV	Murine leukemia virus
MoLV	Molony leukemia virus
MOI	Multiplicity of infection

MpF	Mustela putoris furo (ferret)
MSCV	Murine Stem Cell Virus
NIH	National Institutes of Health
ORF	Open reading frame
P	Passage or promoter
PCR	Polymerase chain reaction
RCA	Replication-competent adenovirus
RCAAV	Replication-competent adeno-associated virus
RCL	Replication-competent lentivirus
RCR	Replication-competent retrovirus
rtTa2^{S}-m2	Reverse transactivator (rtTA2S-M2) of the tetracycline (Tet)
SFM	Serum-free medium
SIN	Self-inactivating (vector)
SV	Simian virus
TetR	Tetracyclin resistance
TK	Thymidine kinase
TNCL	Tn5 cell line
TU	Transducing unit
VSV	Vesicular stomatitis virus
Vg/vg	Vector genome
WCB	Working cell bank
Wt/WT	Wild type

References

1. ICH (1997) ICH Topic Q5 D Quality of Biotechnological Products: Derivation and Charaerisation of cell Substrates Use for Production of Biotechnological/Biological Products. CPMP/ICH/294/95. ICH Technical Coordination, European Medicines Evaluation Agency, London, UK.
2. ICH (1998) Guidance on Quality of Biotechnological/Biological Products: Derivation and Characterization of Cell Substrates Used for Production of Biotechnological/Biological Products. *Fed. Reg.* **63**, 50244–50249.
3. Knezevic, I., Stacey, G., Petricciani, J., Sheets, R., and the WHO Study Group on Cell Substrates. (2010) Evaluation of cell substrates for the production of biologicals: Revision of WHO recommendations. Report of the WHO Study Group on Cell Substrates for the Production of Biologicals, 22–23 April 2009, Bethesda, USA. *Biologicals.* **38**, 162–169.
4. WHO (Expert Committee on Biological Standardisation and Executive Board) (1998) Requirements for the use of animal cells as in vitro substrates for the production of biologicals. (Requirements for biological substances no. 50) WHO Technical Report Series 848, WHO, Geneva, Switzerland.
5. Phillips, H. J. (1973) In: Tissue Culture: Methods and Applications (ed. P.F. Kruse Jr. and M. K. Patterson), pp. 406–408, Academic Press, NJ, USA.
6. Patterson, M. K. (1979) Measurement of growth and viability of cells in culture. *Methods Enzymol.* **58**, 141–152.
7. Freshney, R. I. (2005) Culture of Animal Cells: A Manual of Basic Techniques, 5th edition, Wiley-Liss, NY, USA.
8. Carrier, T., Donahue-Hjelle, L., and Stramaglia, M.J. (2009) Banking parental cells according to cGMP guidelines. *BioProcess International* 7, 20–25.
9. Stacey, G. N., and Masters, J. R. (2008) Cryopreservation and banking of mammalian cell lines. *Nature Protocols* **3**, 1981–1989.
10. Gartler, S.M. (1967) Genetic markers as tracers in cell culture. *Natl. Cancer Inst. Monogr.* **26**, 167–195.
11. Nelson-Rees, W. A., Daniels, D. W., and Flandermeyer, R. R. (1989) Cross-contamination of cells in culture. *Science* **212**, 446–452.

12. MacLeod, R. A., Dirks, W. G., Matsuo, Y., Kaufmann, M., Milch, H., and Drexler, H. G. (1999) Widespread intra-species cross-contamination of human tumor cell lines arising at source. *Int. J. Cancer* **12**, 555–563.
13. Melcher, R., Maisch, S., Koehler, S., Bauer, M., Steinlein, C., Schmid, M., et al. (2005) SKY and genetic fingerprinting reveal a cross-contamination of the putative normal colon epithelial cell line NCOL-1. *Cancer Genet. Cytogenet.* **158,** 84–87.
14. Masters, J. R., Thomson, J. A., Daly-Burns, B., Reid, Y. A., Dirks, W. G., Packer, P., et al. (2001) Short tandem repeat profiling provides an international reference standard for human cell lines. *Proc. Natl. Acad. Sci. USA* **98**, 8012–8017.
15. Parson, W., Kirchebner, R., Mühlmann, R., Renner, K., Kofler, A., Schmidt, S., et al. (2005) Cancer cell line identification by short tandem repeat profiling: power and limitations. *FASEB J.* **19**, 434–436.
16. Hellmann, A. P., Rohleder, U., Eichmann, C., Pfeiffer, I., Parson, W., and Schleenbecker, U. (2006) A proposal for standardization in forensic canine DNA typing: allele nomenclature of six canine-specific STR loci. *J. Forensic. Sci.* **51**, 274–281.
17. Raveendran, M., Harris, R. A., Milosavljevic, A., Johnson, Z., Shelledy, W., Cameron, J., et al. (2006) Designing new microsatellite markers for linkage and population genetic analyses in rhesus macaques and other non-human primates. *Genomics* **88**, 706–710.
18. Smith, D. G., Kanthaswamy, S., Viray, J., and Cody, L. (2000) Additional highly polymorphic microsatellite (STR) loci for estimating kinship in rhesus macaques (Macaca mulatta). *Am. J. Primatol.* **50**, 1–7.
19. Stacey, G. N., Hoelzl, H., Stephenson, J. R. and Doyle, A. (1997) Authentication of animal cell cultures by direct visualisation of DNA, Aldolase gene PCR and isoenzyme analysis. *Biologicals* **25**, 75–83.
20. Folmer, O., Black, M., Hoeh, W., Lutz, R., and Vrijenhoek, R. (1994) DNA primers for amplification of mitochondrial cytochrome c oxidase subunit I from diverse metazoan invertebrates. *Mol. Mar. Biol. Biotechnol.* **3**, 294–299.
21. Stacey, G., Byrne, E. and Hawkins, J.R. (2007) DNA fingerprinting and the characterisation of Animal Cell Lines. In: Animal Cell Biotechnology, 2nd Edition. Ed Poertner, R, Humana press, Totowa, NJ, pp123–145.
22. European Pharmacopeia (2007). European Pharmacopeia section 2.6.1 (Sterility) (6th Edition), Supplement 8, Maisonneuve SA, Sainte Ruffine, pp. 5795–5797 (www.pheur.org).
23. US Food and Drug Administration (2005) Title 21, Code of Federal Regulations, Volume 7, revised April 2005, CFR610.12 (Sterility), FDA, Department of Health and Human Services.
24. European Pharmacopeia (2007). European Pharmacopeia section 2.6.7 (Mycoplasma) (6th Edition), Supplement 6.1, Maisonneuve SA, Sainte Ruffine, pp. 3317–3322.
25. US Food and Drug Administration (2005) Title 21, Code of Federal Regulations, Volume 7, revised April 2005, CFR610.30 (Test for Mycoplama), FDA, Department of Health and Human Services.
26. Lelong-Rebel, I. H., Piemont, Y., Fabre, M., and Rebel, G. (2009) Mycobacterium avium-intracellulare contamination of mammalian cell cultures. *In Vitro Cell. Dev. Biol. Anim.* **45**, 75–90.
27. Stacey, G. N. (2007) Risk assessment of cell culture procedures. In: Medicines from Animal Cell Culture. Eds Stacey G. and Davis J., John Wiley & Sons Ltd., pp. 569–588.
28. Ecker, D. J., Sampath, R., Massire, C., Blyn, L. B., Hall, T. A., Eshoo, M. W., et al. (2008) Ibis T5000: a universal biosensor approach for microbiology. *Nat. Rev. Microbiol.* **6**, 553–558.
29. Merten, O.-W., and Audit, M. (2003) Gene therapy – general safety tests and vector specific safety issues. In. Proceedings of the EDQM-Meeting on "Standardisation and quality control – cell and gene therapy products", pp. 35–57, Strasbourg/F, 24th-25th February, 2003, © Council of Europe.
30. CBER (1998) Guidance for Industry – Guidance for Human Somatic Cell Therapy and Gene Therapy. U.S. Department of Health and Human Services – Food and Drug Administration.
31. Rowe W. P., Pugh, W. E., and Hartley, J. W. (1970) Plaque assay techniques for murine leukemia viruses. *Virology* **42**, 1136–1139.
32. Haapala, D. K., Robey, W. G., Oroszlan, S. D., and Tsai, W. P. (1985) Isolation from cats of an endogenous type C virus with a novel envelope glycoprotein. *J. Virol.* **53**, 827–833.
33. Markowitz, D., Goff, S., and Bank, A. (1988) A safe packaging line for gene transfer: separating viral genes on two different plasmids. *J Virol.* **62**, 1120–1124.
34. Markowitz, D., Goff, S., and Bank, A. (1988) Construction and use of a safe and efficient

amphotropic packaging cell line. *Virology.* **167**, 400–406.

35. Kim, Y.-S., Lim, H. K., and Kim, K. J. (1998) Production of high-titer retroviral vectors and detection of replication-competent retroviruses. *Mol. Cells.* **8**, 36–42.
36. Chen, J., Reeves. L., and Cornetta, K. (2001) Safety testing for replication-competent retrovirus associated with Gibbon Ape Leukemia Virus-pseudotypes retroviral vector. *Hum. Gene Ther.* **12**, 61–70.
37. Audit, M., and Cosset, F. L. (2001) Plasmide chimère comprenant des séquences GAG, POL et enveloppes d'origine rétrovirales et utlisations. French patent. n° 01.14976, date 20.1.01.
38. Farson, D., Witt, R., McGuinness, R., Dull, T., Kelly, M., Song, J., et al. (2001) A new-generation stable inducible packaging cell line for lentiviral vectors. *Hum. Gene Ther.* **12**, 981–997.
39. Smith, K. T., Shepherd, A. J., Boyd, J. E., and Lees, G. M. (1996) Gene delivery systems for use in gene therapy: an overview of quality assurance and safety issues. *Gene Ther.* **3**, 190–200.
40. Ma, D., Newman, A., Lucas, W. T., Meloro, R. N., Rudderow, L., Hughes, J. V., et al. (2002) Methods for detection and evaluation of replication competent adenovirus (RCA). *BioProcessing Fall* 26–30.
41. Koeberl, D. D., Alexander, I. E., Halbert, C. L., Russell, D. W., and Miller, A. D. (1997) Persistant expression of human clotting factor IX from mouse liver after intravenous injection of adeno-associated virus vectors. *Proc. Natl. Acad. Sci. USA* **94**, 1426–1431.
42. Cao, L., During, M., and Xiao, W. (2002) Replication competent helper functions for recombinant AAV vector generation. *Gene Ther.* **9**, 1199–1206.
43. Coecke, S., Balls, M., Bowe, G., Davis, J., Gstraunthaler, G., Hartung, T., et al. (2005), Guidance on Good Cell Culture Practice. A report of the second ECVAM Task Force on Good Cell Culture Practice. *ATLA* **33**, 1–27.
44. Pegg, D. (2007) Fundamentals of cryopreservation. Cryopreservation and Freeze-drying Methods. Eds. Day, D. G. and Stacey, G. N., Humana Press, Totowa, USA.
45. Gray, J. S., Birmingham, J. M., and Fenton, J. I. (2009) Got black swimming dots in your cell culture? Identification of Achromobacter as a novel cell culture contaminant. *Biologicals* **38**, 273–277.
46. McDaniel, L. D. and Schultz, R. A. (1993) Elevation of sister chromatid exchange frequency in transformed human fibroblasts following exposure to widely used aminoglycosides. *Environ. Mol. Mutagen* **21**, 67–72.
47. Graham, F. L., Smiley, J., Russell, W. C., and Nairn, R. (1977) Characteristics of a human cell line transformed by DNA from human adenovirus type 5. *J. Gen. Virol.* **36**, 59–74.
48. Louis, N., Evelegh, C., and Graham, F. L. (1997). Cloning and sequencing of the cellular-viral junctions from the human adenovirus type 5 transformed 293 cell line. *Virology* **233**, 423–429.
49. Lewis, A. M. Jr., Krause, P., and Peden, K. (2001) A defined-risks approach to the regulatory assessment of the use of neoplastic cells as substrates for viral vaccine manufacture. *Dev. Biol.* **106**, 513–535.
50. Shaw, G., Morse, S., Ararat, M., and Graham, F. L. (2002) Preferential transformation of human neuronal cells by human adenoviruses and the origin of HEK 293 cells. *FASEB J.* **16**, 869–871.
51. Park, M. T., Lee, M. S., Kim, S. H., Jo, E. C., and Lee, G. M. (2004) Influence of culture passages on growth kinetics and adenovirus vector production for gene therapy in monolayer and suspension cultures of HEK 293 cells. *Appl. Microbiol. Biotechnol.* **65**, 553–558.
52. Shen, C., Gu, M., Song, C., Miao, L., Hu, L., Liang, D., et al. (2008) The tumorigenicity diversification in human embryonic kidney 293 cell line cultured in vitro. *Biologicals* **36**, 263–268.
53. Jardon, M., and Garnier, A. (2003) pH, pCO_2, and termperature effect on r-adenovirus production. *Biotechnol. Prog.* **19**, 202–208.
54. Kotani, H., Newton, P. B. 3rd, Zhang, S., Chiang, Y. L., Otto, E., Weaver, L., et al. (1994) Improved methods of retroviral vector transduction and production for gene therapy. *Hum. Gene Ther.* **5**, 19–28.
55. Kaptein, L. C., Greijer, A., Valerio, D., and van Beusechem, V.W. (1997) Optimized conditions for the production of recombinant amphotropic retroviral vector preparations. *Gene. Ther.* **4**, 172–176.
56. Le Doux, J. M., Davis, H. E., Morgan, J. R., and Yarmush, M. L. (1999) Kinetics of retrovirus production and decay. *Biotechnol. Bioeng.* **63**, 654–662.
57. Merten, O.-W., Charrier, S., Laroudie, N., Fauchille, S., Dugué, C., Jenny, C., et al. (2011) Large scale manufacture and characterisation of a lentiviral vector produced for

clinical *ex vivo* gene therapy application. *Hum. Gene Ther.* **22**, 343–356.

58. Pear, W. S., Nolan, G. P., Scott, M. L., and Baltimore, D. (1993) Production of high-titer helper-free retroviruses by transient transfection. *Proc. Natl. Acad. Sci. USA.* **90**, 8392–8396.
59. Pensiero, M., et al. Retroviral vectors produced by producer cell lines resistant to lysis by human serum. US Patent 6,329,199 dated Dec 11 2001.
60. Ikeda, Y., Takeuchi, Y., Martin, F., Cosset, F. L., Mitrophanous, K., and Collins, M. (2003) Continuous high-titer HIV-1 vector production. *Nat. Biotechnol.* **21**, 569–572.
61. Throm, R. E., Ouma, A. A., Zhou, S., Chandrasekaran, A., Lockey, T., Greene, M. et al. (2009) Efficient construction of producer cell lines for a SIN lentiviral vector for SCID-X1 gene therapy by concatemeric array transfection. *Blood* **113**, 5104–5110.
62. Stewart, H. J., Leroux-Carlucci, M. A., Sion, C. J., Mitrophanous, K. A., and Radcliffe, P. A. (2009) Development of inducible EIAV-based lentiviral vector packaging and producer cell lines. *Gene Ther.* **16**, 805–814.
63. Segura, M. M., Garnier, A., Durocher, Y., Coelho, H., and Kamen, A. (2007) Production of lentiviral vectors by large-scale transient transfection of suspension cultures and affinity chromatography purification. *Biotechnol. Bioeng.* **98**, 789–799.
64. Durocher, Y., Pham, P. L., St-Laurent, G., Jacob, D., Cass, B., Chahal, P. et al. (2007) Scalable serum-free production of recombinant adeno-associated virus type 2 by transfection of 293 suspension cells. *J. Virol. Meth.* **144**, 32–40.
65. Torrent, C., Bordet, T., and Darlix, J. L. (1994) Analytical study of rat retrotransposon VL30 RNA dimerization in vitro and packaging in murine leukemia virus. *J. Mol. Biol.* **240**, 434–444.
66. Cosset, F., Takeuchi, Y., Battini, J., Weiss, R. A., and Collins, M. K. L. (1995) High-titer packaging cells producing recombinant retroviruses resistant to human serum. *J. Virol.* **69**, 7430–7436.
67. Pensiero, M. N., Wysocki, C. A., Nader, K., and Kikuchi, G. E. (1996) Development of amphotropic murine retrovirus vectors resistant to inactivation by human serum. *Hum. Gene Ther.* 7, 1095–1101.
68. Rigg, R. J., Chen. J., Dando, J. S., Forestell, S. P., Plavec, I., and Böhnlein, E. (1996) A novel human amphotropic packaging cell line: high titer, complement resistance, and improved safety. *Virology* **218**, 190–195.
69. Forestell, S. P., Dando, J. S., Chen, J., de Vries, P., Böhnlein, E., and Rigg, R. J. (1997) Novel retroviral packaging cell lines: complementary tropisms and improved vector production for efficient gene transfer. *Gene Ther.* **4**, 600–610.
70. Sheridan, P. L., Bodner, M., Lynn, A., Phuong, T. K., DePolo, N. J., de la Vega, D. J. Jr., et al. (2000) Generation of retroviral packaging and producer cell lines for large-scale vector production and clinical application: improved safety and high titer. *Mol. Ther.* **2**, 262–275.
71. Patience, C., Takeuchi, Y., Cosset, F.-L., and Weiss, R. A. (2001) MuLV packaging systems as models for estimating/measuring retrovirus recombination frequency. *Dev. Biol.* **106**, 169–179.
72. Duisit, G., Salvetti, A., Moullier, P., and Cosset, F.-L. (1999) Functional characterization of adenoviral/retroviral chimeric vectors and their use for efficient screening of retroviral producer cell lines. *Hum. Gene Ther.* **10**, 189–200.
73. Pizzato, M., Merten, O.W., Blair, E.D., and Takeuchi, Y. (2001) Development of a suspension packaging cell line for production of high titre, serum-resistant murine leukemia virus vector. *Gene Ther.* **8**, 737–745.
74. Jenkins, N. A., Copeland, N. G., Taylor, B. A., and Lee, B. K. (1982) Organization, distribution, and stability of endogenous ecotropic murine leukemia virus DNA sequences in chromosomes of Mus musculus. *J. Virol.* **43**, 26–36.
75. Hatzoglou, M., Hodgson, C. P., Mularo, F., and Hanson, R. W. (1990) Efficient packaging of a specific VL30 retroelement by psi 2 cells which produce MoMLV recombinant retroviruses. *Hum. Gene Ther.* **1**, 385–397.
76. Farson, D., McGuinness, R., Dull, T., Limoli, K., Lazar, R., Jalali, S., et al. (1999) Large-scale manufacturing of safe and efficient retrovirus packaging lines for use in immunotherapy protocols. *J. Gene Med.* **1**, 195–209.
77. Davis, J. L., Witt, R. M., Gross, P. R., Hokanson, C. A., Jungles, S., Cohen, L. K., et al. (1997) Retroviral particles produced from a stable human-derived packaging cell line transduce target cells with very high efficiencies. *Hum Gene Ther.* **8**, 1459–1467.
78. Jolly, D. (1994) Viral vector systems for gene therapy. *Cancer Gene Ther.* **1**, 51–64.
79. Mason, J. M., Guzowski, D. E., Goodwin, L. O., Porti, D., Cronin, K. C., Teichberg, S., et al. (1999) Human serum-resistant retroviral vector, particles from galactosyl containing

nonprimate cell lines. *Gene Ther.* **6**, 1397–1405.
80. Rigg, R. J., et al. Method for obtaining retroviral packaging cell lines producing high transducing efficiency retroviral supernatant. US Patent 6,017,761 dated Jan 25 2000.
81. Ghani, K., Cottin, S., Kamen, A., and Caruso, M. (2007) Generation of a high-titer packaging cell line for the production of retroviral vectors in suspension and serum-free media. *Gene Ther.* **14**, 1705–1711.
82. Schucht, R., Coroadinha, A. S., Zanta-Boussif, M. A., Verhoeyen, E., Carrondo, M. J., Hauser, H., et al. (2006) A new generation of retroviral producer cells: predictable and stable virus production by Flp-mediated site-specific integration of retroviral vectors. *Mol. Ther.* **14**, 285–292.
83. Coroadinha, A. S., Schucht, R., Gama-Norton, L., Wirth, D., Hauser, H., and Carrondo, M. J. (2006) The use of recombinase mediated cassette exchange in retroviral vector producer cell lines: predictability and efficiency by transgene exchange. *J. Biotechnol.* **124**, 457–468.
84. Spitzer, D., Hauser, H., and Wirth, D. (1999) Complement-protected amphotropic retroviruses from murine packaging cells. *Hum. Gene Ther.* **10**, 1893–1902.
85. Loew, R., Meyer, Y., Kuehlcke, K., Gama-Norton, L., Wirth, D., Hauser, H., et al. (2010) A new PG13-based packaging cell line for stable production of clinical-grade self-inactivating gamma-retroviral vectors using targeted integration. *Gene Ther.* **17**, 272–280.
86. Slepushkin, V., Chang, N., Cohen, R., Gan, Y., Jiang, B., Deausen, E., et al. (2003) Large-scale purification of a lentiviral vector by size exclusion chromatography or Mustang Q ion exchange chromatography. *Bioproc. J.* September/October 2003, 89–95.
87. Couture, L. A. (2008) Vector production in support of early clinical trials. Presented at the ASGT Meeting, Boston/USA, 28.5.–1.6.08.
88. Dupont, F. (2008) Large scale manufacturing of a lentiviral vector (ProSavin®) for phase I/II clinical trial. Presented at the CONSERT Labcourse, Evry/F, 29.6.–1.7.08.
89. Negré, O., Denaro, M., Gillet-Legrand, B., Fusil, F., Hehir, K., Dorazio, R., et al. (2008) Long-term correction of murine beta-thalassemia following busulfan conditioning and transplant of bone marrow transduced with clinical-grade lentiviral vector (LentiGlobin™). *Mol. Ther.* **16**, S85.
90. Côté, J., Garnier, A., Massie, B., and Kamen, A.. (1998) Serum-free production of recombinant proteins and adenoviral vectors by 293SF-3 F6 cells. *Biotechnol. Bioeng.* **59**, 567–575.
91. Ansorge, S., Lanthier, S., Transfiguracion, J., Durocher, Y., Henry, O., and Kamen, A. (2009) Development of a scalable process for high-yield lentiviral vector production by transient transfection of HEK293 suspension cultures. *J. Gene Med.* **11**, 868–876.
92. Klages, N., Zufferey, R., and Trono, D. (2000) A stable system for the high-titer production of multiply attenuated lentiviral vectors. *Mol. Ther.* **2**, 170–176.
93. Ni, Y., Sun, S., Oparaocha, I., Humeau, L., Davis, B., Cohen, R., et al. (2005) Generation of a packaging cell line for prolonged large-scale production of high-titer HIV-1-based lentiviral vector. *J. Gene Med.* **7**, 818–834.
94. Xu, K, Ma, H., McCown, T. J., Verma, I. M., and Kafri, T. (2001) Generation of a stable cell line producing high-titer self-inactivating lentiviral vectors. *Mol. Ther.* **3**, 97–104.
95. Brousseau, S., Jabbour, N., Lachapelle, G., Durocher, Y., Tom, R., Transfiguracion, J., et al. (2008) Inducible packaging cells for large-scale production of lentiviral vectors in serum-free suspension culture. *Mol. Ther.* **16**, 500–507.
96. Rohll, J. B., Mitrophanous, K. A., Martin-Rendon, E., Ellard, F. M., Radcliffe, P. A., Mazarakis, N.D., et al. (2002) Design, production, safety, evaluation, and clinical applications of nonprimate lentiviral vectors. *Methods Enzymol.* **346**, 466–500.
97. Kaplan, A. H., and Swanstrom, R. (1991) The HIV-1 gag precursor is processed via two pathways: implications for cytotoxicity. *Biomed. Biochim. Acta.* **50**, 647–53.
98. Karacostas, V., Wolffe, E.J., Nagashima, K., Gonda, M.A., and Moss, B. (1993) Overexpression of the HIV-1 gag-pol polyprotein results in intracellular activation of HIV-1 protease and inhibition of assembly and budding of virus-like particles. *Virology* **193**, 661–671.
99. Kafri, T., van Praag, H., Ouyang, L., Gage, F. H., and Verma, I. M. (1999) A packaging cell line for lentivirus vectors. *J. Virol.* **73**, 576–584.
100. Zufferey, R. (2002) Production of lentiviral vectors. *Curr. Top. Microbiol. Immunol.* **261**, 107–121.
101. Sodroski, J., Goh, W. C., Rosen, C., Campbell, K., and Haseltine, W.A. (1986) Role of the HTLV-III/LAV envelope in

syncytium formation and cytopathicity. *Nature* **322**, 470–474.
102. Imler, J. L., Chartier, C., Dreyer, D., Dieterle, A., Sainte-Marie, M., Faure, T., et al. (1996) Novel complementation cell lines derived from human lung carcinoma A549 cells support the growth of E1-deleted adenovirus vectors. *Gene Ther.* **3**, 75–84.
103. Fallaux, F. J., Bout, A., van der Velde, I., van den Wollenberg, D. J., Hehir, K. M., Keegan, J., et al. (1998) New helper cells and matched early region 1-deleted adenovirus vectors prevent generation of replication-competent adenoviruses. *Hum. Gene Ther.* **9**, 1909–1917.
104. Copeland, N. G., and Cooper G. M. (1979) Transfection by exogenous and endogenous murine retrovirus DNAs. *Cell* **16**, 347–356.
105. Copeland, N. G., Zelenetz, A. D., and Cooper, G. M. (1979) Transformation of NIH/3T3 mouse cells by DNA of Rous sarcoma virus. *Cell* **17**, 993–1002.
106. Markowitz, D., Hesdorffer, C., Ward, M., Goff, S., and Bank, A. (1990) Retroviral gene transfer using safe and efficient packaging cell lines. *Ann. N. Y. Acad. Sci.* **612**, 407–414.
107. Danos, O., and Mulligan, R. C. (1988) Safe and efficient generation of recombinant retroviruses with amphotropic and ecotropic host ranges. *Proc. Natl. Acad. Sci. USA* **85**, 6460–6464.
108. Merten, O.-W., Cornet, V., Petres, S., Couvé, E., and Heard, J. M. (1996) Large scale production of retro-virus vectors. *Cytotechnology* **21**, 8.
109. Miller, A. D., Garcia, J. V., von Suhr, N., Lynch, C. M., Wilson, C., and Eiden, M. V. (1991) Construction and properties of retrovirus packaging cells based on gibbon ape leukemia virus. *J. Virol.* **65**, 2220–2224.
110. Fallaux, F. J., Kranenurg, O. Cramer, S. J., Houweling, A., Van Ormondt, H., Hoeben, R. C., et al. (1996) Characterization of 911: a new helper cell line for the titration and propagation of early region 1-deleted adenoviral vectors. *Hum. Gene Ther.* 7, 215–222.
111. Schiedner, G., Hertel, S., and Kochanek, S. (2000) Efficient transformation of primary human amniocytes by E1 functions of Ad5: generation of new cell lines for adenoviral vector production. *Hum. Gene Ther.* **11**, 2105–2116.
112. Xu, Q., Arevalo, M. T., Pichichero, M. E., and Zeng, M. (2006) A new complementing cell line for replication-incompetent E1-deleted adenovirus propagation. *Cytotechnology.* **51**, 133–140.
113. Nadeau, I., and Kamen, A. (2003) Production of adenovirus vector for gene therapy. *Biotechnol. Adv.* **20**, 475–489.
114. Zhang, S., Thwin, C., Wu, Z., and Cho, T. (1998) An improved method for the production and purification of adenoviral vectors. International Patent, WO 98/22588.
115. Iyer, P., Ostrove, J. M., and Vacante, D. (1999) Comparison of manufacturing techniques for adenovirus production. *Cytotechnology* **30**, 169–172.
116. Nadeau, I., Seanez, G., and Wu, F. (2001) Adenovirus production in 293 cells: a comparative study between a suspension cell and an adherent cell process. Presented at the 17th ESACT Meeting, Tylösand, Sweden, June 10–14, 2001.
117. Irish, T., Baker, W., Fresner, B., Abraham, G., Tvijn, C., Lardenoije, R., et al. (2000) A comparative study of large-scale production strategies used to produce RCA free adenovirus preparations in serum-free media. Research Report. JRH Bioscience, Lenexa, KS, USA.
118. Liu, L. C., and Shoupeng, L. (2001) Method of producing adenoviral vector stocks. US Patent No. 618941 B1, Jan 2001.
119. Chaubard, J. F. (2000) Serum-free media and bioreactor strategies for manufacturing adenoviral gene therapy vectors. Viral vectors. Viral vectors and vaccines, Lake Tahoe, NV, Nov. 6–9, 2000.
120. Garnier, A., Cortin, V., Thibault, J., and Jacob, D. (2002) Production of recombinant adenovirus by 293 cells cultures in perfusion. Cell culture engineering VIII, April 1–6, 2002, Snowmass, CO.
121. Nadeau, I., Seanez, G., and Wu, F. (2002) Optimization of a 293 suspension process for adenovirus production. Cell Culture Engineering VIII, April 1–6, Snowmass, CO.
122. Parks, R. J., Chen, L., Anton, M., Sankar, U., Rudnicki, M. A., and Graham, F. L. (1996) A helper-dependent adenovirus vector system: removal of helper virus by Cre-mediated excision of the viral packaging signal. *Proc. Natl. Acad. Sci. USA.* **93**, 13565–13570.
123. Alba, R., Bosch, A., and Chillon, M. (2005) Gutless adenovirus: last-generation adenovirus for gene therapy. *Gene Ther.* **12**: Suppl 1, S18–S27.
124. Barjot, C., Hartigan-O'Connor, D., Salvatori, G., Scott, J. M., and Chamberlain, J. S. (2002) Gutted adenoviral vector growth using E1/E2b/E3-deleted helper viruses. *J. Gene Med.* **4**, 480–489.

125. Puck, T. T., and Fisher, H. W. (1956) Genetics of somatic mammalian cells: I. Demonstration of the existence of mutants with different growth requirements in a human cancer cell strain (HeLa). *J. Exp. Med.* **104**, 427–434.

126. Darnell, J. E. Jr., Eagle, H., and Sawyer, T. K. (1959) The effect of cell population density on the amino acid requirements for poliovirus synthesis in HeLa cells. *J. Exp. Med.* **110**, 445–450.

127. Forestell, S., Celeri, C., Dang, C., Gong, T., Olsen, M., Sifi, I., et al. (2005) Comparison of host cell lines and production methods for a new generation of oncolytic adenoviral vectors. In: F. Godia and M. Fussenegger teds.), Animal Cell Technology meets genomics, pp. 309–316, Springer, Dordrecht/NL.

128. Yuk, I. H., Olsen, M. M., Geyer, S., and Forestell, S. P. (2004) Perfusion cultures of human tumor cells: a scalable production platform for oncolytic adenoviral vectors. *Biotechnol. Bioeng.* **86**, 637–642.

129. Merten, O.-W., Gény-Fiamma, C., and Douar, A. M. (2005) Current issues in adeno-associated viral vector production. *Gene Ther.* **12**, Suppl 1, S51–S61.

130. Gao, G.-P., Qu, G., Faust, L.Z., Engdahl, R.K., Xiao, W., Hughes, J.V., et al. (1998) High-titer adeno-associated viral vectors from a rep/cap cell line and hybrid shuttle virus. *Hum. Gene Ther.* **9**, 2353–2362.

131. Gao, G.-P., Lu, F., Sanmiguel, J.C., Tran, P.T., Abbas, Z., Lynd, K.S., et al. (2002) Rep/cap gene amplification and high-yield production of AAV in an A549 cell line expressing rep/cap. *Mol. Ther.* **5**, 644–649.

132. Clément, N., Knop, D. R., and Byrne, B. J. (2009) Large-scale adeno-associated viral vector production using a herpesvirus-based system enables manufacturing for clinical studies. *Hum. Gene Ther.* **20**, 796–806.

133. Thomas, D. L., Wang, L., Niamke, J., Liu, J., Kang, W., Scotti, M. M., et al. (2009) Scalable recombinant adeno-associated virus production using recombinant herpes simplex virus type 1 coinfection of suspension-adapted mammalian cells. *Hum. Gene Ther.* **20**, 861–870.

134. Urabe, M., Ding, C., and Kotin, R. M. (2002) Insect cells as a factory to produce adeno-associated virus type 2 vectors. *Hum. Gene Ther.* **13**, 1935–1943.

135. Farson, D., Harding, T. C., Tao, L., Liu, J., Powell, S., Vimal, V., et al. (2004) Development and characterization of a cell line for large-scale, serum-free production of recombinant adeno-associated viral vectors. *J. Gene Med.* **6**, 1369–1381.

136. Chadeuf, G., Favre, D., Tessier, J., Provost, N., Nony, P., Kleinschmidt, J., et al. (2000) Efficient recombinant adeno-associated virus production by a stable rep-cap HeLa cell line correlates with adenovirus-induced amplification of the integrated rep-cap genome. *J. Gene Med.* **2**, 260–268.

137. Samulski, R. J., Chang, L. S., and Shenk, T. (1989) Helper-free stocks of recombinant adeno-associated viruses: normal integration does not require viral gene expression. *J. Virol.* **63**, 3822–3828.

138. Blouin, V., Brument, N., Toublanc, E., Raimbaud, I., Moullier, P., and Salvetti, A. (2004) Improving rAAV production and purification: towards the definition of a scaleable process. *J. Gene Med.* **6**, Suppl 1, S223–S228.

139. Jenny, C., Toublanc, E., Danos, O., and Merten, O. W. (2005) Evaluation of a serum-free medium for the production of rAAV-2 using HeLa derived producer cells. *Cytotechnology* **49**, 11–23.

140. Merten, O.-W., Kierulff, J. V., Castignolles, N., and Perrin, P. (1994) Evaluation of the new serum-free medium (MDSS2) for the production of different biologicals: use of various cell lines. *Cytotechnology* **14**, 47–59.

141. Airenne, K. J., Mähönen, A. J., Laitinen, O. H., and Ylä-Herttuala.S. (2009) Baculovirus-mediated gene transfer: An emerging universal concept, in *Gene and cell therapy: Therapeutic mechanisms and strategies*, (Templeton, N. S. ed.), CRC Press, Boca Raton, pp. 263–307.

142. Volkman, L. E., and Goldsmith, P. A. (1983) In vitro Survey of *Autographa californica* Nuclear Polyhedrosis Virus interaction with nontarget vertebrate host cells. *Appl. Environ. Microbiol.* **45**, 1085–1093.

143. Condreay, J. P., Witherspoon, S. M., Clay, W. C., and Kost, T. A. (1999) Transient and stable gene expression in mammalian cells transduced with a recombinant baculovirus vector. *Proc. Natl. Acad. Sci. USA* **96**, 127–132.

144. Song, S. U., Shin, S. H., Kim, S. K., Choi, G. S., Kim, W. C., Lee, M. H., et al. (2003) Effective transduction of osteogenic sarcoma cells by a baculovirus vector. *J. Gen. Virol.* **84**, 697–703.

145. Hu, Y. C. (2008) Baculoviral vectors for gene delivery: a review. *Curr. Gene Ther.* **8**, 54–65.

146. Cheng, T., Xu, C. Y., Wang, Y. B., Chen, M., Wu, T., Zhang, J., and Xia, N. S. (2004) A rapid and efficient method to express target

genes in mammalian cells by baculovirus. *World J. Gastroenterol.* **10**, 1612–1618.

147. Granados, R. R. (1978) Replication phenomena of insect viruses *in vivo* and *in vitro*, in *Safety Aspects of Baculoviruses as Biological Insecticides*, (Miltenburger, H. G. ed.), Bundesministerium für Forschung und Technologie, Bonn, pp. 163–184.

148. Li, T. C., Scotti, P. D., Miyamura, T., and Takeda, N. (2007) Latent infection of a new alphanodavirus in an insect cell line. *J. Virol.* **81**, 10890–10896.

Chapter 4

Overview of Current Scalable Methods for Purification of Viral Vectors

María Mercedes Segura, Amine A. Kamen, and Alain Garnier

Abstract

As a result of the growing interest in the use of viruses for gene therapy and vaccines, many virus-based products are being developed. The manufacturing of viruses poses new challenges for process developers and regulating authorities that need to be addressed to ensure quality, efficacy, and safety of the final product. The design of suitable purification strategies will depend on a multitude of variables including the vector production system and the nature of the virus. In this chapter, we provide an overview of the most commonly used purification methods for viral gene therapy vectors. Current chromatography options available for large-scale purification of γ-retrovirus, lentivirus, adenovirus, adeno-associated virus, herpes simplex virus, baculovirus, and poxvirus vectors are presented.

Key words: Viral vectors, Gene therapy, Purification, Downstream processing, Centrifugation, Membrane filtration, Chromatography

1. Introduction

Viruses were first introduced as therapeutics more than 200 years ago with the intentional administration of the vaccinia virus to prevent smallpox disease. Whether killed or live attenuated, many more virus-based products were developed since then for vaccination purposes. More recently, the use of viruses as beneficial tools in medicine has regained interest with the emergence of gene therapy. Gene therapy offers great potential for the treatment of many inherited as well as acquired diseases. This relatively new therapeutic approach is likely to play an increasingly important role in health care throughout this century.

Like recombinant proteins and viral vaccines, viral vector therapeutics are scrutinized by regulatory authorities that demand increasingly stringent standards of purity, efficacy, and safety.

Otto-Wilhelm Merten and Mohamed Al-Rubeai (eds.), *Viral Vectors for Gene Therapy: Methods and Protocols*, Methods in Molecular Biology, vol. 737, DOI 10.1007/978-1-61779-095-9_4, © Springer Science+Business Media, LLC 2011

Direct adaptation of methods and technologies already developed for the downstream processing of recombinant proteins and vaccines, although tempting, is not as straightforward as originally envisioned. The different physico–chemical properties of viruses compared to proteins and the need to maintain viral activity as intact as possible throughout the purification process, considering the complex and fragile structure of virus particles, pose new challenges for process developers. Consequently, modifications to traditional purification approaches are required.

Downstream processing of viral vectors comprises a series of steps aimed at increasing the potency and purity of the vector preparation. Strategic design and step by step optimization of the purification process is crucial to maximize yield and quality of the final virus preparation. Considerable progress has been made in the area of downstream processing of viral vectors over the past 10 years. Most research effort has been focused on the development of purification strategies for the widely used adenoviral vectors (1–4), γ-retroviral and lentiviral vectors (5, 6), and adeno-associated viral vectors (AAV) (2). Less effort has been invested in developing and optimizing purification processes for baculovirus, herpes virus, and poxvirus vectors.

This chapter describes various methods available for downstream processing of viral gene therapy products. It intends to aid the reader in selecting the most appropriate methods to be used and define the order in which they should be used to achieve the best purification results. The chapter makes particular emphasis on scalable purification techniques.

2. Purification Strategy

Downstream processing begins with the harvest of viral vector particles from the cell culture. Depending on the virus being considered, viral particles may be enriched in the cellular fraction, the cell culture supernatant, or sometimes in both (Fig. 1). In cases where the virus remains located intracellularly, disruption of the cells is necessary to release viral particles. Clarification of the crude viral vector stocks (supernatant or lysate) typically follows to eliminate remaining producer cells and cell debris (Fig. 2). The concentration of the clarified viral stock at this early stage is often advantageous to reduce the volume of feed and consequently, the size of the equipment required in later operations (pumps, filters, columns, and vessels). This is especially true when dealing with extracellular viruses since the viral product is diluted in the cell culture medium. These initial downstream processing steps are primarily intended to remove cells, cell debris, and water. Some degree of purification may also be accomplished. However, the main

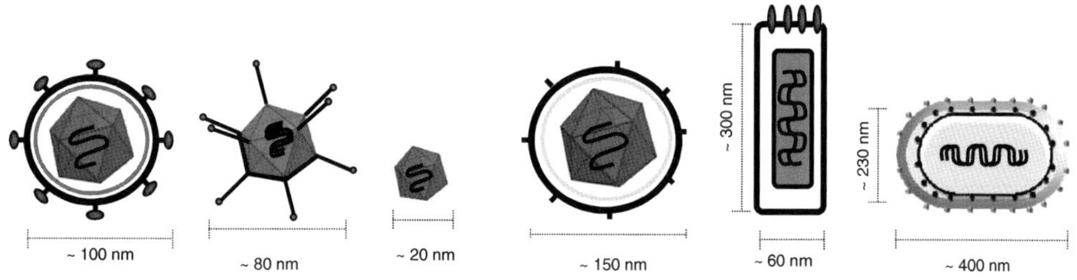

	Retrovirus/ Lentivirus	Adenovirus (Ad5)	AAV (AAV2)	Herpes simplex (HSV-1)	Baculovirus	Poxvirus
Family	Retroviridae	Adenoviridae	Parvoviridae	Herpesviridae	Baculoviridae	Poxviridae
Genome	Plus strand ssRNA	Linear dsDNA	Linear ssDNA	Linear dsDNA	Circular dsDNA	Linear dsDNA
Envelope	yes	no	no	yes	yes	yes
Stability	low	high	high	low	moderate	moderate
Net charge[(1)]	negative	negative	positive			positive
Density (g/cm^3)	1.16 [(2)]	1.34[(3)]	1.39[(3)]	1.26[(3)]		
Fraction	Supernatant	Cell lysate	Cell lysate	Supernatant	Supernatant	Cell lysate/supernatant
Yields (ivp/mL)	10^6-10^7	10^8-10^9	10^7-10^9	10^6-10^7	10^7-10^8	10^7-10^8
Insert size (kb)	~ 7.5	~ 7.5-35	~ 4.7	~ 30-40	~ 30	~ 25

(1) Overall charge at neutral pH
(2) Buoyant density in sucrose gradients
(3) Buoyant density in CsCl gradients

Fig. 1. Viral vector characteristics.

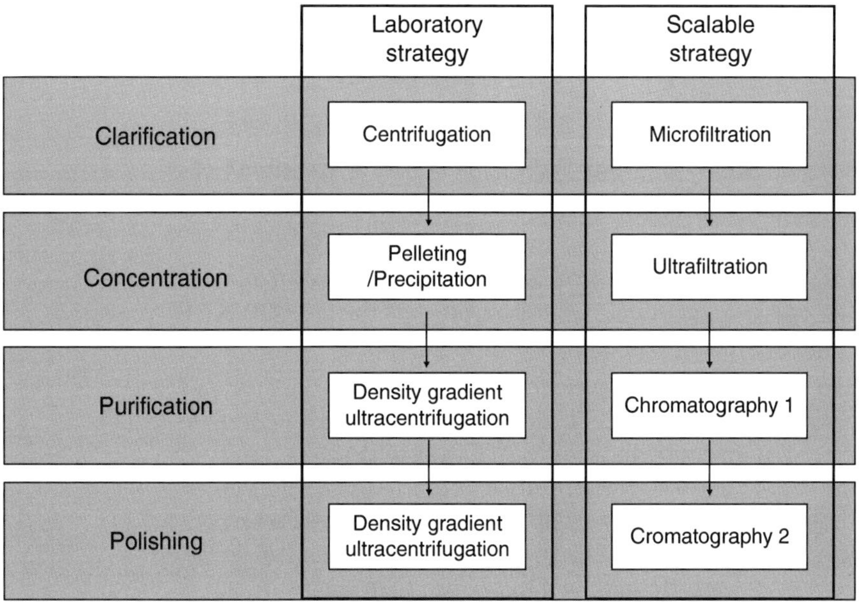

Fig. 2. General purification flow scheme.

purification issues are left to be resolved during the purification stage itself. Typically more than one purification step is required to bring the product to the desired level of purity. In the initial step, viral particles are separated from the most abundant contaminants contained in the vector stock. The polishing step is further introduced to remove remaining impurities and/or closely related species (e.g. defective vector forms and/or cell membrane vesicles).

Before deciding what techniques should be employed, it is key to define purification needs. First, it is important to know how pure the virus needs to be, which will mainly be determined by the end-product final application. For instance, if the viral vector preparation will be used in gene transfer experiments in vitro, it is likely that no purification will be needed or perhaps, at the most, a concentration step to increase the vector potency will be desirable. In contrast, viral vector preparations that will be used in gene transfer experiments ex vivo or in vivo usually need to undergo a series of purification steps to increase the potency and safety of the final product. Nonpurified vector preparations contain contaminating species that are toxic to cells and may reduce transduction efficiencies ex vivo and in vivo. These preparations also induce a systemic immune response and inflammation when injected in vivo. On the other hand, viral vector preparations destined for preclinical studies or clinical studies must attain extremely high levels of purity in accordance with regulatory requirements. Second, it is important to establish how much virus will be needed. The answer to this question will determine the working scale, which in turn will dictate which purification methods are most suitable in each particular case. Figure 2 outlines general purification schemes for viral gene therapy vectors at laboratory scale and large scale. As it can be observed in the schemes, purification of viral vectors at laboratory scale is typically accomplished using ultracentrifugation-based methods whereas, in large-scale purification approaches, scalable technologies such as membrane filtration and chromatography are preferred.

The selection of appropriate virus purification techniques will also depend heavily on the nature of the virus itself. Therefore, it is important to gather as much information as possible about the virus properties before designing a suitable purification scheme. Figure 1 shows some of the characteristics that are important to consider including virus particle size, net charge at neutral pH, relative particle stability, and typical vector yields. The size of viral vectors ranges between 20 nm (AAV vectors) up to 400 nm (poxvirus vectors) and it is significantly larger to that of proteins (typically <5 nm). This feature is extensively exploited for the separation of virus particles from cellular and culture medium derived proteins by employing separation techniques such as ultracentrifugation, membrane filtration, and size exclusion chromatography (SEC).

The net charge at neutral pH will depend on the virion surface composition that will differ for each viral vector type and serotype (Fig. 1). Depending on their net charge, either anion exchange or cation exchange chromatography techniques will be useful for purification. The overall virion charge is affected by the pH of the buffer employed and can be modulated with changes in pH. However, viruses often display a narrow window of pH stability and abrupt changes in pH could lead to virus inactivation. In the case of virus products for gene therapy, it is always important to retain biological activity of the virus preparation. Some viral vectors are more sensitive to changes in pH, buffer composition, temperature, and shear forces than others, with enveloped viruses typically being the most sensitive to such changes (Fig. 1). This will restrict the type of techniques that can be used and the conditions under which purification can be performed. In general, it is important to plan a purification scheme that contains as few steps as possible and minimum changes in buffer composition. Average yields of active virions produced per cell vary considerably among the different types of vectors (Fig. 1). These yields will ultimately determine the titer in crude vector stocks and will provide an idea of how many times a viral stock needs to be concentrated throughout the purification process to attain satisfactory vector potency in the final product.

Finally, crude viral stocks contain contaminants derived from producer cells, cell culture medium, and other substances added throughout vector production (e.g. plasmids, helper viruses, detergents). The composition and abundance of these contaminants will also guide the selection and order of methods used for purification. The harvesting point greatly influences vector yield and also the amount of cellular contaminants that may escape to the supernatant fraction. The choice of cell culture media will also affect the type/amount of contaminants that need to be eliminated, particularly for extracellular viruses. In this sense, several viral vector producer cell lines have been adapted to grow in serum-free media, which greatly facilitates downstream processing.

3. Virus Harvest

At the end of the production phase, virus particles are found enriched either inside or outside the producer cells. Often, naked viruses remain inside the cells until cell lysis occur and thus, can be concentrated in the cellular fraction. In contrast, enveloped viruses usually escape from the host cells by budding through cellular membranes and are diluted in the supernatant. In some cases, active viral particles are found in significant amounts in both the cells and the culture supernatant requiring both fractions to be processed.

This would be the case for the prototypical poxvirus vector (vaccinia virus) that produces four different forms of infectious particles that can be located intracellularly or extracellularly. In addition, adenovirus infection is often allowed to proceed until cell lysis occurs since this practice has shown to improve infectious: total particle ratios (7). In this case again, both the cellular fraction and supernatant are recovered in order to maximize vector recovery. The first downstream processing operation for intracellular viruses is cell lysis. Mammalian cells can be disrupted with relative ease. At laboratory scale repeated freeze–thaw cycles of cell pellets are sufficient to break up the cells. At a larger scale, producer cells are lysed by simply lowering the ionic strength (hypotonic shock) or with the aid of mild pressure changes that can be provoked by a microfluidizer® or cross-flow filtration system. Along with vector particles, vast amounts of cellular DNA, RNA, and proteins are released from the producer cells. In order to reduce the viscosity of the cell lysate, which may cause difficulties in subsequent purification steps, nucleic acids are often eliminated following cell lysis by digestion with nucleases (e.g. Benzonase®). In addition, removal of nucleic acids has been shown to prevent aggregation of adenovirus particles, which would further complicate their purification (3).

4. Clarification Methods

4.1. Centrifugation

Removal of cells and cell debris from harvested supernatants or cell lysates is typically achieved by batch centrifugation at small scale. This simple operation allows separation of viral particles from most cellular debris. Product loss by coprecipitation is not usually an issue when centrifugation force and time are well adjusted, but could be if viral particles associate with cell debris (8). For large volumes of vector stock (>10 L), the use of continuous centrifuges is preferred for practical reasons (1). While low speed centrifugation alone may render a clarified vector stock of sufficient quality for subsequent ultracentrifugation, this step is generally complemented with microfiltration to achieve greater clarification when ultrafiltration or chromatography steps follow in order to avoid filter or column clogging.

4.2. Microfiltration

Microfiltration is widely used for clarification of viral vector stocks either alone or following a centrifugation step. Viral stocks are passed through a membrane that retains cell debris while allowing the recovery of virus particles in the permeate fraction. Membranes with moderately large pore sizes ranging from 0.45 to 0.8 μm are typically employed for virus stock clarification. Smaller pore size filters (0.22 μm) may result in early membrane blockade and the risk of losing active viral particles. Membrane filtration can be

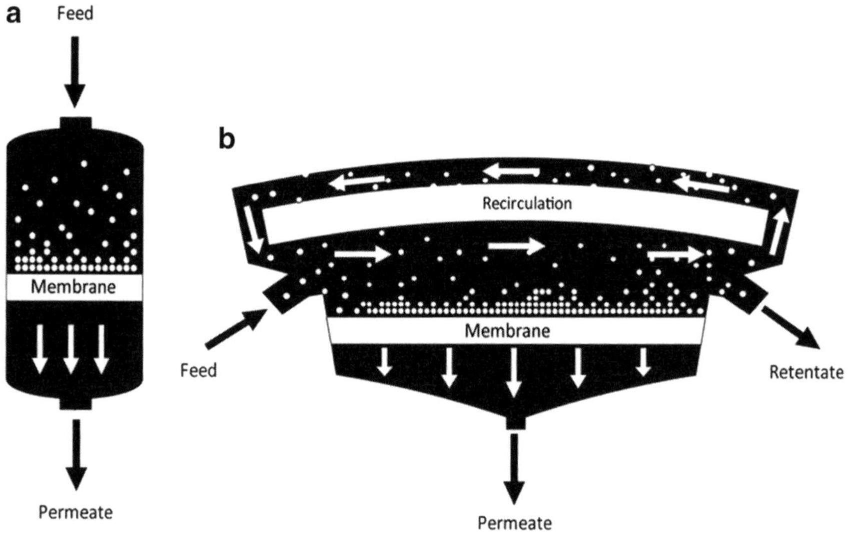

Fig. 3. ***Membrane filtration methods.*** Modes of operation. (**a**) Dead-end filtration system. (**b**) Cross-flow filtration system.

operated in two modes: dead-end and cross-flow filtration (Fig. 3). In cross flow filtration, most of the fluid travels across the surface of the membrane, rather than into the filter, minimizing cake formation and consequently delaying filter blockade. Microfiltration needs to be optimized to minimize the loss of viral product. Clogging of the pores with cell debris over time results in reduction of the actual membrane pore size and consequently virus rejection (9). Therefore, to attain high recovery of virus particles, it is crucial to limit the volume of supernatant to be passed per filter. This volume may vary depending on the initial membrane pore size and quality of the feed, among others. Additionally, selection of an appropriate membrane surface chemistry is important to prevent adsorption and loss of viral particles during filtration. In general, low protein-binding membranes provide satisfactory results. At large scale, a single-step clarification process using only membrane filtration is preferred. To improve filtration performance, in addition to operating in cross-flow filtration mode, crude supernatants are often clarified using a series of membranes with decreasing pore size to further minimize membrane clogging.

5. Concentration Methods

5.1. Virus Pelleting

Virus pelleting is the simplest centrifugation technique traditionally employed to concentrate viruses at a small scale. Both ultracentrifugation and long low-speed centrifugation methods (usually several hours) can be used to efficiently pellet virus particles.

Centrifugation conditions depend on the size and density of the virus. Using centrifugation, high concentration of the virus stocks (over 100-fold) can be easily attained by resuspending viral pellets in small volumes of resuspension buffer. However, transduction efficiencies usually do not increase proportionally with the concentration factor and may not increase at all compared to nonconcentrated virus stocks in the case of labile virus particles. It is not entirely clear whether the loss of active viral particles is due to hydrodynamic pressure at the bottom of the tube, extended processing time, viral aggregation, or shear forces required to disperse the pellet, but it is probably a combination of the above. Coconcentration of viral particles with inhibitors of transduction has also been described as a potential contributor to the loss of active yield (10). In addition, susceptibility of viral vectors to hydrodynamic shear may vary for each type of viral particle or even for each vector pseudotype (11). Recovery of infective particles may be considerably improved by underlaying a cushion of medium with higher density than the virus, where virus particles will band, as this avoids harsh conditions associated with pelleting. Another important limitation of ultracentrifugation procedures is that ultrahigh speed rotors currently in use generally have small volume capacity, although large-scale systems are also available.

5.2. Precipitation with Additives

Concentration of viral particles by precipitation with additives is a commonly used method for the manufacturing of conventional viral vaccines (8). The advantage of using additives is that following the treatment, virus pellets can easily be obtained at low centrifugation speeds in a short time. Using low-speed rotors, larger volumes of supernatant can be processed per run. Several additives have been used to promote precipitation of viral particles including salts and polymers. Salting out of viruses is typically achieved by adding a salt such as ammonium sulfate or calcium phosphate at high concentration (12, 13). This causes partial dehydration of viral particles promoting the formation of aggregates by hydrophobic interactions between particles. Precipitation of viral particles can also be enhanced by the use of nonionic polymers such as polyethylene glycol (PEG) (13). Nonionic polymers occupy large volumes in solution (excluded volume) reducing the volume available for the viral particles, which forces them to interact more. The concentration of such additives required to precipitate virus particles are typically lower than that for individual proteins (14). The use of cationic polymers (poly-L-lysine) able to form lentivirus–polymer complexes that can be pelleted by low-speed centrifugation has also been described (15). Main disadvantages associated with precipitation methods include the low recovery of labile virus particles possibly due to changes in osmotic pressure and the need for dialysis and/or further purification steps to remove salts, polymers, and impurities that coprecipitate with the virus particles due to lack of selectivity of this method. In addition, precipitation processes that include a centrifugation step may be difficult to scale-up.

5.3. Ultrafiltration

Ultrafiltration is the method of choice for large-scale concentration of viral particles because it allows gentle volume reduction of viral stocks in a relatively short time. Furthermore, membrane processes are easily scaled-up and used for GMP manufacturing. Viral particles are enriched in the retentate fraction while water and small molecular weight contaminating molecules pass through the membrane and are removed with the permeate. In contrast to the standard concentration methods discussed above, filtration processes involve no phase change that may be harsh enough to cause virus inactivation. Thus, membrane processes are particularly appealing for labile virus particles, such as γ-retroviral vectors (8). Membranes with moderately large pore sizes ranging from 100,000 to 500,000 molecular weight cut-off (MWCO) are typically employed for ultrafiltration of viral particles. The larger the pore size, the higher the flow rate and purity that can be achieved during the process. However, the limit should be well defined to avoid loss of viral particles either trapped in the membrane pores or in the permeate fraction. Ultra/diafiltration processes offer the possibility of washing off impurities, thus achieving greater levels of purity. More importantly, the retentate could be diafiltered against equilibration buffer used for upcoming chromatography. A relatively low volume of buffer is required for complete buffer exchange compared to SEC and dialysis (1). Ultrafiltration can be carried out using a variety of filtration devices. At small scale, centrifugal filtration devices are a viable option. To process small to medium volumes of vector stocks (up to 2 L) stirred cell tanks are frequently employed. Moderate to large volumes of viral vectors are typically concentrated by cross-flow filtration using either flat-sheet cassette devices or hollow fibers. Membrane fouling is the main problem faced during ultrafiltration since it causes the flow rate to decrease over time. To keep process time within reasonable limits, it is often necessary to restrict the volume reduction. Quantitative recovery of infectious viral particles can be achieved using this concentration technique provided appropriate selection of membrane pore size and chemistry and adjustment of critical operating parameters (i.e. transmembrane pressure, flow rate, process time, volume of feed/cm^2) is performed. However, infectious particle recoveries tend to be lower for semipurified stocks presumably due to lack of protection provided by contaminating proteins present in crude viral stocks or viral aggregation due to high local concentrations of virus particles as proposed for adenoviruses (16).

6. Purification Methods

6.1. Density Gradient Ultracentrifugation

Density gradient ultracentrifugation is the most widely used method for isolation of virus particles in standard virology laboratories. This technique separates virions from other contaminants in solution based on differences in size, shape, and density. Separation of viral particles occurs during their passage through a density gradient enforced by centrifugation at very high speed. Gradients are formed using a dense substance in solution. Caesium chloride (CsCl) and sucrose are the most commonly used gradient forming substances for virus purification. Density gradient ultracentrifugation may be carried out in continuous or discontinuous gradients. In general, two or three purification rounds are required to achieve high levels of purity. The amount of impurities is fairly reduced after the first round, which is often carried out using discontinuous gradients. A continuous density gradient ultracentrifugation often follows to achieve greater levels of purity. Smaller tubes and a narrower density range (closer to the virus buoyant density) are usually employed in the second and third round to attain higher resolution and allow the collection of a well-defined isolated virus band. In cases in which the virus band is not visible (e.g. viruses produced at low yields), collection and analyses of all gradient fractions will be necessary to locate the fraction(s) with high viral activity. Purification of virus vectors by density gradient ultracentrifugation can be accomplished using two different centrifugation modes: equilibrium density or rate zonal ultracentrifugation.

6.1.1. Equilibrium Density Ultracentrifugation

Equilibrium density ultracentrifugation, also known as isopycnic or buoyant density ultracentrifugation, is the most commonly used method for separation of viruses. Using this technique, viral particles are separated according to their buoyant density. A dense solution is layered into the ultracentrifugation tube in such a way that a density gradient is formed, the solution being denser towards the bottom of the tube. The virus stock is usually placed on the top of the continuous gradient and the tube is centrifuged at a very high speed until the virus particles reach a point where the density of the gradient is equal to their buoyant density (equilibrium). Alternatively, the gradient is formed during ultracentrifugation using a self-generating gradient medium such as CsCl, iodixanol, or Nycodenz®. In this case, the virus stock is homogeneously distributed with a medium solution of uniform density and placed on the centrifugation tube. When subjected to a strong centrifugal force, virus particles and contaminants will sediment (or float) until they reach equilibrium. The efficacy of the separation is not limited by the size of the sample, since regardless of their starting position in the gradient, all particles of the same density will band at the same position. However, because centrifugation must be carried out for a time that is sufficient to allow all the particles in the sample to reach equilibrium, the method is usually very time consuming.

6.1.2. Rate Zonal Ultracentrifugation

The buoyant density of viruses ranges between 1.1 and 1.4 g/cm^3 (Fig. 1). While these densities differ notably from that of contaminating nucleic acids (RNA 2.0 g/cm^3 and DNA 1.7 g/cm^3), they may overlap with that of soluble proteins (1.3 g/cm^3) and some cellular organelles (1.1–1.6 g/cm^3) (17), which may copurify with the virus by equilibrium density. In contrast to the latter technique that is independent of particle size, rate zonal ultracentrifugation, also called sedimentation velocity centrifugation, separates virus particles based on their size and density. In rate zonal ultracentrifugation, the virus stock is placed on top of a continuous density gradient as a thin layer. During ultracentrifugation, the particles move through the gradient solution due to their greater density, but at a velocity that is also dependent on their size. Ultracentrifugation must be stopped before the viral particles reach the bottom of the tube or reach equilibrium. Therefore, the process is typically shorter than equilibrium density ultracentrifugation. However, for optimal resolution, it is crucial to layer the sample as a narrow band over a continuous density gradient. Thus, sample volume is usually restricted to 10% of the total gradient volume. Sample concentration prior to ultracentrifugation may be required. This powerful separation technique is underutilized for the purification of viral vectors considering that viruses have a unique size compared to most cellular macromolecules.

Although density gradient ultracentrifugation is extremely helpful for production of small scale viral lots, this purification method is associated with several practical disadvantages that complicates its use at the manufacturing scale. The preparation of density gradients is labor-intensive, time-consuming, and requires technical expertise. In addition, the quality of final vector preparations is variable. Furthermore, the viscous and hyperosmotic nature of commonly used density gradient generating agents (sucrose and CsCl) along with the high shear forces generated in the ultracentrifugation force field can cause disruption of virus particles and thus, loss of virus activity particularly when dealing with labile virus particles such as retroviruses. Other gradient media including iodixanol, Percoll®, and Nycodenz® have been successfully employed for virus purification. Iodixanol is less viscous than CsCl and sucrose and can form iso-osmotic solutions that help preserve virus particle integrity and functionality. In addition, this gradient medium has a much lower toxicity compared to CsCl and allows for direct in vitro or in vivo experimentation directly without prior need of a dialysis or similar desalting steps. The main drawback of density gradient ultracentrifugation is the limited capacity of commonly available laboratory centrifuges, and the very limited number of large-scale systems based on this technology (18, 19). Other disadvantages of this technology include long processing times and potential to generate infectious aerosols. Even though not ideal, ultracentrifugation is the only tool available to date capable of separating viral vectors from contaminating closely related contaminants (see Subheading 6.4).

6.2. Chromatography

Chromatography is the method of choice for selective fractionation of bioproducts in the industrial setting since it overcomes the bottlenecks of common laboratory scale techniques and meets all regulatory requirements. This purification approach is easy to scale-up allowing processing of large volumes of vector stocks in a relatively short period of time. Chromatography-based separations are highly reproducible resulting in a consistent viral vector product in consecutive runs. Modern chromatography systems allow process automation further minimizing human contribution to overall process variation. In addition, the ability to conduct viral vector purification in a closed system that can be sanitized supports aseptic processing. Mild conditions are used to elute viruses from the chromatography column/filter allowing maintenance of the biological activity. In fact, vector yield and potency resulting from chromatography-based purifications often exceeds that obtained by conventional density gradient purification.

Chromatography purifies viral vectors based on the surface properties of the viral particles (adsorptive chromatography) or their size (SEC). In adsorptive chromatography, a clarified viral stock is passed though a solid phase (microparticle, monolith, or membrane) coated with functional groups that capture viral particles while the rest of the solution containing undesired impurities passes through. Retained viral particles are then displaced from the chromatography support using desorption agents and collected in purified fractions. Chromatography adsorbers can also be employed in flow-through mode or negative mode. Using this mode of operation, contaminants bind the chromatography support while the product of interest (virus) passes through the column without binding the matrix and can be collected in the flow-through fraction. In SEC, viral particles are separated from low molecular weight contaminants during their passage through a column packed with microporous particles without binding the solid phase. Virions are excluded from the internal pores of the microparticles due to their large size and elute in the void volume of the column whereas most contaminants present in the virus vector stock are retarded inside the column pores and elute later.

A number of chromatography techniques and supports have been reported for purification of viral vectors. The techniques can be broadly classified into ion-exchange chromatography (IEX), affinity chromatography (AC), hydrophobic interaction chromatography (HIC), and size-exclusion chromatography (SEC) (Table 1). Process development begins with the screening of candidate chromatography supports under various conditions to identify those methods resulting in the highest vector purities and yields. In general, at least two chromatographic steps are required to attain the high levels of purity required for in vivo applications. The best overall purification results are obtained by selecting the steps with the greatest complementarity (orthogonal process design). This is usually achieved by combining steps that are based on distinct separation principles (e.g. one step based on virus charge and another on virus size or hydrophobicity). Purity levels obtained by chromatography are comparable and sometimes higher than those obtained by ultracentrifugation. Since many chromatographic elution buffers are not suitable for in vivo administration, additional purification steps such as dialysis or ultra/diafiltration may be necessary. The latter has the advantage of simultaneously concentrating the final viral vector product.

6.2.1. Ion Exchange Chromatography

Biomolecules differ from one another in their net surface charge. For nucleic acids this charge is always negative due to the phosphate ions in their backbone structure, but for proteins and viruses the charge will depend on the proportion of surface charged amino acids at a particular pH. IEX exploits these differences to separate viruses from contaminating molecules present in vector stocks. This powerful purification tool is widely used in downstream processes since it is efficient and cost-effective (Table 1). Positively charged virions will bind cation exchangers carrying negatively charged ligands such as carboxymethyl (CM) or sulfate (S) groups. In contrast, negatively charged viruses will bind anion exchangers bearing positively charged ligands such as diethylaminoethyl (DEAE) or quaternary ammonium (Q) groups. Viruses are not only eluted from the ion exchangers by increasing the ionic strength (salt concentration) of the buffer, but can also be eluted by changing the pH. Hydroxyapatite chromatography belongs to mixed mode ion-exchange techniques since the resins bear both positively charged functional groups (calcium ions) and negatively charged phosphate groups. Viruses at relatively high salt concentrations can bind hydroxyapatite resins since elution is accomplished using phosphate gradients, which may be advantageous in certain cases (e.g. sample prepurified by IEX). While defining the most suitable binding and elution strategy, one should bear in mind the pH stability range and the susceptibility of the different viral vectors to rapid changes in ionic strength as these changes may result in conformational changes that can be associated with loss of viral activity, aggregation, or disruption of viral particles.

Anion exchange chromatography (AEX) is the most commonly used chromatography purification strategy for viral vectors.

Table 1
Chromatography purification methods used for viral vectors

Purification method	Viral vector	Phase	Chromatography support	Yield (%)[a]	References
Ion-exchange chromatography					
Anion-exchange chromatography	Ad5	Column resin	Fractogel DEAE, Q Sepharose XL, Source Q15, and Streamline Q XL	60–90	(3, 7, 16, 20–24)
		Membrane	Sartobind anion direct Q	62	(25)
	AAV-1	Column resin	POROS HQ and HiTrap Q	42–72	(26, 27)
	AAV-2	Column resin	Q-Sepharose, Source 15Q, POROS HQ, HiTrap Q, POROS PI, and MacroPrep DEAE	53–100	(27–31)
	AAV-4	Column resin	POROS PI	100	(29)
	AAV-5	Column resin	Source 15Q, Mono Q HR, POROS HQ, HiTrap Q, and POROS PI	100	(27, 29, 32–34)
		Membrane	Mustang Q	–	(33)
	AAV-8	Membrane	Mustang Q	43	(33)
	Baculovirus	Membrane	Sartobind D	60–70	(35)
	Lentivirus	Column resin	HiTrap Q and Fractogel TMAE	33–60	(36, 37)
		Membrane	Mustang Q	56–65	(38, 39)
	γ-Retrovirus	Column resin	DEAE Sepharose FF, Q Sepharose FF, and Fractogel DEAE	51–77	(40, 41)
Cation-exchange chromatography	AAV-2	Column resin	UNO S1, Fractogel SO3, POROS 50HS, and SP Sepharose HP	40–100	(28, 30, 32, 42, 43)
	AAV-5	Column resin	SP Sepharose HP	–	(32)
		Membrane	Mustang S	–	(33)
	AAV-8	Membrane	Mustang S	25	(33)
	Baculovirus	Membrane	Mustang S	78	(44)
Hydroxyapatite chromatography	Ad5	Column resin	CHT ceramic hydroxyapatite	71	(24)
	AAV-2	Column resin	CHT ceramic hydroxyapatite	88	(31)
	γ-Retrovirus	Column resin	CHT Ceramic hydroxyapatite	18–31	(45)

Affinity chromatography					
Immunoaffinity chromatography	AAV-1	Column resin	AVB Sepharose HP	50	(46)
	AAV-2	Column resin	A20 Mab coupled to HiTrap-Separose and AVB Sepharose HP	70	(46, 47)
Heparin affinity chromatography	AAV-2	Column resin	POROS HE/P, POROS HE1/M, HiTrap Heparin	73	(27, 42, 48–50)
	γ-Retrovirus	Column resin	Fractogel Heparin	60–63	(51)
	Lentivirus	Column resin	Fractogel Heparin	53	(52)
	Vaccinia virus	Column resin	Toyopearl AF-Heparin	–	(53–55)
		Membrane	Heparinized cellulose membrane	56	(53–55)
Cellufine sulfate affinity chromatography	AAV-2	Column resin	Cellufine sulfate resin	45	(31)
	Vaccinia virus	Column resin	Cellufine sulfate resin	59	(53–55)
		Membrane	Sulfated cellulose membrane	65	(53–55)
Mucin affinity chromatography	AAV-5	Column resin	Mucin-sepharose	–	(56)
Immobilized metal affinity chromatography	Ad5	Column resin	TosoHaas chelate Zinc	49–65	(20, 23)
	His_6 AAV-2	Column resin	Ni-NTA agarose	>90	(57)
	His_6 AAV-8	Column resin	Ni-NTA agarose	>90	(57)
	His_6 Baculovirus	Column resin	Ni-Sepharose fast flow	2–3	(58)
	HAT HSV-1	Column resin	HiTrap IDA-Co	67	(59, 60)
	His_6 γ-Retrovirus	Column resin	Ni-NTA agarose	56	(61, 62)
	His_6 Lentivirus	Column resin	Ni-NTA agarose	>50	(62)
		Monolith	CIM-IDA Ni	–	(63)
Avidin–biotin affinity chromatography	BAP AAV-1	Column resin	Monomeric avidin agarose	>64	(64)
	BAP Ad5	Column resin	Softlink monomeric avidin	17–65	(65)
	Biot γ-Retrovirus	Column resin	Fractogel streptavidin	17	(66)
		Monolith	Streptavidin monolith	8	(67)

(continued)

Table 1
(continued)

Purification method	Viral vector	Phase	Chromatography support	Yield (%)[a]	References
Hydrophobic interaction & reversed-phase chromatography					
Hydrophobic interaction chromatography	Ad5	Column resin	Toyopearl butyl or phenyl, Fractogel propyl	5–30	(20, 23)
	AAV-2	Column resin	Toyopearl butyl and phenyl-Sepharose	48–72	(43, 68)
Reversed-phase chromatography	Ad5	Column resin	Polyflo	94–97	(7)
Size exclusion chromatography					
	Ad5	Column resin	Toyopearl HW-75F, Superdex 200, Sephacryl S-400HR, and Sephacryl S-500	15–90	(3, 16, 20, 22, 23)
	AAV-5	Column resin	Superdex 200	–	(34)
	Baculovirus	Column resin	Sepharose CL-4B	38	(69)
	Lentivirus	Column resin	Sephacryl S-500	80	(38)
	γ-Retrovirus	Column resin	Sepharose CL-4B and Superdex 200	45–70	(40, 70)

His_6, hexahistidine tag; HAT, cobalt affinity peptide; BAP, biotin acceptor peptide; Biot, biotinylated
[a]Refers to maximum and minimum step yields published in terms of either total or infectious viral particles

This is likely related to the fact that most viral vector particles possess an isoeletric point below 7.4. In other words, they are negatively charged at physiological pH. In fact, purification at pH ~ 7.4 is very common since viral vectors are stable under physiological conditions, a prerequisite for gene therapy applications. AEX was found useful for the purification of adenovirus, AAV, γ-retrovirus, lentivirus, and baculovirus vectors (Table 1). In most cases, AEX is used as a first capture chromatography step. Because viral particles are large macromolecules and contain multiple binding sites, they tend to exhibit enhanced binding strength to anion exchangers compared with most contaminating proteins including host cell contaminants as well as free viral components. Therefore, virus loading and binding can usually be carried out at relatively high ionic strength enhancing the selectivity and capacity of the chromatography support. However, nucleic acids often display a similar or higher affinity for AEX supports and can still compete with the vectors for AEX binding sites and often coelute with the vector. Consequently, further purification steps may be required to eliminate similarly charged contaminants including nucleic acids and also salt used for elution.

Some virus particles such as AAVs can withstand important changes in pH without compromising their viral activity. In this case, the net surface charge of the virions can be modified from negative to positive (by adjusting the pH of the viral vector feed) making their binding and separation by cation exchange chromatography possible (Table 1). Hydroxyapatite chromatography was also found useful for the purification of various viral vectors (Table 1). These resins offer an alternative purification approach with distinct selectivity and often result in high vector yields as shown for AAV-2 and Ad5 vectors.

6.2.2. Affinity Chromatography

Affinity chromatography separates biomolecules based on a highly specific interaction between a target molecule and a ligand coupled to a chromatography support. Hydrogen bonds, van der Waals forces, electrostatic and hydrophobic interactions can all contribute to binding. Elution of the viral particles from affinity supports is accomplished by reversing the interaction, either specifically using a competitive ligand, or nonspecifically, by changing the pH or ionic strength in the elution buffer. Due to its exceptionally high selectivity and efficiency, affinity chromatography offers the potential to separate viruses in a single step saving process time and usually allowing high recoveries of viral particles compared to multistep purification processes. In addition, adsorbent binding capacities and concentration effects are generally very high.

Affinity chromatography is the second most widely used chromatography purification method for viral gene therapy vectors (Table 1). The most selective affinity chromatography technique is immunoaffinity chromatography, which relies on the specific interaction between immobilized antibodies and surface viral antigens. This technique was used for the separation of assembled AAV2 vector particles from unassembled capsid proteins using a selective monoclonal antibody (47) as well as the purification of both AAV1 and AAV2 produced in insect cells using the commercially available AVB Sepharose High Performance resin (GE Healthcare) (46). A drawback of immunoaffinity chromatography is that it usually requires stringent elution conditions to break the strong antibody–antigen interactions including low pH, high salt, or the use of denaturing agents. This precludes its application for the purification of labile viral particles. A viable option would be to use immunoaffinity chromatography in negative mode for the specific adsorption of contaminants that may be difficult to eliminate using other purification approaches. For instance, this technique could separate retrovirus particles from cell membrane vesicles provided that a surface protein was found to be exclusively incorporated into the vesicles but not the virions as shown for wild-type HIV-1 (71, 72). In any case, the high costs associated with antibody purification/immobilization and the low stability of these ligands to sanitizing agents make them unattractive for large-scale processes (73).

A biospecific interaction frequently exploited for the purification of viruses is that between a matrix-bound receptor and a viral surface ligand (Table 1). Ubiquitously expressed heparan sulfate proteoglycans are widely utilized by viruses as cell attachment receptors. Viruses known to interact with such receptors can often be purified by heparin (a receptor analog) and/or cellufine sulfate affinity chromatography. Heparin, an inexpensive generic affinity ligand, captures many different types of viruses including AAV2, γ-retrovirus, lentivirus, HSV, and vaccinia (Table 1) (74). Cellufine sulfate affinity chromatography has also been employed for the successful purification of AAV2 and vaccinia viral vectors (Table 1). A further demonstration of the utility of matrix-bound receptors for the purification of viruses was provided by Auricchio et al. (56), who isolated AAV5 vectors by a single-step affinity chromatography using a mucin column. The rationale behind this approach is that widely distributed sialic acids assist AAV5, among other viruses, to enter target cells and mucin is a sialic acid-rich protein.

Alternatively, viral vector particles can be engineered to display affinity tags on their surface in order to facilitate their purification by affinity chromatography. Hexahistidine affinity tags (His_6), frequently used for the purification of recombinant proteins by immobilized metal affinity chromatography (IMAC), have been inserted into the surface of many viral vectors. His_6-tagged viruses show high affinity for immobilized nickel ions and can be purified by Ni-IMAC (Table 1). In addition, HSV-1 vectors bearing a cobalt affinity peptide (HAT) have been generated and successfully purified by Co-IMAC. Interestingly, Ad5 vectors possess a natural affinity for zinc ligands and can be purified by Zn-IMAC with no need for vector engineering. Viral particles can also be purified by exploiting the specific interaction between biotin and avidin (Table 1). In this case, viral particles first need to be biotinylated, either chemically or metabolically. Chemical biotinylation of γ-retrovirus particles was accomplished by exposing producer cells to a biotinylation reagent during the vector production process. In order to metabolically biotinylate viral particles and generate covalently biotinylated virions, a viral surface structural protein must be genetically fused to a biotin acceptor peptide (BAP). A common disadvantage to all vector tagging methods, is that modifying the viral surface structure by inserting tags without reducing or eliminating the virus ability to transduce cells may be a challenging task.

6.2.3. Hydrophobic Interaction and Reversed-Phase Chromatography

HIC and reversed-phase chromatography (RPC) separate biomolecules based on the differences in their surface hydrophobicity. HIC is widely used in preparative protein purification schemes usually complementing other chromatography methods that separate based on charge, size, or bioaffinity. This method has found only modest use for the purification of viral vectors (Table 1). Binding of viral particles to HIC supports is promoted in the presence of moderately high salt concentrations that enhance hydrophobic interactions between the virus surface and the chromatography ligands. Bound virions are typically displaced from hydrophobic supports by gradually lowering the salt concentration in the elution buffer. The use of HIC has been described for AAV and Ad vectors, both in normal and negative modes (Table 1). Low infective recoveries were obtained for Ad5 particles in normal bind-and-elute mode (5–30%). This was attributed to a possible viral particle disruption at the high salt concentrations (1.5 M $[NH_4]_2SO_4$) used to load the virus into the column (20). Another likely explanation is that the virus aggregated inside the column as these concentrations of $[NH_4]_2SO_4$ have been used for virus concentration by precipitation as reported in previous work (13).

On the other hand, due to its outstanding high resolution, RPC is primarily used as an analytical technique for the analysis of disrupted viral particles (structural viral proteins) and for purity checking. RPC has also been employed for preparative purification of Ad5 particles, but only in negative mode in order to preserve viral particle integrity (Table 1). While based on the same separation principle, RPC differs from HIC in that the surface of a RPC support is more hydrophobic leading to stronger interactions. Therefore, elution of tightly bound biomolecules from RPC supports requires the use of organic solvents that can denature proteins. In contrast, HIC separations can be entirely conducted in aqueous solutions.

6.2.4. Size Exclusion Chromatography

SEC, also known as gel filtration chromatography, is a universal purification method for viruses since it takes advantage of the large size of viral particles compared to most contaminating biomolecules. This technique has been widely used for the purification of viral vectors (Table 1), mainly as a final polishing step in multistep purification processes. Importantly, SEC allows simultaneous desalting and buffer exchange. Because no virus binding occurs during the chromatography run and no change in the buffer composition is necessary for virus elution, SEC is a gentle and straightforward approach for viral vector purification. Careful selection of the chromatography media is required since resins with pore sizes in the range of the virus size may lead to virus entrapment inside the pores and consequently, low vector yields (1, 20). Using suitable chromatography supports, high recoveries of active viral particles can be reproducibly obtained.

A few practical disadvantages are associated with the use of SEC for virus purification. Inherent to its nonadsorptive nature, the main limitation of this technique is a low loading capacity (<10% column volume for best peak resolution) that often limits the scalability of the process. If large volumes of vector stock need to be processed, concentration of the starting material by preceding ultrafiltration could be useful. However, the concentration of semipurified high titer vector stocks may result in important loses of vector yield due to virus aggregation (16). Moreover, high resolution SEC separations are carried out at low linear flow rates (~15 cm/h) with long columns, which increases process time and typically results in product dilution of two- to fourfold. On the other hand, it is difficult to separate virus particles from high molecular weight contaminants (e.g. proteoglycans or genomic DNA) that, if present at this stage, would coelute with the virus in the void volume of the column. Alternatively, SEC can be operated in group separation mode at the initial steps of a purification process. This mode of operation has been successfully employed for initial fractionation of plasmid DNA (75) and influenza virus (76). The main advantage of this strategy is that feed volumes of up to 30% of the column volume can be loaded in each run (77). In this mode, separations are faster as linear flow rates can be as high as 80 cm/h and bed heights as low as 25 cm (75). Additionally, (a) no significant product dilution occurs and yields are consistently high, and (b) the buffer can be exchanged to condition the feed for a subsequent chromatography step.

6.3. Chromatography Supports

Most currently available chromatographic matrices were designed to maximize the adsorption of proteins rather than viruses. Consideration of the pore dimensions of conventional microporous chromatography adsorbents (typically 30–80 nm) suggests that adsorption of viruses will be restricted to the bead surface area alone whereas most contaminating proteins will have access to the area inside the pores as well (Fig. 4a). Moreover, large nanoparticles such as viruses and plasmid DNA diffuse more slowly than competing proteins in solution (75, 78). Consequently, both the available virus binding capacity and purification efficiency are compromised using these classical chromatography supports.

Binding capacity is an important variable because it determines the throughput and concentrating potential of a chromatography resin. It can be expressed as static capacity or dynamic capacity. Dynamic capacity values are *more* useful in predicting real process performance since they are carried out under actual flow conditions. The best case scenario is to determine dynamic binding capacities by pumping the actual sample (containing contaminants) into the column as opposed to a purified sample, since the latter approach will result in overestimated values. Reported binding capacity values for viral vectors (10^{11}–10^{12} viral particles per mL of resin) (1) seem to be in line with those reported for plasmid DNA that are usually in the order of hundreds of micrograms of plasmid per mL of chromatographic support in contrast with those commonly obtained for proteins that are in the range of tens to hundreds of milligrams per mL of chromatographic support (79). One way to increase the outer surface area available for virus binding is to decrease bead size since these two parameters inversely correlate. However, small beads generate high column back pressure, which limits the flow rate that can be applied and the bead size that can be used.

Advanced chromatography technologies tailored to improve binding capacities of large particles are rapidly being adopted for virus purification. Among them, tentacle supports, membrane adsorbents, and macroporous monoliths have been tested for the purification of viral vectors (Table 1). Tentacle supports possess sterically accessible ligands available for virus capture (Fig. 4b). The ligands are attached to an inert and flexible spacer arm that separates them from the bead. Therefore, tentacle ligands can access otherwise sterically hindered binding sites and compensate in part for the loss of surface area inside the pores. In addition, since they are no longer exclusively on the surface of the chromatographic bead, larger amounts of ligand are available for binding (80). Tentacle matrices distributed under the trade name Fractogel® (Merck) have been employed for the purification of various viral vectors including AAV-2, Ad5, γ-retrovirus, and lentivirus vectors (Table 1).

More recently, membranes and monoliths are gaining particular interest as alternatives to traditional microporous column packing resins. Owing to their different architecture, mass transport though the pores/channels takes place mainly by convection overcoming virus particle diffusion issues encountered with traditional chromatography supports. This permits the use of higher flow rates at lower pressure drops, which in turn results in higher

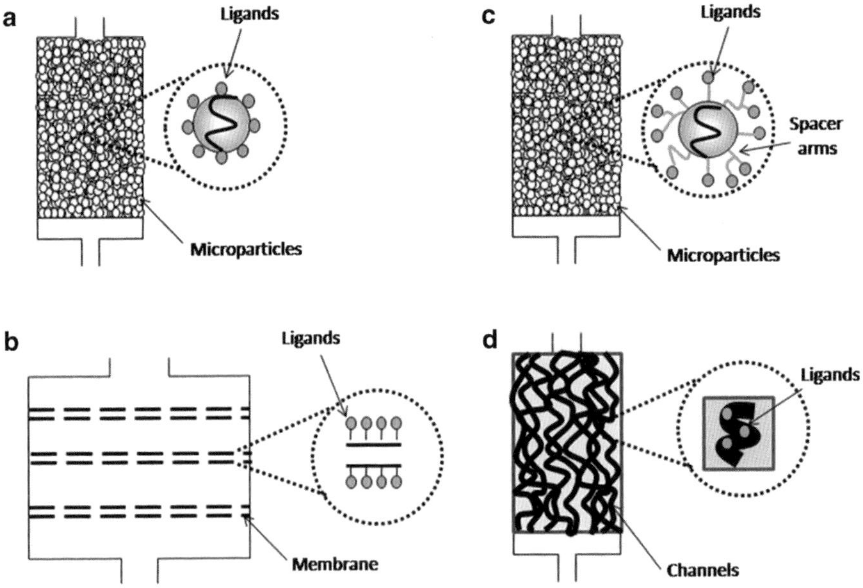

Fig. 4. ***Chromatographic technologies.*** Depiction of chromatography phases available for the purification of virus particles. (**a**) Column packed with conventional porous microparticles, (**b**) column packed with tentacle porous microparticles, (**c**) chromatography membrane device containing several layers of adsorptive membrane, (**d**) monolithic column containing a polymer-based monolith with an uninterrupted, interconnected network of channels.

process productivities. In addition, virus particles have access to the majority of ligands on the adsorber surface, which typically results in increased binding capacities. A further advantage of these modern chromatography technologies is that they do not require packing, thus eliminating packing labor, variation, validation, and risks associated with accidental introduction of air.

Membrane chromatography devices are offered by two major suppliers under the commercial name of Mustang® (Pall) and Sartobind® (Sartorius). They consist of multilayer porous membrane assemblies housed within a capsule. Functional ligands are attached to the membrane surface (Fig. 4c). A wide range of surface chemistries, porous sizes, and formats for different processing scales are available. An additional advantage of membrane chromatography is disposability. Single use membrane adsorbers minimize process validation efforts, time and cost facilitating technology transfer to cGMP operations. The main disadvantage associated with membrane chromatography is the large dead volume of the filter units resulting in peak broadening and decreased separation efficiency. Membrane adsorbers have been used for the purification of AAV, Ad5, baculovirus, lentivirus, and vaccinia vectors showing excellent results (Table 1). Monoliths are continuous beds consisting of a single piece of highly porous solid material, characterized by an uninterrupted, interconnecting network of channels (Fig. 4d). Although only a few viral vector purification studies have been reported so far, monolithic columns offer similar benefits regarding flow rates and capacities as membrane technologies (Table 1). In addition, higher resolution and concentration factors can be attained since, in monolithic columns, the void volume can be decreased to a minimum. Monolithic columns are polymerized directly in a column and can be prepared and derivatized with traditional chromatography ligands in the laboratory using relatively straightforward techniques. Commercial monolithic supports are provided by Dionex (SwiftPro®) (analytical columns), Bio-Rad (UNO™) (analytical and laboratory scale preparative columns), and Bia Separations (CIM®, Convective Interaction Media) (analytical, laboratory, and industrial scale preparative columns).

6.4. Purification Challenges

The main challenge facing researchers working in downstream processing of viral vectors is how to separate the functional viral particles from contaminating closely related viral species such as inactive vector forms (empty capsids), helper viruses, and cell membrane vesicles. Given the structural surface similarity they share with viral vectors, these contaminants pose a serious challenge since they are difficult and sometimes impossible to separate using currently described purification procedures. Minimizing their levels at the production stage is very important, but often not enough to guarantee vector safety.

Levels of contamination with inactive vector forms (including empty viral capsids) vary depending on the specific vector being considered and its production system. Empty capsids are nearly undistinguishable from viral vectors, but they lack the vector genome or contain very little nucleic acid and therefore are inactive. However, they still contribute to the total particle mass and the acute immune response directed against the vector in vivo. Therefore, regulatory authorities have set limits of total-to-infective particle ratios for specific viral vectors types. In the case of Ad vectors, FDA recommends a maximum ratio of 30:1 (81). Separation of empty Ad particles by equilibrium density ultracentrifugation is possible given the significant difference between the buoyant densities of empty and full viral capsids. In contrast, the potential for clearance of empty Ad particles by chromatography remains controversial (21–23). Unlike Ad vectors, the separation of empty and complete AAV particles using a refined column AEX chromatography method has been demonstrated at the lab scale (26, 28) and recently by using membrane based IEX chromatography(82). Although technically challenging, the development of comparable methods for efficient removal of empty capsids from other vector preparations may be possible. Elimination of other inactive vector forms, such as damaged or denatured vector particles, is still rarely reported or discussed in the literature.

A number of vector production systems utilize helper viruses to support viral vector production. Examples include the generation of AAV and helper-dependent Ad vectors using helper adenoviruses, and AAV and lentiviral vectors using recombinant baculoviruses. Chromatography separation of helper viruses from a different species than the viral vector is possible given the differences in chemical and physical properties between the various viruses. In contrast, no reports concerning the separation of helper adenovirus from helper-dependent Ad preparations by chromatography have been published and achieving such chromatographic separation seems unlikely. In this case, the only difference between both types of particles might be represented by merely a few kilobases of DNA between their genomes. By designing helper-dependent Ad and helper Ad virus constructs having a significantly different genome size (and thus, different buoyant densities), separation can be accomplished by equilibrium density ultracentrifugation (83, 84).

Other closely related species that are generated during retroviral vector production and may prove difficult to separate are cell membrane vesicles. Purified retrovirus preparations obtained by equilibrium density ultracentrifugation were described to contain variable amounts of cell membrane vesicles. These are released by producer cells and have a density similar to that of the virus (85, 86). Complete removal of contaminating cell membrane vesicles is difficult to accomplish since these particles show important similarities in morphology, composition, and physical characteristics with the virions. However, since these vesicles show a wider range of size (50–500 nm) than viruses, higher levels of purification can be attained by rate zonal ultracentrifugation. The use of this strategy resulted in highly purified γ-retrovirus preparations with no evident contamination with cell membrane vesicles (87). Another possible way to remove these cellular vesicles is to employ immunoaffinity chromatography as previously mentioned (Subheading 6.2.2) (71, 72).

In order to facilitate the use of ultracentrifugation in a large-scale purification protocol, a possibility would be to combine chromatography and ultracentrifugation. It would be tempting, for instance, to introduce ultracentrifugation as a final polishing step to allow removal of closely related contaminating species at the end of a standard chromatography process when the volume of viral stock is easier to handle.

Acknowledgements

The authors wish to thank Gavin Whissell for careful review of this manuscript. This work was supported by an NSERC Strategic Project grant and the NCE Canadian Stem Cell Network.

References

1. Altaras, N. E., Aunins, J. G., Evans, R. K., Kamen, A., Konz, J. O., and Wolf, J. J. (2005) Production and formulation of adenovirus vectors, *Adv Biochem Eng Biotechnol* **99**, 193–260.
2. Burova, E., and Ioffe, E. (2005) Chromatographic purification of recombinant adenoviral and adeno-associated viral vectors: methods and implications, *Gene Ther* **12** Suppl **1**, S5–17.
3. Lusky, M. (2005) Good manufacturing practice production of adenoviral vectors for clinical trials, *Hum Gene Ther* **16**, 281–291.
4. Segura, M. M., Alba, R., Bosch, A., and Chillon, M. (2008) Advances in helper-dependent adenoviral vector research, *Curr Gene Ther* **8**, 222–235.
5. Rodrigues, T., Carrondo, M. J., Alves, P. M., and Cruz, P. E. (2007) Purification of retroviral vectors for clinical application: biological implications and technological challenges, *J Biotechnol* **127**, 520–541.
6. Segura, M. M., Kamen, A., and Garnier, A. (2006) Downstream processing of oncoretroviral and lentiviral gene therapy vectors, *Biotechnol Adv* **24**, 321–337.
7. Green, A. P., Huang, J. J., Scott, M. O., Kierstead, T. D., Beaupre, I., Gao, G. P., and Wilson, J. M. (2002) A new scalable method for the purification of recombinant adenovirus vectors, *Hum Gene Ther* **13**, 1921–1934.
8. Lyddiatt, A., and O'Sullivan, D. A. (1998) Biochemical recovery and purification of gene therapy vectors, *Curr Opin Biotechnol* **9**, 177–185.
9. Reeves, L., and Cornetta, K. (2000) Clinical retroviral vector production: step filtration using clinically approved filters improves titers, *Gene Ther* **7**, 1993–1998.
10. Le Doux, J. M., Morgan, J. R., Snow, R. G., and Yarmush, M. L. (1996) Proteoglycans secreted by packaging cell lines inhibit retrovirus infection, *J Virol* **70**, 6468–6473.
11. Burns, J. C., Friedmann, T., Driever, W., Burrascano, M., and Yee, J. K. (1993) Vesicular stomatitis virus G glycoprotein pseudotyped retroviral vectors: concentration to very high titer and efficient gene transfer into mammalian and nonmammalian cells, *Proc Natl Acad Sci USA* **90**, 8033–8037.
12. Pham, L., Ye, H., Cosset, F. L., Russell, S. J., and Peng, K. W. (2001) Concentration of viral vectors by co-precipitation with calcium phosphate, *J Gene Med* **3**, 188–194.
13. Schagen, F. H., Rademaker, H. J., Rabelink, M. J., van Ormondt, H., Fallaux, F. J., van der Eb, A. J., and Hoeben, R. C. (2000) Ammonium sulphate precipitation of recombinant adenovirus from culture medium: an easy method to increase the total virus yield, *Gene Ther* **7**, 1570–1574.
14. Pedro, L., Soares, S. S., and Ferreira, G. N. M. (2008) Purification of bionanoparticles, *Chem. Eng. Technol.* **31**, 815–825.
15. Zhang, B., Xia, H. Q., Cleghorn, G., Gobe, G., West, M., and Wei, M. Q. (2001) A highly efficient and consistent method for harvesting large volumes of high-titre lentiviral vectors, *Gene Ther* **8**, 1745–1751.
16. Kamen, A., and Henry, O. (2004) Development and optimization of an adenovirus production process, *J Gene Med* **6 Suppl 1**, S184–192.
17. Koolman, J., and Rohm, K. H. (2004) "Organelles". In *Color Atlas of Biochemistry*, 2nd ed., pp 196–234, Thieme Medical Publishers, Stuttgart-New York.
18. Round, J. J., Liptak, R. A., and McGregor, W. C. (1981) Continuous-flow ultracentrifugation in preparative biochemistry, *Ann N Y Acad Sci* **369**, 265–274.
19. Reimer, C. B., Baker, R. S., Van Frank, R. M., Newlin, T. E., Cline, G. B., and Anderson, N. G. (1967) Purification of large quantities of influenza virus by density gradient centrifugation, *J Virol* **1**, 1207–1216.
20. Huyghe, B. G., Liu, X., Sutjipto, S., Sugarman, B. J., Horn, M. T., Shepard, H. M., Scandella, C. J., and Shabram, P. (1995) Purification of a type 5 recombinant adenovirus encoding human p53 by column chromatography, *Hum Gene Ther* **6**, 1403–1416.
21. Blanche, F., Cameron, B., Barbot, A., Ferrero, L., Guillemin, T., Guyot, S., Somarriba, S., and Bisch, D. (2000) An improved anion-exchange HPLC method for the detection and purification of adenoviral particles, *Gene Ther* **7**, 1055–1062.
22. Peixoto, C., Ferreira, T. B., Carrondo, M. J., Cruz, P. E., and Alves, P. M. (2006) Purification of adenoviral vectors using expanded bed chromatography, *J Virol Methods* **132**, 121–126.
23. Vellekamp, G., Porter, F. W., Sutjipto, S., Cutler, C., Bondoc, L., Liu, Y. H., Wylie, D., Cannon-Carlson, S., Tang, J. T., Frei, A., Voloch, M., and Zhuang, S. (2001) Empty capsids in column-purified recombinant adenovirus preparations, *Hum Gene Ther* **12**, 1923–1936.
24. Konz, J. O., Lee, A. L., Lewis, J. A., and Sagar, S. L. (2005) Development of a purification

process for adenovirus: controlling virus aggregation to improve the clearance of host cell DNA, *Biotechnol Prog* **21**, 466–472.

25. Peixoto, C., Ferreira, T. B., Sousa, M. F., Carrondo, M. J., and Alves, P. M. (2008) Towards purification of adenoviral vectors based on membrane technology, *Biotechnol Prog* **24**, 1290–1296.
26. Urabe, M., Xin, K. Q., Obara, Y., Nakakura, T., Mizukami, H., Kume, A., Okuda, K., and Ozawa, K. (2006) Removal of empty capsids from type 1 adeno-associated virus vector stocks by anion-exchange chromatography potentiates transgene expression, *Mol Ther* **13**, 823–828.
27. Zolotukhin, S., Potter, M., Zolotukhin, I., Sakai, Y., Loiler, S., Fraites, T. J., Jr., Chiodo, V. A., Phillipsberg, T., Muzyczka, N., Hauswirth, W. W., Flotte, T. R., Byrne, B. J., and Snyder, R. O. (2002) Production and purification of serotype 1, 2, and 5 recombinant adeno-associated viral vectors, *Methods* **28**, 158–167.
28. Qu, G., Bahr-Davidson, J., Prado, J., Tai, A., Cataniag, F., McDonnell, J., Zhou, J., Hauck, B., Luna, J., Sommer, J. M., Smith, P., Zhou, S., Colosi, P., High, K. A., Pierce, G. F., and Wright, J. F. (2007) Separation of adeno-associated virus type 2 empty particles from genome containing vectors by anion-exchange column chromatography, *J Virol Methods* **140**, 183–192.
29. Kaludov, N., Handelman, B., and Chiorini, J. A. (2002) Scalable purification of adeno-associated virus type 2, 4, or 5 using ion-exchange chromatography, *Hum Gene Ther* **13**, 1235–1243.
30. Gao, G., Qu, G., Burnham, M. S., Huang, J., Chirmule, N., Joshi, B., Yu, Q. C., Marsh, J. A., Conceicao, C. M., and Wilson, J. M. (2000) Purification of recombinant adeno-associated virus vectors by column chromatography and its performance in vivo, *Hum Gene Ther* **11**, 2079–2091.
31. O'Riordan, C. R., Lachapelle, A. L., Vincent, K. A., and Wadsworth, S. C. (2000) Scaleable chromatographic purification process for recombinant adeno-associated virus (rAAV), *J Gene Med* **2**, 444–454.
32. Brument, N., Morenweiser, R., Blouin, V., Toublanc, E., Raimbaud, I., Cherel, Y., Folliot, S., Gaden, F., Boulanger, P., Kroner-Lux, G., Moullier, P., Rolling, F., and Salvetti, A. (2002) A versatile and scalable two-step ion-exchange chromatography process for the purification of recombinant adeno-associated virus serotypes-2 and -5, *Mol Ther* **6**, 678–686.
33. Davidoff, A. M., Ng, C. Y., Sleep, S., Gray, J., Azam, S., Zhao, Y., McIntosh, J. H., Karimipoor, M., and Nathwani, A. C. (2004) Purification of recombinant adeno-associated virus type 8 vectors by ion exchange chromatography generates clinical grade vector stock, *J Virol Methods* **121**, 209–215.
34. Smith, R. H., Ding, C., and Kotin, R. M. (2003) Serum-free production and column purification of adeno-associated virus type 5, *J Virol Methods* **114**, 115–124.
35. Vicente, T., Peixoto, C., Carrondo, M. J., and Alves, P. M. (2009) Purification of recombinant baculoviruses for gene therapy using membrane processes, *Gene Ther* **16**, 766–775.
36. Scherr, M., Battmer, K., Eder, M., Schule, S., Hohenberg, H., Ganser, A., Grez, M., and Blomer, U. (2002) Efficient gene transfer into the CNS by lentiviral vectors purified by anion exchange chromatography, *Gene Ther* **9**, 1708–1714.
37. Yamada, K., McCarty, D. M., Madden, V. J., and Walsh, C. E. (2003) Lentivirus vector purification using anion exchange HPLC leads to improved gene transfer, *Biotechniques* **34**, 1074–1078, 1080.
38. Slepushkin, V., Chang, N., Cohen, R., Gan, Y., Jiang, B., Deausen, E., Berlinger, D., Binder, G., Andre, K., Humeau, L., and Dropulic, B. (2003) Large-scale purification of a lentiviral vector by size exclusion chromatography or mustang Q ion exchange capsule, *Bioprocessing J.* **2**, 89–95.
39. Kutner, R. H., Puthli, S., Marino, M. P., and Reiser, J. (2009) Simplified production and concentration of HIV-1-based lentiviral vectors using HYPERFlask vessels and anion exchange membrane chromatography, *BMC Biotechnol* **9**, 10.
40. Rodrigues, T., Carvalho, A., Carmo, M., Carrondo, M. J., Alves, P. M., and Cruz, P. E. (2007) Scaleable purification process for gene therapy retroviral vectors, *J Gene Med* **9**, 233–243.
41. Rodrigues, T., Carvalho, A., Roldao, A., Carrondo, M. J., Alves, P. M., and Cruz, P. E. (2006) Screening anion-exchange chromatographic matrices for isolation of onco-retroviral vectors, *J Chromatogr B Analyt Technol Biomed Life Sci* **837**, 59–68.
42. Zolotukhin, S., Byrne, B. J., Mason, E., Zolotukhin, I., Potter, M., Chesnut, K., Summerford, C., Samulski, R. J., and Muzyczka, N. (1999) Recombinant adeno-associated virus purification using novel methods improves infectious titer and yield, *Gene Ther* **6**, 973–985.
43. Chahal, P. S., Aucoin, M. G., and Kamen, A. (2007) Primary recovery and chromatographic purification of adeno-associated virus type 2 produced by baculovirus/insect cell system, *J Virol Methods* **139**, 61–70.

44. Wu, C., Soh, K. Y., and Wang, S. (2007) Ion-exchange membrane chromatography method for rapid and efficient purification of recombinant baculovirus and baculovirus gp64 protein, *Hum Gene Ther* **18**, 665–672.
45. Kuiper, M., Sanches, R. M., Walford, J. A., and Slater, N. K. (2002) Purification of a functional gene therapy vector derived from Moloney murine leukaemia virus using membrane filtration and ceramic hydroxyapatite chromatography, *Biotechnol Bioeng* **80**, 445–453.
46. Smith, R. H., Levy, J. R., and Kotin, R. M. (2009) A simplified baculovirus-AAV expression vector system coupled with one-step affinity purification yields high-titer rAAV stocks from insect cells, *Mol Ther* **17**, 1888–1896.
47. Grimm, D., Kern, A., Rittner, K., and Kleinschmidt, J. A. (1998) Novel tools for production and purification of recombinant adenoassociated virus vectors, *Hum Gene Ther* **9**, 2745–2760.
48. Clark, K. R., Liu, X., McGrath, J. P., and Johnson, P. R. (1999) Highly purified recombinant adeno-associated virus vectors are biologically active and free of detectable helper and wild-type viruses, *Hum Gene Ther* **10**, 1031–1039.
49. Anderson, R., Macdonald, I., Corbett, T., Whiteway, A., and Prentice, H. G. (2000) A method for the preparation of highly purified adeno-associated virus using affinity column chromatography, protease digestion and solvent extraction, *J Virol Methods* **85**, 23–34.
50. Harris, J. D., Beattie, S. G., and Dickson, J. G. (2003) Novel tools for production and purification of recombinant adeno-associated viral vectors, *Methods Mol Med* **76**, 255–267.
51. Segura, M. M., Kamen, A., Trudel, P., and Garnier, A. (2005) A novel purification strategy for retrovirus gene therapy vectors using heparin affinity chromatography, *Biotechnol Bioeng* **90**, 391–404.
52. Segura, M. M., Garnier, A., Durocher, Y., Coelho, H., and Kamen, A. (2007) Production of lentiviral vectors by large-scale transient transfection of suspension cultures and affinity chromatography purification, *Biotechnol Bioeng* **98**, 789–799.
53. Wolff, M. W., Venzke, C., Zimmermann, A., Post Hansen, S., Djurup, R., Faber, R., and Reichl, U. (2007) Affinity Chromatography Of Cell Culture Derived Vaccinia Virus, in *American Institute of Chemical Engineers*, Salt Lake City
54. Wolff, M. W., Sievers, C., Lehmann, S., Opitz, L., Post Hansen, S., Djurup, R., Faber, R., and Reichl, U. (2008) Capturing of Cell Culture Derived Vaccinia Virus by Membrane Adsorbers, in *American Institute of Chemical Engineers*, Philadelphia.
55. Wolff, M. W., Sievers, C., Lehmann, S., Post Hansen, S., Faber, R., and Reichl, U. (2009) Cellufine® sulfate and heparin affinity chromatography to capture cell culture derived Vaccinia Virus particles, in *21st Meeting of the European Society for Animal Cell Technology*.
56. Auricchio, A., O'Connor, E., Hildinger, M., and Wilson, J. M. (2001) A single-step affinity column for purification of serotype-5 based adeno-associated viral vectors, *Mol Ther* **4**, 372–374.
57. Koerber, J. T., Jang, J. H., Yu, J. H., Kane, R. S., and Schaffer, D. V. (2007) Engineering adeno-associated virus for one-step purification via immobilized metal affinity chromatography, *Hum Gene Ther* **18**, 367–378.
58. Hu, Y., Tsai, C., Chung, Y., Lu, J., and Hsu, J. T. (2003) Generation of chimeric baculovirus with histidine-tags displayed on the envelope and its purification using immobilized metal affinity chromatography, *Enzyme and Microbial Technology* **33**, 445–452.
59. Jiang, C., Glorioso, J. C., and Ataai, M. (2006) Presence of imidazole in loading buffer prevents formation of free radical in immobilized metal affinity chromatography and dramatically improves the recovery of herpes simplex virus type 1 gene therapy vectors, *J Chromatogr A* **1121**, 40–45.
60. Jiang, C., Wechuck, J. B., Goins, W. F., Krisky, D. M., Wolfe, D., Ataai, M. M., and Glorioso, J. C. (2004) Immobilized cobalt affinity chromatography provides a novel, efficient method for herpes simplex virus type 1 gene vector purification, *J Virol* **78**, 8994–9006.
61. Ye, K., Jin, S., Ataai, M. M., Schultz, J. S., and Ibeh, J. (2004) Tagging retrovirus vectors with a metal binding peptide and one-step purification by immobilized metal affinity chromatography, *J Virol* **78**, 9820–9827.
62. Yu, J. H., and Schaffer, D. V. (2006) Selection of novel vesicular stomatitis virus glycoprotein variants from a peptide insertion library for enhanced purification of retroviral and lentiviral vectors, *J Virol* **80**, 3285–3292.
63. Cheeks, M. C., Kamal, N., Sorrell, A., Darling, D., Farzaneh, F., and Slater, N. K. (2009) Immobilized metal affinity chromatography of histidine-tagged lentiviral vectors using monolithic adsorbents, *J Chromatogr A* **1216**, 2705–2711.
64. Stachler, M. D., and Bartlett, J. S. (2006) Mosaic vectors comprised of modified AAV1 capsid proteins for efficient vector purification and targeting to vascular endothelial cells, *Gene Ther* **13**, 926–931.

65. Parrott, M. B., Adams, K. E., Mercier, G. T., Mok, H., Campos, S. K., and Barry, M. A. (2003) Metabolically biotinylated adenovirus for cell targeting, ligand screening, and vector purification, *Mol Ther* **8**, 688–700.
66. Williams, S. L., Nesbeth, D., Darling, D. C., Farzaneh, F., and Slater, N. K. (2005) Affinity recovery of Moloney Murine Leukaemia Virus, *J Chromatogr B Analyt Technol Biomed Life Sci* **820**, 111–119.
67. Williams, S. L., Eccleston, M. E., and Slater, N. K. (2005) Affinity capture of a biotinylated retrovirus on macroporous monolithic adsorbents: towards a rapid single-step purification process, *Biotechnol Bioeng* **89**, 783–787.
68. Potter, M., Chesnut, K., Muzyczka, N., Flotte, T., and Zolotukhin, S. (2002) Streamlined large-scale production of recombinant adeno-associated virus (rAAV) vectors, *Methods Enzymol* **346**, 413–430.
69. Transfiguracion, J., Jorio, H., Meghrous, J., Jacob, D., and Kamen, A. (2007) High yield purification of functional baculovirus vectors by size exclusion chromatography, *J Virol Methods* **142**, 21–28.
70. Transfiguracion, J., Jaalouk, D. E., Ghani, K., Galipeau, J., and Kamen, A. (2003) Size-exclusion chromatography purification of high-titer vesicular stomatitis virus G glycoprotein-pseudotyped retrovectors for cell and gene therapy applications, *Hum Gene Ther* **14**, 1139–1153.
71. Esser, M. T., Graham, D. R., Coren, L. V., Trubey, C. M., Bess, J. W., Jr., Arthur, L. O., Ott, D. E., and Lifson, J. D. (2001) Differential incorporation of CD45, CD80 (B7-1), CD86 (B7-2), and major histocompatibility complex class I and II molecules into human immunodeficiency virus type 1 virions and microvesicles: implications for viral pathogenesis and immune regulation, *J Virol* **75**, 6173–6182.
72. Trubey, C. M., Chertova, E., Coren, L. V., Hilburn, J. M., Hixson, C. V., Nagashima, K., Lifson, J. D., and Ott, D. E. (2003) Quantitation of HLA class II protein incorporated into human immunodeficiency type 1 virions purified by anti-CD45 immunoaffinity depletion of microvesicles, *J Virol* **77**, 12699–12709.
73. Andreadis, S. T., Roth, C. M., Le Doux, J. M., Morgan, J. R., and Yarmush, M. L. (1999) Large-scale processing of recombinant retroviruses for gene therapy, *Biotechnol Prog* **15**, 1–11.
74. O'Keeffe, R. S., Johnston, M. D., and Slater, N. K. (1999) The affinity adsorptive recovery of an infectious herpes simplex virus vaccine, *Biotechnol Bioeng* **62**, 537–545.
75. Stadler, J., Lemmens, R., and Nyhammar, T. (2004) Plasmid DNA purification, *J Gene Med* **6 Suppl 1**, S54–66.
76. Kalbfuss, B., Wolff, M., Morenweiser, R., and Reichl, U. (2007) Purification of cell culture-derived human influenza A virus by size-exclusion and anion-exchange chromatography, *Biotechnol Bioeng* **96**, 932–944.
77. Morenweiser, R. (2005) Downstream processing of viral vectors and vaccines, *Gene Ther* **12 Suppl 1**, S103–110.
78. Ljunglof, A., Bergvall, P., Bhikhabhai, R., and Hjorth, R. (1999) Direct visualisation of plasmid DNA in individual chromatography adsorbent particles by confocal scanning laser microscopy, *J Chromatogr A* **844**, 129–135.
79. Urthaler, J., Buchinger, W., and Necina, R. (2005) Improved downstream process for the production of plasmid DNA for gene therapy, *Acta Biochim Pol* **52**, 703–711.
80. Kaufmann, M. (1997) Unstable proteins: how to subject them to chromatographic separations for purification procedures, *J Chromatogr B Biomed Sci Appl* **699**, 347–369.
81. McIntyre, M. (2001) Development of viral vectors for use in gene transfer trials; product characterization and quality concerns., *Adenovirus Reference Material Working Group:* www.wilbio.com.
82. Okada, T., Nonaka-Sarukawa, M., Uchibori, R., Kinoshita, K., Hayashita-Kinoh, H., Nitahara-Kasahara, Y., Takeda, S., and Ozawa, K. (2009) Scalable purification of adeno-associated virus serotype 1 (AAV1) and AAV8 vectors, using dual ion-exchange adsorptive membranes, *Hum Gene Ther* **20**, 1013–1021.
83. Palmer, D., and Ng, P. (2003) Improved system for helper-dependent adenoviral vector production, *Mol Ther* **8**, 846–852.
84. Sakhuja, K., Reddy, P. S., Ganesh, S., Cantaniag, F., Pattison, S., Limbach, P., Kayda, D. B., Kadan, M. J., Kaleko, M., and Connelly, S. (2003) Optimization of the generation and propagation of gutless adenoviral vectors, *Hum Gene Ther* **14**, 243–254.
85. Bess, J. W., Jr., Gorelick, R. J., Bosche, W. J., Henderson, L. E., and Arthur, L. O. (1997) Microvesicles are a source of contaminating cellular proteins found in purified HIV-1 preparations, *Virology* **230**, 134–144.
86. Gluschankof, P., Mondor, I., Gelderblom, H. R., and Sattentau, Q. J. (1997) Cell membrane vesicles are a major contaminant of gradient-enriched human immunodeficiency virus type-1 preparations, *Virology* **230**, 125–133.
87. Segura, M. M., Garnier, A., and Kamen, A. (2006) Purification and characterization of retrovirus vector particles by rate zonal ultracentrifugation, *J Virol Methods* **133**, 82–91.

Chapter 5

Methods to Construct Recombinant Adenovirus Vectors

Miguel Chillon and Ramon Alemany

Abstract

The most efficient system to introduce genes of interest within the adenovirus genome is by homologous recombination in microorganisms. In this chapter, the most popular procedures are described: two for homologous recombination in *Escherichia coli*, and one in yeast. Main differences between procedures are found in the plasmids needed as well as in the selection system used to rapidly identify newly generated recombinant adenovirus. The adenovirus genomes are then analyzed to confirm their identity and integrity, and further linearized to generate a viral pre-stock in permissive human cells. Finally, as a previous step before its amplification at medium or large scale, the viral pre-stock must be analyzed to quantify its potency and infectivity as well as to exclude the presence of unwanted replication competent particles.

Key words: Adenovirus genome, Adenovirus construction, Cloning the gene of interest, Homologous recombination

1. Introduction

Initially, recombinant adenoviruses were generated by direct ligation of the gene of interest into the adenoviral genome. However, direct ligation was technically difficult due to the large adenovirus genome (36 kb), the lack of unique restriction sites for cloning, and the low efficiency of large DNA fragment ligations. Further developments led to a two-step strategy, where the gene of interest was first cloned in a shuttle vector containing part of the adenovirus genome, and then transferred into the vector genome by homologous recombination within an adenovirus packaging cell line. Newly generated recombinants were selected by screening individual plaques in permissive packaging cells (1). However, this strategy needed to be improved due to the low efficiency of homologous recombination, the need for repeated rounds of plaque purification, and the long duration required for completion

Otto-Wilhelm Merten and Mohamed Al-Rubeai (eds.), *Viral Vectors for Gene Therapy: Methods and Protocols*, Methods in Molecular Biology, vol. 737, DOI 10.1007/978-1-61779-095-9_5, © Springer Science+Business Media, LLC 2011

of the viral production process. A third approach recently developed takes advantage of the highly efficient homologous recombination process in microorganisms. In this approach, the gene of interest still needs to be cloned into a shuttle plasmid. However, identification of positive recombinants is facilitated by faster plasmid replication in microorganisms than in mammalian cells, and simpler selection based on antibiotic-resistance markers.

Traditionally, generated adenoviruses were first-generation vectors derived from human serotypes 2 (Ad2) and 5 (Ad5). However, the following methods can also be used to generate vectors derived from other adenovirus serotypes (2); generation of chimeric vectors, which contain viral proteins from different serotypes (3); generation of vectors other than the first generation vectors as oncolytic vectors (4) or helper and helper-dependent adenovirus vectors (5); or even the generation of non-human adenovirus (6). In all cases, the recombination procedures either in BJ5183 bacteria or yeast can be applied directly, though for each particular vector the researcher must use specific plasmids and/or specific permissive cell lines.

Cloning the gene of interest within the adenovirus genome by homologous recombination and further amplification in permissive HEK-293 cells may lead to rearrangements and instability of the viral genome. Therefore, it is highly recommended to analyze the recombinant adenovirus genome (both at the genetic and the functional level), before starting large-scale amplification of the vector. At the genetic level, adenovirus genomes should be digested by a large battery of restriction enzymes and the presence of the gene of interest should be confirmed by PCR (sequencing is also recommended). At a functional level, tests detecting the production of new viral proteins such as the anti-hexon antibody staining method, or the IC50 assays, should be used to confirm infectivity of the vectors produced, as well as to exclude the presence of replication competent adenovirus.

2. Materials

2.1. Adenovirus Construction by Homologous Recombination in Bacteria: Procedure I

1. LB Broth: 2.5 g of Miller's LB in powder in 1 L of ddH_2O. Autoclave.
2. LB + Ampicillin: Add 100 mg of ampicillin to 1 L of LB Broth.
3. LB + Ampicillin plates: Add 15 g of agar to 1 L of LB Broth. Autoclave. Cool down to 50°C and add 100 mg of ampicillin. Poor on plates.
4. *E. coli* strain BJ5183 (endA, *sbcB*$^-$, *recBC*$^-$, *str*^{R}).
5. *E. coli* strains TOP10, DH5α or similar.
6. 0.8% (wt/vol) agarose gel.

2.2. Adenovirus Construction by Homologous Recombination in Bacteria: Procedure II

1. LB + Kanamycin: Add 60 mg of kanamycin to 1 L of LB Broth (see previous protocol).
2. LB + Kanamycin plates: Add 15 g of agar to 1 L of LB Broth. Autoclave. Cool down to 50°C and add 60 mg of kanamycin. Poor on plates.
3. AdEasy kit (Stratagene #240009).
4. *E. coli* strains TOP10, DH5α or similar, and *E. coli* BJ5183.
5. 0.8% (wt/vol) agarose gel.

2.3. Adenovirus Construction by Homologous Recombination in Yeast

Reagents (the basic component can be purchased at Sigma):

1. Transformation mix: 240 μL of PEG (50%, wt/vol), 36 μL of lithium acetate 1.0 M, 10 μL of Boiled SS-Carrier DNA (10 mg/mL), 74 μL [(v) + (i)] DNA plus H_2O (Milli-Q, autoclaved).
2. YPDA^{++} (yeast extract/peptone/dextrose/adenine rich medium): 5 g of Yeast extract, 10 g of Bacto-peptone. Add ddH_2O up to 450 mL. For YPDA^{++} plates add 15 g of bacto agar and autoclave for 20 min. Then, add 50 mL of glucose 20%; 20 mL of adenine 0.5% (previously filtered through a 22 μm filter).
3. SC (basic medium): 3.35 g YNB [yeast nitrogen base without AA, with ammonium sulfate (Difco)]. Add ddH_2O up to 400 mL. For SC plates add 15 g of bacto agar and autoclave for 20 min. Then, add 50 mL of glucose 20% and 50 mL of 10× AA solution (URA- or LEU-, or URA/LEU-).
4. 10× AA Solution: 5.7 g of BSM–His–Leu–Try–Ura + 0.5 g Leucine (do not add for LEU-), 0.2 g tryptophan, 0.1 g histidine, and 0.1 g uracil (do not add for URA-). Add ddH_2O up to 500 mL and autoclave for 15 min. Store at 4°C.
5. SC plates with FOA (5-Fluoroorotic acid):
   (a) In a beaker, mix: 0.63 g of BSM–His–Leu–Try–Ura, 0.04 g of uracil (plasmids that grow in FOA plates must have lost the Ura gene), 0.02 g of Tryp, 0.01 g of His, 0.05 g of Leucine (do not add if the plasmid has CAL), 0.5 g of FOA, 3.5 g of YNB [yeast nitrogen base without AA, with ammonium sulfate (Difco)], 10 g of glucose/dextrose. Add ddH_2O up to 250 mL. Stir and heat on a stir plate to dissolve powders. Try to keep temperature below 45°C. It may take a while to dissolve the 5-FOA. Filter-sterilize when dissolved and keep the solution warm.
   (b) In another flask: Add 10 g of bacto agar plus 250 mL of ddH_2O. Autoclave. Add the filtered mixture (a) and mix thoroughly. Pour in plates (protect from light).

6. Lithium acetate (1.0 M): Dissolve 5.1 g of lithium acetate dihydrate (Sigma) in 50 mL of H_2O, sterilize by autoclaving, and store at room temperature.
7. Polyethylene glycol 3350 (50%, wt/vol): Dissolve 50 g of PEG 3350 (Sigma) in 30 mL of H_2O in a 150-mL beaker on a stirring hot plate. Cool down the solution to room temperature; fill volume up to 100 mL, mix thoroughly by inversion and autoclave. Store, securely capped, at room temperature. Evaporation of water from the solution will increase the concentration of PEG and severely reduce the yield of transformants.

2.4. Generation of Viral Pre-stocks

1. Dulbecco's modified eagle's medium (DMEM) (Gibco/BRL) supplemented with 10% or 1% fetal bovine serum (FBS, Hyclone).
2. *Pac* I restriction enzyme (New England Biolabs).
3. HEK-293 cells or other adenovirus packaging cell lines.

2.5. Purification of Viral Pre-stocks by Banding on CsCl

1. Ultracentrifuge: Beckman Coulter Optima L90K o L100XP and rotor SW40Ti (Beckman Coulter). Polyallomer centrifuge tubes for SW40 rotor (Beckman Coulter ref. 331374).
2. CsCl solutions: 1.4, 1.34, and 1.25 g/mL in PBS 1×.
3. 18-G needles, 2-mL syringes, pipette-aid, and 5 mL pipettes.
4. Amersham/Pharmacia PD-10 columns Sephadex G-25 (ref. 17-0851-01).
5. PBS 1× Ca^{++}/Mg^{++} (Gibco ref. 14080-048), Glycerol, anhydride (Fluka ref. 49769).

2.6. Genome Identity

1. DNase mix: 1 μL of RNase-free DNase (10 U/μL; Roche), 154 μL of nuclease-free water, 18 μL of 10× DNase digestion buffer (500 mM Tris–HCl, 100 mM $MgCl_2$ at pH 7.6).
2. Proteinase K mix: 20 μL of 10× Proteinase K buffer (100 mM Tris–HCl, 100 mM EDTA, and 2.5% SDS at pH 8), 5 μL of Proteinase K (20 mg/mL; Roche Applied Science, Mannheim, Germany).
3. 0.8% (wt/vol) agarose gel.

2.7. Titration Viral Stocks Using Anti-Ad/Hexon Staining

1. Dulbecco's modified eagle's medium (DMEM) (Gibco/BRL) supplemented with 5% FBS (Hyclone).
2. Primary antibody anti-hexon 2Hx-2 from ATCC or similar antibodies.
3. FITC or Alexa488-conjugated secondary antibody.

2.8. Ad Titration Assay for Virus IC50 Determination

1. Dulbecco's modified eagle's medium (DMEM) (Gibco/BRL) supplemented with 0.5% FBS (Hyclone).
2. BCA protein staining (Pierce ref. 23225): Mix 1 part of reactive A in 50 parts of reactive B and vortex the solution. This solution can be stored for 24 h if it is required.

3. Methods

Cloning the genes of interest by homologous recombination in bacteria or yeast is based on a two-step system. In the first step, the gene of interest is cloned into a shuttle vector using adequate restriction enzymes and ligation. The shuttle plasmid contains two fragments of adenovirus sequence (usually 4–5 kb from the 5′ end) flanking the multicloning site. After confirming its presence and orientation by restriction digestion analysis and/or sequence analysis, the second step consists of introducing the gene of interest into the adenovirus genome by homologous recombination between the shuttle plasmid and a large backbone plasmid. This backbone plasmid provides most of the adenovirus genome, but lacks essential genes (usually E1 genes) for virus propagation. Rapid detection of positive recombinants is achieved by antibiotic selection and restriction digestion analysis. The first method below describes the procedure to generate recombinants in *E. coli* by selection with only one antibiotic. The second method describes the commercial system AdEasy, whose cloning plasmids contains resistance for two different antibiotics. The third method describes how to generate recombinant adenovirus efficiently in yeast.

3.1. Adenovirus Construction by Homologous Recombination in Bacteria: Procedure I

In this protocol, the recombination between the shuttle plasmid and the adenovirus genome is performed in the *E. coli* strain BJ5183. Positive recombinants are selected by resistance to only one antibiotic (see Fig. 1). Therefore, to avoid background from undigested plasmids, complete digestions in steps 1 and 3 must be ensured.

1. Linearize backbone plasmid (i.e. pKP1.4 (5, 7) or similar) with a restriction enzyme cutting in the insertion site. In the pKP1.4 plasmid, *Swa*I site is located after the adenovirus packaging signal and marks the insertion point of the gene of interest. Digestion should be made in two steps: First, digest 3 μg of plasmid with 10 U of *Swa*I for 12–18 h. Then, add 10 U more of *Swa*I and digest six additional hours (see Note 1).
2. Check background by transforming BJ5183 bacteria with 100 ng of digested plasmid. After verification, store in aliquots of 200 ng.

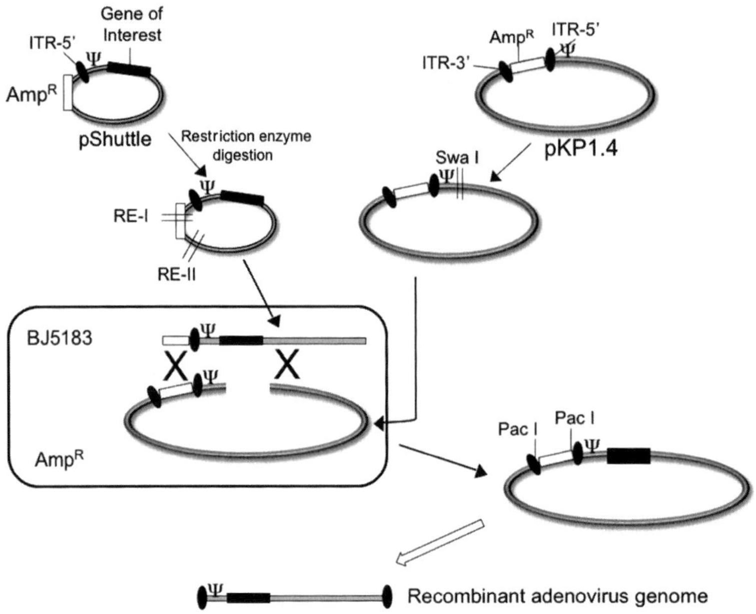

Fig. 1. Schematic outline of the homologous recombination in *E. coli* (procedure I). The shuttle plasmid already contains the gene of interest, which is flanked by the 5′ ITR and packaging signal (ψ) in one end, and adenoviral sequences in the other. The pKP1.4 backbone plasmid contains the adenovirus genome (except the E1 region). First, the pKP1.4 plasmid is linearized by *Swa* I, and the shuttle plasmid is digested by one (RE-I) or two restriction enzymes (RE-II) in the Amp^R gene. Both digested plasmids are co-transfected in BJ5183 bacteria for homologous recombination and only bacteria carrying recombinant plasmids containing the adenoviral genome plus the gene of interest are viable in LB + Amp^R plates. For production of the viral pre-stock, recombinant plasmids are digested with *Pac* I to liberate the vector genome.

3. Digest 2 μg of the shuttle plasmid with one or two appropriate restriction enzymes. Digest for 12–18 h using 2 U of each restriction enzyme (see Note 2).
4. Confirm complete digestion by agarose gel electrophoresis (0.8%). Purify the DNA fragment containing the expression cassette by GENECLEAN® or a similar method (see Note 3).
5. Resuspend in sterile ddH_2O and quantify the DNA by measuring absorbance at 260 nm.
6. Mix 50 ng of linearized pKP1.4 plasmid gently with different amounts of the previously purified DNA fragment (see Note 4). Start with the following molar ratios:

   1:5 pKP:fragment (approximately 50 ng pKP:50 ng of fragment) or

   1:20 pKP:fragment (approximately 50 ng pKP:200 ng of fragment)
7. Transform competent BJ5183 *E. coli* strain, using either heat-shock or electrocompetent standard procedures. Add 1 mL of LB broth and incubate at 37°C for 1 h while shaking at 250 rpm (see Note 5).

8. Culture 500 μL of co-transformed bacteria, in one 10-cm plate of LB + ampicillin. Incubate overnight at 37°C (see Note 6).
9. Pick at least ten isolated small colonies. Inoculate each in 2 mL of LB + ampicillin. Incubate overnight at 37°C shaking at 250 rpm (see Note 7).
10. Purify plasmid DNA with the conventional alkaline lysis procedure (better than with commercial DNA minipreparation kits). Resuspend DNA in 30 μL of Milli-Q H_2O. Check by agarose gel electrophoresis (0.8%). Store at −20°C (see Note 8).
11. Transform competent *E. coli* (strain TOP10, DH5α, or similar). Culture in 1 mL of LB and incubate for 1 h at 37°C while shaking at 250 rpm.
12. Culture 100 μL in one LB + ampicillin plate. Incubate overnight at 37°C. Pick three or four colonies and inoculate 3 mL of LB + ampicillin. Grow overnight at 37°C.
13. Purify plasmid DNA with the conventional alkaline lysis procedure and store at −20°C.
14. Identify positive recombinants and check their genomic integrity by a battery of informative restriction enzymes (see Subheading 3.6).

3.2. Adenovirus Construction by Homologous Recombination in Bacteria: Procedure II

In this protocol, recombination between the shuttle plasmid and the adenovirus genome is also performed in the *E. coli* strain BJ5183. The main difference is that the plasmid used contains resistance to two different antibiotics (8, 9). This strategy is followed by the AdEasy™ Adenoviral Vector System (see Fig. 2).

1. Digest the 2 μg of shuttle plasmid with *Pme*I. Remove the restriction enzyme and buffer, and treat with alkaline phosphatase for 30 min at 37°C. *Pme*I cuts within the flanking adenoviral sequences. This is an important difference with respect to the previous protocol.
2. Confirm complete digestion by agarose gel electrophoresis and purify the linearized shuttle plasmid by GENECLEAN® or similar method (see Note 3).
3. Resuspend in sterile ddH_2O to a final concentration of 1 μg/μL.
4. Mix 100 ng of the plasmid containing the complete adenovirus genome (pAdEasy-1 plasmid in AdEasy kit) and 1 μg of the linearized shuttle plasmid (see Note 9).
5. Transform competent BJ5183 bacteria, using either heat-shock or electrocompetent standard procedures. Add 1 mL of LB broth and incubate at 37°C for 1 h while shaking at 250 rpm (see Note 5). As control use only linearized shuttle plasmid.

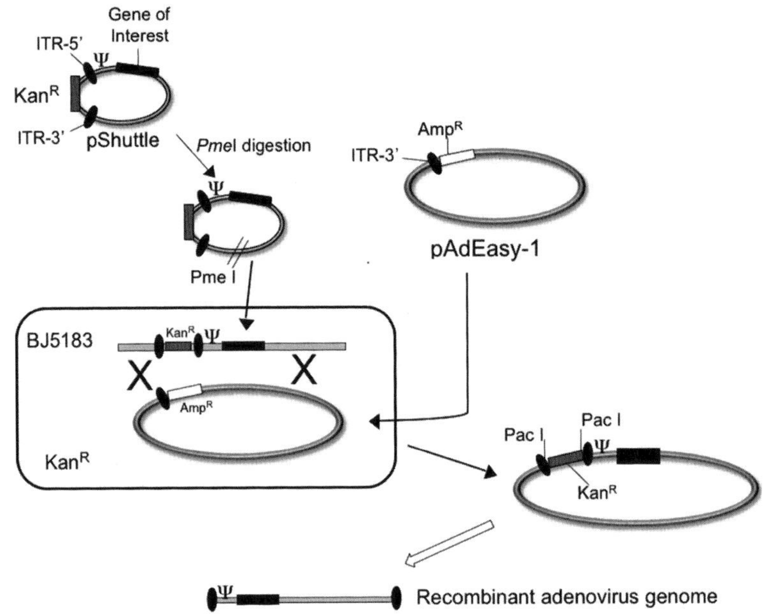

Fig. 2. Schematic outline of the homologous recombination in *E. coli* (procedure II). The shuttle plasmid already contains the gene of interest, which is flanked by the 5′ ITR and packaging signal (Ψ) in one end, and adenoviral sequences plus the 3′ ITR in the other. The AdEasy-1 backbone plasmid contains the adenovirus genome (except for the 5′ ITR, Ψ and the E1 region). First, the shuttle plasmid is digested with *PmeI* into the adenovirus sequence. Both plasmids are co-transfected in BJ5183 bacteria for homologous recombination and only bacteria carrying recombinant plasmids containing the adenoviral genome plus the gene of interest are viable in LB + Kan^R plates. For production of the viral pre-stock, recombinants plasmids are digested with *Pac* I to liberate the vector genome.

7. Culture 100 and 500 μL of co-transformed bacteria, in two 10-cm plates of LB + Kanamycin. Incubate overnight at 37°C (see Note 6).
8. Select at least ten isolated small colonies. Inoculate in 2 mL of LB + Kanamycin. Incubate overnight at 37°C by shaking at 250 rpm (see Note 7).
9. Purify plasmid DNA with the conventional alkaline lysis procedure (better than with commercial DNA minipreparation kits). Resuspend DNA in 50 μL of Milli-Q H_2O. Check by agarose gel electrophoresis. Store at −20°C.
10. Transform competent bacteria (strain TOP10, DH5α or similar) using either heat-shock or electrocompetent standard procedures, only with DNA from colonies with high molecular weight DNA (see Note 8). Culture in 1 mL of LB and incubate for 1 h at 37°C while shaking at 250 rpm.
11. Culture 100 μL in one LB + Kanamycin plate. Incubate overnight at 37°C. Pick three or four colonies and inoculate 3 mL of LB + Kanamycin. Grow overnight at 37°C.
12. Purify plasmid DNA with the conventional alkaline lysis procedure. Store at −20°C.

13. Identify positive recombinants and check their genomic integrity by a battery of informative restriction enzymes (see Subheading 3.6).

3.3. Adenovirus Construction by Homologous Recombination in Yeast

Compared to bacteria, homologous recombination in yeast is more efficient and need much shorter regions of homology (40 bp). This makes yeast a more flexible and efficient system for adenovirus construction. The method requires adapting adenovirus genomic plasmids (i.e. pKP1.4 or pAdEasy-1) to grow in yeast. This means that a centromere "CEN", an autonomously replicating sequence "ARS", and a yeast selection gene (Ura o Leu) has to be inserted in the backbone of such adenovirus plasmid. This is quite straightforward as this sequence can be obtained by PCR from a plasmid with Uracil or Leucine selectable genes (i.e. pRS416 or pRS425, respectively, Stratagene). The Ura gene allows the yeast to grow in media without uracil (URA-) and the Leu gene to grow without Leucine (LEU-). We call this fragment that confers yeast compatibility and selection, CAU or CAL ("C" for centromere, A for autonomous replicating sequence and U or L for URA or LEU). The primers used for this PCR contains 40 nt tails at their 5′ ends that are homologous to the site targeted in the adenovirus plasmid. In addition, the primer contains 20 nt at their 3′ end corresponding to the beginning and end of the CAU or CAL fragment. The PCR product will contain the CAU or CAL with 40-bp flanking regions homologous to the site to be targeted. As the only plasmid that can grow in yeast is the recombination product, it is not necessary to open or linearize the adenovirus genomic plasmid for the recombination. However, if a site with a unique enzyme that does not destroy an essential sequence in the adenovirus genomic plasmid is available then, such a site should be targeted. That is, the 40-bp flanking regions of the CAU or CAL fragments should fall upstream and downstream of this site. Then linearization of the adenovirus genomic plasmid with this unique enzyme increases the rate of homologous recombination.

Once the pAd-CAU or CAL is ready, the modification of this plasmid to generate recombinant adenoviruses follows two general strategies depending on the availability of restriction sites: cut-repair or URA-positive-negative selection. When a restriction site is unique (or partial digestion using a two-cutter) the pAd-CAU or pAd-CAL plasmid ("vector") can be linearized and a fragment of DNA ("insert") with a minimum of 40 bp homology at both sides of the cut can be used to circularize the plasmid and obtain the recombinant with the insert. If no restriction sites exist at the position to be modified, the pAd-CAL genomic plasmid can be used to insert URA without the need to cut it. URA is amplified from pRS416 using primers with 40 nt sequences that flank the desired position in the pAd-CAL plasmid and recombinants

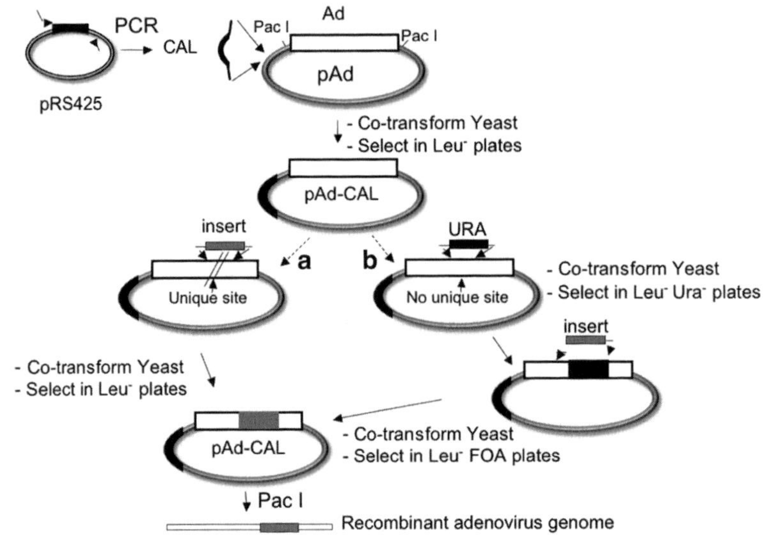

Fig. 3. Main steps involved in the generation of recombinant adenoviruses using homologous recombination in yeast. CAL (Centromere-Autonomously Replicating Sequences – Leucine gene) renders a bacterial plasmid competent for yeast growth. The adenovirus genomic plasmid with CAL can be used to insert a DNA of interest either via cut-repair when a unique restriction site is available at the targeted site (**a**) or via positive–negative selection with the URA3 gene (**b**).

are selected using Uracil and Leucine deficient plates (positive selection of CAU). Then the URA gene is replaced by the desired insert without the need of any restriction using a negative selection against URA-containing plasmids using FOA plates. Alternatively to the negative selection step, the URA gene can be flanked by unique restriction sites and cut it after the positive selection to proceed as in a cut-repair protocol (see Fig. 3). Once yeast colonies are obtained by cut-repair or URA-positive-negative selection, the DNA from the yeast plasmid (low copy) is isolated and transferred to bacteria (high copy) in order to analyze it. A similar system has been published by Hokanson et al. (10).

3.3.1. Preparation of the Insert

There are different types of inserts: the CAU or CAL to adapt a regular bacterial plasmid to grow in yeast, the URA insert to use positive–negative selection, and a regular DNA insert obtained by restriction or PCR to repair (re-circularize) a linearized yeast plasmid. The common requirement is that the 5′ and 3′ ends (a minimum of 40 bp) of the insert are homologous to a region in the receptor vector that will be replaced.

1. Primer design: Oligos to amplify CAU (CAL uses the same oligos but pRS425 instead of PRS416): Forward: 40 bp homology upstream of targeted site + ACCTGGGTCCT TTTCATCAC. Reverse: 40 bp homology downstream of targeted site (reverse orientation) + CATCTGTGCGGTAT TTCACA.

Oligos to amplify URA (from pRS416): Forward: 40 bp homology upstream of targeted site + TCAATTCATCA TTTTT. Reverse: 40 bp homology downstream of targeted site (reverse orientation) + GTAATAACTGATATAA.

2. PCR mix: 0.5 μL template DNA (20 ng) (e.g. pRS416); 25 μL Ex-Taq 2×; 1 μL oligo F (20 μM); 1 μL oligo R (20 μM); and 22.5 μL Milli-Q-autoclaved H_2O.
3. PCR program (PCR fragment 1 kb): 1 min at 95°C/30× (30 s at 95°C – 30 s at 55°C – 2 min 30 s at 72°C)/5 min at 72°C.
4. After the PCR, digest the template plasmid by adding 10 U of *Dpn*I (only cuts the methylated DNA template, not the PCR product) to the PCR tube and incubating for 2 h at 37°C. Then, use phenol–chloroform extraction and clean with gel purification, ethanol precipitation, and resuspend in ddH_2O.

3.3.2. Homologous Recombination in Yeast: Preparing Competent Yeast and Co-transformation with Vector and Insert

Protocols for yeast transformation have been adapted from Gietz and Woods (10).

Day 1

1. In a 50-mL falcon, inoculate 5 mL of YPDA^{++} (Yeast extract/Peptone/Dextrose/Adenine rich medium) with one colony of yeast or 10 μL of yeast glycerol stock (strain YPH857).
2. Incubate O/N at 30°C shaking at 200 rpm.
3. Place a bottle of YPDA^{++} and a 250-mL culture flask in the incubator as well.

Day 2

1. Dilute 1/10 in YPDA^{++} to measure OD_{600} (use YPDA^{++} as blank, 1 $OD_{600} = 1.5 \times 10^7$ cells/mL). Calculate dilution to prepare 50 mL of pre-warmed YPDA^{++} at $OD_{600} = 0.15$ (0.15 $OD_{600} = 2.25 \times 10^6$ cells/mL).
2. Incubate the flask on a rotary shaker at 30°C and 200 rpm until exponential growth is achieved ($OD_{600} = 0.4$–0.9, approximately 5 h).
3. Transfer the 50 mL to a falcon tube and spin at 3,000 × *g* for 5 min at room temperature.
4. Decant supernatant and wash the pellet with 25 mL ddH_2O. Spin again (2,000 rpm, 5 min).
5. Decant supernatant and resuspend the cells in 1 mL of ddH_2O.
6. Boil salmon sperm or Herring DNA (10 mg/mL) in a boiling water bath (use 10 μL for each transformation) for 5 min and then keep on ice while harvesting the cells.

7. Transfer the yeast suspension to an Eppendorf, centrifuge for 30 s at 5,000 × *g* in a microfuge, and discard the supernatant.
8. Add ddH_2O to a final volume of 1 mL and vortex-mix vigorously to resuspend the cells.
9. Pipette 100 μL of samples (10^8 cells) into 1.5-mL microfuge tubes, one for each transformation, centrifuge at 5,000 rpm at room temperature for 30 s, and remove the supernatant.
10. Make up the transformation mix (see Note 10).
11. Add 360 μL of transformation mix to each transformation tube and resuspend the cells by vortex mixing vigorously.
12. Incubate the tubes in a 42°C water bath for 40 min.
13. Microcentrifuge at 5,000 rpm at room temperature for 30 s and remove the supernatant with the micropipette.
14. Pipette 1.0 mL of H_2O (Milli-Q, autoclaved) into each tube, stir the pellet with a micropipette tip and vortex vigorously.
15. Plate appropriate dilutions of the cell suspension onto SC-URA or SC-LEU plates (see Subheading 2.3). Spread gently (few movements). Use plates without URA or LEU or both according to the presence of URA, LEU, or both genes, respectively, in the vector or the insert.
16. Incubate at 30°C for 2–3 days until yeast colonies appear. There should be more colonies in the plates with (v) + (i), than in the (v) and (i)-alone controls.

3.3.3. Yeast Plasmid Extraction After Yeast Transformation

1. Seed 2 mL of liquid SC-URA (or SC-LEU) O/N at 30°C with the desired colony.
2. Transfer 1.5 mL to the Eppendorf tube. Centrifuge at maximum speed for 5 s in a microfuge. Discard supernatant and resuspend the yeast pellet in residual liquid.
3. Add 400 μL of 2% TX-100/1% SDS/0.1 M NaCl/10 mM Tris–HCl at pH 8.0/1 mM EDTA and mix.
4. Add 400 μL phenol:chloroform:isoamyl alcohol (25:24:1).
5. Add 0.3-g glass beads (Sigma). Close Eppendorf with parafilm.
6. Vortex (vertical) for 2 min at 4°C.
7. Spin for 5 min at maximum speed in a microfuge.
8. (Optional) Take 300 μL of the supernatant and extract DNA from the solution following a DNA purification method (e.g. glass milk). Add ddH_2O to a final volume of 300 μL and continue with step 9.
9. Take 300 μL of supernatant and add 600 μL of EtOH/2% NaAc. Invert and leave at −80°C or −20°C to increase DNA precipitation.

10. Centrifuge 20 min max. Speed in microfuge at room temperature.
11. Resuspend in 25 μL of H_2O or TE (H_2O is better when planning to transform this DNA by electroporation).
12. Use 2 μL for electrocompetent DH5α transformation. Next day pick up colonies and purify by miniprep the plasmid DNA.
13. Identify positive recombinants and check their genomic integrity by a battery of informative restriction enzymes (see Subheading 3.6).

3.4. Generation of Viral Pre-stocks

The production of recombinant adenoviruses should be performed in a Biosafety Level 2 laboratory. The requirements include the use of laminar flow hoods and the establishment of proper residue's manipulation. Conventional methods for producing small volumes of viral vectors involve culturing cells in stationary, adherent cultures, such as T-flasks or roller bottles. The principle protocols of this small scale production method are presented below, while methods for large scale production of Ad viral vectors can be found in the literature. Viral pre-stocks obtained at the end of the protocol are ready to be used as starting material for subsequent rounds of larger amplifications.

1. Six to fifteen hours before transfection, plate 10^6 HEK-293 cells in several 25-cm^2 tissue culture flask(s) with DMEM + 10% FBS. Plate at least two flasks. One plate may work as control plate and help to follow vector amplification.
2. Digest 3 μg of recombinant adenoviral plasmid with 30 U of *Pac*I to separate adenovirus genome from bacterial sequences. To ensure complete digestion, 6 h later add another 30 U of *Pac*I. Digest overnight.
3. Precipitate digested DNA with two volumes of ethanol and resuspend in 20 μL of sterile ddH_2O.
4. Perform a standard transfection using 3 μg of *Pac*I digested plasmid per 10^6 HEK-293 cells (see Note 11).
5. Incubate the DNA with the transfection reagent for 30 min at room temperature.
6. Remove growth medium from HEK-293 cells and wash once with serum-free DMEM gently. Remove DMEM and add 2 mL of DMEM + 1%FBS per 10^6 cells.
7. Add DNA complexes dropwise to the cells. Incubate at 37°C and 5% CO_2 for 4–6 h.
8. Remove medium and add 2 mL of fresh DMEM + 10% FBS. Incubate at 37°C and 5% CO_2 for 6 days (see Note 12).
9. Scrape/harvest the medium and cells. Freeze/thaw three times to release adenovirus from cells.

10. Centrifuge at 4°C for 10 min and 500 × *g* and discard the pellet.
11. Use all the centrifuged crude lysate to infect 8×10^6 HEK-293 cells (at 70–80% confluency) in a 10-cm plate.
12. Incubate at 37°C and 5% CO_2 until general cytopathic effect is observed (usually between 3 and 6 days).
13. Harvest medium and cells. Freeze/thaw three times to release adenovirus from cells.
14. Centrifuge at 4°C for 10 min and 500 × *g*. Discard the pellet and keep the supernatant.
15. Use all the previous centrifuged crude lysate to infect 4×10^8 HEK-293 cells (at 70–80% confluency) in twenty 25-cm plate.
16. Incubate at 37°C and 5% CO_2 until general cytopathic effect is observed (usually between 32 and 38 h).
17. Harvest medium and cells. Freeze/thaw three times to release adenovirus from the cells.
18. Centrifuge at 4°C for 10 min and 500 × *g*. Discard the pellet and keep the supernatant (viral-prestock).
19. Titer the viral pre-stock by anti-hexon staining method (see Subheading 3.7).
20. Aliquot and store viral pre-stock at −80°C.

3.5. Purification of Viral Pre-stocks

Although out of the scope of this methods paper, the protocols for purification of AdV vectors have evolved over the last decade. The most classical and easy to acquire for a non-specialized laboratory remain the ultracentrifugation on a CsCl. This method of purification is limited by the capacity of cell lysate volume that can be processed. However, this method is still widely used and most of the time sufficient for fundamental studies and early in vivo preclinical evaluation of the vectors. More complex techniques based on column chromatography and membrane techniques are now well developed for the generation of high purity grade and upscaled production suitable for human clinical applications (11).

3.5.1. Initial Step Gradient

1. In an SW40 polyallomer centrifuge tube, add 2.5 mL of 1.4 g/mL of CsCl.
2. Add 2.5 mL of 1.25 g/mL of CsCl by placing tip of a 5-mL pipette slowly dispensing solution to make two phases.
3. Gently add ~7 mL of cleared vector supernatant on top of 1.25 g/mL CsCl. Leave about 0.5 cm at the top of the tube.
4. Add 0.5 mL of mineral oil on top of cleared vector supernatant.
5. Balance tubes against closest sample and load in rotor.

6. Centrifuge for 90 min at 35,000 rpm, 18°C.
7. Remove tubes from the rotor with forceps.
8. Vector appears as an opaque band at interface of 1.25 g/mL and 1.4 g/mL CsCl. Remove band by piercing the tube about 1 cm below vector with 2-mL syringe loaded with 18-G needle (see Note 13).

3.5.2. Second Isopycnic Gradient

1. Add 5–6 mL of 1.34 g/mL of CsCl to the recovered vector band from previous step into a new polyallomer centrifuge tube. Cover with ~0.5 mL of mineral oil.
2. In a second tube add CsCl (1.34 g/mL) to equilibrate tubes. Centrifuge for 18 h at 35,000 rpm, 18°C.
3. Remove tube from rotor and place in black safety tube holder. Vector appears as opaque band near the center of the tube. Remove band as above in less than 2 mL. Collect vector by keeping the volume to a minimum.

3.5.3. Desalting Column and Storage

1. Prepare PD-10 column following manufacturer's instructions. Load up to 2 mL of vector on column.
2. Collect by adding 0.5 mL of PBS 1× Ca^{++}/Mg^{++}. Repeat the step 9–10 times. Label 0.5-mL tubes.
3. The vector is clearly visible as an opaque elute in a final volume of ~2.0–3.5 mL (from aliquots 4 to 7), depending on the initial volume size.
4. Combine the most opaque tubes (excluding the extremities which can be used to isolate DNA for further analysis) and add glycerol to a final concentration of 10%. Another option is to dialyze the vector in several changes of PBS at 4°C.
5. Aliquot (10, 50, and 100 μL) in 0.5-mL tubes and store at −80°C as quickly as possible.

3.6. Genome Identity

3.6.1. Identity of the Vector Genome by Restriction Enzyme Analysis

Integrity and identity of the vector genome can be quickly analyzed by restriction enzyme digestion. Since, each gene of interest has a specific DNA sequence, informative restriction enzymes must be previously chosen by comparing the expected recombinant adenovirus to the original backbone plasmid with a Sequence Analysis computer program.

1. Digest 1–2 μg of purified plasmid DNA from selected colonies after homologous recombination with 10 U of an informative restriction enzyme, for 6 h.
2. Perform at least seven or eight different digestions and run in a 0.8% agarose gel (see Fig. 4).
3. If only one restriction enzyme pattern does not correspond with the expected pattern, repeat the digestion. If the observed

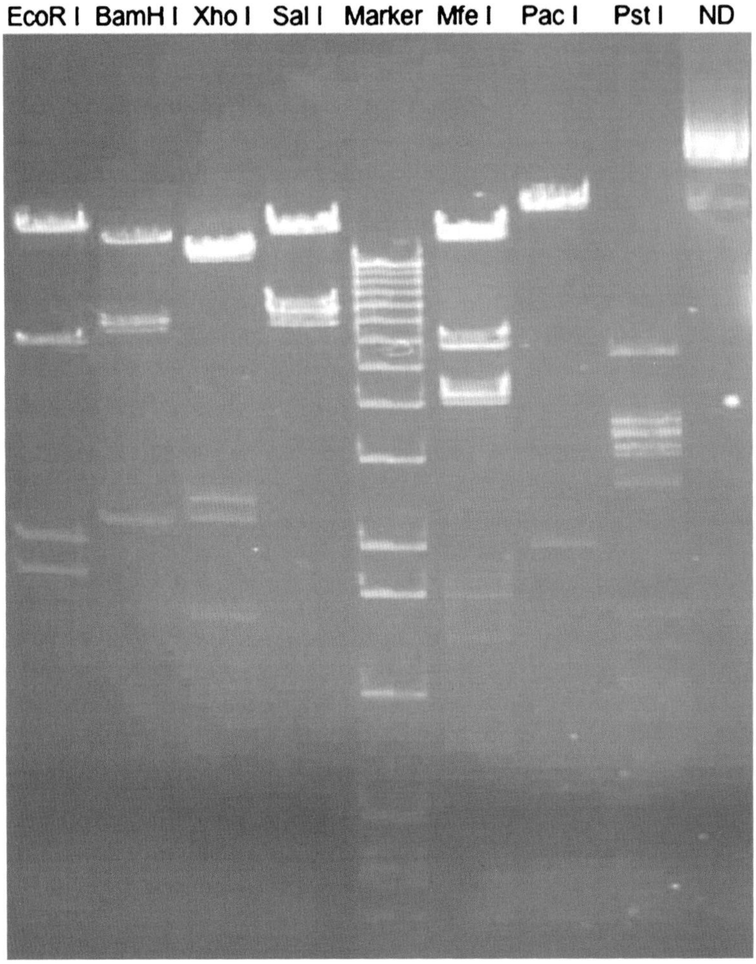

Fig. 4. Analysis of the integrity and identity of the vector genome by multiple restriction enzyme digestion. Informative restriction enzymes must be previously chosen with a Sequence Analysis computer program. Marker, 1-kb marker. *ND* non-digested control.

pattern still does not correspond with the expected pattern or if there are more than one unexpected enzyme patterns, discard the selected DNA.

3.6.2. Detection of the Gene of Interest by PCR

Contaminating cellular DNA from the viral-pre-stock must be removed prior the PCR, especially when the gene of interest is of human origin. To this end, an aliquot of viral pre-stock must be subjected to a pretreatment with DNAse and Proteinase K (11). Once the viral pre-stock has been processed, use specific primers for the gene of interest and previously set-up conditions. In addition, amplified fragment may be sequenced to further confirm the identity of the gene of interest cloned into the viral genome.

1. Incubate 2 μL of viral pre-stock with 173 μL of DNAse mix for 1 h at 37°C. Inactivate DNase by incubation at 75°C for 30 min. Allow viral pre-stock to cool down at room temperature.
2. To disrupt the viral capsids incubate with 25 μL of Proteinase K mix during 1 h at 37°C. Inactivate Proteinase K by incubation at 95°C during 20 min.
3. Use 2–4 μL of treated viral-prestock per PCR.

3.7. Titration of Viral Stocks Using Anti-Ad/Hexon Staining

1. Prepare serial dilutions (typically 1/10) of the stock in 96-well dishes using DMEM + 5% FBS. Final amount for each dilution in each well should be 100 μL. Range of dilutions will be selected according to the estimate concentration of the viral stock to titer.
2. Add 50 μL/well of a cell suspension at 1.10^6 cells/mL (50,000 cells/well) in DMEM + 5% FBS. The cell line must be chosen depending on characteristics of the viral stock to titer (HEK-293 cells are appropriate for most assays) (see Note 14).
3. Incubate virus and cells for 24–48 h (time must be chosen depending on rate of replication of Ad in the cell line in order to avoid secondary infections). For HEK-293 cells incubate for 24–36 h.
4. Remove medium from the wells very carefully to avoid cell loss. Air-dry for 3–5 min.
5. Add 100 μL of 100% ice-cold methanol to each well (fixation). Incubate for 10 min at −20°C.
6. Aspirate methanol. Wash each well containing the cells twice with 100 μL of PBS Ca^{++}/Mg^{++} 1% BSA (PBS without Ca^{++}/Mg^{++} can also be used, but adding bivalent ions can prevent cells detaching from wells).
7. Add 50 μL of primary antibody diluted in PBS Ca^{++}/Mg^{++} 1% BSA to each well. For most hybridomas 1/5 dilution from supernatant is recommended. If using a purified anti-adenovirus or anti-hexon Ab, 1/500 dilution can be initially tested. Avoid bubble formation. Incubate for 1–2 h at 37°C.
8. Wash each well (3×) with 100 μL of PBS Ca^{++}/Mg^{++} 1% BSA.
9. Add 50 μL of FITC or Alexa488-conjugated secondary antibody diluted in PBS Ca^{++}/Mg^{++} 1% BSA (1/300 dilution is suitable for most commercial antibodies) to each well. Using Alexa488 can increase the test sensibility. Avoid bubble formation. Incubate for 1–2 h at 37°C in the dark.
10. Wash each well (3×) with 100 μL of PBS Ca^{++}/Mg^{++} 1% BSA.
11. Quantify green cells in each well using an inverted fluorescence microscope. A cloud of positive cells ("comet effect") suggests secondary infections and consequently it should be

quantified as a single positive cell. For each stock, the mean of different dilutions will be used to calculate concentration. Note that only 100 μL of each dilution has been used for analysis, so a tenfold factor should be applied to obtain transducing units per milliliter (TU/mL).

3.8. Ad Titration Assay for Virus IC50 Determination (Spectrotiter)

1. Seed cells in a 96-well plate at 30,000 cells/well in a total volume of 100 μL/well of DMEM supplemented with 0.5% FBS. To do it, count the total number of cells, centrifuge them to eliminate the supernatant (1,250 rpm during 5 min), and resuspend them in a necessary volume of DMEM + 0.5% FBS to obtain 300,000 cell/mL. In case confluence is needed, seed more cells/well (100,000 cells/well for HEK-293, or 30,000 cells/well for A549).
2. Two days after seeding, infect the cells with serial dilutions of the virus (see Note 15). For cells that get easily infected, use 1/5 dilutions. For cells that need high concentrations to be infected use 1/2 or 1/3 virus's dilutions. Prepare 11 serial dilutions of each virus in an Eppendorf and infect wells by adding 50 μL of each dilution per well. Add to the 12th column of the 96-well plate, 50 μL of DMEM + 0.5% FBS as non-infected controls to each well (see Note 16).
3. Observe the infection daily by comparing the CPE of infected cells with the control cells. Incubate at 37°C until cytopathic effect can be detected in the first seven columns (this is normally occurs 5–8 days after infection).
4. Prepare BCA protein staining. For 96 wells: 96 wells × 200 μL/well = 19.2 mL. Reactive A + B ⇒ 20 mL reagent A + 400 μL reagent B.
5. Shake softly, the 96-well plate, in order to resuspend dead cells and carefully remove the medium from each well. Add 200 μL of reagent A + B to each well as soon as possible with a multi-channel pipette. If DMEM + 5% FBS was used, wash the cells with PBS in order to remove FBS (that could interfere with the results) before adding reagent A+B.
6. Incubate for 30 min at 37°C and read in a spectrophotometer at 540 nm. Use any well from the first or second column (the clearest) as the blank.
7. Introduce the results in an excel file. For each condition (vp/mL or TU/mL), plot the amount of protein as a percentage (%) with respect to the non-infected control (see Fig. 5).
8. Calculate the dose that reduces the protein content to 50% (IC50) for each virus (GraFit software for example).

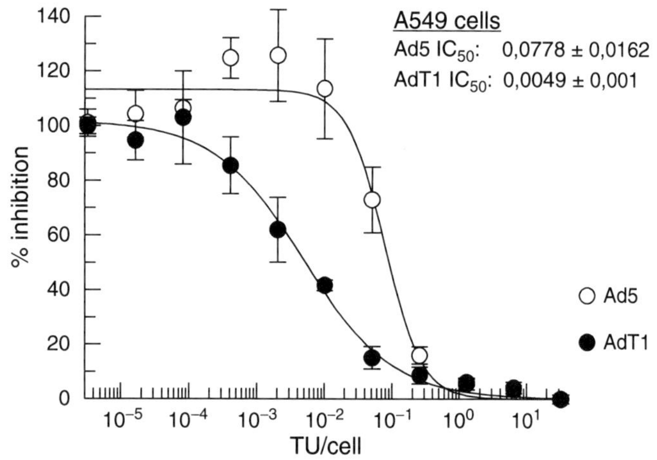

Fig. 5. Spectrophotometric determination of the concentration of the virus that lyses 50% of cells in a given period of time (IC50). This method is very useful to compare the lytic potency of different viruses. Cells are infected in 96-well plates using serial dilutions of the different viruses in triplicates. The researcher arbitrarily stops the experiment when for most viruses, wells with 50% of cytopathic effect (detachment of cells) fall in the middle of the serial dilution. The protein content remaining in the well is measured by colorimetric absorbance as a quantitative indication of the cells that have not been lysed. The more diluted the virus, the highest the protein content, until a 100% is indicated by non-infected control wells. For each triplicate dilution (*x*), the mean% of inhibition achieved (*y*) (100% inhibition being no cell death) and the standard deviation is introduced into a nonlinear regression program (Prism, GraFit, BioDataFit, etc.) to apply an adapted Hill equation that will indicate the mean IC50 and the fitting or standard error of such IC50. Some programs use the log10 of the dilution as (*x*).

4. Notes

1. Digestion should be as complete as possible in order to remove the background from undigested plasmids in the following steps.
2. The enzyme(s) must NOT cut within the expression cassette. To facilitate recombination, it is recommended to leave at least of 1 kb sequences on both sides of the expression cassette. If possible, use two enzymes for digestion, better if at least one is within the resistance gene.
3. Briefly, to isolate DNA from Agarose Gels by Geneclean® Turbo Kit, place gel slice in an Eppendorf tube. Add 100 μL GENECLEAN® Turbo SALT Solution per 0.1 g of gel slice and mix. Incubate at 55°C for 5 min to melt gel. Invert tube to mix. Transfer <600 μL DNA/SALT solution to

GENECLEAN® Turbo Cartridge. Centrifuge until all liquid has passed through the filter. Empty Catch Tube as hended. Add 500 μL of prepared GENECLEAN® Turbo Wash Solution to the filter and centrifuge for 5 s. Empty Catch Tube as hended. Centrifuge GENECLEAN® Turbo Cartridge for an additional 4 min to remove the residual Wash Solution. Remove cap from a new, clean Catch Tube and insert the GENECLEAN® Turbo Cartridge containing the bound DNA. Add 30 μL of GENECLEAN® Turbo Elution Solution directly onto GLASSMILK®-embedded membrane and incubate at room temperature for 5 min. Centrifuge for 1 min to transfer the eluted DNA to a GENECLEAN® Turbo Catch Tube. Discard GENECLEAN® Turbo Cartridge and cap the Catch Tube.

4. Alternatively, mix different molar ratios by changing the amount of purified fragment, for example at a ratio of 1:50 (approximately 50 ng pKP1.4:500 ng fragment).
5. Use as controls (a) Only pKP1.4 linearized with SwaI (optional, if the plasmid has been previously checked as suggested in point 1); and (b) only gel-purified fragment from the shuttle plasmid. It is recommended to use highly competent bacteria since BJ5183 exhibit lower transformation efficiencies than conventional *E. coli* strains.
6. Some authors suggest incubation for at least 24 h, arguing that a shorter incubation time is not sufficient for evident growth of colonies containing recombinant plasmids.
7. Two populations of colonies are expected: large and small size colonies. Large colonies are generally the background from shuttle plasmid, while small colonies will likely contain recombinants plasmids, which are low copy number plasmids. Number of small colonies must be at least three times higher than in control plate (digested pKP1.4 only) to continue the protocol. If number of small colonies is less than three times, start the procedure once again and check for complete *Swa*I digestion of the pKP1.4 plasmid; also use a different ratio in step 5.
8. Do not store the BJ5183 bacteria after overnight growth, as unwanted recombinants might appear. Perform plasmid purification early in the morning. Check by agarose gel electrophoresis to discard colonies containing the shuttle plasmids. Select clones only with high molecular weight DNA. Also check those clones with undetectable DNA since the yield of recombinant DNA is much lower that from background or unwanted rearrangements.
9. It is possible to use competent BJ5813 bacteria previously transformed with the AdEasy-1 plasmid (AdEasier-1 bacteria). Because of the efficient recombination processing in BJ5183

bacteria, perform extensive restriction enzyme analysis to discard unwanted rearrangements, unless commercial AdEasier bacteria are used.

10. Vector (v) is 300 ng linearized plasmids. Insert (i) is 100 ng CAU or CAL PCR fragment or any other insert obtained by PCR or restriction form a donor plasmid. When vector is not linearized because the insert has a selectable gene (Ura or Leu), increase the amount of insert by 100-fold (10 μg). It is recommended to use a mix with vector and insert alone as controls.
11. Usual methods to transfect HEK-293 cells are calcium phosphate precipitation (12), or polyethylenimine (PEI) (6) though other methods based on cationic molecules can also be used. For efficiency, simplicity, and cost, PEI is highly recommended. In the case PEI is used, the following protocol for the preparation of the PEI/DNA complex can be used:
    (a) Prepare PEI and the DNA complexes in 2-mL Eppendorf tubes.
    (b) In a tube labeled A: put 6 μg DNA and 150 μL of sterile 150 mM NaCl. Mix well.
    (c) In a tube labeled B: put 1.35 μL PEI 10× and 150 μL of sterile 150 mM NaCl. Mix well.
    (d) Using a Pasteur pipette, slowly add solution B drop-wise to solution A.
    (e) Incubate for 30 min at room temperature.
    (f) Change cell medium to DMEM + 1%FBS.
    (g) Add PEI–DNA complexes to cells.
12. If vector carries a marker gene, check initial transfection efficiency as well as vector copy during amplification. Though possible, do not expect to observe an evident cytopathic effect (CPE).
13. Carry pierced tube to plastic bottle using loaded syringe, withdraw needle and let drain into 500-mL plastic bottle (an old medium bottle works well).
14. Use non-permissive, A549 human cells to detect replication competent adenovirus. Contrary to the regular first-generation vectors, replication competent adenovirus will be able to produce new viral proteins in A549 cells.
15. Always use pipette tips with filter when pipetting the virus. Before preparing the dilution, choose the vp/cell or TU/cell (viral particle per cell or transducing units per cell) desired to infect the first column. Use 5333.3 vp/cell or 100–300 TU/cell. The vp/mL necessary for the first column is:

$$\begin{aligned}5333.3\ \text{vp/cell} \times 30.000\ \text{cell/well} &= 1.5\text{E}10^{8}\ \text{vp}/50\ \mu\text{L}\\ &= 3\text{E}10^{6}\ \text{vp}/\mu\text{L}\end{aligned}$$

16. When infecting, start adding DMEM + 0.5% of FBS in column number 12. Then, infect column number 11 (the most diluted) and use the same tip while infecting with the same virus. Try to release the 50 μL of infection medium, of each well, without disturbing the monolayer (particularly for HEK293 cells). Try not to make bubbles.

References

1. Graham, F. L., and Prevec, L. (1995) Methods for construction of adenovirus vectors *Mol. Biotechnol.* **3**, 207–220.
2. Barratt-Boyes, S. M., Soloff, A. C., Gao, W., Nwanegbo, E., Liu, X., Rajakumar, P. A., et al. (2006) Broad cellular immunity with robust memory responses to simian immunodeficiency virus following serial vaccination with adenovirus 5- and 35-based vectors. *J. Gen. Virol.* **87**, 139–149.
3. Glasgow, J. N., Kremer, E. J., Hemminki, A., Siegal, G. P., Douglas, J. T., and Curiel, D. T. (2004) An adenovirus vector with a chimeric fiber derived from canine adenovirus type 2 displays novel tropism. *Virology* **324**, 103–116.
4. Cascallo, M., Alonso, M. M., Rojas, J. J., Perez-Gimenez, A., Fueyo, J., and Alemany, R. (2007) Systemic toxicity-efficacy profile of ICOVIR-5, a potent and selective oncolytic adenovirus based on the pRB pathway. *Mol. Ther.* **15**, 1607–1615.
5. Alba, R., Hearing, P., Bosch, A., and Chillon, M. (2007) Differential amplification of adenovirus vectors by flanking the packaging signal with attB/attP-PhiC31 sequences: implications for helper-dependent adenovirus production. *Virology* **367**, 51–58.
6. Kremer, E. J., Boutin, S., Chillon, M., and Danos, O. (2000) Canine adenovirus vectors: an alternative for adenovirus-mediated gene transfer. *J. Virol.* **74**, 505–512.
7. Delenda, C., Chillon, M., Douar, A.-M., and Merten, O.-W. (2007) Cells for gene therapy and vector production. *Methods in Biotechnology: Animal Cell Biotechnology: Methods and Protocols. Humana Press, (Editor: Ralf Poertner)* 24, 23–91.
8. Luo, J., Deng, Z. L., Luo, X., Tang, N., Song, W. X., Chen, J., et al. (2007) A protocol for rapid generation of recombinant adenoviruses using the AdEasy system. *Nat. Protoc.* **2**, 1236–1247.
9. Wu, C., Nerurkar, V. R., Yanagihara, R., and Lu, Y. (2008) Effective modifications for improved homologous recombination and high-efficiency generation of recombinant adenovirus-based vectors. *J. Virol. Methods* **153**, 120–128.
10. Hokanson, C. A., Dora, E., Donahue, B. A., Rivkin, M., Finer, M., and Mendez, M. J. (2003) Hybrid yeast-bacteria cloning system used to capture and modify adenoviral and nonviral genomes. *Hum. Gene Ther.* **14**, 329–339.
11. Burova, E., and Ioffe, E. (2005) Chromatographic purification of recombinant adenoviral and adeno-associated viral vectors: methods and implications. *Gene Ther.* 12 Suppl 1, S5–S17.
12. Umana, P., Gerdes, C. A., Stone, D., Davis, J. R., Ward, D., Castro, M. G., et al. (2001) Efficient FLPe recombinase enables scalable production of helper-dependent adenoviral vectors with negligible helper-virus contamination. *Nat. Biotechnol.* **19**, 582–585.

Chapter 6

Manufacturing of Adenovirus Vectors: Production and Purification of Helper Dependent Adenovirus

Edwige Dormond and Amine A. Kamen

Abstract

Adenoviral vector (AdV) of the third generation also known as helper-dependent adenoviral vector (HDV) is an attractive delivery system for gene therapy applications. However, obtaining high quality-grade HDV in sufficient amount remains a challenge that hampers the extensive use of this vector in preclinical and clinical studies.

Here we review recent progress in the large-scale manufacturing of HDV. The production of HDV is now amenable to large-scale volume with reduced process duration under optimized rescue and co-infection conditions. Also, efficient downstream processing of HDV with acceptable recovery of HDV and minimal contamination by the helper virus is described.

Key words: Gutless adenoviral vectors, Large-scale manufacturing, Reactor culture, Downstream processing

1. Introduction

Manufacturing methods of the first and second generation adenoviral vectors have been extensively reviewed (1) and detailed protocols have been provided in previous editions of this book series. Therefore, this chapter will focus on the most recent progress for large-scale production and purification of the third generation adenoviral vectors.

1.1. Adenoviral Vectors

Human adenoviral vector serotype 5 (AdV) is the most characterized virus among the other 51 serotypes of the same family (2, 3). AdV has been considered as a good candidate for human gene therapy for a number of advantages including its wide cell tropism in quiescent and non-quiescent cells, its inability to integrate the host genome, its high capacity for the therapeutic gene insertion

Otto-Wilhelm Merten and Mohamed Al-Rubeai (eds.), *Viral Vectors for Gene Therapy: Methods and Protocols*,
Methods in Molecular Biology, vol. 737, DOI 10.1007/978-1-61779-095-9_6, © Springer Science+Business Media, LLC 2011

and its high production titre. Within the last two decades, the AdV genome has been progressively modified from the wild-type genome to improve its safety and efficacy in therapeutic applications. A decrease in the immunological response following vector administration has been achieved by a progressive removal of nonessential viral DNA regions. From the first generation of AdV with the deletion of replication necessary genes to the third generation of AdV with the clearance of most viral sequences, an enhanced capacity for a therapeutic gene insertion from ~7 to ~30 kb has been achieved. The therapeutic benefit of the third over the first generation suggests that, in the future, AdV vectors of the third generation will be predominant for clinical approaches.

1.2. Production System

The actual need for large amounts of clinical-grade AdV (10^{12} to 10^{13} viral particles/patient; 10^{10} to 10^{11} plaque-forming units/patient) requires efficient and established processes for large-scale production. While substantial efforts were dedicated to improving the large-scale manufacturing of the first and second generations of AdV, one can access only scarce information on the large-scale production of the third generation, mostly due to the inherent complexity of the production system (4). The third generation AdV genome comprises only cis-acting elements, i.e., the packaging signal (ψ) and the inverted terminal repeats (ITR). Its production in human embryonic kidney 293 cells (HEK293) requires trans-acting elements provided by the first generation of adenoviral vector called helper virus (HV) and by the host cell line (E1 sequences) (5). Therefore, these defective vectors have been referred to as Helper Dependent AdV (HDV). Initial efforts emphasized the necessity to develop cell systems capable of reducing the HV contamination (Fig. 1). This has been achieved through the use of Cre/*loxP* (6–8) or FLP/*frt* recombinase systems (9, 10). Both the Cre/*loxP* and FLP/*frt* systems have shown similar efficiencies in reducing the HV contamination and in amplifying the HDV (9, 10).

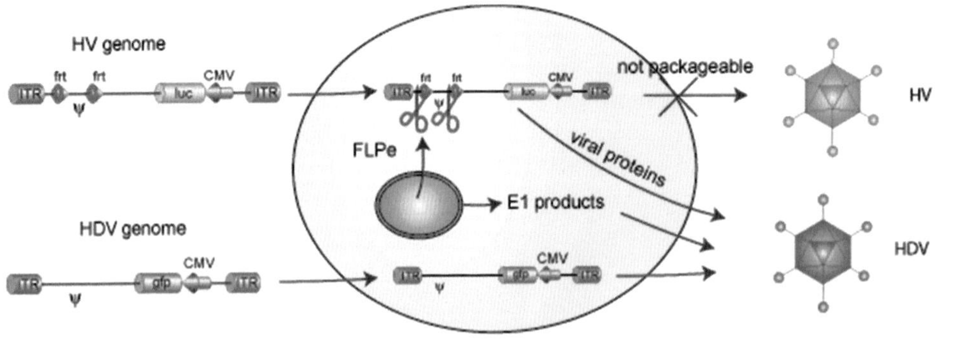

Fig. 1. The FLPe/*frt* recombinase system used for production of HDV in HEK293 derived cell line.

1.3. Production Process

The standard HDV production protocol consists of a multi-step process (Fig. 2). The initial step, commonly known as the rescue step, aims to recover HDV from HDV DNA. Typical rescue protocols involve transfection of adherent producer cells with the linearized HDV genome, i.e., excised from the bacterial sequence, followed by the HV infection 8–18 h post-transfection (11). The viral lysate containing the HDV is recovered when a cytopathic effect is visible, usually 48–72 h post-infection. To overcome the limitations associated with the use of adherent cell culture for large-scale operations, but also to improve the yield of HDV at the rescue step, we have developed a protocol called adenofection (see Subheading 3.1.1) that is easily transferable to large-scale volumes (12).

Because the HDV titer is low at the end of the rescue step [10^2 to 10^5 infectious units (IU) of HDV/mL] (8, 9, 13), further amplification of the HDV is required. To achieve this, typical amplification protocols consist of exhaustive passages of viral lysate on an increasing number of adherent cells using a volume-based method (14, 15). Drawbacks of such a amplification protocol are process time length, fluctuation in titre, and viral recombination (9, 14, 15). An amplification protocol based on the use of infection parameter (see Subheading 3.1.2) has been developed to simultaneously decrease the number of passages (up to two passages), favor the HDV amplification (up to 10^8 infectious units of HDV/mL corresponding to 10^9 total viral genomes of HDV/mL) and limit HV contamination (16).

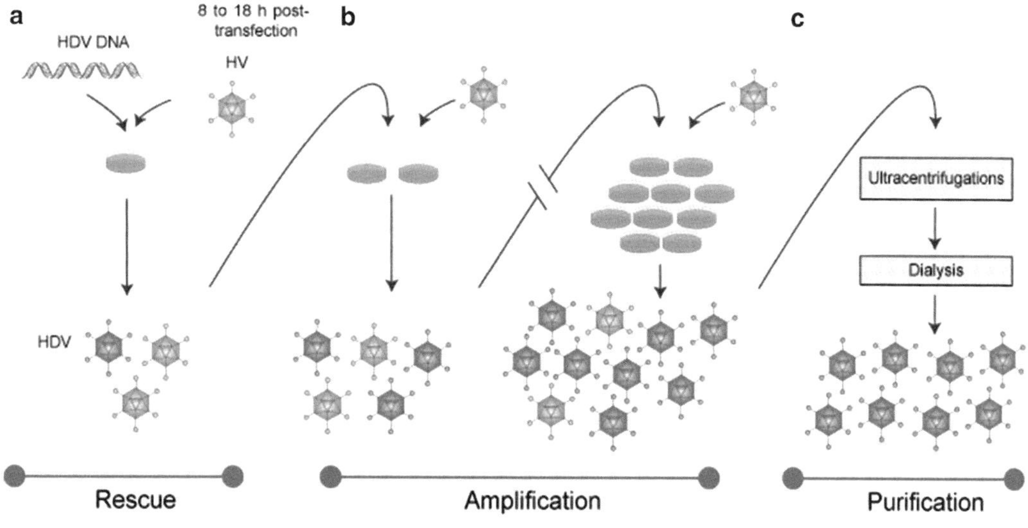

Fig. 2. Standard production process of HDV in adherent cell cultures.

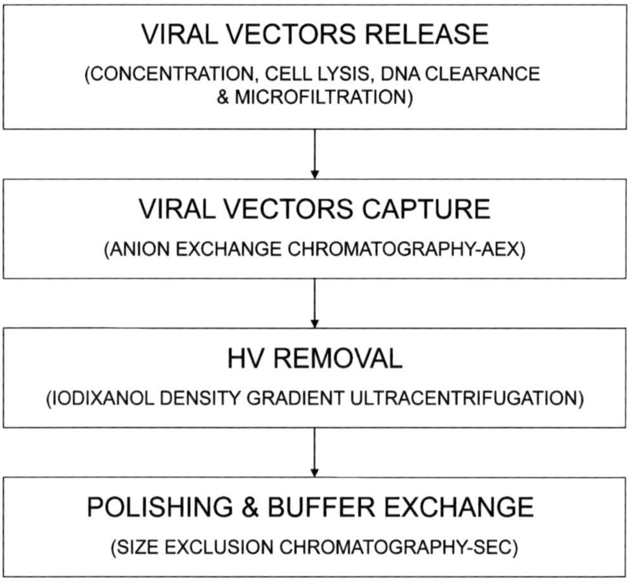

Fig. 3. Downstream processing strategy for the third generation adenoviral vectors.

1.4. Purification Process

The production allows generation of sufficient HDV material at a concentration of 10^8 IU/mL. However, before considering use of this material in preclinical or eventually clinical studies, this bulk material has to be separated from any contaminants (host cell protein, serum, and non-desirable viral species), purified, and characterized. For large-scale applications, the purification of HDV presents a number of challenges that are mostly associated with the residual contamination by the HV. Use of the chromatographic steps as established with the first generation adenoviral vector allows clarification, capture, and purification of the viral material in a scalable manner. Then an improved iodixanol ultracentrifugation procedure allows a rapid separation of the HV from the HDV particles in the final viral preparation (see Subheading 3.2). Figure 3 presents a scheme of the purification process. The overall recovery of infective units of HDV for the complete purification strategy is ~80%. A 10-times diminution of HV contamination ratio from 2 to 0.2% is obtained. A second round of iodixanol ultracentrifugation can be considered if a lower level of contamination is desired (17).

2. Materials

The production data shown in this protocol have been generated using (i) the HDV plasmid, pHCAgfp, a generous gift from Dr. V. Sandig. It carries a *gfp* expression cassette driven by the cytomegalovirus promoter, two ITRs adjacent to the bacterial sequence, the ψ and the E4 promoter region of the adenovirus type 5. pHCA*gfp* is amplified in *E. coli* DH5-α and purified using Giga-Prep columns (Qiagen, Ontario, Canada). The HDV plasmid is PmeI-linearized to liberate the viral sequence (30 kb).

And (ii) the HV, provided generously by Dr. P. Lowenstein. It is an E1/E3 deleted adenoviral vector available in viral form (10). It bears two parallel frt sites flanking the ψ. A high-titer stock of the HV is produced by infecting HEK293SF cells (18, 19) (see Note 1).

Adenoviral vectors must be handled under appropriate biosafety containment by trained personnel using biological safety cabinets and following the guidelines specified by the institution where the experiments are conducted. The nature of the transgene must be taken into account when establishing biosafety requirements. The work described herein has been performed in biosafety level-2 laboratories.

2.1. HDV Production

2.1.1. Adenofection

1. LCSFM is a low calcium formulation of HSFM (Gibco Invitrogen, Grand Island, NY) supplemented with 10 mM HEPES (Calbiochem, Gibbstown, NJ), 0.1% Pluronic F-68 (Sigma-Aldrich, Oakville, ON), and 0.75 μg/mL puromycin.
2. HEK293 FLPe cells (18) in fresh LCSFM medium at a final concentration of 0.5×10^6 cells/mL, 1 h prior to adenofection.
3. HDV DNA PmeI-linearized in 10 mM Tris–HCl, 1 mM EDTA, pH 8.0 at 1 mg/mL (see Note 2). For the preparation of complexes, dilute in HSFM/10 mM HEPES in 1/10 of the cell culture volume for a final concentration of 1 μg/mL of cell culture.
4. Linear 25 kDa polyethylenimine at 1 mg/mL in water, neutralized with HCl. Sterile filtered and stored at −80°C. For the preparation of complexes, dilute in HSFM/10 mM HEPES for a final concentration of 3 μg/mL of cell culture.
5. Purified HV in 10 mM Tris–HCl, pH 7.9, 1 mM $MgCl_2$, 5% sucrose (see Note 3). Dilute in HSFM/10 mM HEPES for a final concentration in cell culture of 5 IU/cell.

2.1.2. Amplification by Infection

1. HEK293 FLPe cells in fresh LCSFM medium at a final concentration of 0.5×10^6 cells/mL, 1 h prior to amplification.
2. HV as viral lysate. Diluted in HSFM/10 mM HEPES in 1/10 of the cell culture volume at a multiplicity of infection (MOI, see Note 4) of 0.5 IU/cell.
3. HDV as viral lysate from rescue by adenofection (see Subheading 2.1.1) and then from amplification by infection. For amplification of passage 1 and passage 2, use non-diluted. For amplification of passage 3, use diluted in HSFM/10 mM HEPES at a MOI of 5.

4. 3.5-L bioreactor in batch mode for the amplification passage 3 (see Note 5).

2.2. HDV Purification

2.2.1. Concentration, Clarification, and Conditioning Step

1. HDV material following amplification at passage 3 in bioreactor.
2. Concentration and lysis buffer: 10 mM HEPES, 2 mM $MgCl_2$ in MilliQ-H_2O at pH 7.5, sterile filtered.
3. Benzonase (Merck KGaA, Darmstadt, Germany).
4. Concentrated conditioning solution: 1.5 M NaCl in MilliQ-H_2O, filtered and degassed.
5. 0.45 μm cellulose acetate membrane in a vacuum filtration unit with a glass fiber pre-filter (Corning Life Sciences, Lowell, MA).

2.2.2. Anion Exchange Chromatography

1. Low-pressure GradiFrac system with UV monitoring at 280 nm (GE Healthcare, Uppsala, Sweden).
2. Fractogel-DEAE beads (Merck KGaA, Darmstadt, Germany) packed into a HR 5/5 glass column (Amersham Biosciences, Pistacaway, NJ) with ~4 mL bead volume.
3. Running AEX Buffer A: 50 mM HEPES, 2 mM $MgCl_2$, 2% sucrose in MilliQ-H_2O at pH 7.5, filtered and degassed.
4. Running AEX Buffer B: 50 mM HEPES, 2 mM $MgCl_2$, 2% sucrose, 1 M NaCl in MilliQ-H_2O, pH 7.5, filtered and degassed.
5. Storage buffer: 150 mM NaCl, 20% ethanol in MilliQ-H_2O, filtered and degassed.
6. Regeneration/sanitization buffer: 1 M NaCl, 0.5 M NaOH in MilliQ-H_2O, filtered and degassed.

2.2.3. Ultracentrifugation

1. Ultracentrifugation medium: OptiPrep medium (60% iodixanol) (Axis-Shield, Oslo, Norway).
2. Ultracentrifugation medium diluent: 10 mM Tris–HCl, 150 mM NaCl, and 1 mM EDTA at pH 7.9.
3. 13.5-mL PA Ultracrimp tube (Thermo Scientific, Milford, MA).
4. Stepsaver 50V39 vertical rotor and a Sorvall discovery ultracentrifuge (Thermo Scientific).

2.2.4. Size-Exclusion Chromatography

1. Low-pressure GradiFrac system with UV monitoring at 280 nm (GE Healthcare, Uppsala, Sweden).
2. Sepharose 4FF resin packed in a XK 16/70 glass column (Amersham Biosciences) with a bead volume of ~30 mL.
3. Running SEC buffer: 10 mM Tris–HCl, 150 mM NaCl, 1 mM EDTA, and 2% sucrose at pH 7.9.

4. Storage buffer: 150 mM NaCl and 20% ethanol in MilliQ-H_2O, filtered and degassed.
5. Regeneration/sanitization buffer: 1 M NaCl and 0.5 M NaOH, filtered and degassed.

2.3. Monitoring and Characterization of HDV Manufacturing

2.3.1. HDV and HV Infectious Particles: Gene Transfer Assay and Cytopathic Assay

The HDV IU are quantified by GFP gene transfer assay (GTA) on target cells in suspension culture.

1. HEK293E cells, seeded at 0.5×10^6 cells/mL in 12-well plates (20).
2. HEK293E medium: HSFM, 10 mM HEPES, 1% BCS, and 50 μg/mL G-418.
3. Dilution of unknown in HEK293E medium (100 μL) to stand around 3–30% of IU/cell.
4. 16% *p*-formaldehyde in phosphate buffered saline.

The HV IU are quantified by a cytopathic effect (CPE) detection following end-point dilution assay (EPDA) on infected target cells.

1. HEK293A cells, seeded at 0.03×10^6 cells/mL, 200 μL, in 96-well plates.
2. HEK293A medium: DMEM+, 2 mM L-glutamine, 1 mM sodium pyruvate, and 10% FBS.
3. 10-Logarithmic dilution of unknown in HEK293A medium (50 μL/well and 8 well/dilution).

2.3.2. HDV and HV Total Particles and HV Contamination: qPCR Assay

1. Benzonase (Merck KGaA).
2. Benzonase activation solution: 50 mM $MgCl_2$.
3. Benzonase inactivation buffer: 0.5 M EDTA.
4. High Pure Viral Nucleic Acid kit (Roche Diagnostics, Laval, QC, Canada).
5. HDV forward and reverse primers (see Table 1) at 5 μM in MilliQ-H_2O.
6. HV forward and reverse primers (see Table 1) at 5 μM in MilliQ-H_2O.
7. Light Cycler Fast-Start SYBR Green I (containing $MgCl_2$ at 25 mM and 10× Master Mix).
8. 10-Logarithmic dilution in MilliQ-H_2O of standard plasmid (see Note 6) containing HDV and HV amplified products (from 10^8 to 10^9 copies/2 μL).
9. Several dilutions of unknown in MilliQ-H_2O in the range of standard plasmid dilution.

Table 1
Description of the duplex real-time qPCR in terms of reaction conditions, quantification, and specificity of reaction

			HDV	HV
qPCR conditions	Primer sequence (5′-3′) forward		AGCTCACAGGCTGTAGTTTG	CCGCAGTTGACAGCATTACC
	Primer sequence (5′-3′) reverse		GGATCACTTGCACGGTTTAG	CGGACCACGTCAAAGACTTC
	Product position and size		**2,551 to 2,570 – 224 bps**	**21,402 to 21,422 – 220 bps**
	Amplification (40 cycles)	Denaturation	95°C for 10 s	
		Annealing	61°C for 7 s	
		Elongation, quantification	72°C for 10 s, single fluorescence acquisition at end	
Quantification	Standard curve	Range	10^8 to 10^3 molecules/2 μL (1:10 dilutions)	10^7 to 10^2 molecules/2 μL (1:10 dilutions)
		Slope	$-3.6 < \text{Slope} < 3.1$	
		R^2	$R^2 > 0.99$	
Reaction specificity	Melting curve	Condition	70°C for 10 s, 70–95°C at 0.10°C/s with continuous fluorescence acquisition	
		Primer dimers[a]	No	No
		Non-specific products[b]	No	No
		Melting of products	$87.2°C < T < 87.5°C$	
	Agarose gel	Primer dimers[c]	No	No
		Non-specific products[d]	No	No
		Bands	Single band at ~200 bps	Single band at ~200 bps

[a] Primer dimers are usually produced when a fluorescence shift is observed at $T < T$ of product on the melting curve
[b] Nonspecific products are usually produced when a fluorescence shift is observed at $T > T$ product on the melting curve
[c] Bands for primer dimers are usually observed above the product band (>200 bps) on the agarose gel
[d] Bands for nonspecific products are usually observed below the product band (<200 bps) on the agarose gel

3. Methods

3.1. Scalable HDV Production

3.1.1. HDV Rescue by Adenofection

The adenofection protocol consists of a combined transfection-infection approach that takes advantage of the HV infection mechanism. Complexes of HDV DNA, PEI, and HV are formed by simple in vitro mixing and then added to the producer cells to deliver the required genetic material and generate new HDV. We have shown that the adenofection protocol outperformed the standard rescue procedures by producing more HDV in a shorter process time length (12). Moreover, this protocol is adaptable to suspension culture, rendering the rescue process scalable.

1. HEK293FLPe cells were maintained as suspension cultures in shake flasks at 37°C in a humidified incubator with 5% CO_2 under orbital agitation (120 rpm). Every 2 or 3 days, cells were sub-cultured to maintain exponential growth.
2. Seed HEK293FLPe cells in fresh medium in a 250-mL shake flask at a final volume of 45 mL, 1 h prior to the adenofection.
3. Prepare a final volume of 5 mL of adenofection complexes by gently mixing 150 μg of prediluted 25 kDa linear PEI with 50 μg of prediluted linearized HDV DNA (see Note 2). Allow it stand for 5 min at room temperature. Add 1.25×10^8 IU of prediluted HV and allow it to stand for 5 min more (see Note 7).
4. Distribute the complexes on the cells and incubate for 48 h.
5. Aliquot the viral lysate and store at –80°C before amplification

3.1.2. HDV Amplification by Infection

Once the first HDV viral particles are obtained, usually two serial amplifications are required to generate a stock with a high HDV titer (1×10^8 IU/mL) and a low HV contamination ratio (around 1%). Amplification is carried out via coinfection using an optimal quantity of HV and HDV viral lysate from the preceding passage. An HDV viral lysate volume of 1:10 of the total cell culture volume is used to minimize addition of toxic elements from infected cell lysate. At this step, because of the low titer of HDV up to passage 2, it is difficult to control the HDV MOI to an optimum MOI of 5. The protocol described below has been designed to produce 3 L of HDV material before purification following a rescue and two amplification steps.

1. Seed 2.5×10^7 HEK293FLPe cells in fresh medium in a 250-mL shake flask, in a final volume of 45 mL, 1 h prior to the coinfection.
2. Coinfect the cell culture with 1.25×10^7 IU of HV and 5 mL of 37°C-thawed viral lysate containing HDV from the rescue step (see Note 8). Incubate the cell culture for 48 h (passage 1).

3. Harvest the viral lysate, aliquot for amplification and quantification, and store at –80°C until further use.
4. Repeat from point 1 to 4 a second time (passage 2).
5. Characterize the HDV material at this point (see Subheading 3.3).

3.1.3. HDV Amplification by Infection in Bioreactor

Once a stock of HDV has been produced and characterized as described in the previous section, a large-scale preparation of HDV can be performed to generate the required quantity of material for in vitro or in vivo studies using the robust MOI-based protocol for amplification. Under these controlled conditions, the HDV material generated should have the required specifications before undergoing the purification process.

1. Seed at least 7.5×10^8 HEK293FLPe cells in the 3.5 L bioreactor.
2. Around 24 h later, count cells using the erythrosine dye exclusion technique. Cell concentration should be at 0.5×10^6 cells/mL before infection.
3. Thaw an aliquot of viral lysate of HV and viral lysate from the passage 2 in a 37°C water bath until the lysates completely melt (see Note 8). The quantity of HV corresponds to 7.5×10^8 IU (MOI of 0.5) and the quantity of HDV corresponds to 7.5×10^9 IU (MOI of 5).
4. Coinfect the cell culture with HV and HDV from the inoculation bottle. Let the cultures run for 48 h. Record the parameters and sample the bioreactor culture every 6 h for cell counts and freeze-store the samples for subsequent analyses at –80°C storage.
5. Harvest the bioreactor viral lysate and proceed immediately to the concentration step in the purification procedure.

3.2. Scalable HDV Purification

At this point, the volume and titer of the HDV bulk material determines the size of the chromatographic column and volume of buffers. For instance, the set-up described herein was designed for the downstream processing of 1 L of crude lysate materials from a 3.5-L bioreactor culture.

3.2.1. Concentration, Clarification, and Conditioning Step

These steps are designed to release the virus from the cells allowing an efficient capture on the AEX column:

1. Centrifuge the viral lysate by spinning the culture at $290 \times g$ for 15 min. Resuspend the viral lysate in lysis buffer in 1:10 of the bioreactor volume.
2. Aliquot and store at –80°C.

3. Quantify HDV in terms of IU and viral genomes (VG) and HV in term of VG before proceeding with the purification protocol.
4. Thaw around 100 mL of the 10× cell lysate in a 37°C water bath (see Note 8).
5. Vortex, triturate with a 5-mL pipet to homogenize this viscous cell lysate solution.
6. Add Benzonase at a final concentration of 100 U/mL and put on a rocking plate for gentle shaking at room temperature for 1 h (see Note 8).
7. Centrifuge at high speed (4,700 ×*g*) for 15 min. Recover the supernatant in a new recipient.
8. Add concentrated NaCl dropwise to reach a final NaCl concentration of 300 mM.
9. Filter the conditioned supernatant using the filtration unit.

3.2.2. AEX Step

The AEX chromatography was used to selectively capture the AdV and remove the majority of protein contaminants.

1. Rinse the column with 10 column volume (CV) of MilliQ-H_2O at 75 cm/h and equilibrate with 5 CV of 30% buffer B at 150 cm/h.
2. Load onto the column at 150 cm/h the clarified conditioned supernatant previously filtered.
3. Apply a stepwise elution strategy consisting of a wash step at 300 mM NaCl with 30% buffer B, an elution step at 450 mM of NaCl with 45% of buffer B in 7 CV and a final high stringency step at 1 M of NaCl with 100% of buffer B.
4. The collected AEX-AdV peak is stored at 4°C and processed immediately in the iodixanol gradient ultracentrifugation step.
5. Clean the column with the regeneration/sanitization buffer at 75 cm/h for 1 h, rinse with 10 CV of MilliQ-H_2O at 150 cm/h, and store in storage buffer in 10 CV at 75 cm/h.

3.2.3. Iodixanol Ultracentrifugation Step

1. The AEX-AdV peaks were processed through an iodixanol gradient ultracentrifugation to isolate HDV from HV.
2. Place horizontal marks every 0.5 cm on the Ultracrimp tube to mark a total of 16 fractions.
3. Distribute the AEX-AdV peak equally among the tubes and weigh the exact quantity distributed.
4. Determine the concentration of iodixanol solution to be added to reach a final 38.60% iodixanol concentration (see Note 9).

Table 2
Characteristics of the self-formed iodixanol density gradients obtained following the 3-h run at 180,000 × g_{av}

HV	Fraction number	6–8
	Iodixanol content (% w/v)	41.6–40.3
	Density (g/mL)	1.224–1.217
HDV	Fraction number	9–12
	Iodixanol content (% w/v)	39.4–37.6
	Density (g/mL)	1.212–1.203

5. Prepare the iodixanol solution and verify the iodixanol concentration by refractive index measurement. Correct eventually by adding iodixanol or buffer. Fill the tube up the neckline. Equilibrate the tubes (see Note 10).
6. Crimp the tube and place them in the Stepsaver 50V39 vertical rotor. Run at 180,000 × *g* for 3 h at 4°C.
7. Puncture the bottom of the tube and collect fractions 1–16. Fractions 6–8 contains HV and fractions 9–12 contains HDV. Pool fractions 9–12 together (see Notes 11 and 12; Table 2).
8. Store the HDV iodixanol-containing fractions at 4°C before proceeding to the size exclusion chromatography (SEC).

3.2.4. Size Exclusion Chromatography Step

SEC was used to remove the iodixanol viscous buffer and the remaining protein contaminants.

1. Rinse the column with 10 CV of MilliQ-H_2O. Equilibrate the column with the running buffer for 5 CV.
2. Load material at a linear flow rate of 90 cm/h. Elute at 135 cm/h. The SEC-AdV peak elutes first in the void volume of the column whereas iodixanol elutes later (see Note 12).
3. Clean the column with the regeneration/sanitization buffer at 75 cm/h for 1 h, rinse with 10 CV of MilliQ-H_2O at 150 cm/h and store in storage buffer in 10 CV at 75 cm/h.
4. Aliquote and store the SEC-AdV peak at –80°C.
5. Characterize the final product.

3.3. Monitoring and Characterization of HDV Manufacturing

3.3.1. HDV Infectious Particles: Gene Transfer Assay

The HDV IU are quantified by GFP gene transfer assay (GTA) on target cells in suspension culture.

1. Seed the HEK293E cells (20) at 0.5×10^6 cells/mL in 12-well plates (see Note 1).
2. Apply at least two dilutions of unknown (100 μL) on the cells.

3. At 24 hpi, estimate the cell concentration by the erythrosine dye exclusion method.
4. Resuspended in 2% *p*-formaldehyde in PBS and allow it stand for 1 h at 4°C.
5. Analyze by flow cytometry at least 10,000 events using the Coulter EPICS™ XL-MCL cytometer and EXPO32 software to determine the percentage of GFP-positive cells. A minimum of two dilutions showing 3–30% GFP-positive cells should be taken into account for the titer calculation.

   HDV titer (IU/mL) = percentage GFP-positive cells × cell concentration (cells/mL)/(dilution of unknown × 0.1 mL)

3.3.2. HV Infectious Particles: Cytopathic Effect Following an End-Point Dilution Assay

The HV IU are quantified by cytopathic effect (CPE) detection following an end-point dilution assay (EPDA) on infected target cells under static conditions This assay is strictly used to determine the quantity of HV to be added during the amplification. Assay should be performed in duplicates.

1. Seed 150 μL of HEK293A cells at 0.03×10^6 cells/mL in 96-well plates.
2. 24 h later, each column of the 96-well plates should receive a logarithmic dilution of unknown viral stock. A noninfected column is considered as a negative control.
3. Place the plates in a humidified Tupperware inside the incubator to limit medium evaporation.
4. 14 days post-infection, the positive infected wells are scored: Infected cells are rounded, form grapes, and might be detached from the bottom surface of the plates. The negative control helps in determining the infected from noninfected wells.
5. The HV titer is estimated according to the calculation by Reed and Muench (21) based on the Median tissue culture infective dose ($TCID_{50}$); amount of IU that will produce pathological change in 50% of cell cultures. Details of the $TCID_{50}$ calculation is given below.

Proportional distance = (% of infected wells at the dilution above 50%–50%)/(% of infected wells at the dilution above 50% − % of infected wells at the dilution below 50%).

$$\text{Log } TCID_{50} = \text{Sum of log of dilutions above } 50\% - \text{proportional distance} \times \log(\text{dilution factor})$$

$$TCID_{50} = 10^{\log TCID50}$$

$$\text{Titer in IU/mL} = TCID_{50}{}^{-1}/\text{infected volume per well in mL}$$

3.3.3. HDV and HV Total Particles and HV Contamination: qPCR

The VG titers of the HDV and the HV and the HV contamination ratio are determined by a duplex SYBR-Green I quantitative PCR assay (qPCR). As the qPCR conditions are similar for both

HDV and HV detection, a single run is performed to quantitate both vectors. We recommend running at least duplicate samples.

1. Thaw 500 μL of cell lysates at 37°C and centrifuge for 2 min at 4,500 ×*g* to remove the cell debris.
2. Treat 200 μL of the supernatants with 1 μL of Benzonase in a final concentration of Benzonase activation buffer of 2 mM $MgCl_2$ for 30 min at 37°C.
3. Add 5 μL of 0.5 M EDTA for DNA inactivation and inactivate RNA by heating the sample at 65°C for 30 min.
4. Extract the viral genomes using the High Pure Viral Nucleic Acid kit following the manufacturer's instructions. Elute the equivalent of 200 μL of supernatant in 50 μL.
5. Prepare separately a HDV mix and a HV mix by mixing per number of samples 12 μL H_2O, 1.6 μL $MgCl_2$ 25 mM, 1.2 μL forward HDV or HV primer 5 μM, 1.2 μL reverse HDV, or HV primer 5 μM.
6. Under limited light exposure, in the HDV or HV mix, add per sample 2 μL of 10× Master Mix Light-Cycler Fast-Start SYBR Green I prepared according to the manufacturer. Mix gently by up and down by pipetting.
7. Distribute 18 μL of the HDV or HV mix among the capillaries previously placed in the cold metallic box. Add 2 μL of samples in the capillaries (serial dilution of standard from 10^8 to 10^2 molecules or unknowns or MilliQ-H_2O for negative control).
8. Spin the capillaries at 300 ×*g* for 1 min to allow the qPCR mix to settle down in the shallow part of the capillary.
9. Perform the run in the Light Cycler according to the program described in Table 1.
10. By highlighting the HDV or the HV samples forming the standard curve, the software displays results for HDV or HV unknowns, respectively.
11. Analyze the data using the Light Cycler 480 software. Specificity of the reaction is confirmed by melting curves analysis and runs of qPCR products on agarose gels (Table 1).
12. Calculate the concentration of HDV and HV and the contamination ratio.

$$\text{VG/mL} = \text{Light cycler value}/0.002\ \mu\text{L}/0.2\ \mu\text{L} \times 0.05\ \mu\text{L}$$

$$\text{TVG (total viral genomes)/mL} = \text{VG of HV/mL} + \text{VG of HDV/mL}$$

$$\%\ \text{Contamination} = (\text{VG of HV/mL/VG of HDV/mL}) \times 100$$

4. Notes

1. The HEK293SF cell line is used to produce HV (18, 19). Briefly, a suspension culture of HEK293SF cultivated in LCSFM medium without puromycin at 0.5×10^6 cells/mL is infected with HV at a MOI of 5. Twenty-four hours later, the viral lysate is recovered. The viral lysate is concentrated 10× by centrifugation (300 ×*g*, 10 min) and stored at −80°C. This lysate might be used in this form for amplification or further purified by ultracentrifugation for adenofection. Quantification of the stocks by EPDA-CPE (see Subheading 3.3.2) allows to determine the viral titer for rescue by adenofection and amplification by infection protocols (see Subheading 3.1).

   In this chapter, references are made to other HEK293 derived clones:

   HEK293 A: A stands for adherent and these cells are used in infectious assays.

   HEK293 E: E stands for EBNA and cells are EBNA transformed cells.

   Both cell lines are maintained and used in this lab based on previously established protocols.
2. Overnight digest at least 50 μg of HDV DNA with PmeI restriction endonuclease. Verify DNA digestion by running a small volume of PmeI-HDV DNA on agarose gel electrophoresis stained with ethidium bromide. Purify the linearized HDV DNA by ethanol precipitation and dilute in TE. Quantify DNA concentration by UV absorbance in 50 mM Tris–HCl at pH 8.0 ensuring that A_{260}/A_{280} stands between 1.80 and 1.95.
3. We recommend using purified HV material to form adenofection complexes. HV might be purified either by ultracentrifugation procedures or chromatographic methods such as described in Subheading 3.2.2. Purification by ultracentrifugation consists of a double CsCl banding (one-step gradient at 100,000 ×*g*, 4°C for 1.5 h and one linear gradient at 100,000 ×*g*, 4°C for 24 h) followed by dialysis against 10 mM Tris–HCl, pH 7.9, 1 mM $MgCl_2$.
4. The multiplicity of infection (MOI) is a commonly used parameter when describing viral infection. It is the quantity of infecting units (IU) per cell and is therefore dependent on the viral quantification method used to assess viral titer. For better clarity in the text, the MOI is given without units.
5. For amplification passage 3, the 3.5 L bioreactor CF-3000 (Chemap, Männedorf, Switzerland) is equipped with surface

baffles and marine impellers for an efficient mass transfer. Temperature is maintained at 37°C by a water jacket. Bottles for virus inoculation and $NaHCO_3$ addition are connected to the bioreactor. The pH is controlled at 7.2 with the addition of 1 M $NaCO_3$ solution or controlling the percentage of CO_2 in the gas inlet. Nitrogen and oxygen are used for controlling the dissolved oxygen concentration at 40% of air saturation. The gas is introduced by surface aeration and its composition was controlled by mass flow controllers. Data acquisition and control is performed using FIX MMI software.

6. For standard plasmid construction, briefly, the HV genome is extracted by using the viral DNA Roche kit. The HDV and HV elongated qPCR products are amplified by standard PCR using the pHCAgfp and the extracted HV genome as templates. Restriction site sequences were added to the 3′-end of each primer sequence (HDV forward: AAAGTTTAAACG CCCAGGTAGTAAATGTCTC containing PmeI sequence, HDV reverse: containing EcoR I sequence, HV forward: AAAAAGCTTGGCCTACCCTGCTAACTTCC containing Hind III sequence, HV reverse AAAGCGGCCGCAG GTACACGGTTTCGATGAC containing Not I sequence). The PCR cycling conditions were 94°C for 5 min; 35 cycles of 94°C for 30 s, 58°C for 30 s, 72°C for 40 s; 72°C for 7 min and final temperature 40°C. Each product is gel purified, digested, desalted, and ligated sequentially into the pTT vector. The resulting standard plasmid pTT3-HDV-HV (7 kb) is amplified in *E. coli* and purified by Qiagen Maxiprep (Qiagen, Mississauga, ON, Canada). The standard stock was quantified with absorbance ratios of 260 nm/280 nm and values greater than 1.8 were considered pure and accepted for further studies.
7. Several key points have to be considered during complex formation: thaw material (PEI, DNA, and virus), mixing and mixing time affect complex formation. For instance, always use material thawed only once, avoid vortex mixing and respect time for complex formation.
8. Do not allow to over warm, as virus is temperature-sensitive. Gently mix by inverting the tubes until complete melting.
9. A filled-up (until the tube neck) 13.5-mL Ultracrimp tube weighs 17.68 g and contains 14.65 mL of material.
10. Because of the infinite difference of density, particular care should be taken in the preparation of the self-forming iodixanol density gradient. For instance, preparation is carried out using volumes and refractive index measurements. The latter is used to measure the exact iodixanol content and correct it accordingly using the following correlation (http://www.axis-shield-density-gradient-media.com/Applic/V01.pdf).

$$\% \text{ Iodixanol } (w/v) = 641.4717\eta - 856.4968$$
$$\% \text{ Iodixanol } (\omega/\varpi) = 190.3302\rho - 191.3011$$

11. From the bottom to top of the tube, the fractions are numbered 1–16. Viral bands are not visible; however, a faint band around fractions 14–15 has been observed and corresponds to aggregated proteins. The refractive index measurement might confirm HDV localization which bands at 39.4–37.6% iodixanol.
12. This protocol aims to separate the 30-kb HDV vector from the 37-kb HV vector. To separate vector of different sizes, we recommend to collect each fraction and to assess vectors presence by gene transfer assays. Here, initial fraction identification was done by assessing each fraction for GFP expression (HDV marker gene) and for luciferase expression on target cells (HV marker gene).
13. A maximum of 20% of the CV should be loaded.

Acknowledgements

Dr. V. Sandig and Dr. P. Lowenstein are also acknowledged for providing the authors with the HDV and the HV constructs. A. Bernier, N. Arcand, P.S. Chahal, D. Jacob, A. Meneses-Acosta,Y. Durocher, S. Perret, R. Tom, R. Gilbert, L. Bourget, and A. Migneault are acknowledged for their technical and scientific support.

References

1. Altaras NE, Aunins JG, Evans RK, Kamen A, Konz JO, Wolf JJ. (2005) Production and formulation of adenovirus vectors *Adv Biochem Eng Biotechnol* **99,** 193–260.
2. McConnell MJ, Imperiale MJ. (2004) Biology of adenovirus and its use as a vector for gene therapy *Hum Gene Ther* **15,** 1022–33.
3. Tatsis N, Ertl HC. (2004) Adenoviruses as vaccine vectors *Mol Ther* **10,** 616–29.
4. Dormond E, Perrier M, Kamen A. (2009) From the first to the third generation adenoviral vector: what parameters are governing the production yield? *Biotechnol Adv* **27,** 133–44.
5. Graham FL, Smiley J, Russell WC, Nairn R. (1977) Characteristics of a human cell line transformed by DNA from human adenovirus type 5 *J Gen Virol* **36,** 59–74.
6. Hardy S, Kitamura M, Harris-Stansil T, Dai Y, Phipps ML. (1997) Construction of adenovirus vectors through Cre-lox recombination *J Virol* **71,** 1842–9.
7. Lieber A, He CY, Kirillova I, Kay MA. (1996) Recombinant adenoviruses with large deletions generated by Cre-mediated excision exhibit different biological properties compared with first-generation vectors in vitro and in vivo *J Virol* **70,** 8944–60.
8. Parks RJ, Chen L, Anton M, Sankar U, Rudnicki MA, Graham FL. (1996) A helper-dependent adenovirus vector system: removal of helper virus by Cre-mediated excision of the viral packaging signal *Proc Natl Acad Sci USA* **93,** 13565–70.
9. Ng P, Beauchamp C, Evelegh C, Parks R, Graham FL. (2001) Development of a FLP/frt system for generating helper-dependent adenoviral vectors *Mol Ther* **3,** 809–15.
10. Umana P, Gerdes CA, Stone D, et al. (2001) Efficient FLPe recombinase enables scalable

production of helper-dependent adenoviral vectors with negligible helper-virus contamination *Nat Biotechnol* **19,** 582–5.

11. Oka K, Chan L. (2005) Construction and characterization of helper-dependent adenoviral vectors for sustained in vivo gene therapy *Methods Mol Med* **108,** 329–50.
12. Dormond E, Meneses-Acosta A, Jacob D, et al. (2009) An efficient and scalable process for helper-dependent adenoviral vector production using polyethylenimine-adenofection *Biotechnol Bioeng* **102,** 800–10.
13. Kumar-Singh R, Chamberlain JS. (1996) Encapsidated adenovirus minichromosomes allow delivery and expression of a 14 kb dystrophin cDNA to muscle cells *Hum Mol Genet* **5,** 913–21.
14. Hartigan-O'Connor D, Barjot C, Crawford R, Chamberlain JS. (2002) Efficient rescue of gutted adenovirus genomes allows rapid production of concentrated stocks without negative selection *Hum Gene Ther* **13,** 519–31.
15. Ng P, Parks RJ, Graham FL. (2002) Preparation of helper-dependent adenoviral vectors *Methods Mol Med* **69,** 371–88.
16. Dormond E, Perrier M, Kamen A. (2009) Identification of critical infection parameters to control helper-dependent adenoviral vector production *J Biotechnol* **142,** 142–50.
17. Dormond E, Chahal P, Bernier A, Tran R, Perrier M, Kamen A. (2009) An Efficient Process for the Purification of Helper-Dependent Adenoviral Vector and Removal of Helper Virus by Iodixanol Ultracentrifugation *J Virol Methods (In press).*
18. Côté J, Bourget L, Garnier A, Kamen A. (1997) Study of adenovirus production in serum-free 293SF suspension culture by GFP-expression monitoring *Biotechnol Prog* **13,** 709–14.
19. Meneses-Acosta A, Dormond E, Jacob D, et al. (2008) Development of a suspension serum-free helper-dependent adenovirus production system and assessment of co-infection conditions *J Virol Methods* **148,** 106–14.
20. Durocher Y, Perret S, Kamen A. (2002) High-level and high-throughput recombinant protein production by transient transfection of suspension-growing human 293-EBNA1 cells *Nucleic Acids Res* **30,** E9.
21. O'Reilly DR, Miller LK, Luckow VA. Baculovirus Expression Vectors: A Laboratory Manual. New York: Oxford University Press; 1994.

Chapter 7

Manufacturing of Retroviruses

Pedro E. Cruz, Teresa Rodrigues, Marlene Carmo, Dagmar Wirth, Ana I. Amaral, Paula M. Alves, and Ana S. Coroadinha

Abstract

Retrovirus vectors derived from moloney murine leukemia virus (MoMLV) were the first class of viral vectors developed for gene therapy. They have been extensively used in clinical trials, particularly in *ex vivo* transduction of hematopoietic stem cells. Although there is a vast experience acquired with retroviruses, their manufacturing is still a difficult task due to the low cell productivities and inherent instability of the infective virus. These viral vectors are most commonly produced using stable producer cell lines in adherent monolayer culture systems. In order to obtain high transduction efficiencies and low toxicity in clinical applications, the viral preparations should be purified, concentrated, and well characterized to attain stringent quality specifications. This chapter describes currently used protocols for manufacturing retroviruses.

Key words: Retrovirus, Gene therapy, Packaging cells, Production, Bioreaction, Purification, Quantification

1. Introduction

Oncoretrovirus or C-type retroviruses derived from Moloney Murine Leukemia Virus (MoMLV) were the first class of viral vectors used in Gene Therapy (1). The retroviruses integrate the transgene in the host-cell chromosome resulting in prolonged expression which makes them particularly suited for inherited diseases (2, 3). Traditionally retroviral vectors have been the vector of choice for *ex vivo* transduction of hematopoietic stem cells, as in the treatment of severe combined immunodeficiencies (SCIDs), where they have shown to be effective in clinical trials (4). Notwithstanding its particular advantage of high transduction efficiency *ex vivo*, retroviruses have also demonstrated promising results in the treatment of other types of diseases, namely cancer (e.g., melanoma, glioblastoma, etc.) (5).

Otto-Wilhelm Merten and Mohamed Al-Rubeai (eds.), *Viral Vectors for Gene Therapy: Methods and Protocols*, Methods in Molecular Biology, vol. 737, DOI 10.1007/978-1-61779-095-9_7, © Springer Science+Business Media, LLC 2011

Table 1
Retrovirus packaging cell lines

	Native cell line	Envelope	Transfer vector	Titers (IP/mL)	References
Modular cell lines					
293 FLEX	Human 293	GaLV	SIN or non-SIN, neo	3×10^6	(8)
Flp293A	Human 293	Amphotropic	SIN or non-SIN, neo	2×10^7	(9)
Classical cell lines					
PA317	Murine NIH/3T3	Amphotropic	Non-SIN, neo	3×10^6	(11), ATCC CRL9078
PG13	Murine NIH/3T3	GaLV	Non-SIN, neo	5×10^6	(12), ATCC CRL10686
FLY A4	Human HT1080	Amphotropic	Non-SIN, LacZ	2×10^7	(13)
FLY RD18	Human HT1080	RD114	Non-SIN, LacZ	1×10^5	(13)
Te Fly Ga18	Human Te671	GaLV	Non-SIN, LacZ	5×10^6	(13, 14)
Phoenix	Human 293	Amphotropic	SIN or non-SIN, neo, puro	1×10^5	(15)
CEM FLY	Human CEM	Amphotropic	Non-SIN, LacZ	2×10^6	(16)

neo neomycin resistance gene, *puro* puromycin resistance gene
Adapted from (6)

Among the major disadvantages of retroviral vectors are the difficulties in its manufacturing, storage, and quality control. Relatively low titers are usually obtained during retroviruses manufacturing due to the low cell productivity and short vector half-life (6). The retroviral vectors are produced in packaging cell lines, generally derived from murine or human origin. These cell lines provide the helper or packaging functions *gag*, *pro*, *pol*, and envelope of the virus in *trans* using molecular constructs that cannot be packaged into the retroviral vectors. The therapeutic gene is supplied by stably transfecting the packaging cell lines with a construct that mimics the viral genome by containing minimal *cis*-acting sequences, allowing its incorporation in the viral particles. Table 1 lists some of the retroviral producer cell lines available (7–9). The production systems used to date to manufacture retrovirus for clinical trials are considered for small scale (10–40 L) and preferably disposable systems, such as T-flasks, cell factories, and roller bottles, although a number of other systems of greater scalability are available (7). Until recently, the purification of retroviruses has been based on centrifugation and ultracentrifugation, but for clinical applications these methods are generally insufficient to meet the quality standards required by the regulatory agencies. The removal of DNA and protein contaminants generally requires the use of chromatographic and membrane technologies (10).

Robust and integrated protocols are needed for efficient manufacturing and characterization of retroviruses. This chapter provides detailed experimental protocols for the production, purification, and quantification of retroviruses.

2. Materials

All materials used in cell culture procedures, media, Fetal Bovine Serum, Dulbecco's Phosphate-Buffered Saline (D-PBS), antibiotics, and other supplements should be of cell culture grade. The chemicals used in all protocols should be of purest grade available from regular commercial sources, unless otherwise specified.

2.1. Establishment of Retroviral Vector Producer Cells

2.1.1. Establishment of Modular Producer Cells by Flp-Mediated Recombination of the Transgene

1. 293 FLEX or Flp293A cells and plasmids pTARFwF5 and pSVFlpe (Table 1, Fig. 1 and Coroadinha et al. 2006 (8) and Schucht et al. 2006 (9)).

   (The cell lines and plasmids are available from the authors.)
2. Cell culture medium: Dulbecco's Modified Eagle's medium (DMEM) supplemented with 10% (v/v) Fetal Bovine Serum (FBS) (see Note 1).
3. Neomycin selection medium: DMEM supplemented with 10% (v/v) FBS and 1,000 μg/mL G418.

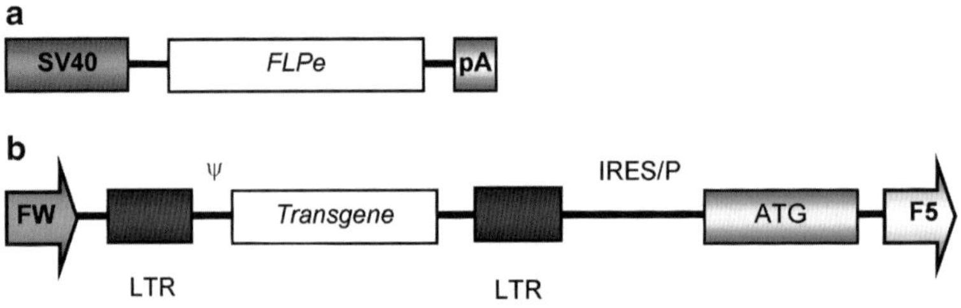

Fig. 1. Schematic representation of pSVFlpe (**a**) and pTARFwF5 (**b**) plasmids used in the establishment of modular retroviral producer cells.

4. Trypsin-EDTA: 0.05% (w/v) Trypsin, 0.53 mM EDTA.
5. Dulbecco's Phosphate-Buffered Saline (D-PBS) w/o calcium and w/o magnesium.
6. Cell culture plates: 96-well, 6-well flat-bottom sterile plates and 100 mm Petri Plates (polystyrene-treated surface).
7. 2.5 M $CaCl_2$ in water, filter sterilized (store at –20°C).
8. 2× HBS: 50 mM Hepes, 280 mM NaCl, 1.5 mM Na_2HPO_4 in water (final pH 7.1), filter sterilized (store at –20°C).

2.1.2. Classical Approach for Establishing Producer Cells by Transfection with the Transgene

1. Packaging cells and retroviral vector plasmids (Table 1 lists a few examples of packaging cell lines available) (see Note 2).
2. Cell culture medium: DMEM supplemented with 10% (v/v) FBS.
3. Selection Antibiotic (see Table 1).
4. Trypsin-EDTA: 0.05% (w/v) Trypsin, 0.53 mM EDTA.
5. D-PBS w/o calcium and w/o magnesium.
6. Cell culture plates: 96-well, 6-well flat-bottom sterile plates and 100 mm Petri Plates (polystyrene-treated surface).
7. 2.5 M $CaCl_2$ in water, filter sterilized (store at –20°C).
8. 2× HBS: 50 mM Hepes, 280 mM NaCl, 1.5 mM Na_2HPO_4 in water (final pH 7.1), filter sterilized (store at –20°C).

2.2. Production of Retrovirus

2.2.1. Production in Small-Scale T-Flasks

1. Stable retrovirus producer cell line (e.g., Table 1).
2. Cell culture medium: DMEM supplemented with 10% (v/v) FBS.
3. Trypsin-EDTA: 0.05% (w/v) Trypsin, 0.53 mM EDTA.
4. D-PBS w/o calcium and w/o magnesium.
5. Cell culture T-flasks 25; 75 or 175 cm^2 (polystyrene-treated surface).
6. Depending on the scale, the filtration at 0.45 μm can be done using either sterile 33 mm Filter Units or Stericup Filter Units (low protein binding – PVDF).

2.2.2. Production in Cell Factories

1. Stable retrovirus producer cell line (e.g., Table 1).
2. Cell culture medium: DMEM supplemented with 10% (v/v) FBS.
3. Trypsin-EDTA: 0.05% (w/v) Trypsin, 0.53 mM EDTA.
4. D-PBS w/o calcium and w/o magnesium.
5. Cell Factory with 10 trays (CF10) corresponding to a total culture area of 6,320 cm^2 and the corresponding accessories, air filter, connector, and white filter adaptor cap from Nunc (Roskilde, Denmark).
6. Four 2 L sterile aspirator bottles mounted with a sterile connector and a clamp (see Note 3).
7. Stericup Filter Unit (0.45 μm) – low protein binding Durapore – PVDF.

2.3. Purification and Storage

2.3.1. Purification by Ultracentrifugation

1. Storage buffer: 10 mM Tris pH 7.2, 2 mM $MgCl_2$, and 0.01% (v/v) Tween 80 (filter sterilized).
2. 20% (w/v) sucrose solution (autoclaved sterilized).
3. 45Ti ultracentrifugation tubes: 70 mL polycarbonate bottle assembly with aluminum caps (Beckman Coulter, Fullerton, CA).
4. 95Ti ultracentrifugation tubes: 10.4 mL polycarbonate bottle with cap assembly (Beckman Coulter).
5. Beckman 45Ti rotor (Beckman Coulter).
6. Beckman 95Ti rotor (Beckman Coulter).
7. Beckman Optima XL-100 ultracentrifuge (Beckman).

2.3.2. Complete Purification Scalable Process by Filtration and Chromatography

1. Buffer 1: 20 mM phosphate buffer with 150 mM of NaCl at a pH value of 7.5, filtered through a 0.22 μm filter.
2. Buffer 2: 20 mM phosphate buffer with 1,500 mM NaCl at a pH value of 7.5, filtered through a 0.22 μm filter.
3. Buffer 3: 20 mM Tris–HCl with 0.5 M sucrose at a pH value of 7.2, filtered through a 0.22 μm filter.
4. 300 mM NaCl solution.
5. 150 mM NaCl solution.
6. 1 M $MgCl_2$ solution.
7. 0.5 M NaOH solution.
8. Benzonase® purity grade II (Merck, Darmstadt, Germany).
9. Sterile containers (Schott bottles and/or disposable bags).
10. 0.8–0.45 μm Sartopore 2 MidiCaps size 7 (Sartorius, Göttingen, Germany).
11. Peristaltic pump.
12. Pressure gauge.

13. Flexible tubing.
14. Hollow fiber cartridge with 500 kDa cutoff, 30 cm length, and 140–420 cm^2 filtration area (GE Healthcare, Uppsala, Sweden).
15. MidGee hollow fiber cartridge with 500 kDa cutoff, 30 cm length, and 16–26 cm^2 filtration area (GE Healthcare).
16. Quick Stand System (GE Healthcare).
17. Advanced MidJet System (GE Healthcare).
18. Anion-exchange chromatography (AEXc) resin Fractogel DEAE EMD 650 (M) media (Merck).
19. XK 26/20 column (GE Healthcare).
20. RK 16/26 packing reservoir (GE Healthcare).
21. ÄKTA™ or FPLC system with a conductivity meter, an UV absorbance detector, pH meter, and a fraction collector.
22. 0.22 μm syringe filters (and a 20 mL sterile syringe).

2.4. Retrovirus Quantification

2.4.1. Infectious Vector Units

2.4.1.1. Quantification of LacZ-Expressing Vectors by Contrast Phase Microscopy

1. Te671 (ATCC CCL-136) target cells (see Note 4).
2. Cell culture medium: DMEM supplemented with 10% (v/v) FBS.
3. Trypsin-EDTA: 0.05% (w/v) Trypsin, 0.53 mM EDTA.
4. D-PBS w/o calcium and w/o magnesium.
5. Cell culture 96 flat-bottomed sterile well plates (polystyrene-treated surface).
6. 12 channel Multichannel micropipettes 10–100 μL and 20–200 μL and reagent reservoir.
7. Phase contrast inverted microscope (100× magnification).
8. Polybrene solution 1 mg/mL (Sigma, Steinheim, Germany) in PBS, filter sterilized.
9. 37% (v/v) formaldehyde solution.
10. 25% (v/v) glutaraldehyde solution.
11. X-gal solution, sterile: 20 mg/mL 5-bromo-4-chloro-3-indolyl-beta-d-galactopyranoside (X-gal, Stratagene, La Jolla, USA) in dimethyl formamide (DMF).
12. 0.5 M $K_3Fe(CN)_6$ solution, filter sterilized.
13. 0.5 M $K_4Fe(CN)_6$ solution, filter sterilized.
14. 0.1 M $MgCl_2$ solution, filter sterilized.

2.4.1.2. Flow Cytometric Analysis of Fluorescent Reporter-Expressing Vectors (or Fluorescence Antibody Staining)

1. Te671 (ATCC CCL-136) target cells (see Note 4).
2. Cell culture medium: DMEM supplemented with 10% (v/v) FBS.
3. Trypsin-EDTA: 0.05% (w/v) Trypsin, 0.53 mM EDTA.

4. D-PBS w/o calcium and w/o magnesium.
5. Cell culture 24-well sterile plates (polystyrene treated surface).
6. 5 mL polystyrene round-bottom tubes.
7. Flow cytometer with Blue Argon Laser (e.g., FACS Calibur, Becton Dickinson).
8. Tabletop Centrifuge.
9. Polybrene solution 1 mg/mL (Sigma) in PBS, filter sterilized.
10. PBS supplemented with 2% (v/v) FBS.
11. 1 mg/mL Propidium iodide solution (Sigma) in PBS.
12. Fix & Perm Cell Permeabilization Reagent B (Invitrogen, Camarillo, CA) when perfoming antibody staining.

2.4.2. Total Vector Units: Viral RNA

1. LightCycler® Systems for Real-Time PCR (Roche, Mannheim, Germany).
2. Thermomixer.
3. LighCycler capillary 20 μL tubes (Roche).
4. 1.5 mL sterile tubes.
5. DNase I (Sigma).
6. PCR-grade water.
7. First-Strand cDNA synthesis kit (Roche).
8. LighCycler-DNA master SYBR Green I (Roche).
9. pSIR standard retroviral vector plasmid (Clontech, Palo Alto, CA) or equivalent plasmid with MoMLV LTRs.
10. Forward Primer: ATT GAC TGA GTC GCC CGG, Tm = 52.4°C, 20 μM (17).
11. Reverse Primer: AGC GAG ACC ACA AGT CGG AT, Tm = 53.6°C, 20 μM (17).

3. Methods

3.1. Establishment of Retrovirus Vector Producer Cells

This section describes the establishment of a stable producer cell line. This requires the stable transfection of a packaging cell line expressing the retrovirus helper functions with retroviral transgene, either by random integration or site-specific flp-mediated integration.

3.1.1. Establishment of Modular Producer Cells by Flp-Mediated Recombination of the Transgene

1. Seed modular packaging cells (293 FLEX or Flp293A) at 6×10^5 cells *per* well in a six-well plate (prepare a cell suspension at 3×10^5 cells/mL and inoculate 2 mL *per* well). Incubate the cells overnight in a humidified incubator at 37°C and 10% CO_2.

2. The next day cells should be 60–80% confluent. Three hours before transfection remove the medium and replace it with 2 mL of fresh cell culture medium.
3. Co-transfect cells with calcium phosphate precipitation method with 4 μg of targeting plasmid (pTARFwF5) and 12 μg of flipase plasmid (pSVFlpe) *per* well. For each transfection, prepare two sterile tubes with solution A and solution B according to Table 2. In tube A, dilute the DNA in Molecular Grade Water, mix well, and add afterwards the 2.5 M $CaCl_2$ and mix again. Perform a negative control replacing the DNA volume by Molecular Grade Water. Prepare the tube B with 150 μL of 2× HBS and to this tube add slowly drop wise tube A solution under vortex mix. Incubate the solution at room temperature for 10–15 min.
4. Vortex transfection solution again and add drop wise to the cells. Swirl the plates and incubate in a humidified incubator at 37°C and 10% CO_2.
5. Between 4 and 16 h post-transfection remove the medium and add 2 mL of fresh growth media.
6. Forty-eight hours after transfection start neomycin selection by transferring the cells to a 100 mm Petri plate with 1,000 μg/mL of G418.
7. Incubate plates in a humidified incubator at 37°C and 10% CO_2.
8. Change the selection medium containing 1,000 μg/mL of G418 twice a week until large drug-resistance colonies (2–3 mm in diameter) are formed (10–20 days).
9. Isolate 10–12 colonies using cloning rings or by suction using a P-200 pipette (see Note 5), place the colonies in a 96-well plates with 200 μL of selection medium.
10. Expand each cell clone; confirm correct recombination and virus production of five to ten clones. The titers should be

Table 2
Solutions for calcium phosphate transfection

	***per* well in six-well plate**	
A	DNA:	
	pTARFwF5	4 μg
	pSVFlpe	12 μg
	2.5 M $CaCl_2$	10 μl
	M.G. Water	To a final volume 150 μl
B	2× HBS	150 μl

M.G. Water (Molecular Grade Water)

homogeneous and around 1–10 × 10^6 IP/mL. The stable producer cell line can now be used for the production of retroviruses according to the following protocols and cryopreserved at −85°C or vapor liquid N_2 according to the general cell culture protocols (for additional details see refs. 8, 9).

3.1.2. Classical Approach for Establishing Producer Cells by Transfection with the Transgene

The classical protocols for establishing a producer cell line are dependent on the helper cell line used (e.g., the transfection method, retroviral plasmids, selection antibiotic) (see Table 1).

1. Seed the classical packaging cells in six-well plates at a cell density that will be between 60 and 80% confluent the next day (depending on the origin of the cell line use between 2 and 6 × 10^5 cells *per* well).
2. Incubate the cells overnight in a humidified incubator at 37°C and 10% CO_2.
3. The next day, 3 h before transfection remove the medium and replace it with 2 mL of fresh cell culture media.
4. Transfect cells with calcium phosphate precipitation method with 5 μg of retroviral vector plasmid *per* well. For each transfection prepare two sterile tubes with solution A and B according to Table 2 but instead of pTARFwF5 and pSVFlpe dilute 5 μg of your transfer retroviral plasmid of interest in the Molecular Grade Water. Mix well and add afterwards the 2.5 M $CaCl_2$ and mix again. Perform a negative control replacing the DNA volume by Molecular Grade Water. Prepare the tube B with 150 μL of 2× HBS and to this tube add slowly drop wise tube A solution under vortex mix. Incubate the solution at room temperature for 10–15 min.
5. Proceed as in steps 4 and 5 in Subheading 3.1.1.
6. Forty-eight hours after transfection, start antibiotic selection by cultivating the cells in a 100 mm Petri dish with the respective resistance antibiotic (see Table 1).
7. Incubate plates in a humidified incubator at 37°C and 10% CO_2.
8. Change the selection medium containing the resistance antibiotic twice a week until large drug-resistance colonies (2–3 mm in diameter) are formed (10–20 days).
9. Isolate around 100 colonies using cloning rings or by suction using a P-200 pipette (see Note 5) and place the colonies in a 96-well plate with 200 μL of selection media (see Note 6).
10. Expand each cell clone and screen for clones producing titers between 1 × 10^5 and 10 × 10^6 IP/mL (see Note 6). The stable producer cell line can now be used for the production of retroviruses according to the following protocols and cryopreserved at −85°C or vapor liquid N_2 according to the general cell culture protocols.

3.2. Production of Retrovirus

The production conditions of retrovirus will depend on the producer cell line used (i.e., cell inoculum, production medium, and harvest time). The protocols below describes the production conditions for 293-derived producer cells 293 FLEX and Flp293A and for Te671-derived producer cells Te Fly Ga18 and Te Fly A7.

3.2.1. Production in Small-Scale T-Flasks

1. Prepare a cell suspension at 2×10^5 or 1×10^5 cells/mL for 293-derived producer cells (293 FLEX or Flp293A) or Te671-derived cells (Te Fly Ga18 or Te Fly A7), respectively, in cell culture medium DMEM supplemented with 10% (v/v) FBS.
2. Inoculate each T-flask with 5, 15, or 35 mL depending if you are using a 25, a 75, or 175 cm^2 T-flask, respectively. This corresponds to an inoculation density of 4×10^4 cells/cm^2 for 293 FLEX and Flp293A and of 2×10^4 cells/cm^2 for Te Fly Ga 18 and Te Fly A7. Place in a humidified incubator at 37°C and 10% CO_2.
3. After 3 days, when cells reached around 60% confluence, exchange the growth medium for fresh new one and re-incubate in a humidified incubator at 37°C and 10% CO_2.
4. The next day, 24 h after medium exchange, harvest the viral supernatant, and filter at 0.45 μm. If not immediately used, the supernatant should be stored at −85°C.

3.2.2. Production in Cell Factories

1. Prepare a cell suspension of 1.5 L at 1.7×10^5 or 8.5×10^4 cells/mL for 293-derived producer cells (293 FLEX or Flp293A) or Te671-derived cells (Te Fly Ga18 or Te Fly A7), respectively, in cell culture medium DMEM supplemented with 10% (v/v) FBS. The suspension should be prepared or transferred to a 2 L sterile aspiration bottle mounted with sterile connector and clamp (additionally see Nunc Cell Factory Instructions).
2. Unpack the Cell Factory CF10 (10 tray – 6,320 cm^2) and place it in the laminar flow cabinet. The following steps should be done under sterile conditions.
3. Remove the seal from one of the white filter adaptor caps, as indicated in the Cell Factory instructions, and immediately insert presterilized air filter (0.22 μm) (see Fig. 2a, b).
4. Remove the second white filter adaptor cap from the Cell Factory and insert the connector from the 2 L aspiration bottle containing the cell suspension (see Fig. 2a, b).
5. Turn the Cell Factory to its side, so that the growth surface is in the vertical position, and raise the aspirator bottle above the Cell Factory level. Gently agitate the aspirator bottle and loosen the clamp, the cell suspension will flow into the Cell Factory (see Fig. 2c).

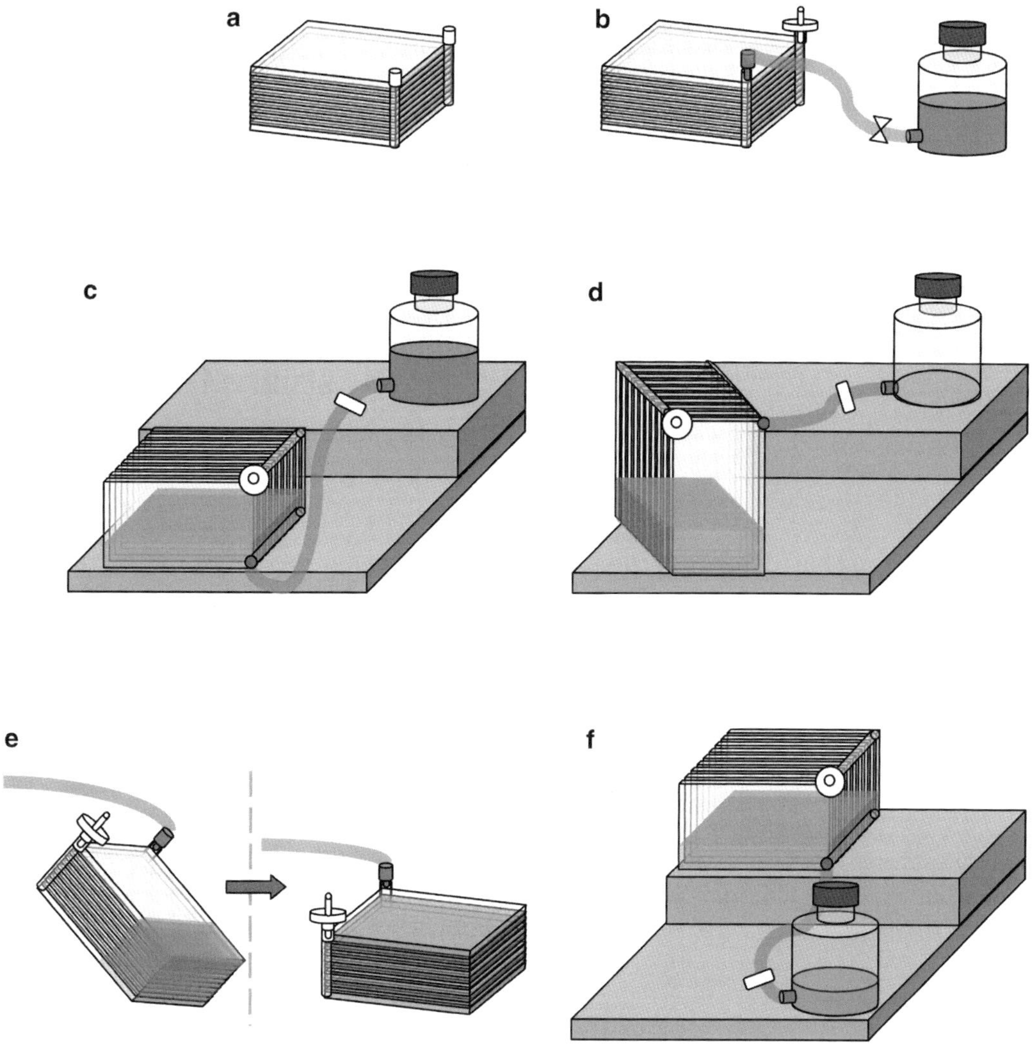

Fig. 2. Schematic representation of manual Cell Factory operation during, inoculation (**a–e**) and supernatant harvesting (**f**).

6. When the filling is complete, allow the levels of the liquid to equalize in all chambers and then turn the Cell Factory 90° in a way that inlet is up and the growth surface is still in the vertical position (see Fig. 2d).
7. Verify that the medium is separated in equal volumes in each chamber and then place the Cell Factory in the horizontal position in order that the growth surface of all trays will be covered by the medium (see Fig. 2e).
8. Remove the connector from the 2 L aspirator bottle and replace it by a white filter adaptor cap (leave the filter on).
9. Incubate the Cell Factory in a humidified incubator at 37°C and 10% CO_2 for 3 days (see Note 7).

10. After 3 days, prepare an empty sterile 2 L aspirator bottle mounted with a sterile connector with clamp. Connect it to the Cell Factory using the adaptor cap port, turn the Cell Factory on its side and raise it above the aspirator bottle. Open the clamp and the supernatant will flow into the aspirator bottle (see Fig. 2f).
11. Prepare a third sterile aspirator bottle with a connector and clamp with 900 mL of fresh cell culture medium previously warmed at 37°C. Connect the aspirator bottle to the adaptor port and raise it above the Cell Factory. Open the clamp and allow the medium to flow into the Cell Factory (see Fig. 2c). Turn the Cell Factory 90°C so that the level of medium is equal in each chamber (see Fig. 2d). Turn the Cell factory in the horizontal position, remove the aspirator bottle connector and replace it by a white filter adaptor cap (see Fig. 2e).
12. Re-incubate the Cell Factory in a humidified incubator at 37°C and 10% CO_2 for 24 hours.
13. The next day prepare a sterile aspirator bottle with a connector with a clamp to harvest the viral supernatant. Perform as described in step 10. Filter the viral supernatant at 0.45 μm and proceed to purification or store at −85°C in appropriate containers.

3.3. Purification and Storage

This section describes a common ultracentrifugation labscale purification procedure and a complete scalable purification process based on filtration and chromatographic techniques (18) of retroviruses.

3.3.1. Purification by Ultracentrifugation

This section describes a general ultracentrifugation purification method for retroviral vectors. Since the principle of purification is based on molecular size, it is applicable to all retroviral vectors pseudotypes, although different yields may be obtained depending on the envelope resistance to shear stress.

1. Place the retrovirus supernatant in 70 mL ultracentrifugation bottles previously sterilized by autoclavation (see Note 8).
2. Concentrate the retrovirus supernatant by ultracentrifugation at 100,000 × *g* for 90 min at 4°C using a Beckman 45Ti rotor.
3. Resuspend the pelleted viruses in a maximum volume of 1.5 mL of storage buffer.
4. Fill sterile ultracentrifugation 10.4 mL bottles with 7 mL of 20% (w/v) sucrose solution.
5. Place up to 1.5 mL of the concentrated pelleted viruses on top of the sucrose solution.
6. Ultracentrifuge at 200,000 × *g* for 120 min at 4°C using a 90Ti rotor.

7. Resuspend the final pellet in the desired volume of storage buffer (or other buffer desired, accordingly to the subsequent use of the purified retroviral preparation).
8. If not for immediate use, store the viral vector preparation at −85°C (see Note 9).

3.3.2. Complete Purification Scalable Process by Filtration and Chromatography

This method describes a complete purification process for retroviral vectors relying on the retrovirus size and charge. The following protocol was tested for Amphotropic pseudotyped retroviruses, although it is applicable for all envelope pseudotypes (due to the different envelope charge differences minor optimization for different envelopes may be desirable for achieving good yields). The process includes six steps: (1) an initial microfiltration clarification of the vector supernatant followed by (2) benzonase treatment and (3) concentration using 500 kDa MWCO tangential flow hollow fiber Ultra/Diafiltration membrane, (4) purification by anion-exchange chromatography (AEXc) using a tentacle matrix bearing DEAE functional ligands, (5) concentration and buffer exchange of the vector into a storage buffer by ultrafiltration and, finally, (6) sterile filtration of the purified vector using a 0.22-μm filter. To maintain viral stability and achieve higher transducing unit yields, it is recommended to perform the purification protocol at low temperature (between 4 and 6°C).

3.3.2.1. Step 1: Microfiltration

3.3.2.1.1. Preparation and Conditioning of the Microfiltration Capsule

The dimensions of the micro and ultrafilters to be used depend on the initial volume of the supernatant to be processed. The filters herein described have been tested for processing 2–4 L of initial retroviral supernatant (18).

1. Mount a pressure gauge close to the filter's inner port and a flexible tubing upstream the pressure gauge (add enough tubing in order to be able to filter in a vertical position into a reservoir) following Fig. 3. Use clamps to ensure the tube is tight to the filter and the pressure gauge.
2. Adjust and secure the tubing to the pump head and introduce the tubing extremity into a reservoir containing Buffer 1.
3. Remove the top plug of the filter and start the pump at 50 mL/min. This will allow removal of the air inside the capsule.
4. Close the top plug of the filter when the capsule is filled with liquid and rinse with 500 mL of Buffer 1.
5. Stop the pump and leave the buffer to soak the membrane till further use.

3.3.2.1.2. Microfiltration Purification Procedure

1. Remove the tubing from the Buffer 1 reservoir and start the pump to drain the Sartopore 2 filter.

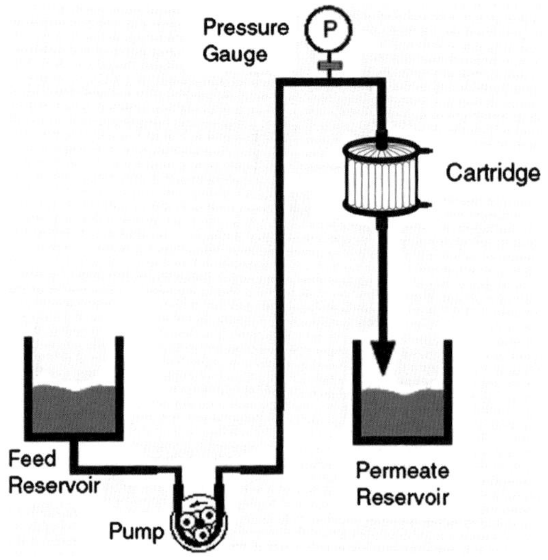

Fig. 3. Schematic representation of the microfiltration setup.

2. Place the extremity of the tubing into the reservoir containing the retroviral supernatant, remove the top plug of the filter, and start the pump at 50 mL/min.
3. Close the top plug of the filter when the capsule is filled with liquid and rinse with 500 mL of Buffer 1. Filter through in a vertical position.
4. Close the top plug and increase the flow rate to 100–300 mL/min ensuring the inlet pressure is below 2 bar (decrease flow rate if necessary).
5. Collect the filtrate into a sterile Schott bottle (see Note 10).

3.3.2.2. Step 2: Benzonase Treatment

1. Add 200 units of benzonase and 2 mL of 1 M $MgCl_2$ *per* L of clarified supernatant (19).
2. Incubate the supernatant overnight at 4°C or at 37°C for 1 h.
3. Condition the supernatant at the appropriate temperature for further purification (see Notes 11 and 12)

3.3.2.3. Step 3: First Ultra/Diafiltration

3.3.2.3.1. Preparation of the Ultrafiltration Cartridges

1. Install and connect the cartridges to the appropriate system: QuickStand for the larger cartridge and the Advanced MidJet System for the MidGee cartridge. See Fig. 4 for guidance.
2. Connect the retentate and the permeate lines to a waste container.
3. Fill the feed reservoir with warm ultrapure water.
4. Start the pump at a low flow rate and adjust the feed pressure to 0.3 bar (5 psi).
5. Adjust the pump speed and retentate valve such that the retentate flow rate is approximately one-tenth of the permeate flow (see Note 13).

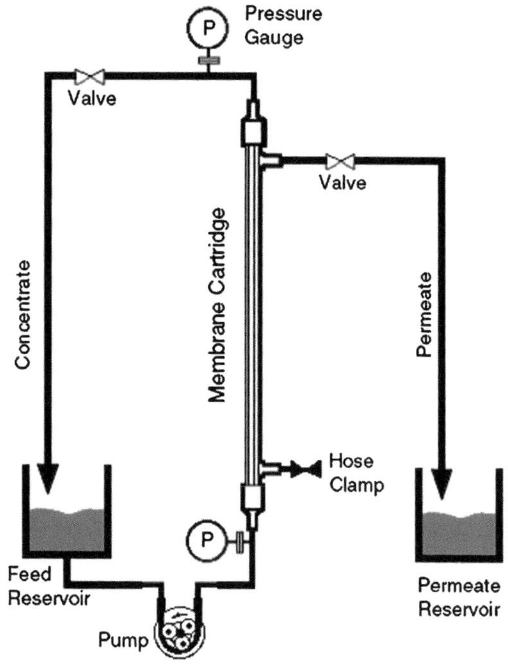

Fig. 4. Schematic representation of a tangential flow filtration setup using a hollow fiber (courtesy of GE Healthcare).

6. Continue rinsing for 90 min adding more water to the reservoir as necessary.
7. Stop the pump and drain the system.
8. Direct the retentate and permeate lines to the feed reservoir.
9. Recirculate a solution of 0.5 M of NaOH at 30°C for 30–60 min.
10. Drain the system.
11. Rinse the cartridge with ultrapure water as described above.
12. Drain the system.
13. Add Buffer 1 (first ultra/diafiltration) or Buffer 3 (second ultra/diafiltration) to the feed reservoir (5–10 L of buffer *per* m^2 of filter surface area).
14. Open the retentate and permeate valves. Start the pump slowly and increase the feed rate until solution flows from the retentate and permeate lines.
15. Adjust transmembrane pressure to 0.3 bar (5 psi).
16. Open the retentate valve and close the permeate valve. Increase the retentate flow rate to 15 mL/(cartridge fiber. min), e.g., a cartridge containing six fibers should run at a recirculation flow rate of 90 mL/min.

17. Open the permeate valve and adjust the retentate valve to the transmembrane pressure noted above.
18. Recirculate the buffer solution for 30 min.
19. Drain the buffer from the feed reservoir, leaving a small amount in the bottom of the reservoir to prevent introduction of air into the system.

3.3.2.3.2. First Ultra/Diafiltration Purification Procedure

1. Allow the clarified supernatant and the Quick Stand System to reach the temperature at which the process will be run (4°C or room temperature).
2. Introduce the feed and the concentrate lines into the feed reservoir containing the clarified supernatant. Keep the retentate returning to the feed reservoir below the liquid level to avoid splashing, foaming and excess air entrainment (Fig. 4).
3. Place the permeate line in a waste reservoir.
4. Open the concentrate valve and close the permeate valve.
5. Start the pump and increase the recirculation flow rate slowly. Set the recirculation flow rate to 12 mL/(cartridge fiber.min).
6. Open the permeate valve and adjust the inlet pressure to 1.5 bar by closing the concentrate valve. Adjust the concentrate valve to keep the inlet pressure constant.
7. Add 0.1× initial supernatant volume of Buffer 1 when the volume of concentrate reaches 10% of the initial supernatant volume.
8. Concentrate further till approximately 100 mL of concentrate remain in the retentate reservoir.
9. To maximize the recovery of concentrate from the system, place the retentate line above liquid level and start the pump at a low flow rate in the reverse mode. Most of the retentate in the membrane and the system will be drawn back to the retentate reservoir.
10. Stop the pump and carefully drain the retentate out (see Note 14).

3.3.2.4. Step 4: Anion-Exchange Chromatography

The dimensions of the AEXc column to be used depend on the initial volume and biological titer of the supernatant to be processed. Herein, we describe the packing of a column with approximately 100 mL volume (CV = 100 mL), 20 cm height, and 2.6 cm diameter suitable for processing up to 4 L of initial retroviral supernatant (18).

3.3.2.4.1. AEXc Column Packing and Cleaning

1. Equilibrate all materials at room temperature.
2. Remove the storage solution of approximately 150 mL of resin slurry by decanting and wash with 500 mL of ultrapure water. Decant again and wash two times more with 500 mL of 300 mM NaCl. Decant and add 150 mL of 300 mM NaCl.

3. Mount the XK 16/20 column end pieces. Eliminate air by flushing column end pieces with 150 mM NaCl. Ensure no air is trapped under the column net. Close column outlet leaving 1–2 cm of buffer in the column.
4. Adjust the packing reservoir to the top of the column and level the column.
5. Gently resuspend Fractogel DEAE medium and pour into the column.
6. Immediately fill the column and packing reservoir with 150 mM NaCl.
7. Close the packing reservoir and connect it to the chromatography system.
8. Open the column outlet and pump 20 mL/min of 150 mM NaCl through the column. Ensure the backpressure does not exceed 0.5 MPa.
9. Stop the pump when the bed height no longer decreases and close the column outlet. Remove the packing reservoir and carefully fill the rest of the column with buffer to form an upward meniscus at the top.
10. Insert the adaptor into the column at an angle; ensuring that no air is trapped under the net, and slide the adaptor slowly down the column (the outlet of the adaptor should be open). Lock the adaptor in position.
11. Flush the chromatography system with 0.5 M NaOH.
12. Clean the column with 1.5 CV of 0.5 M NaOH at a flow rate of 5 mL/min.
13. Flush the chromatography system with Buffer 1.
14. Equilibrate the column with Buffer 1 at a flow rate of 10 mL/min. Monitor the conductivity and pH of the outflow. Stop when both reach stable values.

3.3.2.4.2. AEXc Purification Procedure

1. Equilibrate the AEXc column with 1 CV of Buffer 1 at 10 mL/min.
2. Reset the UV detector.
3. Load the concentrated supernatant into the AEXc column at a flow rate of 7 mL/min.
4. Wash the column with 2 CV of Buffer 1 at a flow rate of 10 mL/min (maintain this flow rate till the end of the process).
5. Start elution of contaminant proteins with a mixture of 25% (v/v) of Buffer 2 and 75% (v/v) of Buffer 1 (prepare the buffer solutions previously if the chromatography system has no buffer mixer). Elution is accompanied by an increase in absorption at 280 nm.

6. After 2 CV of buffer (or when absorbance reaches baseline again) start elution of the viral vectors with a mixture of 60% (v/v) of Buffer 2 and 40% (v/v) of Buffer 1.
7. The viral peak should start eluting at 40–50 mL after starting to pump the elution buffer through the column. Collect the viral peak using the fraction collector (the elution volume will be approximately 30–40 mL).
8. Regenerate the column with 2 CV of buffer 2 and 5 CV of buffer 1 afterwards.

3.3.2.5. Step 5: Second Ultra/Dialfiltration

The second concentration step is similar to the first but performed at a smaller scale. Preparation of the ultrafiltration cartridges follows as described in step 3 "Preparation of the Ultrafiltration Cartridges."

1. Allow the AEXc purified vector and the Advanced MidJet System to reach the temperature at which the process will be run (4°C or room temperature).
2. Introduce the feed and the concentrate lines into the feed reservoir containing the AEXc purified vector (as in Fig. 4). Keep the retentate returning to the feed reservoir below the liquid level.
3. Place the permeate line in a waste container.
4. Open the concentrate valve and close the permeate valve.
5. Start the pump and increase the recirculation flow rate slowly. Set a recirculation flow rate of 12 mL/(cartridge fiber.min).
6. Open the permeate valve and adjust the inlet pressure to 1.0 bar by closing the concentrate valve. Adjust the concentrate valve to keep the inlet pressure constant.
7. To start diafiltration, add 10 mL of Buffer 3 to the concentrate reservoir when the volume of concentrate reaches approximately 10 mL.
8. Concentrate further till reaching 10 mL of concentrate and repeat the previous action two more times. The virus can be concentrated down to 5 mL.
9. To maximize the recovery of concentrate from the system place the cartridge inlet line above liquid level and start the pump at a low flow rate. Most of the retentate in the membrane will be pumped through system into the retentate reservoir.
10. Stop the pump and recover the retentate.

3.3.2.6. Step 6: Sterile Filtration

Due to the small volume of purified vector obtained in the end of the process, the best way to sterile filter the vector is to use a low binding 0.22 μm MiniSart® NML syringe filter. It is advisable not to exert too much pressure during filtration. Filter under sterile conditions.

If not for immediate use, store the viral vector preparation at −85°C (see Note 9).

3.4. Retrovirus Quantification

The quantification of infectious retroviruses depends on the gene expressed, herein are described two protocols for retrovirus expressing either the marker genes LacZ or fluorescent reporter proteins. The latter protocol can be adapted to genes for which an antibody is available. This section also describes the quantification of total viruses by quantifying by real-time RT-PCR the viral RNA, it can be applied for all retrovirus possessing an MoMLV LTR (17).

3.4.1. Infectious Vector Units: Quantification of LacZ-Expressing Vectors by Contrast Phase Microscopy

1. Prepare a cell suspension of target cells Te671 at 1.65×10^5 cells/mL in DMEM supplemented with 10% (v/v) FBS (see Note 4).
2. Inoculate 100 μL *per* well of the cell suspension in 96-well plates (this corresponds to an initial cell density of 5×10^4 cells/cm^2). Incubate overnight in a humidified incubator at 37°C and 10% CO_2.
3. The next day perform serial dilutions of the viral samples (generally between 10^{-1} and 10^{-7} for titers between 10^3 and 10^{10} I.U./mL). These dilutions should be performed in 96-well plates, using a multichannel micropipette, by diluting 20 μL of viral suspension successively in 180 μL of DMEM supplemented with 10% (v/v) FBS and polybrene at 8 μg/mL.
4. Remove the supernatant from the Te671 target cells, that should be 60–80% confluent, and infect cells in triplicate with 50 μL of viral suspension of the several dilutions performed. Incubate cells in a humidified incubator at 37°C and 10% CO_2.
5. After 4 h of incubation, for virus adsorption, add 150 μL of DMEM supplemented with 10% (v/v) of FBS. Incubate 48 h in a humidified incubator at 37°C and 10% CO_2.
6. Two days after infection the cells are fixed: the medium is removed from the target cells Te671, cells are washed with PBS 100 μL *per* well, and 100 μL of the fixing solution is added and incubated for 3 min at room temperature. The fixing solution should be prepared freshly by adding (for one 96-well plate): 675 μL of glutaraldehyde 25% (v/v) and 100 μL of formaldehyde 37% (v/v) to 12.5 mL of PBS.
7. After fixing the cells, wash with 100 μL of PBS *per* well.
8. Stain the cells by adding 100 μL of staining solution *per* well. The staining solution should be prepared freshly by adding (for one 96-well plate): 125 μL of x-Gal 20 mg/mL, 125 μL of 0.5 M $K_3Fe(CN)_6$, 125 μL of 0.5 M $K_4Fe(CN)_6$, 125 μL of 0.1 M $MgCl_2$ to 12 mL of PBS.
9. Incubate for 24 h at 37°C (see Note 15).

10. Count the number of stained blue cells (corresponding to the infected cells expressing β-Galactosidase) *per* well, using a phase contrast inverted microscope. Only wells with 20–200 blue stained cells should be considered.
11. The titer is calculated by multiplying by the dilution factor according to the equation:

$$[\text{Titer}]\,(\text{I.U.}/\text{mL}) = \frac{n^{\circ}\ \text{Blue cells}}{0.05\ \text{mL}} \times \text{Viral dilution}.$$

3.4.2. Infectious Vector Units: Flow Cytometric Analysis of Fluorescent Reporter-Expressing Vectors (or Fluorescent Antibody Staining)

1. Prepare a cell suspension of target cells Te671 at 1×10^5 cells/mL in DMEM supplemented with 10% (v/v) FBS (see Note 4).
2. Inoculate 0.5 mL of Te671 target cells *per* well in 24-well plates and incubate cells in a humidified incubator at 37°C and 10% CO_2 overnight.
3. The next day determine the cell concentration *per* well at the time of infection, in duplicate, by tripsinizing the Te671 cells in two wells.
4. Perform serial dilutions of the viral samples (generally between 10^{-1} and 10^{-6} for titers between 10^5 and 10^{10} I.U./mL) in DMEM supplemented with 10% (v/v) FBS and polybrene at 8 μg/mL.
5. Remove the supernatant from the Te671 target cells, that should be at 60–80% confluent, and infect cells in triplicate with 200 μL of viral suspension of the several dilutions performed. Perform a negative control with cells not infected.
6. Incubate the cells for 4 h in a humidified incubator at 37°C and 10% CO_2 for virus attachment.
7. Add 0.8 mL of DMEM supplemented with 10% (v/v) FBS and re-incubate for 48 h in a humidified incubator at 37°C and 10% CO_2.
8. Two days after infection harvest cells: remove the supernatant from the wells, wash with 0.5 mL of PBS, trypsinize with 200 μL of trypsin, and after cell detachment resuspend cells with 300 μL of PBS supplemented with 2% FBS.
9. Centrifuge cells at $200 \times g$ (either in Eppendorf tubes or FACS polysterene tubes), for 5 min at 4°C. For fluorescent marker genes, remove the supernatant and resuspend each pellet in 500 μL of PBS with 2% (v/v) FBS and 2 μg/mL of PI (Propidium Iodide). For fluorescence antibody staining of the gene expressed, wash by removing the supernatant, resuspend each pellet in 500 μL of PBS and centrifuge again. Resuspend pellet with 100 μL of the fluorescent-labeled primary antibody (FITC- or PE-conjugated) diluted in

reagent B of the Fix & Perm Cell Permeabilization (at the antibody manufacturer recommended dilution). Wash twice in PBS by centrifugation and resuspend in 500 μL of PBS with 2% (v/v) FBS and 2 μg/mL of PI (see Note 16).

10. Analyze samples for fluorescent-positive viable cells using a flow cytometer (see Note 17).
11. The titer is calculated by multiplying the percentage of fluorescent-positive cells by the number of cells *per* well at the time of infection and by the dilution factor according to the equation:

$$[\text{Titer}](\text{I.U / mL}) = \frac{\%\text{Fluorescent cells}}{0.20\text{ mL}} \times \text{Viral Dilution} \times n^{\circ}\text{cells / well}$$

(see Note 18)

3.4.3. Total Vector Units: Viral RNA

3.4.3.1. Part I: Pretreatment of Samples and cDNA Synthesis

1. Incubate 50 μL of viral samples (previously filtered at 0.45 μm) at 75°C for 10 min in a Thermomixer to release the viral RNA.
2. Let the samples cool down to room temperature and spin down the tubes in a bench top centrifuge.
3. Add 1 μL of DNase I (1 U/μL) and 1.8 μL of 25 mM $MgCl_2$, vortex and incubate the mixture for 30 min at 25°C in order to destroy any DNA from the cell lysates.
4. Incubate the mixture at 75°C for 10 min in the Thermomixer to inactivate the DNase I.
5. Prepare 11.5 μL of the cDNA synthesis mix to the indicated end-concentrations (see Table 3 and First-Strand cDNA synthesis kit instructions from Roche).

Table 3
cDNA synthesis mix solution

cDNA synthesis mix	
1× reaction buffer	2 μL
5 mM $MgCl_2$	4 μL
1 mM dNTP	2 μL
1 μM reverse primer	1.7 μL
50 U RNase inhibitor	1 μL
20 U AMV reverse transcriptase	0.8 μL
Total volume	11.5 μL

6. Add 8.5 μL of viral sample to 11.5 μL of cDNA synthesis mix, vortex briefly, and spin down the mixture.
7. Incubate the reaction at 25°C for 10 min and subsequently for 60 min at 42°C.
8. Inactivate the AMV (avian myeloblastosis virus) reverse transcriptase by heating for 5 min at 99°C. Store the sample either at 4°C for 1–2 h or -20°C for longer periods.

3.4.3.2. Part II: Real-time PCR

1. Prepare 10 μL *per* sample of the LightCycler Real-Time SYBR Green reaction mastermix to the indicated end-concentrations in a sterile 1.5 mL tube (see Table 4 and LightCycler Fast Start DNA master SYBR Green I manual instructions from Roche).
2. Vortex the mastermix and distribute 10 μL *per* each LighCycler capillary reaction tube.
3. To the negative control capillary tube add 10 μL of PCR-grade water and close it. To the sample capillary tubes add the samples of cDNA previously synthesized and diluted in PCR-grade water (generally a dilution of 10^{-1} is adequate for titers between 10^{6}–10^{12} T.U./mL).
4. Prepare the standard curve of retroviral plasmid (e.g., pSIR) at 10^{2}, 10^{3}, 10^{4}, 10^{5}, 10^{6}, 10^{7}, and 10^{8} copies/mL in PCR-grade water and add 10 μL of each to the respective capillary tube.
5. Centrifuge the capillary tubes to spin down the reaction, place them in the LightCycler rotor and run the PCR: denaturation program (95°C for 10 min); amplification and quantification program repeated 45 times (60°C for 10 min; 72°C for 10 min with a single fluorescence measurement); melting curve program (65–95°C for 10 min with continuous fluorescence measurement); and cooling step to 40°C.
6. Analyze the data using the second derivative maximum method (see Note 19). The titer of Total Units is calculated

Table 4
Real-time SYBR Green PCR mastermix solution

Real-time SYBR Green PCR mastermix	
Fast start DNA master SYBR Green I	2 μL
4 mM $MgCl_2$	3.2 μL
0.5 μM forward primer	0.5 μL
0.5 μM reverse primer	0.5 μL
PCR-grade water	3.8 μL
Total volume	10 μL

taking into account: an efficiency of 30% of the reverse transcriptase, two copies of LTR *per* retroviral RNA (if the plasmid has two LTRs such as pSIR this term is eliminated); two copies of RNA *per* viral particle and the dilution factor accordingly to the equation:

$$[\text{Total Units}](\text{T.U.}/\text{mL}) = [\text{DNA}_{\text{PCR}}](\text{copies}/\mu\text{L}) \times \text{Dilution factor} \times 1000 \times \frac{100}{30} \times \frac{1}{2}.$$

4. Notes

1. All cell culture media used are high glucose (4.5 g/L) and high glutamine (4 mM).
2. The choice of transfer vector depends both on the packaging cell line and target cell for gene delivery. A review on the state of the art transfer vector design can be found at Schambach et al. (2008) (20).
3. The 2 L aspirator bottles should be mounted and sterilized 20 min at 121°C in a autoclave.
4. There are several target cell lines available that can be used for the titration of infectious retrovirus namely, NIH 3T3 (murine cell line), HT1080, HCT 116, HEK293, and Te671 (human cell lines). The choice of target cell depends mainly on the envelope used in the retrovirus (although other factors such as the transgene promoter in the case of SIN vectors should be accounted for). Te671 cells are a suitable target cell line for both GaLV and Amphotropic envelope pseudotyped retrovirus.
5. After selecting and marking the location of colonies in the Petri plate, aspirate the medium and add small volume of PBS sufficient to cover the plate. Place the pipette tip, with the plunger fully depressed, directly over the colony and slowly release the plunger to aspirate the colony. Place the colony under trypsin in a 96-well plate a few minutes and resuspend cells with cell culture medium.
6. Isolation of a transfected cell clone expressing high-titers of infectious retrovirus is a low-yield process due to the random integration of the plasmid. Hundreds of clones have to be screened in order to find a high-performance clone. For non-SIN vectors, since they contain an active promoter element at the LTR (original U3 sequence or a heterologous promoter inserted at the deleted U3 region of the 3′LTR) driving the expression of primary transcripts, viral transduction by retroviral

infection has been preferred for their stable integration in the packaging cell line. This procedure generally leads to a higher yield of high-performance clones; however, requires a previous transient transfection of a packaging cell in order to produce the virus.

7. When a microscope allowing the monitoring of Cell Factory with 10 trays is not available, it is recommended to inoculate simultaneously one 175 cm^2 T-flask using the same inoculum and the same cells' density (i.e., inoculate 41 mL in order to obtain 4×10^4 cells/cm^2 for 293 FLEX and Flp293A and of 2×10^4 cells/cm^2 for Te Fly). Monitor the 175 cm^2 T-flask, and if cells do not grow as expected, the medium exchange can be delayed or anticipated.
8. Use retroviral supernatant previous filtered at 0.45 μm.
9. To increase the vector half-life at −85°C several stabilizers can be added to the storage buffer. Examples of such are sucrose, ectoin, and firoin (generally added at 0.5 M) (21) or alternatively recombinant proteins, like BSA or HSA (0.4 mg/mL) (22).
10. The filter can be rinsed with 200 mL of Buffer 1 collected into the filtrate to maximize recovery.
11. If the inactivation kinetics of the vectors is fast the benzonase treatment can be performed during the production phase, i.e., benzonase and $MgCl_2$ can be added to the medium used for the last medium exchange.
12. A second benzonase treatment can be performed after the first ultra/diafiltration step if necessary.
13. The transmembrane pressure is given by (Pin + Pout)/2-Ppermeate.
14. Performing this step at 4°C results in a decrease of the permeate flux through the membrane thus, significantly increasing processing time.
15. After incubation at 37°C the plates can be stored at 4°C up to 1 week if sealed with parafilm (prevents evaporation).
16. If the primary antibody is not conjugated with a fluorescent probe, after the first incubation with the antibody, wash twice with PBS and incubate for 1 h with 100 μL of a fluorescent-labeled secondary antibody diluted in reagent B of the Fix & Perm kit (at the antibody manufacturer recommended dilution). After incubation, wash twice in PBS by centrifugation and resuspend in 500 μL of PBS with 2% (v/v) FBS and 2 μg/mL of PI.
17. Homogenize cell samples immediately before flow cytometry analysis.

18. Consider only the dilutions of viral samples giving a linear response (dilution vs. % of positive fluorescent cells transduced).
19. Examine the melting curves for the presence of specific amplification and the absence of primer dimers. The amplicon peak corresponds to an 84°C melting temperature. The calibration curve should have a slope between –3.3 and –3.9 and an error below 0.1.

References

1. Blaese, R. M., K. W. Culver, A. D. Miller, C. S. Carter, T. Fleisher, M. Clerici, G. Shearer, L. Chang, Y. Chiang, P. Tolstoshev, J. J. Greenblatt, S. A. Rosenberg, H. Klein, M. Berger, C. A. Mullen, W. J. Ramsey, L. Muul, R. A. Morgan and W. F. Anderson (1995) T lymphocyte-directed gene therapy for ADA-SCID: initial trial results after 4 years. *Science* **270**, 475–480.
2. Andrew Mountain (2000) Gene Therapy: the first decade. *TIBTECH* **18,** 119–128.
3. Thomas, C.E., Ehrhardt, A., and Kay, M.A. (2003) Progress and problems with the use of viral vectors for gene therapy. *Nat Rev Genet* **4**, 346–358.
4. Aiuti, A., Cattaneo, F., Galimberti, S., Benninghoff, U., Cassani, B., Callegaro, L., Scaramuzza, S., Andolfi, G., Mirolo, M., Brigida, I., Tabucchi, A., Carlucci, F., Eibl, M., Aker, M., Slavin, S., Al-Mousa, H., Al Ghonaium, A., Ferster, A., Duppenthaler, A., Notarangelo, L., Wintergerst, U., Buckley, RH., Bregni, M., Marktel, S., Valsecchi, MG., Rossi, P., Ciceri, F., Miniero, R., Bordignon, C., Roncarolo, M.G. (2009) Gene therapy for immunodeficiency due to adenosine deaminase deficiency. *N Engl J Med.* **360**, 447–58.
5. Roth, J. A. and R. J. Cristiano (1997) Gene therapy for cancer: what have we done and where are we going? *J Natl Cancer Inst* **89**, 21–39.
6. Cruz, P.E., Coroadinha, A.S., Rodrigues, T., and Hauser, H. (2008) Production of Retroviral Vectors: from the producer cell to the final product, In Gene and Cell Therapy: *Therapeutic Mechanisms and Strategies Taylor & Francis eds. (CRC Press)* pp. 17–32.
7. Merten, O.-W. (2004) State-of-the-art of the production of retroviral vectors. *J Gene Med* **6**, S105–124.
8. Coroadinha, A.S., Schucht, R., Gama-Norton, L., Wirth, D., Hauser, H., and Carrondo, M.J.T. (2006) The use of recombinase cassette exchange in retroviral vector producer cell lines: predictability and efficiency in transgene exchange. *J. Biotechnol.* **124**, 457–468.
9. Schucht, R., Coroadinha, A.S., Zanta-Boussif, M.A., Verhoeyen, E., Carrondo, M.J., Hauser, H., and Wirth, D. (2006) A new generation of retroviral producer cells: predictable and stable virus production by Flp-mediated site-specific integration of retroviral vectors. *Mol Ther* **14**, 285–292.
10. Rodrigues, T., Carrondo, M.J., Alves, P.M., Cruz, P.E. (2007) Purification of retroviral vectors for clinical application: biological implications and technological challenges. *J Biotechnol* **127**, 520–41.
11. Miller, A.D., Buttimore, C. (1986) Redesign of retrovirus packaging cell lines to avoid recombination leading to helper virus production. *Mol. Cell. Biol.* **6**, 2895–2902.
12. Miller, A.D., et al. (1991) Construction and properties of retrovirus packaging cells based on gibbon ape leukemia virus. *J. Virol.* **65**, 2220–2224.
13. Cosset, F.L., Takeuchi, Y., Battini, J.L., Weiss, R.A., Collins, M.K. (1995) High-titer packaging cells producing recombinant retroviruses resistant to human serum. *J Virol* **69**, 7430–7436.
14. Swift, S., Lorens, J., Achacoso, P., Nolan, G.P. (2001) Rapid production of retroviruses for efficient gene delivery to mammalian cells using 293T cell-based systems. *Curr Protoc Immunol.*, Chapter 10: Unit 10.17C.
15. Duisit, G., Salvetti, A., Moullier, P., Cosset, F.-L. (1999) Functional characterization of adenoviral/retroviral chimeric vectors and their use for efficient screening of retroviral producer cell lines. *Hum Gene Ther*, **10**, 189–200.
16. Pizzato, M., Merten, O.W., Blair, E.D., Takeuchi, Y. (2001) Development of a suspension packaging cell line for production of high titre, serum-resistant murine leukemia virus vectors. *Gene Ther.*, **8**, 737–45.
17. Carmo, M., Peixoto, C., Coroadinha, A.S., Alves, P.M., Cruz, P.E., Carrondo, M.J.T. (2004). Quantitation of MLV-based retroviral

vectors using real-time RT-PCR. *J Virol Methods* **119**, 115–119.

18. Rodrigues, T., Carvalho, A., Carmo, M., Carrondo, M.J.T., Alves, P.M., Cruz, P.E. (2007) Scalable purification process for gene therapy retroviral vectors. *J Gene Med* **9**, 233–243.
19. Transfiguracion, J., Coelho, H., Kamen, A. (2004) High-performance liquid chromatographic total particles quantification of retroviral vectors pseudotyped with vesicular stomatitis virus-G glycoprotein. *J Chromatogr B Analyt Technol Biomed Life Sci* **813**, 167–173.
20. Schambach, A., Maetzig, T. and BAUM, C. (2008) Retroviral Vectors for Gene and Cell Therapy, In Gene and Cell Therapy: *Therapeutic Mechanisms and Strategies Taylor & Francis eds. (CRC Press)* pp. 3–15.
21. Cruz, P.E., Silva, A.S., Roldão, A., Carmo, M., Carrondo, M.J.T., Alves. P.M. (2006) Screening of Novel Excipients for Improving the Stability of Retroviral and Adenoviral Vectors. *Biotech. Pro.* **22**, 568–576.
22. Carmo, M., Alves, A., Rodrigues, A.F., Coroadinha, A.S., Carrondo, M.J.T., Alves, P.M., Cruz, P.E. (2009) Stabilization of gammaretroviral and lentiviral vectors: from production to gene transfer. *J Gene Med* **11**, 670–678.

Chapter 8

Lentiviral Vectors

Marc Giry-Laterrière, Els Verhoeyen, and Patrick Salmon

Abstract

Lentiviral vectors have evolved over the last decade as powerful, reliable, and safe tools for stable gene transfer in a wide variety of mammalian cells. Contrary to other vectors derived from oncoretroviruses, they allow for stable gene delivery into most nondividing primary cells. In particular, lentivectors (LVs) derived from HIV-1 have gradually evolved to display many desirable features aimed at increasing both their safety and their versatility. This is why lentiviral vectors are becoming the most useful and promising tools for genetic engineering, to generate cells that can be used for research, diagnosis, and therapy.

This chapter describes protocols and guidelines, for production and titration of LVs, which can be implemented in a research laboratory setting, with an emphasis on standardization in order to improve transposability of results between laboratories. We also discuss latest designs in LV technology.

Key words: Lentivirus, Vector, Lentivector, Gene transfer, Gene therapy, Genetic engineering, Cell engineering, Cell therapy

1. Introduction

1.1. From Lentiviruses to Lentivectors

Retroviral vectors have three characteristics of a highly attractive gene delivery system. First, they integrate their genetic cargo into the chromosome of the target cell, a likely prerequisite for long-term expression. Second, they have a relatively large capacity, close to 10 kb, allowing for the delivery of most cDNAs. Finally, they do not transfer sequences that encode for proteins derived from the packaging virus, thus minimizing the risk that vector-transduced cells will be attacked by virus-specific cytotoxic T lymphocytes. Conventional retroviral vectors, however, are of limited usefulness for many applications because they are derived from oncoretroviruses such as the mouse leukemia virus (MLV), and, as a consequence, cannot transduce nondividing cells. In contrast

Otto-Wilhelm Merten and Mohamed Al-Rubeai (eds.), *Viral Vectors for Gene Therapy: Methods and Protocols*, Methods in Molecular Biology, vol. 737, DOI 10.1007/978-1-61779-095-9_8, © Springer Science+Business Media, LLC 2011

to oncoretroviruses, lentiviruses, such as the human immunodeficiency virus (HIV), are a subfamily of retroviruses that can infect both growth-arrested and dividing cells.

An infectious retroviral particle comprises an RNA genome that carries *cis*-acting sequences necessary for packaging, reverse transcription, nuclear translocation and integration, as well as structural proteins encoded by the gag and env genes, and the enzymatic products of the pol gene. The assembly of these components leads to the budding of the virion at the plasma membrane of the producer cell. In lentiviruses, the efficient expression of Gag and Pol requires a virally-encoded post-transcriptional activator called Rev.

The envelope protein (Env) mediates the entry of the vector particle into its target. HIV-1 Env specifically recognizes CD4, a molecule present on the surface of helper T cells, macrophages, and some glial cells. Fortunately, as with all retroviruses, the HIV-1 envelope protein can be substituted by the corresponding protein of another virus. This process, which alters the tropism of the virion, is called pseudotyping. The envelope of the amphotropic strain of MLV was used in some early experiments to pseudotype HIV-derived vectors (1). Its receptor, Pit-2, however, is only present at very low level on hematopoietic stem cells, an important target for gene therapy. Very often, the G protein of vesicular stomatitis virus (VSV-G) is used to pseudotype lentiviral as well as oncoretroviral vector particles, because it is highly stable, allowing for the concentration of the vector by ultracentrifugation, and because its phospholipid receptor is ubiquitously expressed in mammalian cells. Moreover, the association of the VSV-G glycoprotein with viral cores derived from lentiviruses results in vector pseudotypes that can integrate into non-proliferating target cells (2). More selective tropisms were achieved by taking advantage of the natural tropisms of glycoproteins (gps) from other membrane-enveloped viruses (see Table 1).

For instance, the use of surface glycoproteins derived from viruses that cause lung infection and infect via the airway epithelia, like Ebola virus or Influenza virus, may prove useful for gene therapy of the human airway (3). Exclusive transduction of retinal pigmented epithelium could be obtained following subretinal inoculations of some vector pseudotypes in rat eyes (4). Importantly, several viral gps target lentiviral vector to the central nervous system (CNS) such as rabies, mokola, lymphocytic choriomeningitis virus envelope (LCMV) or Ross River viral gps that permit even transduction of specific cell types in the CNS (Table 1). Some other envelope gps have been proven specifically efficient for LV transduction of hepatocytes or skin (Table 1). Likewise, screening of a large panel of pseudotyped vectors established the superiority of the Gibbon Ape Leukemia virus (GALV) and the cat endogenous retroviral glycoproteins (RD114) for

Table 1
Pseudotyping of lentiviral vectors with heterologous envelope glycoproteins relying on the natural tropism of these glycoproteins (after (22))

Glycoprotein	Virus of origin	Targeted cells – tissues	Reference
VSV-G	Vesicular stomatitis virus	Broad tropism (mouse and human cells)	(2)
MLV-10A1 gp	Murine leukemia virus – amphotropic strain	Broad tropism (mouse and human cells)	(5)
MLV-E gp	Murine leukemia virus – ecotropic strain	Broad tropism (mouse cells)	(6)
Rabies gp	Rabies virus	Neurons	(23–26)
Mokola gp	Mokola virus	Neurons Retinal pigment epithelium	(23) (25, 27, 28)
LCMV gp	Lymphocytic choriomeningitis virus	Glioma and neural stem cells	(27, 29, 30)
Ross River gp	Ross River virus	Glial cells	(31)
Ebola gp	Ebola virus	Airway epithelium Skin	(3, 32, 33) (34)
GP64	Baculovirus	Hepatocytes	(35)
HCV gp	Hepatitis C virus	Hepatocytes	(36)
F protein	Sendai virus	Hepatocytes	(37)
RD114 modified gp	Feline endogenous retrovirus	Hematopoietic cells	(7)
GALV modified gp	Gibbon ape leukemia virus	Hematopoietic cells	(5, 7, 38)
HA gp	Hemagglutinin – influenza A virus	Broad tropism – retinal epithelium	(4)
H and F measles gps	Measles virus H (hemagglutinin) and F (fusion protein)	Resting B cells and T cells	(8, 39)

transduction of progenitor and differentiated hematopoietic cells (5–7). Importantly, replacement of the cytoplasmic tail of RD114 and GALV gps with that of MLV-A glycoprotein resulted in strongly increased incorporation of these chimeric gps as well as high titers (5). Measles virus (MV) gps also require a modification of their cytoplasmic tails to allow efficient in corporation onto lentiviral vectors. Interestingly, lentivectors pseudotyped with such modified MV gps can transduce quiescent T and B cells more efficiently than VSV-G pseudotyped LVs (8). Although many different pseudotyped vectors have been generated as described above, pseudotyping with VSV-G gp provides lentiviral vectors with the highest titers and the most robust particles.

This technique is thus widely and routinely used in basic research as well as in clinical research. Therefore this chapter focuses on production of the VSV-G-pseudotyped vectors.

When producing vector stocks, it is mandatory to avoid the emergence of replication-competent recombinants (RCRs). In the retroviral genome, a single RNA molecule that also contains critical *cis*-acting elements carries all the coding sequences. Biosafety of a vector production system is therefore best achieved by distributing the sequences encoding its various components over as many independent units as possible, to maximize the number of recombination events that would be required to recreate a replication-competent virus. In the lentiviral vector systems described here, vector particles are generated from three or four separate plasmids (Fig. 1). This ensures that only replication-defective viruses are produced, because the plasmids would have to undergo multiple and complex recombination events to regenerate a replication-competent entity.

HIV is a human pathogen. However, its pathogenic potential stems from the presence of nine genes that all encode for important virulence factors. Fortunately, six of these genes (namely Env, Vif, Vpr, Vpu, Nef, and Tat, see Fig. 1) can be deleted from the HIV-derived vector system without altering its gene-transfer ability. The resulting multiply-attenuated design of HIV vectors ensures that the parental virus cannot be reconstituted.

Because lentiviruses can infect both dividing and nondividing cells, vectors were developed from this subgroup of retroviruses with the hope that they would be able to transduce cells that proliferate very little or not at all. The proof-of-principle of this concept was first provided with vectors derived from HIV-1, using the adult rat brain as an in vivo paradigm. Since then, gene delivery systems based on animal lentiviruses such as the simian and feline immunodeficiency viruses (SIV and FIV) and the equine infectious anemia virus (EIAV) have been described. This chapter presents exclusively the HIV1-based vector system

Fig. 1. Evolution in the design of HIV-1 based LV vectors. HIV-1-based LV vectors are derived from wild-type HIV-1 (**a**) by dissociation of the *trans*-acting components (*blue* boxes) coding for structural and accessory proteins (gag, pol, env, tat, rev, vif, vpr, vpu, nef) and the *cis*-acting sequences required for packaging and reverse transcription of the genomic RNA (LTR U3-R-U5, psi, RRE) (*yellow* boxes). (**b**) First generation system. The pHR vector genome has intact 5′LTR and 3′LTR. The R8.2 packaging plasmid expresses all HIV-1 proteins except Env. (**c**) Second generation system. The pSIN vector genome has a self-inactivating (SIN) deletion in the U3 sequence of the 3′LTR. The R8.91 packaging plasmid expresses only the structural and regulatory proteins of HIV-1. (**d**) Third generation system. The pCCL vector genome has a chimeric 5′LTR that is independent of the Tat protein. The packaging system is composed of 2 plasmids, pMDLg/pRRE coding of the structural proteins of HIV-1 and pRSV-Rev providing the Rev protein. Note that all vector systems need the presence of complementary plasmid providing the env gene. *CMV* human cytomegalovirus immediate-early promoter, *RRE* rev-responsive element, *RSV* Rous sarcoma promoter, *polyA* polyadenylation site, *U3-R-U5* HIV-1 LTR, *psi* HIV-1 packaging signal, *PRO* promoter of the internal expression cassette, *GOI* transgene of interest, Δ*U3* self-inactivating deletion of the U3 part of the HIV-1 LTR.

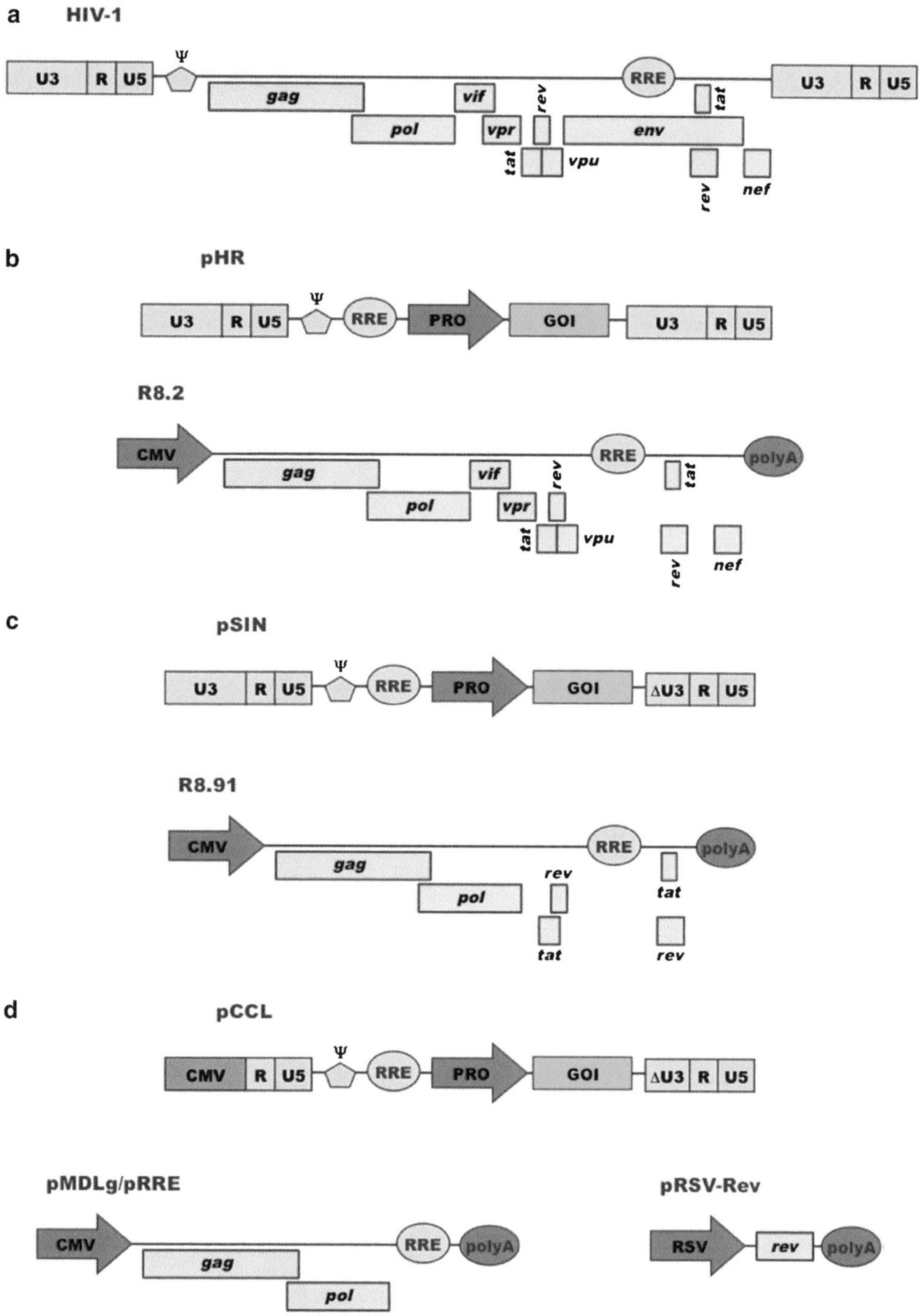
a
HIV-1
U3
R
U5
Ψ
gag
pol
vif
vpr
rev
tat
vpu
RRE
env
tat
rev
nef
U3
R
U5
b
pHR
U3
R
U5
Ψ
RRE
PRO
GOI
U3
R
U5
R8.2
CMV
gag
pol
vif
vpr
rev
tat
vpu
RRE
tat
rev
nef
polyA
c
pSIN
U3
R
U5
Ψ
RRE
PRO
GOI
ΔU3
R
U5
R8.91
CMV
gag
pol
rev
tat
RRE
tat
rev
polyA
d
pCCL
CMV
R
U5
Ψ
RRE
PRO
GOI
ΔU3
R
U5
pMDLg/pRRE
CMV
gag
pol
RRE
polyA
pRSV-Rev
RSV
rev
polyA

(VSV-G pseudotyped LV vectors) because it is presently the most advanced, and because, in its latest version, it offers a level of biosafety that matches, if not exceeds, that of the MLV-derived vectors currently used in the clinic.

1.2. Evolution and Design of Lentivectors

The potential of lentiviral vectors was first revealed in 1996 through the demonstration that they could transduce neurons in vivo (2). Since then, many improvements have been brought to achieve high levels of efficiency and biosafety. The principle, however, remains the same and consists in building replication-defective recombinant chimeric lentiviral particles from three different components, the genomic RNA, the internal structural and enzymatic proteins, and the envelope glycoprotein. The genomic RNA contains all the *cis*-acting sequences, whereas the packaging plasmids contain all the *trans*-acting proteins, necessary for adequate transcription, packaging, reverse transcription, and integration. A diagram of the evolution of HIV1-based systems is depicted in Fig. 1.

The first generation of lentiviral vectors was manufactured using a packaging system that comprised all HIV genes but the envelope (2). In a so-called second generation system, five of the nine HIV-1 genes were eliminated, leaving the gag and pol reading frames, which encode for the structural and enzymatic components of the virion, respectively, and the tat and rev genes, fulfilling transcriptional and post-transcriptional functions (9). Sensitive tests have so far failed to detect replication-competent-recombinants (RCRs) when this system is used. This good safety record, combined with its high efficiency and ease of use, explains why the second generation lentiviral vector packaging system is utilized for most experimental purposes. In a third generation system, geared up towards clinical applications, only gag, pol, and rev genes are still present, using a chimeric 5′ LTR (long terminal repeat) to ensure transcription in the absence of Tat.

The genetic information contained in the vector genome is the only one transferred to the target cells. Early genomic vectors were composed of the following components. The 5′ LTR, the major splice donor, the packaging signal (encompassing the 5′ part of the gag gene), the Rev-responsive element (RRE), the envelope splice acceptor, the internal expression cassette containing the transgene, and the 3′ LTR. In the latest generations, several improvements have been introduced. The Woodchuck Hepatitis Virus Posttranscriptional Regulatory Element (WPRE) has been added to increase the overall levels of transcripts both in producer and target cells, hence increasing titers and transgene expression (10). The central polypurine tract of HIV has also been added back in the central portion of the genome of the transgene RNA (11, 12). This increases titers at least in some targets. The U3 region of the 3′ LTR is essential for the replication of a wild-type retrovirus, since it contains the viral promoter in its RNA genome. It is dispensable for a replication-defective vector and has been deleted to remove all transcriptionally active sequences, creating the so-called self-inactivating (SIN) LTR (13). SIN vectors are thus unable to reconstitute their promoter and are safer than their counterparts with full-length LTRs. Finally, chimeric 5′ LTRs have been constructed, in order to render the LV promoter Tat-independent. This has been achieved by replacing the U3 region of the 5′ LTR with either the CMV enhancer (CCL LTR) or the corresponding Rous sarcoma virus (RSV) U3 sequence (RRL LTR) (14). Vectors containing such promoters can be produced at high titers in the absence of the Tat HIV transactivator. However, the Rev-dependence of these third generation LV has been maintained, in order to maximize the number of recombination events that would be necessary to generate an RCR. This latest generation represents the system of choice for future therapeutic projects. In the laboratory, however, this third generation is not mandatory, and the second generation system offers a high level of safety for P2 conditions. For most research applications, it is thus easier to use only three plasmids, i.e., an envelope plasmid, a second generation plasmid providing Gag, Pol, Tat, and Rev proteins, and any vector genome plasmid (second generation with native 5′ LTR or third generation with chimeric 5′ LTR) since the presence of Tat is required for optimal activity of the native LTR and does not affect the activity of the chimeric LTRs. Thus, for in vitro and vivo research, we advise to use an all-purpose packaging plasmid, such as the psPAX2, which encodes for the HIV-1 Gag, Gag/Pol, Tat, and Rev proteins.

The vector plasmid represented in Fig. 2 provides several desirable features. It contains a gene switch, the TET promoter/rTTA system (15), under the control of the highly and ubiquitously active ubiquitin promoter (16). Transduced cells can also be live-sorted using GFP. The gene of interest can be easily cloned using the Gateway® system, and is expressed in a drug-controlled fashion. When the gene product is toxic, one can thus control its expression in target cells, and also prevent its expression in producer cells, hence avoiding titer drop due to the death of LV-producing cells. The LoxP sequence is duplicated during reverse transcription, and allows the proviral cassette to be excised upon Cre expression (17). Note that, although lentiviral vectors can theoretically accommodate up to 9 kb of transgenic sequence, some inserts can induce a rapid and important titer drop. This is the case, for example, for the powerful chimeric CAG promoter (CMV enhancer/beta-actin promoter, beta-globin intron) (18) in our hands. Also, the UBI promoter (ubiquitin gene promoter) (16) can be replaced by other ubiquitously active promoters,

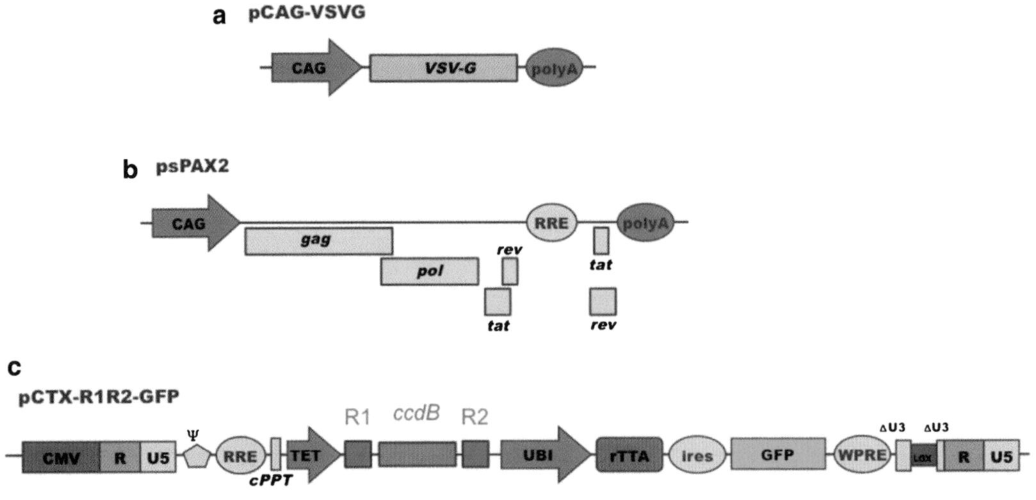

Fig. 2. Example of plasmids used for HIV-1 based LV production. (**a**) The pCAG-VSVG plasmid (courtesy of A. Nienhuis, (6)), providing the envelope of the LV particles is composed of the CAG chimeric promoter, the coding sequence of the Vesicular Stomatitis Virus Envelope protein (VSV-G), and the polyadenylation signal from the rabbit beta-globin gene. (**b**) The second-generation psPAX2 packaging plasmid (P. Salmon, unpublished), providing the structural and enzymatic proteins of the LV particle is composed of the CAG chimeric promoter, the gag, pol, tat and rev genes, the Rev-responsive element of HIV-1 (RRE) and the polyadenylation signal from the rabbit beta-globin gene. (**c**) The third generation pCTX-R1R2-GFP vector plasmid, providing the genome of the LV particles is depicted here as an example of the latest development in LV design. The 5′LTR is composed of the CMV promoter, and the R and U5 regions of HIV-1. This renders it tat-independent. *psi* HIV-1 packaging signal, *RRE* rev-responsive element, *cPPT* central polypurine tract, *R1-ccdB-R2* att-flanked cassette for Gateway® cloning of genes of interest, *UBI* ubiquitin promoter, *rTTA* reverse TET-transactivator, *ires* EMCV internal ribosome entry site, *GFP* green fluorescent protein, *WPRE* Woodchuck Hepatitis Virus Posttranscriptional Regulatory Element, Δ*U3* self-inactivating deletion of the U3 part of the HIV-1 LTR, *lox* Cre recombinase LoxP target sequence.

such as EF1 (19) or EFs (20), or tissue-specific promoters. In that latter case, the gene-switch will be active only in a specific cell type.

Detailed informations (maps, sequences, etc.) as well as other LV backbones are available at our institutional website: http://medweb2.unige.ch/salmon/lentilab/.

1.3. Safety Issues

The system presented here contains numerous safeguards as compared to the first-generation HIV vectors, in which genes encoding all HIV-1 proteins, except for Env, were present. A second generation was characterized by the exclusion of four accessory genes (vif, vpr, vpu, and nef). These deletions improved considerably the safety of the vector because they excluded major determinants of HIV-1 virulence. In the third-generation system, described in this unit, Gag, Pol, and Rev are the only HIV-1 proteins still present. Vectors with self-inactivating (SIN) LTR and produced with the third generation packaging system have been tested for RCR. Thus far, no RCR have been detected amongst a total of 1.4×10^{10} transducing units (21).

In general, transduced cells must always be fixed (using formaldehyde or paraformaldehyde as described below) before being taken out of the P2 laboratory. If a live sorting is needed outside of the P2 laboratory, a careful handling and decontamination of the equipment used must be performed afterward.

Given the very broad tropism of VSV-G-pseudotyped lentiviral vectors both in vitro and in vivo, biosafety precautions need to take into account the nature of the transgene. A P2 laboratory, P2 standard equipment, and P2 safety procedures are required. In dependence of the country and the local legislation, procedures using lentiviral vectors must be reviewed and approved by the local biosafety committee of the institution where they are conducted or need authorization from the competent authority. Extra precautions must be taken when working with transgenes that are themselves potential biohazards. For instance, working in a P3 laboratory is recommended for the lentivector-mediated transfer of genes involved in cell proliferation.

2. Materials

All solutions and equipment coming into contact with living cells must be sterile, and aseptic techniques should be used accordingly. All maps and sequences of plasmids described here are available at http://medweb2.unige.ch/salmon/lentilab/. Common plasmids for the generation of HIV1-based lentivectors can be obtained from http://www.Addgene.org. Use ultrapure or double-distilled water in all recipes.

2.1. Production of HIV-1 Based Lentiviral Vectors by Transient Transfection of 293T Cells

1. Producer cells: 293T/17 cells (from ATCC Cat# CRL-11268).
2. D10 medium: Dulbecco's modified Eagle medium (with 4.5 g/l glucose, glutamine, and pyruvate, Invitrogen Cat# 41966052 or equivalent) supplemented with antibiotics and 10% FBS.
3. Serum-free medium: Advanced DMEM (Invitrogen Cat# 12491015) supplemented with 2 mM glutamine.
4. TE buffer: 10 mM Tris–HCl – 1 mM EDTA, pH 8.0. Used to redissolve all plasmids.
5. Envelope plasmid: pCAG-VSVG dissolved at 1 μg/μl in TE buffer.
6. Packaging plasmid: psPAX2 (encoding HIV-1 Gag, Pol, Tat, and Rev proteins) dissolved at 1 μg/μl in TE buffer.
7. Vector plasmid: pFUGW dissolved at 1 μg/μl in TE buffer.

8. 0.5 M $CaCl_2$: Dissolve 36.75 g of $CaCl_2 \times 2H_2O$ (SigmaUltra Cat# C5080) into 500 ml of H_2O. Filter sterilize through a 0.22-µm nitrocellulose filter. Store at −70°C in 50 ml aliquots. Once thawed, the $CaCl_2$ solution can be kept at +4°C for several weeks without observing significant change in the transfection efficiency.
9. 2×HeBS (HEPES-buffered saline): Dissolve 16.36 g of NaCl SigmaUltra Cat# S7653 (0.28 M final), 11.9 g of HEPES SigmaUltra Cat# H7523 (0.05 M final), and 0.213 g of Na_2HPO_4, anhydrous SigmaUltra Cat# S7907 (1.5 mM final) into 800 ml of H_2O. Adjust pH to 7.00 with 10 M NaOH. Be careful, obtaining a proper pH is very important. Below 6.95, the precipitate will not form, above 7.05, the precipitate will be coarse and transfection efficiency will be low. Add H_2O to 1000 ml, and make the final pH adjustment. Filter sterilize through a 0.22-µm nitrocellulose filter. Store at −70°C in 50 ml aliquots. Once thawed, the HeBS solution can be kept at +4°C for several weeks without observing significant change in the transfection efficiency.
10. 75% Ethanol in a spray bottle.
11. PBS, pH 7.4.
12. PBS-Ca^{2+},Mg^{2+} at pH 7.4.
13. 20% Sucrose: Dissolve 20 g of Sucrose SigmaUltra in 100 ml of PBS-$Ca^{2+}Mg^{2+}$. Filter sterilize through a 0.22-µm nitrocellulose filter. Store at +4°C.
14. 0.25% Trypsin/EDTA.
15. 13–14% Bleach solution (w/v).
16. 10-cm tissue culture dishes.
17. 37°C humidified incubators, 5% CO_2.
18. 1.5-ml microcentrifuge tubes, sterile, disposable.
19. 15- and 50-ml conical centrifuge tubes, sterile.
20. 50 ml syringes and 0.45-µm pore size PVDF filters.
21. 30-ml Beckman Konical tubes (Cat# 358126, Beckman-Coulter) for ultracentrifuge.
22. Ultracentrifuge (such as Beckman Optima™ L-90K) with SW 28 rotor.

2.2. Titration by FACS

1. Target cells: HT-1080 cells (Cat# CCL-121, ATCC).
2. D10 medium: same as above.
3. Trypsin/EDTA: same as above.
4. MW6 tissue culture plates (Cat# 353224, BD Biosciences).

5. PBS: same as above.
6. 1% Formaldehyde (w/v) in PBS: Mix 1 ml of 37% formaldehyde (w/v) in 36 ml of PBS. Store at +4°C.
7. Fluorescence-activated cell sorter (FACS; Becton Dickinson with 488 nm excitation laser and green filter) and appropriate tubes.

2.3. Titration by qPCR

1. Target cells: HT-1080 cells (same as above).
2. D10 medium: same as above.
3. Trypsin/EDTA: same as above.
4. MW6 tissue culture plates (same as above).
5. PBS: same as above.
6. Real-time PCR machine (ABI PRISM® 7900HT Real Time PCR System, Applied Biosystems or equivalent, with a dedicated analysis program, SDS2.2.2, Applied Biosystems or equivalent).
7. Genomic DNA extraction kit (DNeasy Blood & Tissue Kit, Qiagen GmbH, Germany).
8. 2× Reaction buffer (Cat# RT-QP2X-03 Eurogentec, Belgium).
9. 96-well Optical Reaction plate (Cat# 4306737, Applied Biosystems).
10. Optical caps (Cat# N801-0935, Applied Biosystems).
11. Filter tips (1000, 100, and 10 μl).
12. Primers and probe for quantification of HIV sequences (10× GAG set, see Subheading 2.5 below).
13. Primers and probe for quantification of human genomic sequences (10× HB2 set, see Subheading 2.5 below).

2.4. RCR Assay

1. Target cells: HT-1080 cells (same as above).
2. Full HIV-1 genome-containing cells: 8E5 cells (Cat# CRL-8993, ATCC).
3. D10 medium: same as above.
4. Trypsin/EDTA: same as above.
5. MW6 tissue culture plates (same as above).
6. PBS: same as above.
7. Real-time PCR machine (same as above).
8. Genomic DNA extraction kit (same as above).
9. 2× Reaction buffer (same as above).
10. 96-well Optical Reaction plate (same as above).
11. Optical caps (same as above).

12. Filter tips (1000, 100, and 10 μl).
13. Primers and probe for quantification of HIV sequences (10× GAG set, see Subheading 2.5 below).
14. Primers and probe for quantification of human genomic sequences (10× HB2 set, see Subheading 2.5 below).
15. Primers and probe for quantification of HIV packaging sequences (10× PRO set, see Subheading 2.5 below).

2.5. Oligos

2.5.1. Human Beta-Actin Taqman® Probe and Primers

These oligos are used to normalize for the amount of genomic DNA and are specific for the human beta-actin gene.

1. HB2-P: (probe, sense) 5′-(FAM)-CCTGGCCTCGCTGTC CACCTTCCA-(TAMRA)-3′.
2. HB2-F: (forward primer)5′-TCCGTGTGGATCGGCGGCT CCA-3′.
3. HB2-R: (reverse primer)5′-CTGCTTGCTGATCCACAT CTG-3′.

2.5.2. GAG Taqman® Probe and Primers

These oligos are used for amplification of HIV-1 derived vector sequences and are specific for the 5′ end of the gag gene (GAG). This sequence is present in all HIV-1 vectors for it is part of the extended packaging signal.

1. GAG-P: (probe, antisense) 5′-(FAM)-ACAGCCTTCTGAT GTTTCTAACAGGCCAGG-(TAMRA)-3′.
2. GAG-F: (forward primer)5′-GGAGCTAGAACGATTCGCA GTTA-3′.
3. GAG-R: (reverse primer)5′-GGTTGTAGCTGTCCCAGTA TTTGTC-3′.

2.5.3. PRO Taqman® Probe and Primers

These oligos are used for amplification of sequences present in RCRs are specific for the region of the pol gene coding for the HIV-1 protease (PRO).

1. PRO-P: (probe, sense) 5′-(FAM)-ACAATGGCAGCAATTT CACCAGT-(TAMRA)-3′.
2. PRO-F: (forward primer) 5′-AGCAGGAAGATGGCCAGT AA-3′.
3. PRO-R: (reverse primer) 5′-AACAGGCGGCCTTAACT GTA-3′.

Oligos can be ordered on-line from several companies such as Eurogentec, Invitrogen, or Sigma. FAM fluorescent dye can be replaced by other equivalent molecule, and TAMRA can be replaced by other quenchers.

3. Methods

3.1. Production of LV Stocks

1. Maintain 293T cells in D10 medium, in 10-cm tissue culture dish in a 37°C humidified incubator with a 5% CO_2 atmosphere, and split them at ratio 1:10 using Trypsin/EDTA, three times per week (e.g., every Monday, Wednesday, and Friday). Frequent passages and keeping the 293T as individual cells will ensure high transfection efficiency.
2. The day before the transfection, seed 1–10 dishes at 1.5 to 2.5 million cells per dish (10 cm). Cells must be approximately 1/2 to 2/3 confluent on the day of transfection. Incubate overnight in a 37°C humidified incubator with a 5% CO_2 atmosphere. On the following day, co-transfect the cells according to the following recipes.
3. For one plate of 10 cm, mix in a sterile 1.5-ml microcentrifuge tube.

Envelope plasmid	pCAG-VSVG	4 μg
Packaging plasmid	psPAX2	8 μg
Vector plasmid	pFUGW	8 μg

4. The vector plasmid (pFUGW given as example above) can be second or third generation since the psPAX2 plasmid provides Tat protein.
5. Adjust to 250 μl with sterile buffered water and mix well by pipetting
6. Add 500 μl of 2× HeBS and mix well by pipetting
7. Put 250 μl of 0.5 M $CaCl_2$ in a 15-ml sterile conical tube
8. To each 15-ml tube containing the $CaCl_2$ solution, slowly transfer, dropwise, the 750 μl of DNA/HeBS mixture, while vigorously vortexing. Vigorous vortexing will ensure the formation of a fine precipitate that can be taken up efficiently by cells.
9. Leave the precipitates (1 ml final volume per tube) at room temperature for 5–30 min.
10. Add the 1-ml precipitate dropwise to the cells in 10 ml of medium in one culture dish prepared as above. Mix by gentle swirling until the medium has recovered a uniformly red color.
11. Place the dish overnight in a 37°C humidified incubator with a 5% CO_2 atmosphere.
12. Early the next morning, aspirate the medium, wash with 10 ml of prewarmed PBS, and gently add 15 ml of fresh

Advanced DMEM, prewarmed to 37°C. Incubate for 24 h. If 293T cells adhere poorly, washing with PBS can be omitted.

13. Transfer the supernatant from each plate to one 50-ml centrifuge tube. Close the tubes, and spray them with 70% ethanol before taking them out of the hood. Store the supernatant at +4°C. Add another 15 ml of fresh Advanced DMEM, prewarmed to 37°C. Incubate for another 24 h with the cell monolayer.
14. Pool the supernatants of day 1 and 2 and centrifuge for 5 min at 500 g, at 4°C, to pellet detached cells and debris.
15. Filter the 30 ml of pooled supernatant (total harvest from 2 days: 30 ml/dish) with a 50 ml syringe connected to a 0.45 μm PVDF disk filter.

The LV stocks can be stored at +4°C for 1–4 days without significant titer loss, before they are used for transduction of target cells or further processing such as concentration. For longer storage, LV stocks must be kept at −80°C.

The transfection can be started late in the afternoon and the medium changed early the next morning. If you notice cell toxicity, you can transfect early in the morning and change the medium late in the afternoon the same day. The transfection procedure can be scaled up to ten culture dishes of 10 cm, or other cell culture systems with equivalent or larger surface.

3.2. Concentration of LV Stocks

1. For concentration, use 30-ml Beckman conical tubes (Cat# 358126, Beckman-Coulter), in a SW 28 rotor in an ultracentrifuge. Put 4 ml of 20% sucrose on the bottom of the tube. Very slowly pour the supernatant on the surface of the sucrose cushion until the tube is full (allow a 3–5 mm dry zone to the top of the tube). Spin at 50,000 g for 120 min at +16°C.
2. Aspirate the medium with a sterile pipette down to the sucrose interface.
3. Aspirate the sucrose until you have 1–2 ml of colorless sucrose solution and then invert the tube while aspirating the remaining sucrose. Never touch the bottom of the tube where the vector pellet is.
4. Place the conical tube in a 50-ml Falcon tube and quickly add 30–100 μl of PBS-$Ca^{2+}Mg^{2+}$ on the pellet (not always visible). Do not leave the pellet dry for more than 5 min or it may result in significant titer decrease. Close the Falcon tube. You can resuspend the vector pellet of one tube in a minimal volume of 30 μl. In this case, you will achieve a ~1000-fold concentration.
5. Vortex at half-speed for 2 s.

6. Leave the vector pellet to resuspend for 1–2 h at room temperature or 2–4 h at +4°C.
7. Vortex at half-speed for 2 s.
8. Pipet up and down 20 times and freeze at −80°C in aliquots for long-time storage (see Notes 1–8).

3.3. Titration of LV Stocks

Titers of viruses in general and lentivectors in particular, critically depend on the methods and cells used for titration. The quantification of vector particles capable of achieving every step from cell binding to expression of the transgene depends on both vector and cell characteristics. First, the cell used as target must be readily permissive to all steps from viral entry to integration of the vector genetic cargo. Second, the expression of the foreign gene must be easily monitored and rapidly reach levels sufficient for reliable quantification. Early vectors had the lacZ bacterial gene as reporter, under the control of the CMV promoter. Current vectors now have the green fluorescent protein (GFP) gene as a reporter, under the control of promoters that are active in most primary cells.

Measured titers can also vary with the conditions used for titration, i.e., volume of sample during vector-cell incubation, time of vector-cell incubation, number of cells used, etc. For several years now, numerous laboratories have been using HeLa cells as target cells for LVs. Although these cells are easy to grow and 100% susceptible to transduction by VSV-G-pseudotyped LVs, they are very unstable in terms of morphology and karyotype. For this reason, we are now using HT-1080 cells, which are stable, of human origin and give titers identical to HeLa cells.

Physical titration based on the quantification of HIV-1 capsid p24 antigen is not used anymore in our lab. Instead, our current standard procedure relies on determination of infectious titers by transduction of HT-1080 target cells. Also, we always produce a test batch of a standard GFP lentivector alongside all LV productions. This test batch is used to monitor the overall efficiency of the procedure and detects any anomaly in producer cells or reagents that will result in titer drop.

Here we described a procedure that is used on a weekly basis in our lab for several years, and that has been standardized in order to compare titers from one batch to another one or from one lab to another one. Changes in this procedure can be made, but one must keep in mind that, for example, reducing the cell culture surface or increasing the number of target cells will result in an increase of the final calculated titer, from the exact same vector batch.

3.3.1. General Procedure

1. On day 0, seed HT-1080 cells at 50,000 cells per well in MW6 plate in D10. Make sure that HT-1080 cells are well separated and uniformly distributed in the well.

2. On day 1, put into three independent wells 500, 50, or 5 µl of the vector suspension (either pure from unconcentrated supernatants or diluted in complete medium if it comes from a concentrated stock, i.e., 1/100 if the vector is concentrated 100-fold).
3. Polybrene can be omitted for transduction with VSV-G pseudotyped vectors since this compound does not influence permissivity of cells to VSV-G pseudotyped vectors.
4. On day 2, remove the supernatant and replace by 2 ml of fresh D10.
5. On day 5, wash the cells with 2 ml of PBS; detach them with 250 µl of Trypsin/EDTA for 1 min at 37°C.
6. Add 250 µl of D10 and mix well to resuspend the cells. This step inactivates the trypsin and EDTA.
7. Spin cells in a microcentrifuge for 2 min at 200 g. Note that if you need to run a FACS analysis and a qPCR analysis on the same sample, you must split your cells in two separate microcentrifuge tubes.

3.3.2. Titration of Lentivectors by FACS

This method can only be used to titer stocks of vectors that carry a transgene that is easily monitored by FACS (such as GFP, or any living colors, or any membrane protein that can be detected by flow cytometry), and whose expression is governed by a promoter that is active in HT-1080 cells (tissue-specific promoter-containing vector must be functionally assayed in specific cells, and titered by QPCR in HT-1080 cells (see below). We describe here the titration of an Ubiquitin promoter-GFP vector (pFUGW, see above).

1. Add 500 µl of 1% formaldehyde in PBS to the cell pellet obtained at step 7 above. This step will fix the cells and inactivate the vector particles. Samples can thus be taken out of the P2 laboratory.
2. Resuspend the cells thoroughly in the well and transfer them to a FACS tube.
3. Analyze the cells in a flow cytometer. If you are not familiar with flow cytometry, you must seek help from your institutional FACS specialist.
4. Once chosen the appropriate dilution (as described in Fig. 3 and Notes 9–11), apply the following formula: Titer (HT-1080-TU/ml) = 100,000 (target HT-1080 cells) × (% of GFP-positive cells/100)/volume of supernatant (in ml).

3.3.3. Titration of Lentivectors by Quantitative PCR

When lentivectors contain DNAs coding for genes other than GFP or LacZ, or promoters which are active only in specific primary cells and tissue, FACS titration cannot be used. Therefore,

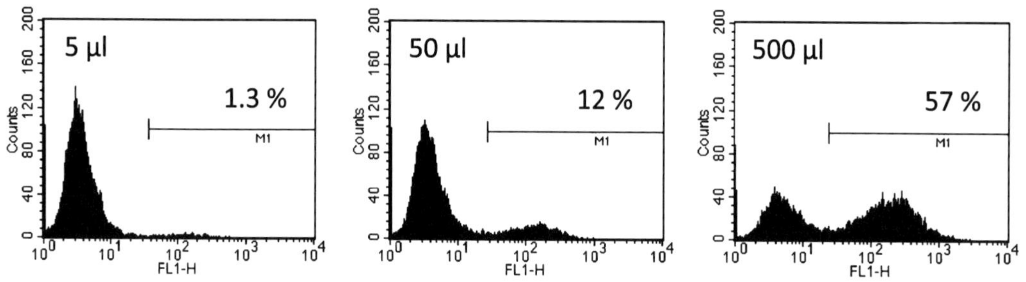

Fig. 3. A representative FACS analysis of HT-1080 cells used for titration of GFP-coding LV. HT-1080 cells (10^5) were incubated with increasing volumes of a supernatant containing a LV expressing GFP under the control of the human Ubiquitin promoter (pFUGW) as described above. After 5 days, cells were detached, fixed and analyzed by FACS for GFP fluorescence (*x* axis, 4-decade log scale, FL1) versus number of cells (*y* axis, linear scale). The percentage of GFP-expressing cells was measured by placing a marker discriminating between GFP-negative (mean of fluorescence intensity 3-4) and GFP-positive cells (mean of fluorescence intensity 200).

most new LVs will need an alternative method to measure the number of copies of LV stably integrated in HT-1080 target cells, after transduction as described above for GFP vectors. This assay, however, only measures the number of LV copies integrated in the target cell genome. The overall functionality of the vector must be tested at least once in cells in which the promoter is active and/or with appropriate techniques to detect the expression of the transgene product. The QPCR assay proceeds as follows, using a real-time PCR machine. HT-1080 cells are transduced as for FACS analysis. Then, one half can be used if FACS analysis is performed in parallel, or target cells can be lysed directly in the plate and the DNA is extracted using a genomic DNA extraction kit (such as Qiagen DNeasy). Then, a fraction of the total DNA is analyzed for copy number of HIV sequences using the following real-time PCR protocol.

1. Extract target cell DNA from each individual well of a MW6 plate (see general titration procedure above) using the genomic DNA extraction kit, following manufacturer's recommendations. For the DNA elution step, use 100 µl of AE buffer (component of the DNeasy tissue kit) instead of 200 µl.
2. Perform qPCR or store DNA at –20°C until use.
3. Prepare a mix containing everything but the sample DNA for the number of wells needed for the QPCR analysis, including all samples and standards in duplicates or triplicates, according to the following recipe (for one well):

2× Reaction buffer	7.5 µl
10× Oligo mix (GAG or HB2, see below)	1.5 µl
DNA sample	1 µl
H_2O	5 µl

4. Distribute 14 μl of this mix into the wells of a 96-well Optical Reaction plate.
5. Add sample DNAs.
6. Close with optical caps.

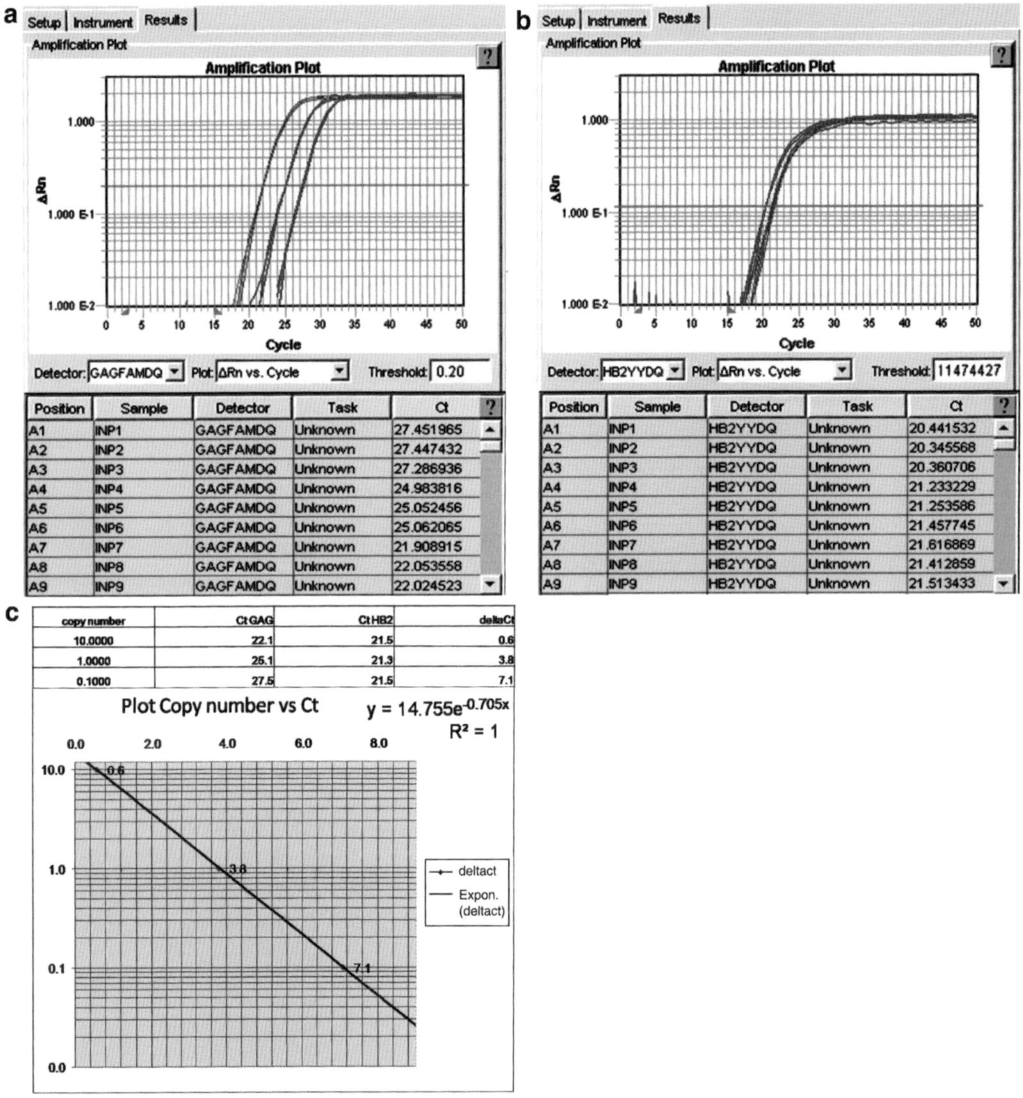

Fig. 4. A representative qPCR analysis used for titration of HIV-1 based LVs. DNA from HT-1080 cells transduced with serial 10-fold dilutions of pFUGW vectors was subjected simultaneously to qPCR titration analysis and FACS analysis as described above. A sample of each dilution was submitted to qPCR amplification and monitoring using an ABI PRISM® 7900HT Real Time PCR System (Applied Biosystems), and sets of primers and probes specific for HIV gag sequences (GAG-FAM, panel **A**) or beta-actin sequences (HB2-FAM, panel **B**). Amplification plots were displayed and cycle threshold values (Ct) were set as described in text. Values of GAG Ct and HB2 Ct were exported in an Excel worksheet to calculate ΔCt values (*x* axis, linear scale) and plot them against copy number values (*y* axis, log scale) (panel **C**). A sample giving 10% of GFP-positive cells was set as cells containing 0.1 copy of HIV sequences per cell. The regression curve can then be used to calculate GAG copy numbers (*Y* value) of unknown samples by applying the formula to ΔCt values (*X* values) of the sample (see Note 18).

7. Centrifuge the plate at 200 g for 1 min to bring all liquid on the bottom of the wells.
8. Place the 96-well Optical Reaction plate in the real-time PCR machine and run the appropriate program depending on the fluorochromes and quenchers used in your Taqman probes (see Notes 12–22).
9. Analyze results and calculate titer using the SDS2.2.2 program (Applied Biosystems). An example of amplification profiles of HIV sequences in human DNA is given in Fig. 4.
10. Ask SDS2.2.2 to analyze the amplification reactions.
11. Set the threshold values (Ct) where the amplification curve is the steepest, both for the gene of interest (GAG-FAM, panel A) and for the internal control (HB2-FAM, panel B). These Ct values are the number of cycles required for the amplification curve to cut the absorbance threshold values.
12. Export the results as a Microsoft Excel sheet.
13. Using standards of cells containing 10, 1, and 0.1 copy of LV per cell (see Note 16), ask excel to calculate the ΔCt values (Ct GAG minus Ct HB2).
14. Ask Excel to display an exponential formula giving the copy number as a function of ΔCt.
15. Apply the formula to unknown samples, to calculate their corresponding copy number of HIV sequences.
16. Calculate the titers by applying the following formula: Titer (HT-1080-TU/ml) = 100,000 (target HT-1080 cells) × number of copy per cell of the sample/volume of supernatant (in ml).

3.4. Quantitative PCR Assay of Replication-Competent-Recombinants

The absence of Replication-Competent Recombinants (RCRs) is essential to downgrade the biohazard level of cells that have been transduced by retroviral vectors, including LVs. We propose here a test based on the detection (or absence of detection) in the chromosomal DNA of transduced cells, of HIV sequences that are absent in the vector plasmid (vector genome), but are present in the packaging plasmid and are essential for HIV (or RCR) replication. The target sequence chosen in our assay is located in the sequence coding for the viral protease that is present in the packaging plasmid, essential for virus replication and absent in the vector genome. Although the assay described here is performed on a small number of cells, at least 3 weeks after initial transduction, it can be scaled up to meet requirements for the detection of RCR in preclinical vector batches. Other RCR tests have been described in the literature. One earlier paper describes a true RCR assay, which failed to detect any RCR in vector batches produced from third generation packaging systems (21). Several other tests have been described, but they detect biological entities that need trans complementation to replicate. Although these assays can measure the level of recombination during the production of lentivectors, they are not suitable to detect genuine RCR that may represent a biological hazard due to potential dissemination within primary human cells.

1. At least 3 weeks prior to assay, transduce HT-1080 cells with lentiviral vector (LV) of interest and with standards (see below). This extended growth period allows for dilution of packaging DNA carried over from vector production steps (see Note 23).
2. After ≥3 weeks of cell growth, extract DNA from the transduced cells using a DNeasy kit according to the manufacturer's instructions. Store DNA at −20°C until use. The number of cells and final volume should be such that 1 μl of the final DNA solution corresponds to 10^4 cells.
3. For each sample or standard, prepare three independent mixes containing everything but the sample DNA for the number of wells needed for the qPCR reaction, including all samples and standards in duplicates, according to the following recipe (for one well):

2× Reaction buffer	7.5 μl
10× Oligo mix (GAG, PRO or HB2, see below)	1.5 μl
DNA sample	1 μl
H_2O	5 μl

4. Distribute 14 μl of this mix into the wells of a 96-well Optical Reaction plate.
5. Add sample DNAs.
6. Close with optical caps.
7. Centrifuge the plate at 200 g for 1 min to bring all liquid on the bottom of the wells.
8. Place the MW96 in the real-time PCR machine and run the appropriate program depending on the fluorochromes and quenchers used in your Taqman probes (see Notes 12–14).
9. Analyze as described in the qPCR titration section. In this case, however, two types of standards are used. One standard corresponds to cells containing vector sequences only (LV standard, target for GAG oligo set), and one corresponds to cells containing all HIV sequences (HIV standard, target for GAG and PRO oligo sets). The first is provided by cells transduced with LV as described above. The second is provided by cells having one copy of full-length HIV genome, such as

8E5 cells (see Note 24 and ATCC website for details about these cells). In the case of 8E5, the DNA will contain 1 copy of HIV per genome. Serial tenfold dilutions of 8E5 DNA into human DNA (up to 10^{-3} copy per genome) can be performed to provide a HIV DNA standard curve. A negative control both for LV sequences and HIV sequences will be provided by HT-1080 cells.

10. Results are expressed as Ct values for each oligo set, i.e., GAG-HB2ΔCt and PRO-HB2ΔCt. The sample DNA will be considered negative for PRO sequences and hence negative for RCR if its PRO-HB2ΔCt value is similar to the PRO-HB2ΔCt value of HT-1080 cells, with a GAG-HB2ΔCt value above the range corresponding to 1 copy of LV sequence per genome.

3.5. Plasmid Preparation

Plasmids containing retroviral long terminal repeats (LTRs) are prone to undergo deletion in some *Escherichia coli* strains. The Top10 or HB101 strains are strongly recommended for propagating the plasmids used in this section. We also recommend CcdB Survival 2 T1R strain (Invitrogen) for Gateway® clonings. We recommend JetStar Kits (GENOMED GmbH, Germany) to prepare DNAs for transfection. The last step of the DNA prep should be an additional precipitation with ethanol and resuspension in TE. Do not treat DNA with phenol/chloroform as it may result in chemical alterations. Also to avoid salt co-precipitation, do not precipitate DNA below +20°C.

3.6. Troubleshooting Lentivector Production

Transfection efficiency is the most critical parameter affecting vector titer. 293T/17 cells are highly transfectable using a variety of protocols. When establishing vector production procedures, it is highly recommended that the transfection protocol be optimized using a plasmid encoding GFP. Transfection efficiency should not be assessed solely on the basis of the percentage of GFP-positive cells, but also on the mean fluorescence intensity, which reflects the number of plasmid copies taken up by the cells. This makes FACS analysis of the transfected cells mandatory. FACS can be done as soon as 15 h after the transfection, allowing many variables to be tested rapidly. The factors most likely to impact on the transfection efficiency are the pH of the 2×HeBS solution, the quality of the batch of fetal bovine serum used, the cell density, the total amount of DNA per plate, and the quality of DNA. A coarse precipitate will give poor transfection whereas a fine precipitate (barely visible after application on cells) will give good transfection. As a rule of thumb, the precipitate will be coarser as the pH of 2×HeBS increases, the DNA quantity decreases, the temperature or the incubation time for precipitate formation increases.

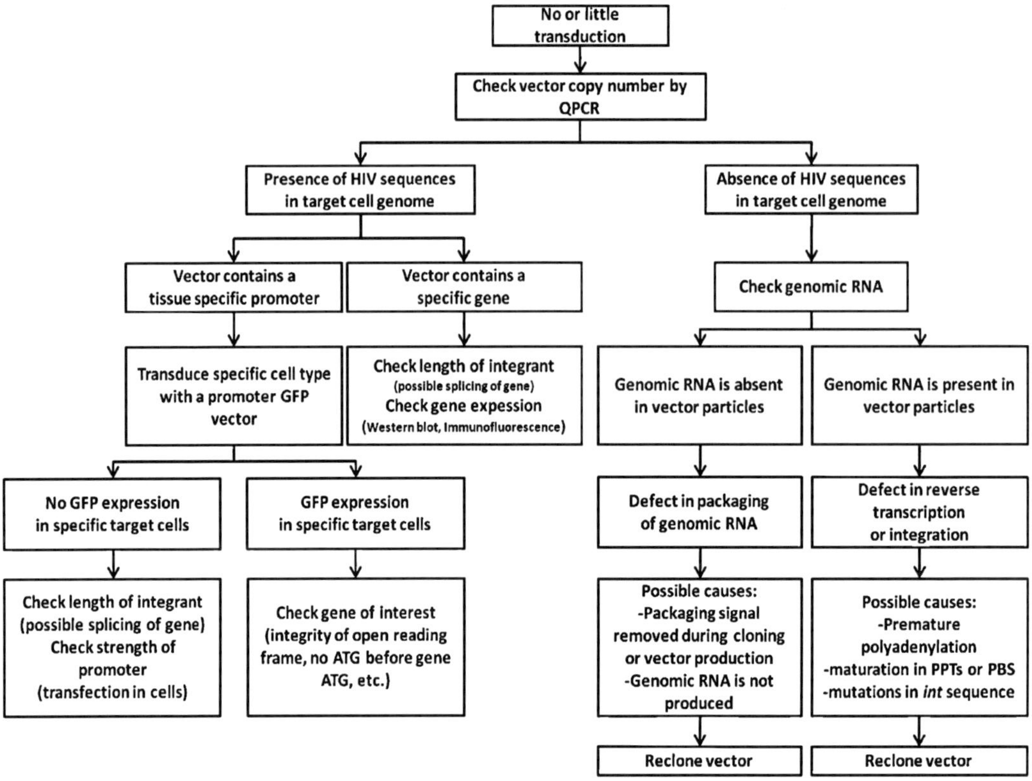

Fig. 5. Troubleshooting diagram for lentiviral vector production and transduction.

In the case of lack of transduction of a specific cell type with a specific lentiviral vector, a synoptic diagram is provided in Fig. 5 to help addressing most of the problems that could account for it.

3.7. Anticipated Results

When applied optimally, the procedure described here yields crude unconcentrated vector titers between 1×10^6 and 1×10^7 TU/ml. After centrifugation, a yield of at least 50% is expected. A similar 50% yield is also expected after one freeze/thaw cycle. The cells produce equally during the 48 h post transfection. You maximize the total yield by harvesting twice.

Note that there is no current procedure for purification per se of infectious particles. The only methods available (ion exchange, centrifugation, etc.) will only concentrate the vector particles and/or wash soluble material. One must keep in mind that all other particulates generated by the producer cells, such as defective vector particles and exosomes, are also coated with VSV-G proteins and will co-sediment or copurify with infectious vector particles. This implies that there is no current way to enrich in infectious particles a vector stock displaying a poor infectivity index. Defective particles will be enriched alongside causing an increase in cell toxicity.

4. Notes

1. P2 practices require that open tubes always be handled in the laminar flow hood. Tubes can be taken out of the laminar flow only when they are closed, and sprayed with 75% ethanol.
2. All solid waste and plasticware must be discarded in a trash bin in the laminar flow hood and all liquids must be aspirated into a liquid waste bottle containing fresh concentrated bleach. Refill the liquid waste bottle with fresh bleach when the color of the liquid is no longer yellow.
3. When full, bags are closed inside the laminar flow hood, then autoclaved.
4. When full, and at least 15 min after neutralization with fresh bleach, the liquid waste bottle can be emptied into a regular sink.
5. In case of a major spill of vector-containing liquid, absorb liquid with paper towels and neutralize with fresh concentrated bleach prior to disposal.
6. In case there is a leak in the SW 28 buckets, remove the tubes in the hood, fill the buckets with 75% ethanol, and invert them several times. Leave under the hood for ≥20 min. Discard the 75% ethanol and remove the conical adapters under the hood. Spray the adapters with 75% ethanol and leave them under the hood for >20 min.
7. When resuspending the pellets, try to avoid bubbles since it will result in decrease of final volume and hence decrease of yield.
8. Try to avoid repeated freeze-thaw cycles of stored vectors. This may result in drop of titer, although the VSV-G pseudotyped particles are more resistant to this procedure than particles pseudotyped with retrovirus-derived envelopes.
9. A reliable measure of the fraction of GFP+ cells relies on the level of GFP expression. In the example shown in Fig. 3, GFP-positive and GFP-negative cells can be readily discriminated when GFP is expressed from a human Ubiquitin promoter, and allowed to accumulate in cells for 4–5 days.

A marker can then be set to measure the fraction of transduced versus total cells.

10. Cells fixed with formaldehyde can be stored in the dark at +4°C for several hours. A final 0.5% formaldehyde concentration is enough to fix the cells and inactivate the vectors. Increasing formaldehyde concentration (up to 4% final) will increase the autofluorescence of cells and decrease GFP fluorescence.
11. In a typical titration experiment, only dilutions yielding to 1–20% of GFP-positive should be considered for titer calculations. Below 1%, the FACS may not be accurate enough to reliably determine the number of GFP-positive cells. Above 20%, the chance for each GFP-positive target cell to be transduced twice significantly increases, resulting in underestimation of the number of transducing particles.
12. The precise settings of a qPCR protocol depend on the real-time PCR machine used. This aspect is beyond the scope of this protocol. If you are not familiar with qPCR techniques, you should seek advice from your local qPCR expert or from the technical assistance of your real-time PCR machine.
13. Standard concentrations in 10× oligo sets are 1 μM of probe and 3 μM of each primer in water.
14. Stocks of probes and primers usually come lyophilized and are stored at 100 μM in water.
15. DNA typically comes from 2×10^6 HT-1080 cells (one confluent well of a MW6 plate), extracted and resuspended in 100 μl of Buffer AE (DNeasy Tissue Kit).
16. Standards of HT-1080 cells containing 10, 1, and 0.1 copy of LV per cell can be prepared from HT-1080 cells transduced with a GFP vector, using serial tenfold dilutions. The 0.1 copy per cell standard will be provided by the sample displaying 10% of GFP-positive cells.
17. It is advisory to run a dual titration (FACS plus qPCR) using one GFP vector alongside the other vectors, for each experiment of qPCR titration. This will help comparing the FACS titration with the qPCR titration.
18. A prototypic excel worksheet for calculation of qPCR titers can be downloaded from the following link: http://medweb2.unige.ch/salmon/lentilab/QPCRtitration.html
19. Using standard DNA extraction procedures in a laboratory context where HIV sequences are often handled, you can expect a level of background contamination with HIV sequences corresponding to cells containing 1 copy per 1000 or 100 genomes. In this case, consider higher copy numbers for calibration.

20. Vector stocks failing to give higher than 0.01 copy per genome in a qPCR assay, using the highest titration dose must have experienced one or several problems during their design, packaging, and/or production. You must then refer to the Subheading 3.6 to solve this issue.
21. Using careful DNA extraction procedures and standardization as described above, you can expect a reproducibility within a twofold range. Ask your local qPCR expert if you need a more stringent quantification qPCR procedure.
22. Always use pipet tips containing aerosol-barrier filters when preparing solutions, mixes, samples, and plates for qPCR, to prevent cross-contamination.
23. Cells being analyzed for the absence of RCR must be confined cells in a culture flask with vented cap until result of RCR analysis. If the result is negative, the biohazard level of the cells can be downgraded; after spraying the flask with 75% ethanol, it can be transferred outside of the culture laboratory.
24. ATCC recommends that 8E5 cells be handled in a P3 laboratory. Indeed, although they contain a full copy of noninfectious HIV, they can form syncytia with uninfected CD4+ cells.

Acknowledgements

We thank Ophélie Cherpin and David Suter for their help in the construction and design of Gateway® lentivectors.

References

1. Russell, S. J., and Cosset, F. L. (1999) Modifying the host range properties of retroviral vectors, **J Gene Med 1**, 300–311.
2. Naldini, L. et al. (1996) In vivo gene delivery and stable transduction of nondividing cells by a lentiviral vector, **Science 272**, 263–267.
3. Kobinger, G. P., Weiner, D. J., Yu, Q. C., and Wilson, J. M. (2001) Filovirus-pseudotyped lentiviral vector can efficiently and stably transduce airway epithelia in vivo, **Nat Biotechnol 19**, 225–230.
4. Duisit, G. et al. (2002) Five recombinant simian immunodeficiency virus pseudotypes lead to exclusive transduction of retinal pigmented epithelium in rat, **Mol Ther 6**, 446–454.
5. Stitz, J. et al. (2000) Lentiviral vectors pseudotyped with envelope glycoproteins derived from gibbon ape leukemia virus and murine leukemia virus 10A1, **Virology 273**, 16–20.
6. Hanawa, H. et al. (2002) Comparison of various envelope proteins for their ability to pseudotype lentiviral vectors and transduce primitive hematopoietic cells from human blood, **Mol Ther 5**, 242–251.
7. Sandrin, V. et al. (2002) Lentiviral vectors pseudotyped with a modified RD114 envelope glycoprotein show increased stability in sera and augmented transduction of primary lymphocytes and CD34+ cells derived from human and nonhuman primates, **Blood 100**, 823–832.
8. Frecha, C. et al. (2008) Stable transduction of quiescent T cells without induction of cycle progression by a novel lentiviral vector pseudotyped with measles virus glycoproteins, **Blood 112**, 4843–4852.
9. Zufferey, R., Nagy, D., Mandel, R. J., Naldini, L., and Trono, D. (1997) Multiply attenuated lentiviral vector achieves efficient gene delivery in vivo, **Nat Biotechnol 15**, 871–875.

10. Zufferey, R., Donello, J. E., Trono, D., and Hope, T. J. (1999) Woodchuck hepatitis virus posttranscriptional regulatory element enhances expression of transgenes delivered by retroviral vectors, **J Virol 73**, 2886–2892.
11. Follenzi, A., Ailles, L. E., Bakovic, S., Geuna, M., and Naldini, L. (2000) Gene transfer by lentiviral vectors is limited by nuclear translocation and rescued by HIV-1 pol sequences, **Nat Genet 25**, 217–222.
12. Zennou, V., Petit, C., Guetard, D., Nerhbass, U., Montagnier, L., and Charneau, P. (2000) HIV-1 genome nuclear import is mediated by a central DNA flap, **Cell 101**, 173–185.
13. Zufferey, R. et al. (1998) Self-inactivating lentivirus vector for safe and efficient In vivo gene delivery, **J Virol 72**, 9873–9880.
14. Dull, T. et al. (1998) A third-generation lentivirus vector with a conditional packaging system, **J Virol 72**, 8463–8471.
15. Gossen, M., Freundlieb, S., Bender, G., Muller, G., Hillen, W., and Bujard, H. (1995) Transcriptional activation by tetracyclines in mammalian cells, **Science 268**, 1766–1769.
16. Lois, C., Hong, E. J., Pease, S., Brown, E. J., and Baltimore, D. (2002) Germline transmission and tissue-specific expression of transgenes delivered by lentiviral vectors, **Science 295**, 868–872.
17. Salmon, P., Oberholzer, J., Occhiodoro, T., Morel, P., Lou, J., and Trono, D. (2000) Reversible immortalization of human primary cells by lentivector- mediated transfer of specific genes, **Mol Ther 2**, 404–414.
18. Niwa, H., Yamamura, K., and Miyazaki, J. (1991) Efficient selection for high-expression transfectants with a novel eukaryotic vector, **Gene 108**, 193–199.
19. Mizushima, S., and Nagata, S. (1990) pEF-BOS, a powerful mammalian expression vector, **Nucleic Acids Res 18**, 5322.
20. Kostic, C. et al. (2003) Activity analysis of housekeeping promoters using self-inactivating lentiviral vector delivery into the mouse retina, **Gene Ther 10**, 818–821.
21. Escarpe, P. et al. (2003) Development of a sensitive assay for detection of replication-competent recombinant lentivirus in large-scale HIV-based vector preparations, **Mol Ther 8**, 332–341.
22. Frecha, C., Szecsi, J., Cosset, F. L., and Verhoeyen, E. (2008) Strategies for targeting lentiviral vectors, **Curr Gene Ther 8**, 449–460.
23. Mochizuki, H., Schwartz, J. P., Tanaka, K., Brady, R. O., and Reiser, J. (1998) High-titer human immunodeficiency virus type 1-based vector systems for gene delivery into nondividing cells, **J Virol 72**, 8873–8883.
24. Mazarakis, N. D. et al. (2001) Rabies virus glycoprotein pseudotyping of lentiviral vectors enables retrograde axonal transport and access to the nervous system after peripheral delivery, **Hum Mol Genet 10**, 2109–2121.
25. Wong, L. F. et al. (2004) Transduction patterns of pseudotyped lentiviral vectors in the nervous system, **Mol Ther 9**, 101–111.
26. Azzouz, M. et al. (2004) Lentivector-mediated SMN replacement in a mouse model of spinal muscular atrophy, **J Clin Invest 114**, 1726–1731.
27. Watson, D. J., Kobinger, G. P., Passini, M. A., Wilson, J. M., and Wolfe, J. H. (2002) Targeted transduction patterns in the mouse brain by lentivirus vectors pseudotyped with VSV, Ebola, Mokola, LCMV, or MuLV envelope proteins, **Mol Ther 5**, 528–537.
28. Bemelmans, A. P. et al. (2005) Retinal cell type expression specificity of HIV-1-derived gene transfer vectors upon subretinal injection in the adult rat: influence of pseudotyping and promoter, **J Gene Med 7**, 1367–1374.
29. Miletic, H. et al. (2004) Selective transduction of malignant glioma by lentiviral vectors pseudotyped with lymphocytic choriomeningitis virus glycoproteins, **Hum Gene Ther 15**, 1091–1100.
30. Stein, C. S., Martins, I., and Davidson, B. L. (2005) The lymphocytic choriomeningitis virus envelope glycoprotein targets lentiviral gene transfer vector to neural progenitors in the murine brain, **Mol Ther 11**, 382–389.
31. Kang, Y. et al. (2002) In vivo gene transfer using a nonprimate lentiviral vector pseudotyped with Ross River Virus glycoproteins, **J Virol 76**, 9378–9388.
32. Medina, M. F. et al. (2003) Lentiviral vectors pseudotyped with minimal filovirus envelopes increased gene transfer in murine lung, **Mol Ther 8**, 777–789.
33. Silvertown, J. D., Walia, J. S., Summerlee, A. J., and Medin, J. A. (2006) Functional expression of mouse relaxin and mouse relaxin-3 in the lung from an Ebola virus glycoprotein-pseudotyped lentivirus via tracheal delivery, **Endocrinology 147**, 3797–3808.
34. Hachiya, A. et al. (2007) Gene transfer in human skin with different pseudotyped HIV-based vectors, **Gene Ther 14**, 648–656.
35. Kang, Y. et al. (2005) Persistent expression of factor VIII in vivo following nonprimate lentiviral gene transfer, **Blood 106**, 1552–1558.

36. Bartosch, B., Dubuisson, J., and Cosset, F. L. (2003) Infectious hepatitis C virus pseudo-particles containing functional E1-E2 envelope protein complexes, **J Exp Med 197**, 633–642.

37. Kowolik, C. M., and Yee, J. K. (2002) Preferential transduction of human hepatocytes with lentiviral vectors pseudotyped by Sendai virus F protein, **Mol Ther 5**, 762–769.

38. Christodoulopoulos, I., and Cannon, P. M. (2001) Sequences in the cytoplasmic tail of the gibbon ape leukemia virus envelope protein that prevent its incorporation into lentivirus vectors, **J Virol 75**, 4129–4138.

39. Frecha, C. et al. (2009) Efficient and stable transduction of resting B lymphocytes and primary chronic lymphocyte leukemia cells using measles virus gp displaying lentiviral vectors, **Blood 114**, 3173–3180.

Chapter 9

Adeno-Associated Viruses

Mauro Mezzina and Otto-Wilhelm Merten

Abstract

Adeno-associated virus (AAV) vectors have evolved over the past decade as a particularly useful gene vector for in vivo applications. In contrast to oncoretro- and lentiviral vectors, this vector stays essentially episomal after gene transfer, making it safer because of the absence of insertional mutagenesis. AAV's non-pathogenicity is a further advantage. For decades, this vector could only be produced at a small scale for research purposes and, eventually, used at very small doses for clinical studies, because only transfection methods were available, which have limited scalability. However, since the development of scalable production methods, this bottleneck is resolved and, from a technical point of view, large quantities of AAV vectors can be produced, opening the possibility of using AAV vectors for whole body treatments in gene therapy trials. This chapter presents the basic principles of small- and large-scale production procedures as well as detailed procedure of small-scale production, purification, and analytical protocols for AAV vectors. In Chapter 10, the reader will find a large-scale production method based on the use of the insect cell/baculovirus system.

Key words: rAAV vectors, Triple-transfection method, PEI, Purification, Ultracentrifugation, Dot blot, ELISA, Gene transfer, Gene therapy

1. Introduction

1.1. Biology of AAV Vectors

AAVs are small non-enveloped single-stranded (ss) DNA viruses with a diameter of 20–25 nm. They belong to the Parvoviridae family and are classified in the *Dependovirus* genus. Eleven strains have been isolated and characterized from humans and primates, and new serotypes are continuously discovered (1–5). Phylogenetic and functional analyses have revealed that primate AAVs are segregated into six clades whose members are closely related phylogenetically and share functional and serological similarities (4, 6). All serotypes share similar structure, genome size, and organization, i.e. the structures and locations of the open reading frames (*orfs*), promoters, introns, and poly-adenylation site. At the

Otto-Wilhelm Merten and Mohamed Al-Rubeai (eds.), *Viral Vectors for Gene Therapy: Methods and Protocols*, Methods in Molecular Biology, vol. 737, DOI 10.1007/978-1-61779-095-9_9, © Springer Science+Business Media, LLC 2011

biological level, they are all dependent on the presence of a helper virus for their replication and gene expression. The most divergent serotype is AAV5 with notable differences of the inverted terminal repeat (ITR) size (167 nucleotides for AAV5 compared to 143–146 for AAV1 to 4 and AAV6) and function (7).

AAVs are frequently found in human populations, 70–80% of individuals having been exposed to an infectious event (8, 9). No known adverse clinical consequences are associated with AAV infection or latency in humans.

The viral particle is composed of an icosahedral capsid and a single-stranded (ss) DNA molecule of the viral genome of either polarity (3) has a density of 1.41 g/cm^3. AAVs are very resistant to extreme conditions of pH, detergent and temperature, making them easy to manipulate. Finally, wild-type (wt) AAV has the unique property of integrating into the human genome at a specific site (the S1 site located in the long arm of chromosome 19), which has been extensively described for AAV2 (10–13). However, this property is not maintained in recombinant vectors because of viral genome manipulation which removes the Rep proteins in AAV vectors.

As the first serotype used to generate vectors, AAV type 2 is so far the best characterized prototype and the majority of gene transfer studies have been based on the use of this serotype (13–15). However, for the last few years, a whole set of recombinant (r)AAVs have been developed from alternative serotypes which present very similar structure but different tropism and immunological properties. The physical characteristics of the different serotypes are close enough to allow handling them under very similar conditions to AAV2 for production and purification. Only those purification systems that are based on the capsid structure (e.g. isoelectric point/range, specific ligands, ...), which may differ from a serotype to another one, has to be adapted for each serotype. Many groups are actively evaluating the in vivo performances of these serotypes in various animal and disease models for a recent and extensive review on the serotypes [see Gao et al. (6), Grimm and Kay (16), and Grimm et al. (17)].

The 4.7 kb genome of AAV2 (Fig. 1) contains two *orfs* which encode four regulatory proteins, the Rep proteins and the three structural Cap proteins, (18, 19). The compacted genome of the AAV is framed by two ITRs, which are base-paired hairpin structures of 145 nucleotides in length. The ITRs contain the only necessary regulatory *cis*-acting sequences required by the virus to complete its life cycle, namely the origin of replication of the genome, the packaging and the integration signals. The two major Rep proteins, Rep78 and Rep68, are involved in viral genome excision, rescue, replication and integration (20) and also regulate gene expression from AAV and heterologous promoters (21, 22). The minor Rep proteins, Rep52 and Rep40, are involved in replicated ssDNA genome accumulation and packaging (23). The cap *orf* is initiated at the p40 promoter and encodes the three structural capsid proteins VP1,

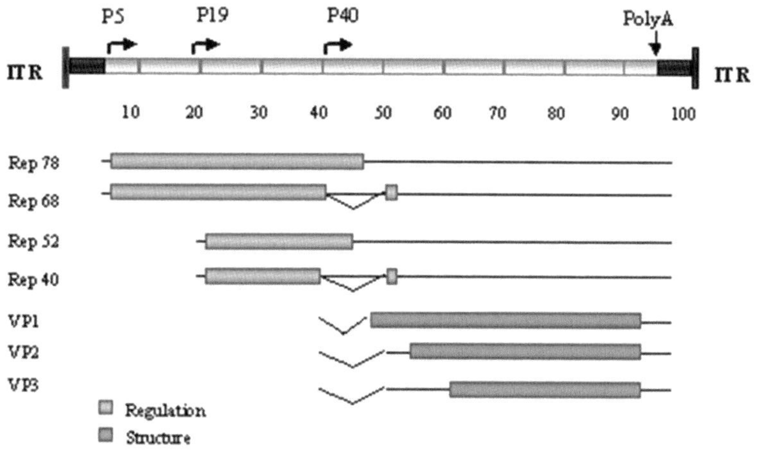

Fig. 1. ***AAV-2 genome organization.*** General organization of the genome and genetic elements of AAV type 2 (a scale of 100 map units is used, 1 map unit being equivalent to approximately 47 nucleotides). The general organization of the other serotypes is similar; T-shaped boxes indicate the ITRs. The ***horizontal arrows*** indicate the three transcriptional promoters. The ***solid lines*** indicate the transcripts; the introns are shown by the ***broken lines.*** A polyadenylation signal at position 96 is common to all transcripts. The first orf (corresponding to the promoter 5 sequences) encodes the four regulatory proteins arising from the promoters p5 and p19 and the alternative splicing. The second orf (promoter p40) encodes the three capsid proteins from two transcripts. VP-1 is initiated from the first cap transcript, and VP-2 and VP-3 are initiated at two different codon sites from the second cap transcript. Note that the initiation site of VP2 (start codon) is an ACG.

VP2, and VP3. Their stoichiometry in the assembled particle is 1:1:10. Finally, all transcripts share the same polyadenylation signal and equal amounts of virions are found containing strands of plus or minus polarity. Details on the encapsidation of AAV genomes were presented in a recent review by Timpe et al. (24).

AAVs are naturally replication defective viruses, making them dependent on the presence of an auxiliary virus in order to achieve their productive cycle. This feature is reflected in the name of the *Dependovirus* genus. Several viruses can ensure the auxiliary functions: adenovirus, herpes simplex virus, and vaccinia virus (25, 26) as well as cytomegalovirus (27) and human papillomavirus (28 and 29 as comprehensive description of the state of the art). The infection scheme is fairly simple. Once AAV has entered the cell and has been conveyed to the nucleus, its ss genome is converted into a replicative double-stranded (ds) form, required for gene expression. In the absence of an auxiliary virus, the AAV genome integrates into the host genome and latently persists in a proviral form. In the presence of a helper virus (by either concomitant or super infection), the AAV genome can undergo the process of active replication, during which capsid proteins are synthesized and DNA packaged (30). Therefore, since AAV does not possess a lytic capability by itself in the natural infection process, the liberation of AAV virions usually relies on the lytic effect of the helper virus.

1.2. AAV Vector Design and Vector Production

The design of AAV-based vectors is straightforward, the ITRs being retained and the exogenous sequences to be transferred being cloned in between. Therefore the Rep and Cap functions have to be supplied in *trans*. Similarly, the helper functions from the auxiliary virus have to be provided. Figure 2 depicts the general principle of rAAV design and production.

Several production systems exist, with their own advantages and drawbacks. In laboratory scale, most of the current methods of producing viral particles are still based on transient transfection of human embryonic kidney (HEK) 293 cells with minor variations.

The traditional transfection methods employ HEK293, A549, or HeLa cells that have to be co-transfected with two plasmids containing the rAAV vector (pAAV) and the *rep* and *cap* genes (pHelper), followed by an infection with helper virus (in most cases, adenovirus [wtAd]) to induce the replication of rAAV. The major drawback is the co-production of rAAV and helper virus. Thus, the helper virus (wtAd) was replaced by a helper plasmid bringing in the adenoviral functions necessary for the replication of the rAAV (this helper plasmid [e.g. pXX6] provides the following functions: E2A, E4, and VA; the other essential

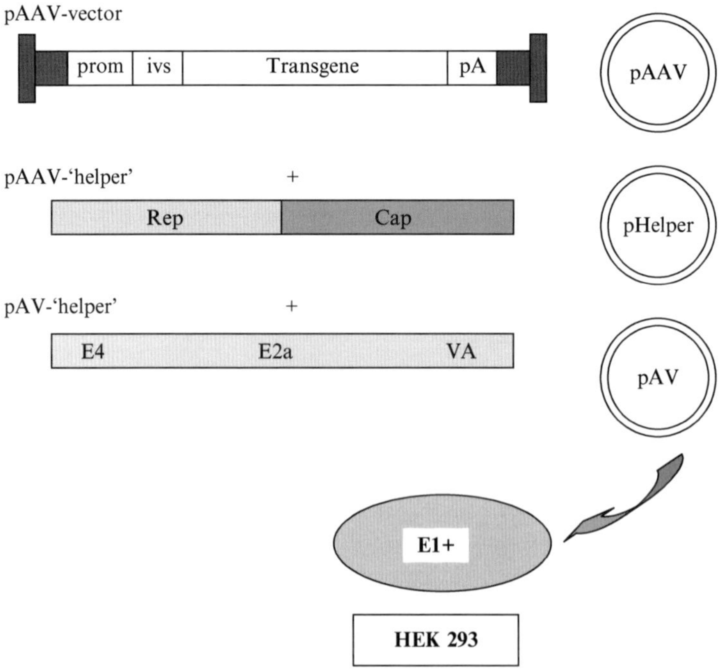

Fig. 2. *Classical transfection production method by using HEK293 cells.* In a transient production system, the pAAV-"helper," (carrying *rep* and *cap* functions), the pAV-"helper" pXX6 mini-plasmid (carrying the AdV helper functions E4, E2a and VA), and the pAAV-vector plasmid carrying the ITRs, the gene of interest, the regulatory sequences (prom), intervening sequence (ivs) and the polyadenylation site (pA) will be brought to the producer cells (HEK293) by transfection. This method leads to the generation of about 10^3 particles per cell.

adenoviral functions E1A and E1B are provided by the HEK 293 cells). When using HeLa or A549 cells, the helper plasmid must provide all essential functions, including adenoviral E1A and E1B. This production method is generally based on the co-transfection of HEK293 cells with three plasmids that contain the rAAV vector, the *rep* and *cap* genes, and the adenovirus helper genes (31, 32), as indicated in Fig. 2.

With the discovery and development of new AAV serotypes, the idea emerged rapidly to develop vectors from these serotypes (2, 3, 33–35). In the vast majority of the studies, a pseudotyping strategy has been adopted for simplicity. Basically, all recombinant genomes are based on AAV2 ITRs and rep2 function, and only the capsid is of the serotype of interest. The production strategies for pseudotyped particles are exactly the same as for the classical AAV2.

This transient production system is largely used for research and developmental purposes due to its high flexibility (the easy change of the plasmids; thus, the transgene as well as the AAV serotype can be adapted rapidly to specific needs). Due to its interests for R&D purposes, this production protocol will be presented in detail in what follows.

However, the system's main drawbacks are its limited scalability – although roller bottle (36) and CellCube-based processes (37) have been established – and the relatively high incidence of recombination events between the plasmids used, leading to rep + rAAVs and rcAAV production.

Thus, to improve the scalability, many attempts have been undertaken to develop producer cells. The use of producer cell lines has mainly focused on HeLa cells, although some investigators have evaluated the use of HEK293 and A549 cells. These cells contain the rAAV vector and the *rep* and *cap* genes of AAV. To induce the production of rAAV, the cells have to be infected with the helper virus. Thus, generation of stable lines as cell factories for rAAVs is better suited to large-scale production than transient transfection. Nevertheless, generation of such cells is hard, tedious, and time consuming. Furthermore, the highest virus titers for these production methods are typically about 10^7 IP/ml (infectious particles/ml). Since an estimated number of 10^{12} to 10^{14} rAAV particles is required for clinical human use (38), 10^2 to 10^4 liters of medium with this vector concentration should be necessary. As no really satisfying cell line is available to date, this approach will not be further developed in these protocols, and the reader is referred to the specific literature for more details (for review, see 39).

To overcome these limitations, recent studies have focused on producing rAAV vectors in insect cell cultures, using the recombinant baculovirus system (40) derived from the *Autographa Californica* nuclear polyhedrosis virus (*Ac*NPV). Production of rAAV particles is achieved by coinfecting Sf9 cells with three

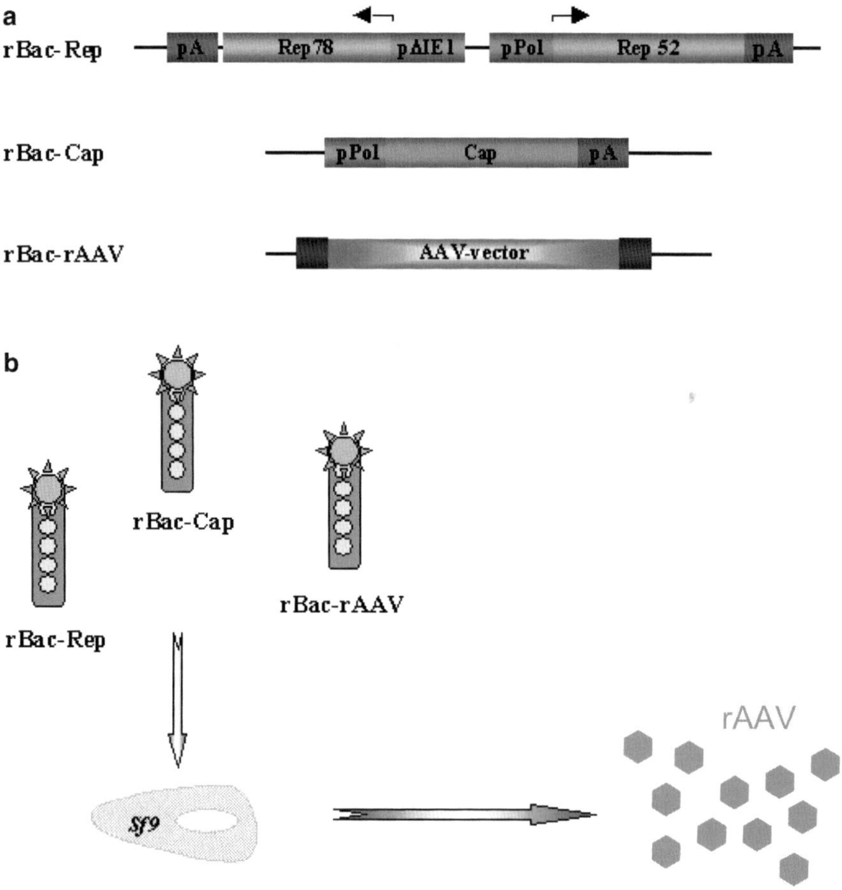

Fig. 3. ***The baculovirus/insect cell production system for rAAV.*** (**a**) Genetic constructions of the recombinant baculoviruses. The 2-split *rep orf* are driven by two insect promoters, the polyhedrine promoter (pPol) of AcMNPV and a truncated form of the immediate-early 1 gene promoter (pΔEI1) of *Orgyia pseudotsugata* nuclear polyhedrosis virus. The difference in the promoter strength allows high expression of the small Rep and a reduced expression of the toxic large Rep. rBac-Cap expresses capsid proteins. The three proteins are directly translated from one transcript. rBac-GFP carries an rAAV-GFP genome. The presence of a cytomegalovirus (CMV) and p10 promoter allows green fluorescent protein (GFP) expression in both mammalian and insect cells. (**b**) AAV production is done by a triple infection of Sf9 cells. The three recombinant baculoviruses are used at a ratio of 1:1:1 with an MOI of 5 per recombinant baculovirus. Three days postinfection rAAVs are harvested from cells and supernatant and are purified.

baculovirus vectors, BacRep, BacCap, and Bac-rAAV (Fig. 3) which encode the respective components of the rAAV production machinery. This system lends itself to large-scale production under serum-free conditions, as Sf9 cells grow in suspension. Infection of Sf9 cells at a multiplicity of infection (MOI) of 5 and a ratio of 1:1:1 for all three baculoviruses has yielded a total of 2.2×10^{12} IP in a 3 L bioreactor.

This system has been optimized to improve the stability of the baculovirus constructs (41–43) and to simplify the system by reducing the number of baculoviruses to two by joining the rep and cap functions on only one recombinant baculovirus (43) or by developing stable Sf9-based packaging cell lines containing stably integrated *rep* and *cap* functions thus needing only the infection with one recombinant baculovirus with the vector construct (44).

The process was scaled up to 125 L (45) and 50 L (O.-W. Merten, presented at the Clinigene session on industrial vector production at the ESGCT Meeting in Hannover/D in 2009) without loss of productivity and demonstrated that quantities sufficient to meet clinical demand can be achieved. Further optimization of this system has been performed to improve productivity at high cell densities (46); however, this production system provides the possibility of easily producing large quantities of rAAVs with the advantages of a rather elevated flexibility to change the transgene (using another Bac-rAAV construct) or to modify the AAV serotype (using of another BacCap construct). The establishment of a new baculovirus construct (including cloning, selection, and amplification included) takes about 6 weeks. These advantages make the Sf9/baculovirus system the most interesting and promising production system for rAAV and protocols are presented in the Chapter 10.

Table 1 compares the different production systems. Further details on rAAV production issues can be found in a review by Merten et al. (39).

The protocols for the purification of rAAV have evolved over the last decade. The protocols that are most classical and easy to

Table 1
Production yields of rAAV using different production systems

Production method	Yield (vg/cell)	Scale-up	References
293, triple *transfection*	10–10^3	Small scale	(32, 57, 61–63)
HeLa-based *producer cell*, rAAV production induced by infection with wt Ad5	10^4 to 10^6	Reactor scale possible	(52, 64)
A549-based *producer cell*, rAAV production induced by infection with a Ad ts	10^5	Reactor scale: 15 L, larger scale possible	(65, 66)
Baculovirus system: Sf9 infected with 3 different rec. baculoviruses	10^4 to 10^5	Reactor scale: 50 L, larger scale possible	Merten, Clinigene session at the ESGCT Meeting in Hannover/D in 2009
Sf9 infected with 2 different rec. baculoviruses		Reactor scale: 125 L, larger scale possible	(45)

acquire for a non-specialized laboratory remain ultracentrifugation on a CsCl (47) or iodixanol gradient (48). These purification methods are limited by the capacity of cell lysate volume that can be processed and by the low purity achieved. However, they are still widely used and most of the time they are sufficient for fundamental studies and early in vivo pre-clinical evaluation of the vectors: therefore, the CsCl ultracentrifugation protocol is also presented here (Subheadings 2.2 and 3.2). More complex techniques based on ion-exchange column chromatography and membrane techniques are now well developed for the generation of high purity grade and up-scaled production suitable for clinical applications (49–54). In addition, the recently developed immunoaffinity chromatography (AVB Sepharose High Performance, GE Healthcare) allows the highly efficient purification of different AAV serotypes (1–3, 5, 6, 8, 43).

The most important endpoint of all vector production protocols is the vector titer and the total vector quantity produced. The dot blot is the classical method for the quantification of the physical particles by hybridization of the packaged rAAV genomes by using DNA probes specific for the transgene cassette. A positive signal in this assay indicates that rAAV virions were produced, and quantification yields a particle number in virions per milliliter. However, this assay will not indicate whether the virus is infectious or the expression cassette is functional. This test is described under Subheadings 2.3 and 2.5, solutions and gradients section and under Subheadings 3.3 and 3.5.

The determination of infectious particles has more relevance, in particular for in vivo applications. However, these tests are rather specific, when marker genes, such as *GFP* or *lacZ* are used, and the reader is referred to the literature (55). For transgenes of clinical interest, other tests have been employed, based on quantitative PCR to detect the sequence of the transgene in the transduced target cells: the reader is referred, for instance, to the paper by Farson et al. (56) for more information as well as to Chapter 11 in this book.

To help worldwide standardization of quantification methods and to facilitate the interpretation of pre-clinical and clinical data, reference materials for the AAV2 vector were generated (Richard Snyder, Director of Biotherapeutic Programs, University of Florida, ICBR, Building 62, South Newell Drive, PO Box 110580, Gainesville FL 32611-0580). This reference material is now available in the ATCC repository (cat# 37216™, http://www.atcc.org/ATCCAdvancedCatalogSearch/ProductDetails/tabid/452/Default.aspx?ATCCNum=37216&Template=vectors).

This chapter presents small scale production (based on the transient transfection of HEK293 cells) and purification methods as well as traditional titration methods of AAV vectors.

2. Materials (see Note 1)

2.1. rAAV Production by Transient Transduction of 293 Cells

1. pAAV vector cloning: a helper free system can be purchased from Stratagene, or the psub201 plasmid (for cloning the transgene between the AAV termini) is available at the ATCC (#68065), or a model AAV2 vector plasmid (pAAV-CMV(nls) lacZ) can be obtained from Dr. F. Wright (University of Pennsylvania School of Medicine, Clinical Vector Core, Center for Cellular and Molecular Therapeutics, The Children's Hospital of Philadelphia, ARC1216C, 3615 Civic Center Blvd, Philadelphia PA 19104, wrightf@email.chop.edu).
2. pXX6 plasmid: the adenoviral helper plasmid (can be obtained from the UNC Vector Core Facility (Gene Therapy Center, Division of Pharmaceutics, University of North Carolina at Chapel Hill, NC27599, USA)) (32).
3. pRepCap4 plasmid (57): the AAV2 helper plasmid (Dr. F. Wright, University of Pennsylvania School of Medicine, Clinical Vector Core, Center for Cellular and Molecular Therapeutics, The Childrens Hospital of Philadelphia, ARC1216C, 3615 Civic Center Blvd, Philadelphia PA 19104, wrightf@email.chop.edu).
4. HEK293 tissue culture cell line (ATCC #CRL 1573), derived from a controlled frozen cell stock.
5. Cell culture media/solutions:
   (a) DMEM (4.5 g/l glucose) (Invitrogen #41966), supplemented with 10% foetal calf serum (FCS).
   (b) DMEM (4.5 g/l glucose) (Invitrogen #41966), supplemented with 1% FCS.
   (c) DMEM (1 g/l glucose) (Invitrogen #31885), supplemented with 10% FCS (see Note 2).
6. Trypsine/EDTA (0.05%/0.2 g/l) (Invitrogen).
7. T-flask (175 cm^2, Corning).
8. Trypan blue solution.
9. Disposable 15 ml or 50 ml polystyrene and polypropylene centrifuge tubes (Falcon).
10. Polyethyleneimine (PEI, 25 kDa) (10 mM) (Aldrich #40872-7) (Preparation: dissolve 4.5 mg of pure PEI in 8 ml of deionized water (mix well), neutralize with HCl (pH 6.5–7.5), adjust the volume to 10 ml, sterilize by filtration through a 0.22 μm filter). The solution is equivalent to 10 mM expressed in nitrogen.
11. NaCl (150 mM) (Sigma).
12. Low-speed tabletop centrifuge (e.g. Jouan CL412).

2.2. Viral Vector Harvest and Purification Using Caesium Chloride Gradient Ultracentrifugation

1. Cell scrapers (Corning).
2. Phosphate-buffered saline (PBS) (Ca^{++}, Mg^{++}) (Invitrogen).
3. Lysis buffer (50 mM Hepes, 150 mM NaCl, 1 mM $MgCl_2$, 1 mM $CaCl_2$) (Sigma).
4. Disposable 50 ml polystyrene and polypropylene centrifuge tubes (Falcon).
5. Low-speed tabletop centrifuge (e.g. Jouan CL412).
6. Dry ice/ethanol bath.
7. Water bath (37°C).
8. Benzonase (250 U/μl) (Merck).
9. Saturated Ammonium sulphate $(NH_4)_2SO_4$ solution (Merck), pH 7.0, 4°C (add 450 g of $(NH_4)_2SO_4$ to 500 ml water. Heat on a stir plate until $(NH_4)_2SO_4$ dissolves completely. Filter through Whatman paper while still warm and allow to cool (upon cooling crystals will form which should not be removed). Adjust the pH to pH 7.0 with ammonium hydroxide. Store up to 1 year at 4°C.
10. Beckman High-speed centrifuge and JA17 rotor or equivalent.
11. CsCl (Fluka) gradient solutions:
    (a) 1.35 g/ml density CsCl (47.25 g CsCl filled up to 100 ml with PBS).
    (b) 1.5 g/ml density CsCl (67.5 g CsCl filled up to 100 ml with PBS).
    (c) 1.4 g/ml density CsCl (54.5 g CsCl filled up to 100 ml with PBS) (see Note 3).
12. Beckman ultracentrifuge with 90Ti rotor.
13. 8.9-ml Beckman Optiseal tubes for the 90Ti rotor.
14. Beckman Fraction Recovery System.
15. Refractometer (Abbé A.B986).
16. Pierce Slide-A-Lyzer dialysis cassettes (MWCO 10000; PIERCE).
17. 21-gauge needles.

2.3. Titration: Determination of Physical Particles by Dot-Blot Hybridization

1. DMEM (Invitrogen #31966).
2. 2× Proteinase K buffer (0.5% SDS; 10 mM EDTA; (Sigma)).
3. Proteinase K (10 mg/ml) (Merck).
4. DNAse I (Invitrogen).

5. Phenol/chloroform/isoamyl alcohol: 25:24:1 (v/v/v).
6. Sodium acetate 3 M.
7. Glycogen (Boehringer M901393) (20 mg/ml).
8. 100% and 70% ethanol.
9. 0.4 N NaOH/10 mM EDTA.
10. 2× SSC (Amresco 0804-41; 0.3 M NaCl/0.03 M Na-citrate).
11. rAAV plasmid used to make recombinant viral vector (see Subheading 2.1).
12. Water bath (37°C).
13. 100°C heating block.
14. 0.45-μm N^+ nylon membrane (Hybond RPN119B).
15. Dot-blot device (BioRad 170 3938).
16. Non-radioactive labelling system (Alk Phos Direct, Amersham: see manufacturer's instructions).
17. ECF substrate detection (Amersham, see manufacturer's instructions).

2.4. Titration: Determination of Transducing Particles (Example: GFP)

1. HEK293 cells, seeded at 0.5×10^5 cells/ml in 24 well-plates.
2. HEK293 medium: DMEM, 10% FCS.
3. Dilution of unknown in HEK293 medium (100 μL) to stand around 3–30% of TU/cell.
4. 16% p-formaldehyde in PBS.

2.5. Titration: Determination of Physical Particle Titer by ELISA

Such tests have been developed for AAV1 (cat number: **PRAAV1**), AAV2 (cat number: **PRATV**), and AAV5 (cat number: **PRAAV5**) and are available from Progen Biotechnik GmbH (Maaßstraße 30, D-69123 Heidelberg, Germany: Internet: http://www.progen.de).

2.6. Solutions Required for the Supplementary Protocols Needed for Dot-Blot Hybridization

1. Hybridization buffer: Add NaCl to the hybridization buffer to give a concentration of 0.5 M. Add blocking reagent to a final concentration of 4%. Immediately, mix thoroughly, to get the blocking reagent into a fine suspension. Continue mixing at room temperature for 1–2 h on a magnetic stirrer. This buffer can be used immediately or stored in suitable aliquots at −15°C to −30°C.
2. Primary wash buffer (200 ml): Make a solution of 24 g of Urea, 1 ml of SDS (0.2 g/ml), 20 ml of 0.5 M sodium phosphate (pH 7.0), 1.74 g of NaCl, 2 ml of 1.0 M $MgCl_2$, and of 0.4 g Blocking reagent (Amersham RPN3680) (= 0.2 %). The primary wash buffer can be kept for up to 1 week at 2–8°C.

3. Secondary wash buffer – 20× stock: Make a solution of 121 g of Tris base (1 M) and of 112 g of NaCl (2 M) in 900 ml of water. Adjust pH to 10.0. Make up to 1 L with water. This buffer can be kept for up to 4 months at 2–8°C.
4. Secondary wash buffer – working dilution: Dilute stock 1:20 and add 2 ml/l of 1 M $MgCl_2$ to give a final concentration of 2 mM Mg^{++} in the buffer. This buffer should not be stored.

3. Methods

The scheme of the transient production protocol is presented in Fig. 4.

3.1. rAAV Production by Transient Transfection of HEK293 Cells: (see Notes 4 and 5)

3.1.1. Thawing and Preparation of HEK293 Cells

1. Thawing of an ampoule from the cell stock prepared in DMEM supplemented with 10% FCS and 10% DMSO (5×10^6 c/ampoule): after rapid thawing of the cells (incubation of the cryovial in a water batch (37°C)), the content (1 ml) is transferred to a 15 ml or 50 ml Falcon tube and 9 ml of fresh medium (DMEM + 10% FCS) (preheated to 37°C) is added dropwise.
2. Centrifugation at 1,200 rpm (=140 × *g*) at 20°C for 5 min (Jouan) and elimination of the supernatant.
3. The cell pellet is taken up in 30 ml of DMEM + 10% FCS, and plated in a T-flask (175 cm^2).
4. Incubate the T-flask in a CO_2 incubator (37°C, 5% CO_2). The medium is eventually changed 3 days post-inoculation.
5. At confluence, the medium is eliminated and the cells are washed once with 10 ml of PBS (without Ca^{++}/Mg^{++}).
6. The cells are trypsinized by adding 4 ml of trypsine/EDTA.
7. 30 s to 5 min later, the T-flask is agitated to detach all cells and 16 ml of DMEM + 10% FCS is added.
8. The cell suspension is transferred to a 50 ml tube (Falcon) and centrifuged (1,200 rpm – 140 × *g*) (Jouan) for 5 min.
9. The cells are taken up in 10 ml of fresh medium, the suspension is homogenized, and a sample is counted (using trypan blue).
10. The subcultures are started with about 30,000 c/cm^2, using 30 ml of DMEM + 10% FCS per T-flask (175 cm^2).

3.1.2. Transfection of HEK293 Cells in View of rAAV Production: (see Note 6)

1. Seed 3×10^7 cells in 15 cm dishes and maintain the cells in complete DMEM (4.5 g/l Glucose) + 10% FCS medium. Incubate overnight in a humidified 37°C, 5% CO_2 incubator (see Note 7).

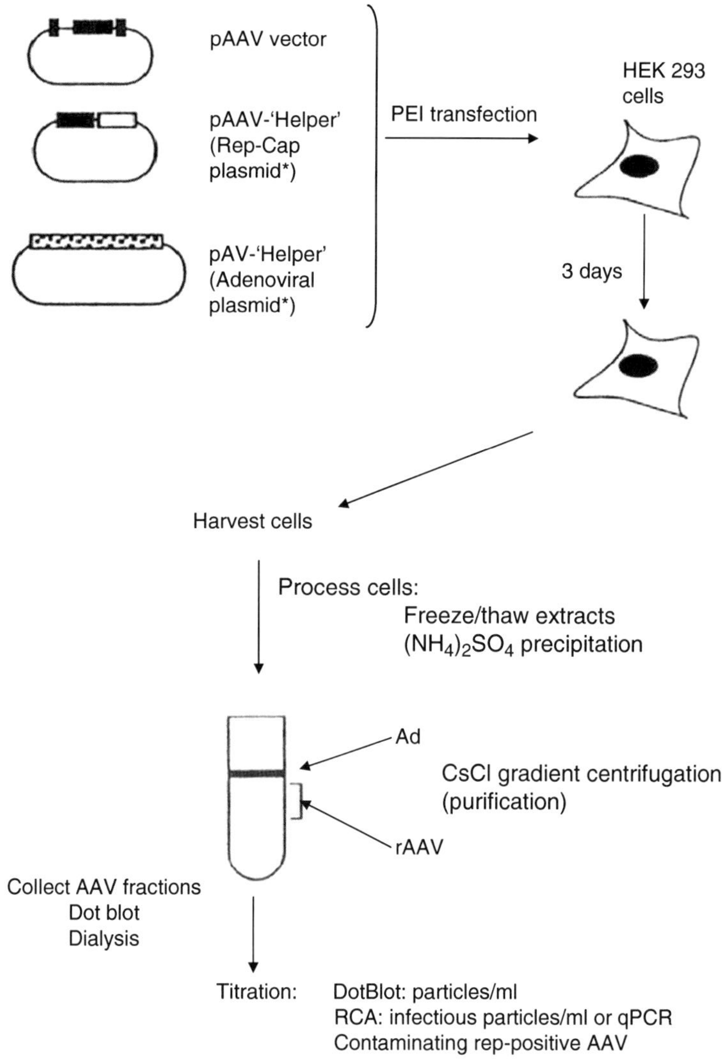

Fig. 4. ***Triple-transfection protocol for rAAV production.*** HEK293 cells are transfected with the pAAV-vector plasmid (the rep-cap pAAV-"helper" plasmid), and the adenoviral pAV-"helper" plasmid (1:1:2 ratio). Fresh medium is added 5–7 h after transfection, and the cells are collected 3 days later and processed (freezing–thawing and $(NH_4)_2SO_4$ precipitation). After CsCl gradient centrifugation, rAAV-containing fractions are pooled, dialyzed, and titered by dot blot, ELISA, RCA (Replication Centre Assay) (57), or qPCR ((59, 60), see also Chapter 11). Notice that an adenovirus band is only present. After CsCl gradient centrifugation when real adenovirus helper virus is used. This band is not present when the pAV-"helper" plasmid is used.

2. The cell culture should be at 70–80% of confluence for optimal transfection.
3. Change medium 1 h before transfection (see Note 8).
4. For transfection of five 15 cm dishes, combine first the following in a disposable 50-ml polystyrene tube:

   125 μg pXX6 helper plasmid (adenoviral [AV] helper genes).

62.5 μg rAAV vector plasmid.

62.5 μg pRepCap4 helper plasmid (AAV helper genes).

Fill up to 5 ml with a NaCl solution (150 mM).

(*This refers to the DNA solution. The total DNA is equal to 50 μg per plate*)

Use set of four tubes for twenty 15 cm dishes.

(see Note 9).

For transfection of five 15-cm dishes, combine the following in a disposable 50 ml polystyrene tube:

562.5 μl PEI 25 kDa (10 mM).

Fill up to 5 ml with a NaCl solution (150 mM)

(*This refers to the PEI solution*).

Use a set of four tubes for twenty 15-cm dishes.

5. Add dropwise the PEI solution into the DNA solution. Incubate 15–20 min at room temperature to allow the formation of the PEI/DNA complexes.
6. Optional: meanwhile, rinse the cells with 10 ml of DMEM (4.5 g/l Glucose) + 1% FCS.
7. Add 12 ml of DMEM (4.5 g/l glucose) + 1% FCS.
8. Add 2 ml of the transfection complex dropwise to the medium in each of four 15-cm plates of cells (from step 5). Swirl the plates to disperse homogenously.
9. Repeat steps 6–8 to transfect each set of five dishes at a time until all 20 dishes have been transfected.
10. Incubate cells for 5–7 h.
11. Add 12 ml of medium (DMEM (1 g/l Glucose) + 10% FCS) to each 15-cm plates (see Note 10).
12. Incubate the cells until 72 h post-transfection (CO_2 incubator, 37°C).

3.2. Viral Vector Harvest and Purification Using Caesium Chloride Gradient Ultracentrifugation

In this procedure, ammonium sulphate precipitation removes most of the cellular debris from the virions before the CsCl gradient is performed. This allows a greater number of cells to be processed and results in more concentrated stocks of rAAV following the CsCl gradient. This recipe is developed for 20–40 15-cm plates, but can be scaled up two to four times.

1. Collect the cells and the supernatant from the tissue culture plates by scraping the cells with a cell scraper to collect cells and medium. Transfer the cell suspension to 50 ml polypropylene centrifuge tubes (ten 50 ml tubes will be necessary for a 20 dishes preparation).

2. Centrifuge the tubes for 10 min at 700 × *g* (1,500 rpm, Jouan CL412), 8°C. Decant the supernatant (see Note 11).

   Dissolve the pellet with 2 ml of lysis buffer per 50-ml tube (equivalent to 1 ml per initial 15-cm plate) and vortex vigorously. Pool the pellets in one 50-ml tube.

3. Freeze and thaw the cell suspension four times by transferring the tubes between a dry ice/ethanol bath and a 37°C water bath. Vortex vigorously between each cycle (see Note 12).

   Centrifuge the tubes for 15 min at 1,500 × g (2,600 rpm, Jouan CL412), 4°C. Decant the supernatant into fresh 50-ml polypropylene centrifuge tubes.

4. Treat the lysate with Benzonase (250 U/μl) at a final concentration of 25 U/ml (0.1 μl per ml of lysate) and incubate for 15 min at 37°C in a water bath.

5. Centrifuge the tubes for 20 min at 10,000 × *g* (8,500 rpm Beckman High-Speed JA17), 4°C. Decant the supernatant into fresh 50-ml polypropylene centrifuge tubes.

6. Add 1 volume of ice-cold saturated ammonium sulphate to the supernatant. Mix thoroughly and precipitate on ice for 1 h (see Note 13).

   Centrifuge the tubes 30 min at 12,000 × *g* (10,000 rpm Beckman High-Speed JA17 rotor) 4°C.

7. Slowly decant the supernatant into a container, avoiding touching the pellet (see Note 14).

8–16. CsCl gradient purification:

8. Dissolve pellet in 2.5 ml of PBS (Ca^{++}, Mg^{++}).

9. Add 3 ml of a 1.5 g/ml CsCl solution to 8.9-ml Beckman Optiseal centrifuge tubes. Gently overlay with 3 ml of a 1.35 g/ml CsCl solution. Then overlay with the 2.5 ml-sample. Fill up the tube with PBS (Ca^{++}, Mg^{++}) (about 0.4 ml).

10. Centrifuge the samples at 385,000 × *g* (67,000 rpm for 6 h in a Beckman 90Ti rotor), 8°C (see Note 15).

11. Decelerate with brake to 1,000 rpm, and then turn the brake off (see Note 16).

12. Using a Beckman Fraction Recovery System, collect 10 fractions of 500 μl (equivalent to 30 drops) from the bottom of the tube.

13. Using a refractometer, analyze the fractions by measuring the refraction index (RI). Pool the positive fractions (see Note 17).

14. Add 1.40 g/ml CsCl solution to the pooled fractions to attain a final volume of 8.9 ml and transfer the virus solution to an Optiseal tube to re-band the virus. Re-band, drip and assay the gradient as in steps 12–15 (see Note 18).

15. Dialyze the rAAV in MWCO 10000 Slide-A-Lyzer dialysis cassettes (Pierce) against three 500-ml changes of sterile 1× PBS (Ca^{++}, Mg^{++}) for at least 3 h each at 4°C. For in vivo application, five changes of sterile PBS are recommended for the dialysis (see Note 19).
16. Transfer the virus suspension into convenient aliquots (typically 100–200 μl) to avoid repeated freezing and thawing (see Note 20).

3.3. Titration: Determination of Physical Particles by Dot-Blot Hybridization

This assay detects the packaged rAAV genomes using DNA probes specific for the transgene cassette. A positive signal in this assay indicates that rAAV virions were produced, and quantification yields a particle number in virions per millilitre. However, this assay will not indicate whether the virus is infectious or whether the expression cassette is functional.

1–8. *Digest virus particles to release DNA*

1. Place samples (usually 2 and 10 μl) of the viral solution in microtubes. Complete to 200 μl with DMEM.
2. Add 10 U of DNAse I and incubate for 30 min at 37°C in a water-bath (see Note 21).
3. Release viral DNA by adding 200 μl of 2× proteinase K buffer and 10 μl of 10 mg/ml proteinase K (0.25 mg/ml final concentration). Incubate for 60 min at 37°C in a water bath.
4. Extract with 1 vol of phenol/chloroform/isoamyl alcohol (25:24:1 (v/v/v)).
5. Precipitate the viral DNA by adding 40 μg of glycogen as a carrier, 0.1 vol of 3 M sodium acetate, and 2.5 vol of 100% ethanol. Incubate for 30 min at −80°C.
6. Centrifuge at 10,000 ×*g* for 20 min at 4°C. Discard the supernatant and wash the pellet with 70% ethanol, then centrifuge at 10,000 ×*g* for 10 min, 4°C. Air-dry the pellet for 10 min and re-suspend it in 400 μl of 0.4 N NaOH/10 mM EDTA.
7. Prepare a twofold serial dilution of the pAAV plasmid used to generate the rAAV stock in 20 μl. Add 400 μl of 0.4 N NaOH/10 mM EDTA (see Note 22).
8. Denature samples by heating at 100°C for 5 min, then chill it on ice.

9–13. *Dot Blot*

9. Set up the dot-blot device with a pre-wetted 0.45 μm nylon membrane.
10. Wash the wells with 400 μl of deionized water. Apply vacuum to empty the wells.

11. Add the viral DNA and the plasmid range samples without vacuum. Apply the vacuum until each well is empty.
12. Rinse wells with 0.4 N NaOH/10 mM EDTA, apply vacuum to dry the membrane.
13. Disassemble the device and rinse the membrane in 2× SSC.

14 and 15. *Hybridization*

14. Probe the membrane with a non-radioactive labelled probe (see Subheading 3.6) specific for the transgene sequence.
15. Expose the membrane following the manufacturer's instruction, and quantify the signal using a PhosphorImager associated with the appropriate software.

3.4. Titration: Determination of Transducing Particles (Example: GFP)

The transducing titer is highly dependent on the cell type used and the sensitivity of the assay, making the titers of rAAV with various transgenes difficult to compare. However, transducing titers can be used to compare rAAV preps of the same transgene. The choice of the target cell depends on the AAV serotype. For AAV1 and AAV2, for instance, HEK293 cells can be used (see Note 23).

1. Seed appropriate target cells (HEK293, 50,000 c/well) into multi-well tissue culture plates (e.g. 24-well plates) with appropriate medium (DMEM + 10% FCS).
2. Infect cells by adding rAAV (dilutions) directly to the medium of the cells or mix rAAV with fresh medium immediately before adding it to the cells. For assaying transducing titer, cells should be infected with serial fivefold dilutions of the rAAV stock to titer (see Note 24).
3. Optionally, adenovirus type 5 can be added with rAAV at an MOI of 5.
4. Perform detection of transduced cell with the appropriate method and deduce the transducing unit (t.u.) titer:
5. At 24 or 48 hpi, estimate the cell concentration by the erythrosine dye exclusion method.
6. Resuspended in 2% p-formaldehyde in PBS and let stand for 1 h at 4°C.
7. Analyze at least 10,000 events by flow cytometry, using the Coulter EPICS™ XL-MCL cytometer and EXPO32 software to determine the percentage of GFP-positive cells. A minimum of two dilutions showing 3–30% GFP-positive cells should be taken into account for the titer calculation.
8. Calculation of titer (t.u./mL) = percentage GFP-positive cells times the cell concentration (cells/mL)/(dilution of unknown × 0.1 mL).

3.5. Titration: Determination of Physical Particle Titer by ELISA

The company Progen Biotechnik GmbH has developed commercial ELISA kits for the titration of the physical particle titer (of intact capsides) in cell culture supernatants and purified virus preparations. Such tests have been developed for AAV1 (cat number: **PRAAV1**), AAV2 (cat number: **PRATV**), and AAV5 (cat number: **PRAAV5**).

The procedure to be employed is presented in detail in the description added to the test kits.

3.6. Supplementary Protocols Needed for Dot-Blot Hybridization: Non-radioactive Labelling of DNA Probe (AlkPhos Direct, RPN 3680, Amersham)

1–9. *Preparation of labelled probe*:

1. Dilute 2 μl of the cross-linker solution with 8 μL of the water supplied to give the working concentration.
2. Dilute DNA to be labelled to a concentration of 10 ng/μL using the water supplied.
3. Place 10 μl of the diluted DNA sample in a micro-centrifuge tube and denature by heating for 5 min in a vigorously boiling water bath.
4. Immediately cool the DNA on ice for 5 min. Spin briefly in a micro-centrifuge to collect the contents at the bottom of the tube.
5. Add 10 μl of reaction buffer to the cooled DNA. Mix thoroughly but gently.
6. Add 2 μl labelling reagent. Mix thoroughly but gently.
7. Add 10 μl of the cross-linker working solution. Mix thoroughly. Spin briefly in a micro-centrifuge to collect the contents at the bottom of the tube.
8. Incubate the reaction for 30 min at 37°C.
9. The probe can be used immediately or kept on ice for up to 2 h. For long-term storage, labelled probes may be stored in 50% (v/v) glycerol at –15°C to –30°C for up to 6 months.

10–13. *Hybridization*:

10. Pre-heat the required volume of prepared AlkPhos Direct hybridization buffer to 55°C. The volume of the buffer should be equivalent to 9 ml for blot hybridized in bottles.
11. Place the blot into the hybridization buffer and pre-hybridize for at least 15 min at 55°C in a hybridization oven.
12. Add the labelled probe to the buffer used for the pre-hybridization step. Typically use 100 ng per 9 ml of buffer.
13. Hybridize at 55°C overnight in a hybridization oven.

14–18. *Posthybridization stringency washes*:

14. Pre-heat the primary wash buffer to 55°C. This is used in excess at a volume of 100 ml for a blot.
15. Carefully transfer the blot to this solution and wash for 10 min at 55°C in a hybridization oven.
16. Perform a further wash in fresh, primary wash buffer at 55°C for 10 min.
17. Place the blot in a clean container and add an excess of secondary wash buffer (150 ml). Wash under gentle agitation for 5 min at room temperature.
18. Perform a further wash in fresh, secondary wash buffer at room temperature for 5 min.

19–21. *Detection with ECF substrate*;

19. Pipette 2 ml of ECF substrate onto the blot and incubate for 1 min. Transfer the blot directly to a fresh detection bag. Fold the plastic over the top of the blot and immediately spread the reagent evenly over the blot.
20. Incubate at room temperature in the dark, for the required length of time.
21. Scan the blot.

3.7. Anticipated Results

In these protocols, twenty 15-cm tissue culture plates should yield 10^{12} to 10^{13} particles as assayed by dot blot. This would translate into 10^9 to 10^{10} rAAV particles that are capable of delivering and expressing the transgene (approximately 1,000 particles per cell). The typical ratio of particles to transducing units is 100:1 to 1,000:1 when assayed on HEK293 or HeLa cells (which are well transduced). Cells that are less well transduced will give a higher ratio. The particle preparations should be free of contaminating cellular and adenovirus proteins and free of infectious adenovirus.

3.8. Time Considerations

Transfection of the cells requires 5 days. On day 1 the cells are split. On day 2 the cells are transfected. On day 5 the cells are harvested.

The purification requires 3–4 days: 1 day to perform cell lysis and fractionation leading to the overnight CsCl gradient centrifugation: 1 day to locate the rAAV by reading the refraction index and overnight re-banding of the virus: and 1 day for dialysis (the last round can be performed overnight).

The dot-blot assay requires 2 days: 1 day to process the samples, bind to the filter, and hybridize, and another day to wash the filter and expose.

4. Notes

1. All solutions are prepared by using double-distilled water or equivalent.
2. The media can be stored at 4°C in the dark for up to 4 weeks, cell culture media are not stable because of the inactivation of glutamine (if not, ala-gln or gly-gln are used), some other amino acids, and some vitamins (B vitamins).
3. For all CsCl gradient solutions, check the density of each solution by weighing 1 ml. Then filter the solutions through 0.22 μm filters. Store the gradient solutions up to 1 year at room temperature.
4. Remark: AAV vectors must be handled under appropriate biosafety containment by trained personnel using biological safety cabinets and following the guidelines specified by the institution where the experiments are conducted. The nature of the transgene must be taken into account when establishing biosafety requirements. The work described herein has been performed in biosafety level-2 laboratories. In addition, all cell culture work should be performed according to GCCP (Good Cell Culture Practice). For details, see Coecke et al. (58) and Chapter 3.
5. The equipment and reagents dedicated to tissue culture cells and vector purification should be sterile. All tissue culture incubations are performed in a humidified 37°C, 5% CO_2 incubator unless otherwise specified.
6. The production procedure described in this protocol is for twenty 15-cm dishes but can be scaled up four times.
7. Alternatively, cells can be seeded 2 days prior transfection at 1.2×10^7 cells in 15-cm dishes.
8. This step can be omitted if the cells appear too fragile, which often occurs with HEK293 cells.
9. Transfection is performed at an optimized ratio of PEI 25 kDa to DNA [$R = 2.25$ = vol (μl) PEI/weight (μg) DNA].
10. The final glucose concentration will be about 2.75 g/l, allowing cells to divide at a slow rate, thereby producing more viruses.
11. The pellet can be stored at this stage at −20°C.
12. Freezing and thawing liberates most of the virus particles from the cells. This suspension can be stored at −20°C for up to 6 months.

13. The ammonium sulphate will precipitate the virus in the supernatant.
14. Do not use bleach to decontaminate ammonium sulphate supernatants because a noxious odour will be produced.
15. It is convenient at this stage to perform the ultracentrifugation step overnight.
16. AAV should form a diffuse band in the middle of the tube; however, most of the time the AAV band is not visible.
17. The RI for positive fractions containing rAAV usually ranges between 1.376 and 1.368. Alternatively, a positive fraction can be determined by dot-blot hybridization using an rAAV-specific probe (see titering protocol).
18. Re-banding the fractions will increase the purity and concentration of the rAAV, but some loss will occur with these supplementary manipulations.
19. Overnight dialysis will not result in any loss of titer.
20. The virus can be stored at –20°C or –80°C for more than 1 year.
21. This step digests any DNA that may be present and that has not been packaged into virions.
22. The dilution range should be from 40 ng to 0.3125 ng of plasmid.
23. Some researchers find convenient to assay transducing titer in the presence of the helper adenovirus, which increases the transduction efficiency. Although the addition of adenovirus gives a better indication of the number of particles that are competent to transduce a cell by inducing an optimal environment for AAV infection, one might find its use not relevant, a titer derived with the use of adenovirus coinfection in vitro may not accurately reflect the in vivo competency.
24. Incubation time depends on the assay. Most often, 24–48 h incubation time will suffice.

Acknowledgements

These protocols are based on the protocols used at the 3rd Eurolabcourse: Advanced industrial methods for production, purification and characterisataion of gene vectors (see http://www.easco.org/adminarea/upload/files/0306_0720_EVRY2004Report.pdf) (organizers: O.-W. Merten, M. Mezzina and G. Waksmann), Evry/F, 14–26 June 2004.

References

1. Berns, K. I. (1996) Parvoviridae: The virus and their replication. In: Fields BN, Knipe DM, Howley PM (eds). *Fields in Virology*. Lippincott – Raven, Philadelphia, pp. 2173–2197.
2. Rutledge, E.A., Halbert, C. L., and Russell, D. W. (1998) Infectious clones and vectors derived from adeno-associated virus (AAV) serotypes other than AAV type 2. *J. Virol.* **72**, 309–319.
3. Gao, G. P., Alvira, M. R., Wang, L., Calcedo, R., Johnston, J., and Wilson, J. M. (2002) Novel adeno-associated viruses from rhesus monkeys as vectors for human gene therapy. *Proc. Natl. Acad. Sci. USA* **99**, 11854–11859.
4. Gao, G., Vandenberghe, L. H., Alvira, M. R., Lu, Y., Calcedo, R., Zhou, X., and Wilson, J. M. (2004) Clades of adeno-associated viruses are widely disseminated in human tissues. *J. Virol.* **78**, 6381–6388.
5. Mori, S., Wang, L., Takeuchi, T., and Kanda, T. (2004) Two novel adeno-associated viruses from cynomolgus monkey: pseudotyping characterization of capsid protein. *Virology* **330**, 375–383.
6. Gao, G., Vandenberghe, L. H., and Wilson J. M. (2005) New recombinant serotypes of AAV vectors. *Curr. Gene Ther.* **5**, 285–297.
7. Qiu, J. and Pintel, D. J. (2004) Alternative polyadenylation of adeno-associated virus type 5 RNA within an internal intron is governed by the distance between the promoter and the intron and is inhibited by U1 small nuclear RNP binding to the intervening donor. *J. Biol. Chem.* **279**, 14889–14898.
8. Erles, K., Sebokova, P., and Schlehofer, J. R. (1999) Update on the prevalence of serum antibodies (IgG and IgM) to adeno-associated virus (AAV). *J. Med. Virol.* **59**, 406–411.
9. Tobiasch, E., Burguete, T., Klein-Bauernschmitt, P., Heilbronn, R., and Schlehofer, J. R. (1998) Discrimination between different types of human adeno-associated viruses in clinical samples by PCR. *J. Virol. Methods* **71**, 17–25.
10. Kotin, R. M., Siniscalco, M., Samulski, R. J., Zhu, X. D., Hunter, L., Laughlin, C. A., *et al.* (1990) Site-specific integration by adeno-associated virus. *Proc. Natl. Acad. Sci. USA* **87**, 2211–2215.
11. Samulski, R. J., Zhu, X., Xiao, X., Brook, J. D., Housman, D. E., Epstein, N., *et al.* (1991) Targeted integration of adeno-associated virus (AAV) into human chromosome 19. *Embo J.* **10**, 3941–3950.
12. Shelling, S. N. and Smith, M. G. (1994) Targeted integration of transfected and infected adeno-associated virus vectors containing the neomycin gene. *Gene Ther.* **1**, 165–169.
13. Samulski, R. J., Berns, K. I., Tan, M., and Muzyczka, N. (1982) Cloning of adeno-associated virus into pBR322: rescue of intact virus from the recombinant plasmid in human cells. *Proc. Natl. Acad. Sci. USA* **79**, 2077–2081.
14. Samulski, R. J., Chang, L. S., and Shenk, T. (1987) A recombinant plasmid from which an infectious adeno-associated virus genome can be excised in vitro and its use to study viral replication. *J. Virol.* **61**, 3096–3101.
15. Tratschin, J. D., Miller, I. L., Smith, M. G., and Carter, B. J. (1985) Adeno-associated virus vector for high-frequency integration, expression, and rescue of genes in mammalian cells. *Mol. Cell. Biol.* **5**, 3251–3260.
16. Grimm, D. and Kay, M. A. (2003) From virus evolution to vector revolution: use of naturally occurring serotypes of adeno-associated virus (AAV) as novel vectors for human gene therapy. *Curr. Gene Ther.* **3**, 281–304.
17. Grimm, D., Lee, J. S., Wang, L., Desai, T., Akache, B., Storm, T. A., *et al.* (2008) In vitro and in vivo gene therapy vector evolution via multispecies interbreeding and retargeting of adeno-associated viruses. *J. Virol.* **82**, 5887–5911.
18. Tratschin, J. D., Miller, I. L., and Carter, B. J. (1984) Genetic analysis of adeno-associated virus: properties of deletion mutants constructed in vitro and evidence for an adeno-associated virus replication function. *J. Virol.* **51**, 611–619.
19. Srivastava, A., Lusby, E. W., and Berns, K. I. (1983) Nucleotide sequence and organization of the adeno-associated virus 2 genome. *J. Virol.* **45**, 555–564.
20. Weitzman, M. D., Kyostio, S. R., Kotin, R. M., and Owens, R. A. (1994) Adeno-associated virus (AAV) Rep proteins mediate complex formation between AAV DNA and its integration site in human DNA. *Proc. Natl. Acad. Sci. U S A* **91**, 5808–5812.
21. Horer, M., Weger, S., Butz, K., Hoppe-Seyler, F., Geisen, C., and Kleinschmidt, J. A. (1995) Mutational analysis of adeno-associated virus Rep protein-mediated inhibition of heterologous and homologous promoters. *J. Virol.* **69**, 5485–5496.
22. Pereira, D. J., McCarty, D. M., and Muzyczka, N. (1997) The adeno-associated virus (AAV) Rep protein acts as both a repressor and an

activator to regulate AAV transcription during a productive infection. *J. Virol.* **71**, 1079–1088.

23. King, J. A., Dubielzig, R., Grimm, D., and Kleinschmidt, J. A. (2001) DNA helicase-mediated packaging of adeno-associated virus type 2 genomes into preformed capsids. *Embo J.* **20**, 3282–3291.
24. Timpe, J., Bevington, J., Casper, J., Dignam, J. D., and Trempe, J. P. (2005) Mechanisms of adeno-associated virus genome encapsidation. *Curr. Gene Ther.* **5**, 273–284.
25. Schlehofer, J. R., Ehrbar, M., zur Hausen, H. (1986) Vaccinia virus, herpes simplex virus, and carcinogens induce DNA amplification in a human cell line and support replication of a helpervirus dependent parvovirus. *Virology* **152**, 110–117.
26. Weindler, F. W. and Heilbronn, R. (1991) A subset of herpes simplex virus replication genes provides helper functions for productive adeno-associated virus replication. *J. Virol.* **65**, 2476–2483.
27. McPherson, R. A., Rosenthal, L. J., and Rose, J. A. (1985) Human cytomegalovirus completely helps adeno-associated virus replication. *Virology* **147**, 217–222.
28. Ogsten, P., Raj, K., and Beard, P. (2000) Productive replication of adeno-associated virus can occur in human papillomavirus type 16 (HPV-16) episome-containing keratinocytes and is augmented by the HPV-16 E2 protein. *J. Virol.* **74**, 3494–3504.
29. Geoffroy, M.-C. and Salvetti, A. (2005) Helper functions required for wild type and recombinant adeno-associated virus growth. *Curr. Gene Ther.* **5**, 265–271.
30. Wistuba, A., Kern, A., Weger, S., Grimm, D., and Kleinschmidt, J. A. (1997) Subcellular compartmentalization of adeno-associated virus type 2 assembly. *J. Virol.* **71**, 1341–1352.
31. Grimm, D., Kern, A., Rittner, K., and Kleinschmidt, J. A. (1998) Novel tools for production and purification of recombinant adenoassociated virus vectors. *Hum. Gene Ther.* **9**, 2745–2760.
32. Xiao, X., Li, J., and Samulski, R.J. (1998) Production of high-titer recombinant adeno-associated virus vectors in the absence of helper adenovirus. *J. Virol.* **72**, 2224–2323.
33. Chiorini, J. A., Yang, L., Liu, Y., Safer, B., and Kotin, R. M. (1997) Cloning of adeno-associated virus type 4 (AAV4) and generation of recombinant AAV4 particles. *J. Virol.* **71**, 6823–6833.
34. Chiorini, J. A., Kim, F., Yang, L., and Kotin, R. M. (1999) Cloning and characterization of adeno-associated virus type 5. *J. Virol.* **73**, 1309–1319.
35. Xiao, W., Chirmule, N., Berta, S. C., McCullough, B., Gao, G., and Wilson, J. M. (1999) Gene therapy vectors based on adeno-associated virus type 1. *J. Virol.* **73**, 3994–4003.
36. Qu, G., McClelland, A., and Wright, J. F. (2000) Scaling-up production of recombinant AAV vectors for clinical applications. *Curr. Opin. Drug Discov. Devel.* **3**, 750–755.
37. Brown, P., Barrett, S., Godwin, S., Trudinger, M., Marschak, T., Norboe, D., *et al.* (1998) Optimization of production of adeno-associated virus (AAV) for use in gene therapy, Presented at: Cell Culture Engineering VI, San Diego/CA.
38. Clark, K. R. (2002) Recent advances in recombinant adeno-associated virus vector production. *Kidney Int.* **61**, 9–15.
39. Merten, O.-W., Gény, C., and Douar A. M. (2005) Current issues in adeno-associated viral vectors production. *Gene Ther.* **12**, S51–S61.
40. Urabe, M., Ding, C., and Kotin, R. M. (2002) Insect cells as a factory to produce adeno-associated virus type 2 vectors. *Hum. Gene Ther.* **13**, 1935–1943.
41. Urabe, M., Ozawa, K., Haast, S. J. P., and Hermens, W. T. J. M. C. (2007) Improved AAV vectors produced in insect cells. WO 2007/046703 A2.
42. Chen, H. (2008) Intron splicing-mediated expression of AAV Rep and Cap genes and production of AAV vectors in insect cells. *Mol. Ther.* **16**, 924–930.
43. Smith, R. H., Levy, J. R., and Kotin, R. M. (2009) A simplified baculovirus-AAV expression vector system coupled with one-step affinity purification yields high-titer rAAV stocks from insect cells. *Mol. Ther.* **17**, 1888–1896.
44. Aslanidi, G., Lamb, K., and Zolotukhin, S. (2009) An inducible system for highly efficient production of recombinant adeno-associated virus (rAAV) vectors in insect Sf9 cells. *Proc. Natl. Acad. Sci. U S A.* **106**, 5059–5064.
45. Cecchini, S., Virag, T., Negrete, A., and Kotin, R. M. (2009) Production and processing of rAAV-U7smOPT in 100 L bioreactors for canine models of Duchenne muscular dystrophy. *Mol. Ther.* **17**, S1
46. Mena, Y. A., Aucoin, M. G., Chahal, P. S., and Kamen, A. A. (2008) Improvement of adeno-associated vector titers in high density insect cell cultures by combined feeding and asynchronous infection. Poster P12 presented at the ATGQ (Association de thérapie génique

du Québec) Meeting, 26–27 May 2008, Montreal/Quebec.

47. Snyder, R. O., Xiao, X., and Samulski, R. J. (1996) Production of recombinant adeno-associated viral vectors. In: N. Dracopoli JH, B. Krof, D. Moir, C. Morton, C. Seidman, J. Seidman, and D. Smith (ed). *Current Protocols in Human Genetics*. John Wiley and Sons Publisher: New York, pp. 12.11.11–24.
48. Zolotukhin, S., Byrne, B. J., Mason, E., Zolotukhin, I., Potter, M., Chesnut, K., *et al.* (1999) Recombinant adeno-associated virus purification using novel methods improves infectious titer and yield. *Gene Ther.* **6**, 973–985.
49. Zolotukhin, S., Potter, M., Zolotukhin, I., Sakai, Y., Loiler, S., Fraites, T. J. Jr., *et al.* (2002) Production and purification of serotype 1, 2, and 5 recombinant adeno-associated viral vectors. *Methods* **28**, 158–167.
50. Brument, N., Morenweiser, R., Blouin, V., Toublanc, E., Raimbaud, I., Chérel, Y., *et al.* (2002) A versatile and scalable two-step ion-exchange chromatography process for the purification of recombinant adeno-associated virus serotypes-2 and -5. *Mol. Ther.* **6**, 678–686.
51. Kaludov, N., Handelman, B., and Chiorini, J. A. (2002) Scalable purification of adeno-associated virus type 2, 4, or 5 using ion-exchange chromatography. *Hum. Gene Ther.* **13**, 1235–1243.
52. Blouin, V., Brument, N., Toublanc, E., Raimbaud, I., Moullier, P., and Salvetti, A. (2004) Improving rAAV production and purification: towards the definition of a scalable process. *J. Gen. Med.* **6**, S223–S228.
53. Burova, E. and Ioffe, E. (2005) Chromatographic purification of recombinant adenoviral and adeno-associated viral vectors: methods and implications. *Gene Ther.* **12**, S5-S17.
54. Duffy, A. M., O'Doherty, A. M., O'Brian, T., and Strappe, P. M. (2005) Purification of adenovirus and adeno-associated virus: comparison of novel membrane-based technology to conventional techniques. *Gene Ther.* **12**, S62–S72.
55. Drittanti, L., Jenny, C., Poulard, K., Samba, A., Manceau, P., Soria, N., *et al.* (2001) Optimised helper virus-free production of high-quality adeno-associated virus vectors. *J. Gene Med.* **3**, 59–71.
56. Farson, D., Harding, T. C., Tao, L., Liu, J., Powell, S., Vimal, V., *et al.* (2004) Development and characterization of a cell line for large-scale, serum-free production of recombinant adeno-associated viral vectors. *J. Gen. Med.* **6**, 1369–1381.
57. Salvetti, A., Oreve, S., Chadeuf, G., Favre, D., Cherel, Y., Champion-Arnaud, P., *et al.* (1998) Factors influencing recombinant adeno-associated virus production. *Hum.Gene Ther.* **9**, 695–706.
58. Coecke, S., Balls, M., Bowe, G., Davis, J., Gstraunthaler, G., Hartung, T., *et al.* (2005) Guidance on Good Cell Culture Practice. A report of the second ECVAM task force on Good Cell Culture Practice. *ATLA* **33**, 261–287.
59. Rohr, U. P., Wulf, M. A., Stahn, S., Steidl, U., Haas, R., and Kronenwett, R. (2002) Fast and reliable titration of recombinant adeno-associated virus type-2 using quantitative real-time PCR. *J. Virol. Methods* **106**, 81–88.
60. Rohr, U. P., Heyd, F., Neukirchen, J., Wulf, M. A., Queitsch, I., Kroener-Lux, G., *et al.* (2005) Quantitative real-time PCR for titration of infectious recombinant AAV-2 particles. *J. Virol. Methods* **127**, 40–45.
61. Matsushita, T., Elliger, S., Elliger, C., Podsakoff, G., Villarreal, L., Kurtzman, G.J., *et al.* (1998) Adeno-associated virus vectors can be efficiently produced without helper virus. *Gene Ther* **5**, 938–945.
62. Grimm, D., Kern, A., Pawlita, M., Ferrari, F., Samulski, R., Kleinschmidt, J. (1999) Titration of AAV-2 particles via a novel capsid ELISA: packaging of genomes can limit production of recombinant AAV-2. *Gene Ther* **6**, 1322–1330.
63. Collaco, R.F., Cao, X., and Trempe, J.P. (1999) A helper virus-free packaging system for recombinant adeno-associated virus vectors. *Gene* **238,** 397–405.
64. Jenny, C., Toublanc, E., Danos, O., and Merten, O.-W. (2005) Serum-free production of rAAV-2 using HeLa derived producer cells. *Cytotechnology* **49**, 11–23.
65. Gao, G.P., Lu, F., Sanmiguel, J.C., Tran, P.T., Abbas, Z., Lynd, K.S., *et al.* (2002) Rep/Cap gene amplification and high-yield production of AAV in an A549 cell line expressing Rep/Cap. *Mol Ther* **5**, 644–649.
66. Farson, D., Harding, T.C., Tao, L., Liu, J., Powell, S., Vimal, V., *et al.* (2004) Development and characterization of a cell line for large-scale, serum-free production of recombinant adeno-associated viral vectors. *J Gene Med.* **6**, 1369–1381.

Chapter 10

Manufacturing of Adeno-Associated Viruses, for Example: AAV2

Haifeng Chen

Abstract

Adeno-associated virus (AAV) is one of the most promising vectors for gene therapy. There are several ways of producing AAV vectors but large-scale production of this vector remains a major challenge. Virovek developed a novel method of expressing the AAV Rep and Cap genes in insect cells mediated by intron-splicing mechanism and producing AAV vectors with these Rep and Cap sequences containing the artificial intron. The recombinant baculoviruses harboring these artificial intron-containing Rep and Cap sequences are very stable and the AAV vectors produced in insect cells with these recombinant baculoviruses are very infectious.

Key words: Adeno-associated virus, Manufacturing, Insect cells, Purification, Cesium chloride, Ultracentrifugation

1. Introduction

Adeno-associated virus (AAV) has emerged in recent years as a preferred viral vector for gene therapy because of its ability to infect efficiently both nondividing and dividing cells, and pose a relatively low pathogenic risk to humans (1–3). In view of these advantages, recombinant AAV has been used in gene therapy clinical trials for hemophilia B, malignant melanoma, cystic fibrosis, Parkinson's disease, and other diseases (4). The most commonly used protocol for AAV vector production utilizes transient transfection of production plasmids into adherent 293 cells. However, this method is not easily scaled up and requires a high degree of operator skill. Since the first report by Urabe et al. (5) the technology using insect cells to produce AAV vectors has gained more and more attention and has been significantly improved (6–8). Recently, Virovek developed a novel method of AAV production in insect cells utilizing the

Otto-Wilhelm Merten and Mohamed Al-Rubeai (eds.), *Viral Vectors for Gene Therapy: Methods and Protocols*, Methods in Molecular Biology, vol. 737, DOI 10.1007/978-1-61779-095-9_10, © Springer Science+Business Media, LLC 2011

intron-splicing mechanism (9). It utilized a single Rep coding sequence to express both AAV Rep78 and Rep52 and avoided the use of repeat sequences of Rep78 and Rep52 so that the recombinant baculovirus was made more stable. Furthermore, it used the authentic AUG start codon for VP1 expression and significantly improved the infectivity of AAV vectors produced in insect cells due to the increased expression level of VP1 protein, which contains a phospholipase A2 domain that is essential for the infectivity of AAV. The protocol provided here describes the detailed steps of this novel method for AAV vector production in insect cells.

2. Materials

2.1. Preparation of Recombinant Bacmid DNA

1. Plasmids pFB-inRep, pFB-inCap2, and pFB-GFP from Virovek, Inc. (Hayward, CA) diluted in TE buffer at 1 ng/μL.
2. MAX Efficiency DH10Bac competent cells from Invitrogen (Carlsbad, CA, USA).
3. LB medium containing 50 μg/mL kanamycin, 7 μg/mL gentamycin, and 10 μg/mL tetracycline.
4. LB plates containing 50 μg/mL kanamycin, 7 μg/mL gentamycin, 10 μg/mL tetracycline, 100 μg/mL X-gal, and 40 μg/mL IPTG (isopropyl-γ-D-thiogalactopyranoside).
5. IPTG stock solution prepared in molecular biology grade water at 100 mg/mL and X-gal stock solution prepared in DMSO at 40 mg/mL.
6. Resuspension buffer (P1): 25 mM Tris–HCl, pH 8.0, 50 mM Glucose, 10 mM EDTA, and 100 μg/mL RNase A.
7. Lysis buffer (P2): 0.2 M NaOH and 1% SDS.
8. Neutralizing buffer (P3): 5 M potassium acetate at pH 4.8.
9. 100% Isopropanol and 70% ethanol.

2.2. Generation of Recombinant Baculoviruses and Manufacturing of rAAV in Shaker Flasks

1. *Spodoptera frugiperda* Sf9 cells from Invitrogen are cultured in Sf-900 III SFM medium (Invitrogen) or ESF921 protein free medium (Expression Systems, LLC, Woodland, CA, USA) containing 100 U/mL of penicillin and 100 μg/mL of streptomycin. The cell density usually can reach from 8×10^6 to 1×10^7 cells/mL. Split the cells to 1×10^6 cells/mL for maintenance.
2. Cellfectin II Reagent from Invitrogen (Carlsbad, CA, USA).
3. BacEZ Baculovirus Titer Kit (Virovek, Inc., Hayward, CA).

2.3. Preparation of Cell Lysates and Purification of rAAV Vectors Through CsCl-Gradient Ultracentrifugation

1. Sf9 cell lysis buffer [50 mM Tris–HCl, pH 7.8, 50 mM NaCl, 2 mM $MgCl_2$, 1% DOC (Sodium Deoxycholate), and 0.5% CHAPS (3-[(3-Cholamidopropyl)dimethylammonio]-1-propanesulfonate)].
2. 5 M NaCl.
3. Dulbecco's Phosphate Buffered Saline (DPBS) and Benzonase from Sigma-Aldrich (St. Louis, MO, USA).
4. CsCl solutions of 1.32 g/cc, 1.38 g/cc, and 1.55 g/cc prepared in DPBS buffer.

2.4. Salt Removal and Buffer Exchange

1. Disposable PD-10 desalting columns and LabMate PD-10 buffer reservoirs from GE Healthcare Bio-Sciences Corp. (Piscataway, NJ, USA).
2. Equilibration buffer (DPBS buffer containing 0.001% pluronic F-68, sterile filtered).
3. Vac-Man Vacuum Manifold from Promega (Madison, WI, USA).
4. 0.2 μm Low-protein binding Syringe filters from Pall Corporation (Ann Harbor, MI, USA).

2.5. Large-Scale rAAV Manufacturing

1. Wave bioreactor system 20/50 from GE HealthCare Bio-Sciences Corp. (Piscataway, NJ, USA).
2. Other items see Subheading 2.3.

2.6. Large-Scale rAAV Purification

1. Two Beckman ultracentrifuges with 50.2 Ti rotors.
2. Other items see Subheading 2.4.

3. Methods

There are several ways of generating recombinant baculoviruses, such as using a homologous recombination procedure to insert foreign genes into the baculovirus genome, direct cloning into the viral genome, in vivo *Cre–LoxP* recombination, and using a bacmid form of baculovirus genome that can be maintained in *Escherichia coli*. Because of its ease to use and less time-consuming, this instruction assumes the use of Bac-to-bac Baculovirus Expression System, which employs the bacmid form of baculovirus genome, to generate recombinant baculoviruses. The shuttle plasmids used in this protocol for generation of the recombinant baculoviruses are pFB-inRep, an artificial intron-containing Rep coding sequence that expresses AAV2 Rep78 and Rep52 proteins, pFB-inCap2, an artificial intron-containing Cap coding sequence that expresses AAV2 VP1, VP2, and VP3 proteins, and pFB-GFP that contains the green fluorescent protein under control of CMV promoter and flanked by inverted terminal repeats of AAV2 (9). The GFP gene can be replaced with other genes of interest depending on research purposes. In addition, other serotypes of AAV capsids (e.g. AAV serotypes 1, 5, 6, 8, and 9, etc.) can be used to replace the AAV2 capsid coding sequence to produce specific serotypes of rAAV vectors. One can also use a dual Rep–Cap baculovirus together with the transgene containing baculovirus to produce AAV vectors in a dual infection method as described (9). The recombinant baculoviruses can then be amplified once and used as working bank for rAAV production. If repeated amplification of recombinant baculoviruses is needed, plaque purification is required so that the rAAV production yield is not compromised.

There are two basic ways to purify rAAV vectors, one employing gradient ultracentrifugation procedures, and the other column chromatographic methods. While column chromatographic methods have been reported to facilitate large-scale vector purification, such methods require optimized chromatography resins and conditions for individual AAV serotypes and result in co-purification of AAV empty capsids, which increases the total particle to infectious unit ratio of the resulting purified product. Though the gradient ultracentrifugation procedures using cesium chloride to purify AAV vectors are more time consuming and somewhat limited by scale, they are easy to perform and can be applied to all serotypes. In addition, high titer, purity, and recovery of rAAV vectors can be obtained with these procedures (see Fig. 2 and Table 1). By using larger capacity rotors, it is possible to purify more than 10^{15} vg scale of AAV vectors with conventional ultracentrifuges in common research labs. This instruction describes the detailed steps of rAAV vector manufacturing with the baculovirus expression system, and purification with CsCl-gradient ultracentrifugation, as well as

Table 1
Examples of rAAV vector production yields in Sf9 cells

Exp. No.	Types and transgene	Yield (vector genome/L of Sf9 culture)
1	AAV1–GFP	8.5×10^{13}
2	AAV9–GFP	4.1×10^{13}
3	AAV9–R65	8.3×10^{13}
4	AAV8–GFP	2.5×10^{14}
5	AAV2–GFP	1.6×10^{14}
6	AAV2–RFP	1.7×10^{14}

desalting and buffer exchange with PD-10 desalting columns. Furthermore, it also provides detailed descriptions of manufacturing AAV vectors up to 1×10^{15} vg and purification of the vectors with the conventional CsCl-gradient ultracentrifugation method that does not require the investment in expensive equipments.

3.1. Preparation of Recombinant Bacmid DNA

1. Label three microcentrifuge tubes, one for pFB-inRep, second for pFB-inCap2, and the third for pFB-GFP and put in the ice bucket. Thaw one vial of MAX Efficiency DH10 Bac competent cells on ice and dispense 20 μL into each tube. Aliquot the rest of competent cells into 20 μL/vial and store at −80°C for later use. Add 2 μL (1 ng/μL) of the recombinant shuttle plasmid to the corresponding tube and gently mix the DNA with the competent cells. Incubate on ice for 30 min.
2. Heat shock the competent cells for 30 s at 42°C and chill on ice for 2 min. Add 500 μL of SOC to the competent cells and incubate at 37°C with agitation for at least 4 h to allow recombination happen.
3. Take 2.5 μL and 25 μL of the transformation mixture after 4-h incubation and dilute each into 100 μL with SOC medium. Spread each onto one LB-plate containing kanamycin, gentamycin, tetracycline, X-gal, and IPTG. Incubate at 37°C for 2 days (40–48 h) to allow the development of color.
4. Pick several white colonies (usually three white colonies are sufficient) from each transformation and culture them overnight in 3 mL of LB-medium containing kanamycin, gentamycin, and tetracycline.
5. The next morning, transfer 2 mL of the overnight culture to a microcentrifuge tube and centrifuge at full speed in a tabletop microcentrifuge for 30 s. Resuspend pelleted bacterial cells in 300 μL of Buffer P1 through vortexing. Add 300 μL of Buffer P2 and gently invert the tube ten times to mix. Do not vortex or it may shear the large bacmid DNA molecules. Incubate at room temperature for 5 min. Then add 300 μL of Buffer P3 and gently invert the tube ten times to mix.
6. Centrifuge for 5 min at full speed at room temperature and carefully pour the supernatant to a 2-mL microcentrifuge tube containing 650 μL of isopropanol. Mix by gently inverting the tube ten times. Centrifuge at full speed (>13,000 × *g*) for 15 min at room temperature to pellet the Bacmid DNA. Carefully pour out the supernatant immediately after centrifugation. Add 1 mL of 70% ethanol to rinse the pellet and centrifuge at full speed for 1 min. Work under sterile environment and carefully pour out the ethanol. Centrifuge again briefly and remove the remaining ethanol with a pipette tip. Add 60 μL of sterile TE buffer to each tube. Carefully dissolve the pellet using a pipette. Avoid vortexing and harsh pipetting!

7. Verify the presence of gene inserts in the bacmids by PCR assay using M13 Forward (5′-GTTTTCCCAGTCACGC-3′) and M13 Reverse (5′-CAGGAAACAGGTATGAC-3′) primers (see Note 1). Pick the correct bacmids and proceed to the next step for generation of recombinant baculoviruses or store the bacmids at −20°C; if not use immediately.
8. Prepare a bacmid glycerol stock by mixing 0.8 mL of overnight culture from step 5 and 0.2 mL of 50% glycerol and store at −80°C for future bacmid DNA isolation.

3.2. Generation of Recombinant Baculoviruses and Manufacturing of rAAV in Shaker Flasks

1. Seed Sf9 cells in a 6-well plate at 1×10^6 cells/well in 2-mL medium and allow the cells to attach for at least 30 min before transfection.
2. For each well of a 6-well plate, dilute 10 μL of the Bacmid DNA into 90 μL medium without antibiotics in a 1-mL microcentrifuge tube and 6 μL CellFectin into 94 μL medium without antibiotics in a second 1-mL microcentrifuge tube. Transfer the diluted transfection reagent solution to the tube containing Bacmid DNA and mix gently by slowly pipetting up and down three times. Incubate at room temperature for 30–45 min.
3. Add 0.8 mL of medium without antibiotics to each tube that contains the transfection mixture. Now aspirate the medium from the cells and add the transfection mixture on to the cells (1 mL total for each well). Incubate for 5 h at 28°C. At the end of incubation remove the transfection solution and add 2 mL of medium containing the antibiotics (100 U/mL of penicillin and 100 μg/mL of streptomycin) and incubate for 3–4 days at 28°C (see Note 2).
4. Tilt the plates and collect the media that contains the recombinant baculoviruses (P_0 stock) in 15-mL tubes and centrifuge for 5 min at 2,000 rpm (900 × *g*) to remove cell debris. Store the baculoviruses at 4°C under dark (see Note 3).
5. In a 500-mL shaker flask, seed 200 mL of Sf9 cells at 2×10^6 cells/mL and incubate at 28°C and 140 rpm until the cell density reaches about 8×10^6 cells/mL. Dilute the cells to 2×10^6 cells/mL with prewarmed fresh medium containing antibiotics in the shaker flasks. Prepare three 500-mL shaker flasks with 200 mL of diluted Sf9 cells, one for Bac-inRep, a second for Bac-inCap2, and a third for Bac-GFP. Add 1 mL of the P_0 stock baculovirus, respectively to the three flasks and incubate for 3 days (72 h) at 28°C and 140 rpm in the shaker incubator.
6. Harvest the supernatants by centrifugation at 2,000 rpm (900 × *g*) for 10 min in a Beckman GS-6 Centrifuge. These are the amplified passage 1 (P_1) baculoviruses. Store at 4°C under dark. Titer the baculoviruses according to Virovek's BacEZ Baculovirus Titer Kit protocol. The P_1 baculoviruses work very well for rAAV production and are used as working stock. However,

plaque purification is required if repeated amplifications are needed to obtain more baculoviruses (see Note 4).

7. Seed 200 mL of Sf9 cells with a density of at least 2×10^6 cells/mL in a 1,000-mL shaker flask and culture at 28°C and 140 rpm in a shaker incubator until the density reaches about 8×10^6 cells/mL (see Note 5). Dilute the cells 1:1 with prewarmed medium to obtain about 400-mL volume with cell density of about 4×10^6 cells/mL. Calculate the required volume of each recombinant baculovirus for infection of a 400-mL culture with 1 moi each of the three recombinant baculoviruses.
8. Mix the recombinant baculoviruses together and add to the diluted Sf9 cells. Incubate at 28°C and 140 rpm in the shaker incubator for 3 days. Harvest cell pellets by centrifugation at 3,500 rpm (2,750 × *g*) for 10 min in the Beckman GS-6 Centrifuge and discard the supernatant. Proceed to next step for purification or store the cell pellets at −20°C if not used immediately.

3.3. Purification of rAAV Vectors Through CsCl-Gradient Ultracentrifugations

1. Resuspend each cell pellet collected from a 400-mL culture into 17 mL of Sf9 lysis buffer containing 125 U/mL of Benzonase. Set the Output Control of a Branson Sonifier 250 to 7 (Micro Tip Limit) and sonicate for 20–30 s to thoroughly break the cells.
2. Incubate the lysate at 37°C for 30 min to digest the nucleic acids. Then add 0.4 mL 5 M NaCl to the lysate, mix and incubate for another 30 min. At the end of incubation add 1.4 mL of 5 M NaCl to the lysate to adjust the NaCl concentration to about 500 mM. Mix and centrifuge at 8,000 rpm (7,750 × *g*) for 30 min in a Beckman Avanti-J25 Centrifuge to pellet the cell debris. Collect the supernatant.
3. Set up step CsCl-gradient in ultra-clear centrifuge tubes for SW28 rotor by carefully adding 10 mL of 1.32 g/cc CsCl solution first and then 5 mL 1.55 g/cc CsCl solution to the bottom. Mark the interface between the CsCl solutions. Carefully load the supernatant (21–22 mL) onto the top layer of the CsCl and balance the centrifuge tubes. Centrifuge at 28,000 rpm (141,000 × *g*) overnight (17–20 h) at 15°C in a Beckman Optima LE-80K Ultracentrifuge.
4. Carefully take out the tubes and assemble one tube at a time in a stand. Aim a beam light at the bottom of the tube to assist visualization of the AAV bands. An example of AAV vector banding after first CsCl ultracentrifugation is shown in Fig. 1a.
5. Insert a syringe needle (18G1 gauge) slightly below the interface mark to draw the rAAV band. Collect about 6 mL samples and proceed to next step or store the samples at 4°C.
6. Transfer the samples into ultraclear centrifuge tubes for SW41 rotor and add CsCl solution (1.38 g/cc) to near the top of the tube. Carefully balance the tubes.

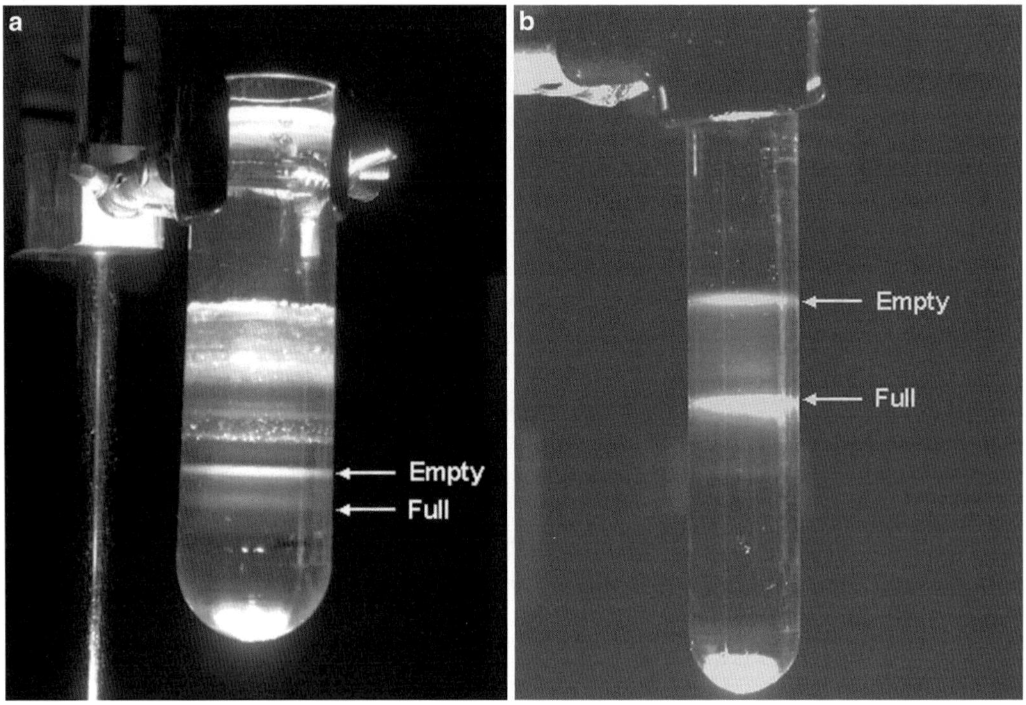

Fig. 1. Examples of rAAV banding after CsCl ultracentrifugation. (**a**) First round of CsCl ultracentrifugation with SW28 rotor. (**b**) Second round of CsCl ultracentrifugation with SW41 rotor. *Arrows* indicate the empty and the full rAAV particles.

7. Centrifuge at 41,000 rpm (288,000 × *g*) for 2 days. Carefully take out the tubes and assemble one tube at a time in a stand. Aim a beam light at the bottom of the tube to assist visualization of the AAV bands. An example of the AAV vector banding after second CsCl ultracentrifugation is shown in Fig. 1b. Draw the viral band (lower band) using a syringe. Collect about 2.5 mL of rAAV samples from the centrifuge tubes. Store the samples at 4°C until desalting.

3.4. Salt Removal and Buffer Exchange Using PD-10 Desalting Columns

1. Cut tips of the columns and pour out the storage solution. Assemble the columns with reservoir in the Vac-Man Vacuum Manifold and add 25 mL of equilibration buffer to each column (the final buffer for the rAAV vectors). Turn on the vacuum and control the flow rate at 10–15 mL/min until all the solution runs out.
2. Assemble the columns in 50-mL centrifuge tubes and centrifuge the columns at 2,000 rpm (900 × *g*) for 2 min in a Beckman GS-6 Centrifuge to remove excess buffer from the columns.
3. Transfer 2.5 mL of rAAV sample to each column and assemble in a new 50-mL centrifuge tube. Centrifuge at 2,000 rpm (900 × *g*) for 2 min to collect the rAAV solution. More than 99% of the salts and detergents are removed by this step.
4. Assemble the used PD-10 columns in the Vac-Man Vacuum Manifold. Add 25 mL of equilibration buffer to each column

and apply gentle vacuum (10–15 mL/min) to draw all buffer through. Centrifuge the columns at 2,000 rpm (900 × *g*) for 2 min to remove excess buffer.

5. Assemble the columns in fresh 50-mL centrifuge tubes. Transfer the rAAV solutions to the columns and spin at 2,000 rpm (900 × *g*) for 2 min. Collect the rAAV solutions. These repeated desalting steps ensure that trace of the salts and detergents are removed and the rAAV vectors are exchanged to the final buffer. Sterilize the rAAV vectors using syringe filters with low-protein binding properties.
6. Determine the rAAV genome copy numbers by quantitative PCR (qPCR) method, aliquot the vectors, and store at −80°C until use.
7. Examine the purity of rAAV vectors with SDS–PAGE followed by silver-staining. An example of silver-staining gel is shown in Fig. 2.

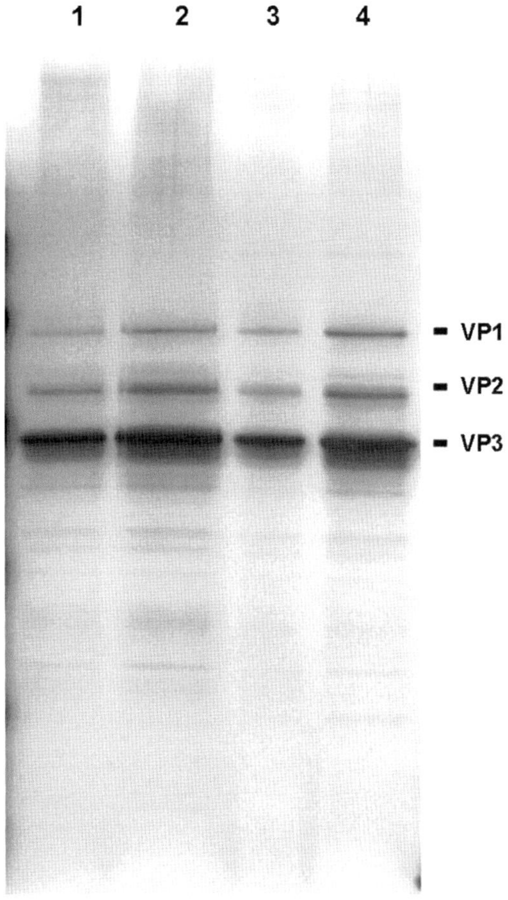

Fig. 2. An example of silver-staining gel showing the purity of rAAV samples purified by double CsCl ultracentrifugations. *Lanes 1* and *2*, AAV2-GFP Lot#1, 1×10^{10} vg and 2×10^{10} vg loaded, *lanes 3* and *4*, AAV2-GFP lot#2, 1×10^{10} vg and 2×10^{10} vg loaded.

3.5. Large Scale Manufacturing of rAAV Vectors

If more than 5×10^{14} vg of rAAV vectors are required for experiments, one can easily scale up to 10 L either with 25 shaker flasks of 1,000-mL volume that can hold 400-mL culture or a 20-L Wave bioreactor that has 10 L culture volume for the manufacturing. These instructions assume the use of the 20/50 Wave Bioreactor System for the 10-L scale rAAV manufacturing. It can be linearly scaled up to 100 L using a 200-L Wave bioreactor and easily adapted to other bioreactor systems with minor modifications.

1. Prepare 200 mL of each of the three P_1 recombinant baculoviruses at titers of 2.5×10^8 pfu/mL or above. Freshly amplified recombinant baculoviruses are key to success.
2. Culture 2×400-mL Sf9 cells to about 8×10^6 cells/mL in a shaker incubator at 28°C and 140 rpm (see Note 5).
3. Transfer the cells into a 20-L Wave Cellbag with a peristaltic pump. Add about 3.2 L of prewarmed fresh medium to obtain cell density of about 2×10^6 cells/mL. Set the angle to 7 and speed to 18 of the Wave bioreactor system. Culture the cells at 28°C until it reaches between 8×10^6 and 1×10^7 cells/mL in the 4-L volume. You may need to batch-feed the cells with small volume (0.25 vol.) of fresh medium to boost cell growth after 48 h of culturing if they do not reach 8×10^6 cells/mL.
4. Calculate the required amount of recombinant baculoviruses (Bac-inRep, Bac-inCap2, and Bac-GFP) based on the baculovirus titers and the total cell number in the Cellbag. One moi each of the three baculoviruses is required for the triple infection. Mix the three recombinant baculoviruses first and then transfer to the cell culture with the peristaltic pump. Add prewarmed fresh medium to the culture to make a total volume of about 10 L. Start O_2 pump and set to 30% oxygen, increase speed to 25 rpm and angle to 9°.
5. Infect for 3 days (72 h) at 28°C. Collect about 800 mL in each centrifuge tube and spin at 3,500 rpm ($2,750 \times g$) for 10 min. Discard the supernatant and store the cell pellets (a total of 12 pellets) at −20°C if not used immediately.
6. Proceed to the next step for rAAV purification.

3.6. Large-Scale Purification of rAAV Vectors Through CsCl-Gradient Ultracentrifugations

To purify rAAV vectors from a 10-L culture volume, two ultracentrifuges with 50.2 Ti rotors are preferred. If only one ultracentrifuge is available, just repeat the centrifugation to process all the samples. This instruction assumes the use of two ultracentrifuges with the 50.2 Ti rotors.

1. Prepare 420 mL of Sf9 lysis buffer and add benzonase to a final concentration of 125 U/mL right before use. Resuspend each cell pellet in 34-mL of Sf9 lysis buffer by vortexing and then sonicate for 30–40 s in an ice bucket to break the cells.

Thoroughly lysis of the cells is very important to get good AAV recovery. Incubate the samples at 37°C for 30 min to digest the nucleic acids. Then add 0.8 mL of 5 M NaCl to each tube, mix and incubate for another 30 min. At the end of incubation, add 2.8 mL of 5 M NaCl to each tube to adjust the NaCl concentration to about 500 mM. Mix and centrifuge at 8,000 rpm (7,750 × *g*) for 30 min to pellet the cell debris. Harvest the supernatants.

2. Setup the step CsCl-gradient in 24 ultraclear centrifuge tubes for 50.2 Ti rotor in the following order: first transfer the cleared cell lysates to the centrifuge tubes through syringes attached to an 18 gauge 3½ in. needle and then add 10 mL 1.32 g/cc followed by 7 mL 1.55 g/cc CsCl solutions. Fill the tubes with PBS and carefully balance the tubes. Seal the tubes and assemble in the rotor with caps.
3. Set both acceleration and deceleration to low and temperature to 15°C. Centrifuge at 50,000 rpm (302,000 × *g*) overnight (>20 h).
4. Carefully take out the tubes and assemble one tube at a time in a stand. Aim a beam light at the bottom of the tube to assist visualization of the AAV bands. The upper band contains the empty AAV particles and the lower band the full rAAV particles. Draw the lower band with a syringe needle (18G1 gauge). Collect about 6 mL samples.
5. Combine all the samples in a 250-mL storage bottle and transfer the samples equally into four ultraclear centrifuge tubes for 50.2 Ti rotor. Fill to top with CsCl solution of 1.38 g/cc and balance the tubes. Seal the tubes and assemble with caps in the rotor.
6. Set deceleration at low and temperature to 15°C. Centrifuge at 50,000 rpm (302,000 × *g*) overnight (>20 h).
7. Collect the lower band with a syringe by drawing 4–5 mL samples from each tube.
8. Proceed to desalting steps as described in Subheading 3.4.
9. Determine the rAAV genome copy numbers by quantitative PCR (qPCR) method, aliquot the vectors, and store at −80°C until use.

4. Notes

1. PCR amplification of non-recombinant bacmid will yield a fragment of ~300 bp and of recombinant bacmid a fragment of ~2,300 bp + size of insert (see Bac-to-Bac Baculovirus Expression System User Manual).

2. Transfection for 4 days instead of 3 days yields higher titers of baculoviruses. Although the protocol from Invitrogen indicated that 4-day amplification could yield lower quality of baculoviruses, lower quality of recombinant baculoviruses for AAV production was not observed.
3. Baculoviruses are sensitive to light and need to be stored under dark.
4. Plaque purification is required for repeated amplifications of baculoviruses. Otherwise the rAAV production yield will be compromised.
5. Sf9 cell usually takes 24 h to double its number. If after 48-h incubation the cell density does not reach 8×10^6 cells/mL, add 0.5 vol. of fresh medium and continue to incubate for another 24 or more hours. Adding fresh medium will boost the cell growth.

References

1. Grimm, D. and Kay, .M.A., (2003) From virus evolution to vector revolution: use of naturally occurring serotypes of adeno-associated virus (AAV) as novel vectors for human gene therapy. *Curr. Gene Ther.* **3**, 281–304.
2. High, K.A. (2003) Theodore E. Woodward Award. AAV-mediated gene transfer for hemophilia. *Trans. Am. Clin. Climatol. Assoc.* **114**, 337–352.
3. Hildinger, M. and Auricchio, A. (2004) Advances in AAV-mediated gene transfer for the treatment of inherited disorders. *Eur. J. Hum. Genet.* **12**, 263–721.
4. Daya, S. and Berns, K.I. (2008) Gene Therapy Using Adeno-Associated Virus Vectors. *Clin. Microbiol. Rev.* **21**, 583–593.
5. Urabe, M., Ding, C., and Kotin, R.M. (2002) Insect Cells as a Factory to Produce Adeno-Associated Virus Type 2 Vectors. *Hum. Gene Ther.* **13**, 1935–1943.
6. Urabe, M., Nakakura, T., Xin, K.Q., Obara, Y., Mizukami, H., Kume, A., et al. (2006) Scalable generation of high-titer recombinant adeno-associated virus type 5 in insect cells. *J Virol.* **80**, 1874–1885.
7. Kohlbrenner, E., Aslanidi, G., Nash, K., Shklyaev, S., Campbell-Thompson, M., Byrne, B.J., et al. (2005) Successful production of pseudotyped rAAV vectors using a modified baculovirus expression system. *Mol. Ther.* **12**, 1217–1725.
8. Merten, O.-W., Geny-Fiamma, C., and Douar, A.M. (2005) Current issues in adeno-associated viral vector production. *Gene Ther.* **12 Suppl 1**, S51–S61.
9. Chen, H. (2008) Intron Splicing-mediated Expression of AAV Rep and Cap Genes and Production of AAV Vectors in Insect Cells. *Mol. Ther.* **16**, 924–930.

Chapter 11

Vector Characterization Methods for Quality Control Testing of Recombinant Adeno-Associated Viruses

J. Fraser Wright and Olga Zelenaia

Abstract

Adeno-associated virus (AAV)-based vectors expressing therapeutic gene products have shown great promise for human gene therapy. A major challenge for translation of promising research to clinical development is the establishment of appropriate quality control (QC) test methods to characterize clinical grade AAV vectors. This chapter focuses on QC testing, providing an overview of characterization methods appropriate for clinical vectors prepared for early phase clinical studies, and detailed descriptions for selected assays that are useful to assess AAV vector safety, potency, and purity.

Key words: Recombinant AAV, Clinical trials, Quality control, Vector characterization

1. Introduction

Recombinant adeno-associated viruses (rAAVs) have demonstrated significant promise as DNA-delivery vectors to treat serious human diseases. Clinical studies using AAV vectors for alpha1-antitrysin deficiency, Alzheimer's disease, arthritis, Batten's disease, Canavan's disease, cystic fibrosis, Hemophilia B, HIV infection, Leber congenital amaurosis, muscular dystrophy, Parkinson's disease, prostate cancer, and malignant melanoma have been initiated (1–5). The manufacture and certification for use of recombinant AAV as an investigational biologic product requires knowledge of the complex methods needed to generate, purify, and characterize AAV vectors, combined with the implementation of current Good Manufacturing Practice (cGMP), comprehensive procedural controls to ensure clinical product purity, potency, safety, and consistency. A general description of cGMP for biologic products (6) and specific guidelines for

Otto-Wilhelm Merten and Mohamed Al-Rubeai (eds.), *Viral Vectors for Gene Therapy: Methods and Protocols*, Methods in Molecular Biology, vol. 737, DOI 10.1007/978-1-61779-095-9_11,

Chemistry, Manufacturing and Controls information for gene therapy products are available (7, 8).

1.1. General Remarks

AAV vectors manufactured for use in clinical research should be extensively characterized, and quality control (QC) test results must meet the predetermined specifications pertaining to vector safety, potency, purity, identity, and stability (7, 9). The quality of the vector product is established by factors that include (1) appropriate selection and control of raw materials, reagents, and components used in manufacturing process; (2) use of optimized manufacturing process steps for vector generation and purification; (3) rigorous environmental controls during aseptic processing to prevent inadvertent contamination; and (4) optimized final product formulation and storage. Implementation of rigorous QC testing supports quality by design by ensuring the high quality of manufacturing raw materials and components, aseptic performance of

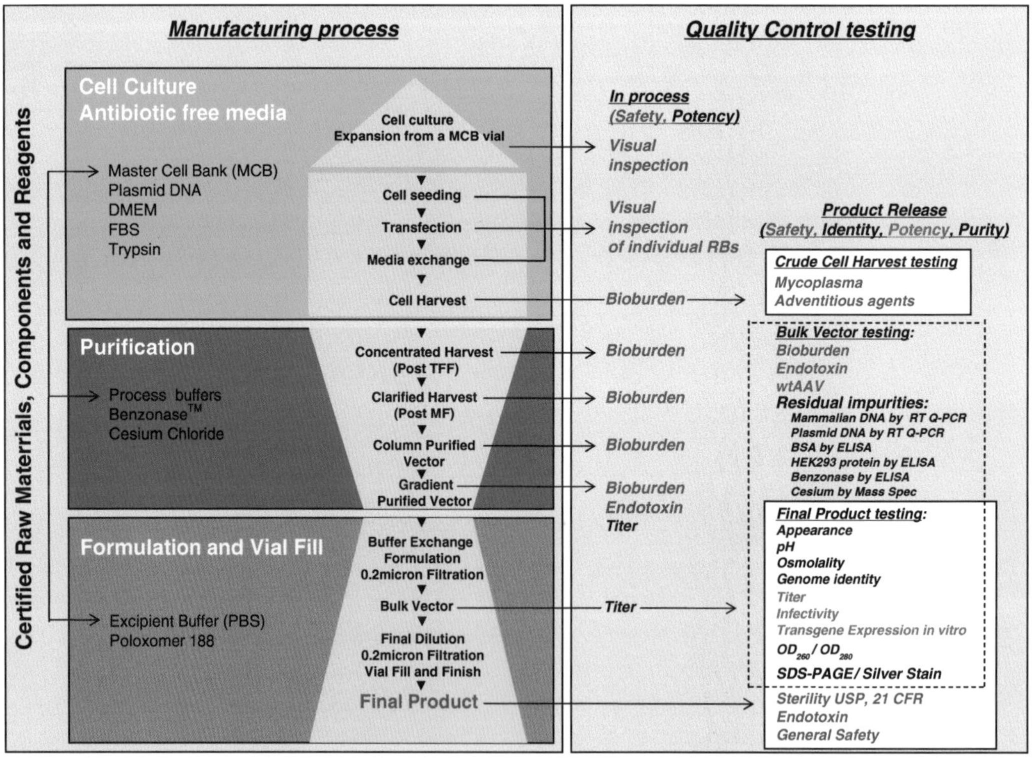

Fig. 1. An example of a quality control testing strategy for characterization of a clinical AAV gene therapy vector. The vector is manufactured using appropriately certified raw materials, components, and reagents using a well-defined manufacturing process designed to consistently achieve optimal vector product purity, potency, and safety. Quality Control tests are performed at each step of the manufacturing process, including process assessment of all product intermediates, and additional testing at key stages, including Crude Cell Harvest, Bulk Vector, and Final Product. In the vector characterization testing strategy shown here, several QC tests are performed at both the Bulk Vector and Final Product stages as indicated by the overlapping boxes. *MCB* Master Cell Bank, *TFF* tangential flow filtration, *PBS* phosphate-buffered saline, RT Q-PCR Real Time Quantitative PCR.

manufacturing process steps, and thorough characterization of the vector product at intermediate and final stages of manufacture. One example of a strategy for QC testing in support of clinical-grade AAV vector manufacturing is illustrated in Fig. 1. A set of AAV vector product release assays and suggested specifications, listing QC tests that might be included in a Certificate of Analysis for a vector prepared for a Phase 1 clinical study, is provided in Table 1. Assays used for QC testing of clinical products must employ appropriate standards and references to ensure that results measured for a given Test Article are valid.

Table 1
Suggested quality control testing for an AAV vector investigational product

Test	Method	Specification
Harvest		
Safety: Viral contaminants	21CFR, in vitro assay	Not detected
Safety: Mycoplasmas	21CFR	Negative
Bulk drug substance		
Appearance	Visual inspection	Clear, colorless solution
pH	Potentiometry	7.3 ± 0.5
Osmolality	Osmometry	300–400 mOsm/kg
Potency: VG titer	Q-PCR	Product specific
Potency: Infectivity	Limiting dilution in C12 cells[59]	<100 vg/IU
Potency: In vitro expression	In vitro transduction/ELISA	Product specific
Purity: Protein	SDS-PAGE	Comparable to Reference
Purity: OD_{260}/OD_{280}	Spectrophotometry[78]	≥1.2
Purity: Residual host cell DNA	Q-PCR	<10 ng/dose
Purity: Residual plasmid DNA	Q-PCR	<100 pg/10^9 vg
Purity: Residual BSA	ELISA	Report result
Purity: Residual HEK293	ELISA	Report result
Purity: Residual Benzonase™	ELISA	<1.0 pg/10^9 vg
Purity: Residual cesium	Mass spectrometry	< 0.1 μg/10^9 vg
Safety: Endotoxin	LAL	<10 EU/mL
Safety: Sterility	Bioburden	Negative
Safety: wt AAV	Infectious Center Assay	<1 rcAAV/10^8 vg
Final product		
Appearance	Visual inspection	Clear, colorless solution
pH	Potentiometry	7.3 ± 0.5
Osmolality	Osmometry	300–400 mOsm/kg
Vector genome identity	DNA sequencing	Matches Reference
Potency: VG titer	Q-PCR	Product specific
Potency: Infectivity	Limiting dilution in C12 cells[59]	<100 vg/IU
Potency: In vitro expression	In vitro transduction/ELISA	Product specific
Purity: Protein	SDS-PAGE	Comparable to Reference
Safety: Sterility <USP>, 21CFR	21CFR	Negative
Bacteria and fungistatic activity	21CFR	No B and F activity
Safety: Endotoxin	LAL	<10 EU/mL

1.2. Safety

Quality control tests used to demonstrate the absence of microbial contamination in the vector product are critical to ensure parenteral product safety. Such testing should include assays for sterility, mycoplasma, endotoxin, general safety, and adventitious viral agents (AVA). Sensitive, validated tests to demonstrate the absence of mycoplasma and AVA must be performed on samples of crude cell harvest, the point at which such contaminants would be most readily detected. Sterility testing must be performed on the Final Product. The sterility assay must be qualified by the assessment of the bacteriostatic and fungistatic activity of the vector product, i.e., to ensure that the Test Article itself does not interfere with the detection of microbial contaminants. Endotoxin testing should be performed on the Final Product using a validated assay, and acceptance criteria must be appropriate for the planned route of administration. Bioburden testing should be performed on samples obtained at intermediate stages of the manufacturing process, for example, following each purification step, to document aseptic conditions throughout the manufacturing process. The formation of wild-type or replication competent AAV (wtAAV or rcAAV), by mechanisms including homologous or non-homologous recombination between AAV ITRs with *rep* and *cap* sequences present in trans in helper components or producer cells, may occur during vector generation (10, 11). Vector manufacturing components should be designed to eliminate the potential for wtAAV generation (11–14), and QC testing for wtAAV is required for clinical AAV vector products. Testing for wtAAV can be performed using an infectious center assay such as that described in this chapter. Development of wtAAV assays for AAV serotypes other than Type 2 is a challenge due to the low efficiency of infection of cultured cells observed for most non-Type 2 AAV serotypes. The use of real-time quantitative PCR (Q-PCR) using primers designed to detect amplicons expected to be present only in wtAAV, for example, an amplicon spanning from the AAV ITR to AAV Rep, is a strategy that may be useful.

1.3. Potency

Quality control assays to measure the concentration and functional activity of AAV vectors are required to define investigational product dosing and assure consistent lot-to-lot functional activity. The concentration of vector genomes (vg) can be determined by dot blot hybridization or Q-PCR. Assays to quantify vg must include a nuclease step performed prior to denaturation of the vector so that only AAV packaged DNA (i.e., nuclease-resistant DNA) is measured. Real-time Q-PCR is amenable to standardization and validation, and an example of Q-PCR-based vector titer determination is described in this chapter. For measurement of functional activity, two steps required for manifestation of rAAV functional activity include (1) vector-mediated delivery of therapeutic DNA to the nucleus of target cells ("infection"), followed by; (2) expression, i.e., transcription and translation of the therapeutic DNA ("transduction"). Examples of measurement of vector infectivity for AAV2 vectors by quantification of vector genome DNA replication in target cells, and vector transduction by quantification of transgene expression by ELISA, are described in this chapter. Development of functional assays for AAV serotypes other than Type 2 may be a challenge because of the lower efficiency of infection and transduction of cultured cells observed for most non-Type 2 AAV serotypes. While infectivity and transduction assays measure vector-mediated delivery and expression of therapeutic DNA, and may be sufficient for characterization of vectors prepared for early phase clinical trials, they typically do not measure the functional activity of the therapeutic molecule produced in the target cell. Late phase clinical studies and product licensure require a bioassay that measures the functional activity of the actual therapeutic product (15).

1.4. Purity

As a guideline, purity specifications for recombinant AAV manufactured for clinical studies can be based in part on those developed for existing licensed biologic products, such as vaccines and recombinant proteins. However, unlike vaccines, AAV vectors generally aim to avoid immune responses. Rather, most AAV vectors are intended to establish long-term expression of the vector genome-encoded therapeutic product, a mechanism more akin to repeat/periodic administration of recombinant proteins for protein replacement, or therapeutic monoclonal antibodies. Therefore, a high level of purity should be attained, and to the degree possible, any impurities that may contribute to immune responses following in vivo administration and thereby limit transgene expression should be removed. Quality control tests should be established to measure levels of process impurities including residual host cell constituents, residual production plasmid DNA, and other reagents that are added to the product as part of the manufacturing process. Useful methods to assess protein impurities in purified recombinant viral vectors include sodium dodecyl sulfate polyacrylamide gel electrophoresis (SDS-PAGE), reverse phase high-pressure liquid chromatography, enzyme linked immunoassays (ELISA) to measure residual production cell protein and bovine serum albumin (BSA) (e.g., if bovine serum used for cell culture-based vector generation), and spectrophotometry. Examples of the use of SDS-PAGE and spectrophotometry for the assessment of purity are provided in this chapter. A useful method to assess DNA impurities is Q-PCR, using specific primers and probes to quantify specific target sequences of interest. Examples of specific methods for Q-PCR-based measuring of residual production plasmid DNA and residual host cell DNA are provided in this chapter.

A particular challenge for AAV vector manufacture is that vector-related impurities are generated concurrent with vector generation during cell culture. Such impurities, including empty capsids and AAV-encapsidated nucleic acid impurities, represent unnecessary extra viral antigen if they are not removed. AAV empty capsids are generally produced in amounts corresponding to >50% of total AAV particles generated in cell culture (16, 17). Vector-related impurities also include AAV-encapsidated nucleic acids, which are fragments of production cell- or helper component-derived nucleic acids contained within AAV capsids. Packaged residual DNA impurities derived from production plasmids in vectors prepared by transient transfection or by helper virus infection of stably transfected mammalian cells have been reported to range from 1 to 8% of total DNA in purified vector particles (18–20). Packaging of fragments of mammalian producer cell genomic DNA has been reported to range from ~1% of total DNA in highly purified vectors to ~3% in vectors that are co-purified with empty capsids (21). Levels of residual DNA impurities in gene therapy products for human use is a topic of ongoing discussion (22), in any case such impurities should be reduced to the lowest levels achievable. Design of manufacturing processes used to prepare clinical-grade vectors should aim to minimize generation of vector-related impurities during cell culture, and maximize their removal during purification, and QC methods established to measure vector-related impurities in the Final Product. Total AAV particles (i.e., bona fide vectors plus vector-related impurities) can be measured using capsid-specific ELISA (16) or by a nonspecific absorbance assay (17). Levels of AAV-encapsidated DNA impurities can be assessed by Q-PCR using primers and probes, and distinction can be made between encapsidated vs. “naked” DNA impurities based on nuclease sensitivity.

1.5. Stability

Quality control testing is required to document vector stability as part of the characterization of vectors prepared for clinical studies (7, 23, 24). Formal stability studies are required to verify that purified clinical vectors stored under designated storage conditions maintain their purity, potency, and safety characteristics. For early phase clinical studies, it may be acceptable to perform stability studies concurrent with vector use in clinical trials. However, by late stage clinical studies, the stability of a vector product should be well defined in order to provide accurate product expiry information.

1.6. Assay Validation

Quality control assays that are used to characterize IND-supporting preclinical studies (e.g., product efficacy and toxicity studies in animals) and early phase clinical studies should be developed and characterized so that data demonstrating assay sensitivity, linearity, accuracy, precision, and specificity are obtained and documented. For QC assay validation, an incremental approach may be acceptable based on recognition that some of the information and data required to validate analytical methods are obtained during clinical product development (25). However, QC tests used to ensure the absence of microbial and adventitious viral contaminants in starting materials, intermediates, and final vector product should be validated from the outset of clinical development. All QC assays must be finalized and validated for Biologics License Application.

The analytical methods described below represent current practices employed by our laboratory for QC testing of recombinant AAV generated by transient transfection of HEK293 cells and purified by combined column chromatography and gradient ultracentrifugation. Eight specific assays are described here in detail; however, additional methods are required for comprehensive clinical product characterization. Quality control methods that are considered as current practices should be expected to evolve and improve.

2. Materials

2.1. Replication Competent AAV (rcAAV)

1. Low passage HEK293 cells (ATCC).
2. Growth medium: DMEM supplemented with 10% FBS/Pen/strep (GIBCO/BRL, Bethesda, MD).
3. Adenovirus with titer $>1.5 \times 10^8$ vg/mL.
4. wtAAV2 with titer $>1 \times 10^9$ vg/mL.
5. 37°C, 5% CO_2 incubator.
6. 37°C water bath.
7. Class 100 biosafety cabinet.
8. Filtration apparatus, Kontes.
9. Hemacytometer.
10. Hybridization oven and bottles.
11. UV cross-linker.
12. X-ray cassette, Fisher.
13. Film, BioMax MR, Kodak.
14. X-ray film developer.
15. Nitrocellulose membrane filters (Millipore).
16. Pipettes 2–20, 20–200, 200–1,000 μL, and tips.
17. Cell culture dishes: 100 mm, Falcon (PN 353003), and six-well, Corning (PN 3506).
18. Conical tubes, 15 and 50 mL.

19. Dry ice.
20. Forceps.
21. Microcentrifuge tubes.
22. Saran wrap.
23. Sterile 10 mL serological pipettes.
24. 150 mL bottle.
25. Filter paper.
26. ^{32}P-labeled AAV2 Cap probe. (see Note 1).
27. Hybridization buffer, QuikHyb® (Stratagene).
28. 10 N NaOH.
29. Sodium chloride.
30. Phosphate-buffered saline.
31. 20× SSC (BioWhittaker).
32. 10% SDS solution.
33. Tris–HCl.
34. Trypsin–EDTA (Gibco BRL).

2.2. Preparation of DNA Standards for Real-Time Q-PCR

1. Gel electrophoresis box and power supply.
2. Microcentrifuge.
3. Spectrophotometer.
4. Vortex.
5. 37°C water bath.
6. Agarose, ultrapure (Invitrogen).
7. Ethanol.
8. 10 mg/mL ethidium bromide (Invitrogen).
9. Microcentrifuge tubes.
10. Nuclease-free water.
11. Salmon sperm DNA (Stratagene).
12. 3 M sodium acetate, pH 6.0.
13. Tris–EDTA, pH 8.0.

2.3. Vector Genome Titer by Real-Time Q-PCR

1. Sequence detection system, ABI Model 7500.
2. Centrifuge with 96-well reaction plate adaptor.
3. 65°C heating block for microfuge tubes.
4. Microcentrifuge.
5. 0.1–2, 2–20, 20–200, and 200–1,000 μL pipettes, with tips.
6. Repeater pipette.
7. 37°C water bath.

8. 96-Well reaction plate (ABI).
9. Conical tubes, 50 mL (Corning).
10. Microcentrifuge tubes.
11. Optical adhesive cover (ABI).
12. Rack for 96-well reaction plate (ABI).
13. DNase I and 10× DNase buffer (Invitrogen).
14. 5 mM EDTA pH 8.0.
15. Nuclease-free water.
16. Q-PCR dilution buffer.
17. Salmon sperm DNA.
18. Universal PCR Master Mix (ABI).
19. Target-specific reagents: TaqMan Probe (100 μM, TE), Forward primer (50 μM, TE).
20. Reverse primer (50 μM, TE), Q-PCR Standard stock (linearized plasmid DNA 2×10^{11} copies/mL, TE).
21. Reference Vector.
22. Standard diluent: 2 ng/μL Salmon Sperm DNA in nuclease-free water.

2.4. Vector Infectivity by Limiting Dilution with Q-PCR Readout

1. Low passage C12 cells (ATCC).
2. Culture medium: DMEM/10% FBS supplemented with G418 Sulfate and Pen/strep (GIBCO BRL).
3. ABI Model 7500 Sequence Detection System.
4. Target-specific reagents: TaqMan forward and reverse primers (50 μM, TE), TaqMan probe (100 μM, TE), and plasmid DNA standard.
5. TaqMan Universal PCR Master Mix (ABI).
6. 2–20, 20–200, and 200–1,000 μL pipettes, and tips.
7. 1–20, 5–50, and 50–300 μL 12-channel pipettes, and tips.
8. 37°C, 5%, CO_2 incubator.
9. Heating block.
10. Centrifuge.
11. Vortex.
12. A 1-L, 0.2-μm filter unit.
13. Sterile 96-well U-bottom tissue culture plates.
14. Sterile reservoirs.
15. 50 mL conical tube.
16. Sterile microcentrifuge tubes.
17. Optical adhesive cover (ABI).

18. Rack for 96-well reaction plate (ABI).
19. 10% SDS solution.
20. 10 mM Tris, 2 mM EDTA, pH 8 (TE).
21. Tris–HCl 1 M, pH 8.0.
22. 500 mM EDTA.
23. Adenovirus.
24. Milli-Q water.

2.5. Vector Potency by Transduction and Transgene ELISA

1. HepG2 cells (ATCC).
2. DMEM supplemented with 10% FBS (Hyclone).
3. Pen/strep (Gibco BRL).
4. Trypsin/EDTA (Gibco BRL).
5. PBS.
6. Class 100 biosafety cabinet.
7. Cell culture microscope.
8. 37°C water bath.
9. 37°C, 5% CO_2 incubator.
10. Hemocytometer.
11. Cryo-vial rack.
12. 24-Well tissue culture dishes (Falcon).
13. Sterile microcentrifuge tubes.
14. Pipette aid (Drummond).
15. 2–20 and 20–200 μL pipettes, and tips.
16. T75 cm^2 tissue culture flasks.
17. 15 mL conical tubes (Corning).
18. Balance.
19. Magnetic plate stirrer and stir bar.
20. 37°C 5% CO_2 incubator.
21. 12-Channel pipette, 5–50 μL, and tips.
22. 12-Channel pipette, 50–300 μL, and tips.
23. Pipette, 20–200 μL, and tips.
24. Pipette aid (Drummond).
25. Plate Reader (Molecular Device, Spectramax 190).
26. Timer.
27. Vortex.
28. Aluminum foil.
29. 1,000 mL beaker.
30. Bucket of crushed ice.

31. 50 mL conical tubes.
32. 500 and 1,000 mL graduated cylinders.
33. 500 and 1,000 mL glass bottles.
34. Microcentrifuge tubes.
35. Parafilm.
36. Sterile 10 mL pipettes.
37. Immuno-plate, Immuno-plate (Maxisorp, Nunc).
38. 50 mL reservoir.
39. Sealing Tape (Nunc).
40. ABTS buffer (Roche).
41. ABTS tablets (Roche).
42. BSA (Sigma).
43. 0.5 M EDTA pH 8.0.
44. Monoclonal mouse anti-human Factor IX Clone HIX-1 (Sigma).
45. Goat anti-hFIX antibody HRP-conjugated (Enzyme Research Laboratory).
46. Milli-Q water.
47. Nonfat milk powder.
48. 10× PBS.
49. Sodium bicarbonate buffer 7.5% (Gibco BRL).
50. TriniCHECK Level 1, (Trinity Biotech).
51. 20% Tween 20 (Fisher Biotech).
52. ELISA Wash buffer: PBS supplemented with 0.5% Tween 20.
53. ELISA dilution buffer: PBS, 0.5% Tween 20, 1 mM EDTA pH 8.0, 6% BSA.
54. ABTS solution: 1 ABTS tablet in 50 mL ABTS buffer. Store at 2–8°C in the dark.

2.6. Vector Purity by SDS-PAGE/Silver Staining

1. Reference Vector.
2. Heating block for microfuge tubs.
3. Orbital shaker.
4. Pipette-Aid.
5. 2–20, 20–200, and 200–1,000 μL pipettes, and tips.
6. Microcentrifuge.
7. Electrophoresis power supply.
8. XCell SureLock Mini-Cell (Invitrogen).
9. 100 and 1,000 mL graduated cylinders.
10. PBS (Gibco BRL).

11. DryEase Mini-Gel dryer apparatus (Invitrogen).
12. DryEase Mini cellophane (Invitrogen).
13. Gel Knife.
14. Microcentrifuge tubes.
15. 4–12% NuPae Bis-Tris Gel, 1.0 mm ten-well (Invitrogen).
16. 5, 10, 25, and 100 mL serological pipettes.
17. StainEase Gel Staining Tray (Invitrogen).
18. Glacial acetic acid.
19. Gel-Dry drying solution (Invitrogen).
20. LDS Sample buffer 4×, NuPAGE (Invitrogen).
21. Mark12 unstained standards (Invitrogen).
22. Methanol.
23. Milli-Q water.
24. 20× MOPS SDS running buffer (NuPAGE, Invitrogen).
25. 10× sample reducing agent (NuPAGE Invitrogen).
26. SilverXpress silver staining kit (Invitrogen).
27. Solutions for silver staining (per kit instructions, freshly prepared).
28. Fixing solution: 90 mL Milli-Q water + 100 mL methanol + 20 mL glacial acetic acid.
29. Sensitizing solution: 105 mL Milli-Q water + 100 mL methanol + 5 mL sensitizer (the sensitizer is a component of the SilverXpress kit).
30. Staining solution: 95 mL Milli-Q water + 5 mL developer (the developer is a component of the SilverXpress kit).
31. Stopper solution: Add 5 mL of stopper solution directly to the staining solution (the stopper is a component of the SilverXpress kit) when proteins bands corresponding to the AAV VP bands have reached sufficient intensity, i.e., VP1, 2, and 3 are distinct. At this point, additional bands corresponding to protein impurities may or may not be visible, depending on the purity of the vector preparation being characterized.
32. Protein molecular weight marker: Add 5 μL 4× LDS sample buffer in a fresh microcentrifuge tube, add 14 μL Milli-Q water, add 1 μL Mark12 protein MW standard. and mix well. Ensure that crystalline precipitate in the protein MW standard tube is dissolved before using. Prepare fresh for each assay.
33. Cellophane (Invitrogen DryEase Mini Cellophane).

2.7. Vector Purity by Optical Density

1. 2–20, 20–200 μL pipettes, and tips.
2. Spectrophotometer (Beckman DU800 UV/Vis).

3. Lint-free disposable wipes.
4. Quartz cuvette (Beckman Coulter Micro Cell 8 mm).
5. Spray bottle containing DI water.
6. Diluent: 180 mM NaCl, 20 mM sodium phosphate, 0.001% Pluronic F68, pH 7.3.
7. Reference Vector: Requires a total of 0.3 mL, 0.1 mL for each of three replicate determinations.
8. Test Article: Requires a total of 0.3 mL, 0.1 mL for each of three replicate determinations.

2.8. Residual Mammalian DNA by Real-Time Q-PCR

1. 2–20, 20–200, 200–1,000 μL pipettes, and tips.
2. 96-Well reaction plate (ABI).
3. Optical adhesive cover (ABI).
4. Rack for 96-well reaction plate (ABI).
5. TaqMan Universal PCR Master Mix (ABI).
6. 20× 18S rRNA Assay Reagent (ABI).
7. Nuclease-free water.
8. Q-PCR dilution buffer.
9. Microcentrifuge tubes.
10. 50 mL conical tubes.
11. HEK293 cell genomic DNA (Roche).
12. Restriction enzyme *Eco*RI (New England Biolabs).
13. Phenol–chloroform–isoamylalcohol: 25:24:1 (Fisher).
14. Chloroform–isoamylalcohol: 24:1.
15. Tris–EDTA (Fisher).
16. Ethanol, 100%.
17. 3 M sodium acetate, pH 6.0.
18. BSA (New England Biolabs).
19. Microcentrifuge.
20. Sequence Detection System (ABI Model 7500).

2.9. Residual Plasmid DNA by Real-Time Q-PCR

1. 2–20, 20–200, 200–1,000 μL pipettes, and tips.
2. Microcentrifuge.
3. 96-Well reaction plate (ABI).
4. Optical adhesive cover (ABI).
5. Rack for 96-well reaction plate (ABI).
6. Sequence detection system, ABI Model 7500.
7. Microcentrifuge tubes.

8. 50 mL conical tubes.
9. Universal PCR Master Mix (ABI).
10. Q-PCR dilution buffer: 0.001% Pluronic F68, Sigma, CAT. # P5556, in nuclease-free H_2O, Fisher (PN BP2484).
11. Target-specific reagents: TaqMan forward and reverse primers, TaqMan probe, standard and Reference Vector.

3. Methods

3.1. Replication Competent AAV

The cell-based semiquantitative assay described here detects wild-type AAV2 (wtAAV) defined here as species that can replicate in HEK293 cells in the presence of helper adenovirus. This assay is designed to show that wtAAV in vector product is below a certain level as detected in a control that has been spiked with known amount of wtAAV (see Note 2).

1. Seed HEK293 cells in 15×100 mm dishes at density 6×10^6 cells per dish.
2. The next day, dilute Test Article, wtAAV2, and adenovirus stocks in fresh medium to final concentrations of 10^{11} vg/mL, 10^5 vg/mL, and 1.2×10^6 vg/mL, respectively.
3. Aspirate the medium from cells and perform triplicate transductions by adding virus dilutions to cells as described below:
   (a) *Negative control # 1, "wt AAV2 only":* Add 10 μL wtAAV2 (10^3 vg) in 10 mL medium.
   (b) *Negative control # 2, "Ad only":* Add 10 mL Ad dilution.
   (c) *Positive control:* Add 10 mL Ad dilution + 10 μL wtAAV2 (10^3 vg).
   (d) *Spike:* 10 mL Ad dilution + 10 μL Test Article (10^9 vg) + 10 μL wtAAV (10^3 vg) (see Note 3).
   (e) *Test:* 10 mL Ad dilution + 10 μL Test Article (10^9 vg).
4. Incubate cells at 37°C, 5% CO_2 for 48 h.
5. Trypsinize and harvest cells from each dish into a separate prelabeled 15 mL conical tube. The harvested cells can be stored at −80°C if not processed immediately.
6. To release the virus, lyse the cells by three–freeze/thaw rounds using dry ice and ethanol.
7. Centrifuge the tubes at $3{,}000 \times g$ for 5 min. Transfer the supernatant from each tube into a new 15-mL tube. Label appropriately. These supernatants will be used in the second round of transduction. The supernatants may be used immediately or stored at 2–8°C for up to 24 h prior to the second round of transduction.

8. The day before the second transduction prepare 5 six-well plates. Seed three wells on each plate with 1×10^6 HEK293 cells per well. Each plate represents an individual sample and each well of the plate an individual replicate in the first round of transduction.
9. Dilute adenovirus in culture medium to final concentration of 1.5×10^6 vg/mL.
10. Add 2 mL of medium + 250 μL of "Negative control # 1" supernatants to triplicate wells.
11. Add 2 mL of Ad dilution + 250 μL of supernatant from first transduction to all other wells (one well per tube). Label wells accordingly.
12. Incubate plates in 37°C, 5% CO_2 incubator for 24 h.
13. Trypsinize and harvest cells from each well into a separate 15 mL tube. Keep the tubes on ice until transferred onto filter membranes.
14. Using vacuum and filtration apparatus transfer the cell harvest from each tube to a filter (one tube per filter). Label the filter with sample ID. Make sure to rinse the apparatus thoroughly with PBS between samples to avoid cross-contamination.
15. Lyse the cells by placing the filters on Whatman paper soaked in 1.5 M NaCl, 0.5N NaOH for 5 min. Follow by neutralization in 3 M NaCl, 0.5 M Tris–HCl, pH 7.4, two times for 5 min each.
16. Soak filters in 2× SSC for 5 min and immobilize DNA by cross-linking.
17. To detect wtAAV particles, hybridize filters with ^{32}P-labeled AAV2 Cap probe (26). Use the following hybridization and wash conditions: prehybridization (QuikHyb®, 30 min, 65°C), hybridization (1 h, 65°C); wash with 1× SSC 1% SDS (65°C, 2 × 50 min); wash with 0.1× SSC 1% SDS (65°C, 2 × 20 min).
18. Expose filters to X-ray film and compare three Test Article filters with positive and negative control filters. Test filters may be negative or positive. If Test filters are positive, compare with spike filters with known concentration of wtAAV2 spike to determine the levels of contamination.

3.2. Preparation of DNA Standards for Real-Time Q-PCR

Linearized plasmid DNA standards (or other linear DNA templates) should be prepared and used to ensure accuracy and precision of Q-PCR for the quantification of target DNA amplicons (see Note 4).

1. Purified plasmid DNA carrying the sequence of interest.
2. Select single-cutting restriction enzyme with site located outside of the target sequence.

3. Linearize 20–50 μg of selected plasmid by restriction digest. Take an aliquot of digest and run on 1% agarose gel to verify that digest is complete.
4. Add 1/10th of remaining reaction volume of 3 M sodium acetate pH 6.0 and vortex.
5. Add 2.5 volumes of 100% EtOH, vortex.
6. Precipitate DNA at −20°C overnight. Centrifuge for 30 min at 14,000 ×*g*.
7. Decant the supernatant, wash DNA pellet with 500 μL of 70% EtOH and let air dry.
8. Dissolve the pellet in 200 μL TE. Make sure pellet is dissolved completely.
9. Determine the DNA concentration and purity (A_{260}/A_{280}) by spectrophotometry (26)
10. One microgram of a 1-kb dsDNA is equivalent to 9.1×10^{11} molecules (copies) or 1.82×10^{12} copies of ssDNA.
11. Calculate the linearized plasmid concentration in copies/mL using the formula given below:

$$\text{Concentration [copies / mL]} = \frac{\text{Conc.}[\mu\text{g/mL}] \times 9.1 \times 10^{11} [\text{copies} / \mu\text{g}]}{\text{Plasmid size [kb / 1kb]}}$$

12. Prepare Q-PCR Standard stock by diluting linearized plasmid in TE to final concentration of 2×10^{11} copies/mL.
13. Prepare 25 μL aliquots of Q-PCR Standard stock 2×10^{11} copies/mL. Store at −20°C. Use one aliquot to prepare standard curve dilutions for each Q-PCR assay.

3.3. Vector Genome Titer by Real-Time Q-PCR

This method utilizes target sequence-specific primers to define an amplicon within the target DNA sequence, and fluorescent probes within this amplicon to quantify the generation of fluorescence due to increasing concentration of the amplicon as the PCR reaction proceeds. The amplification curves are compared to curves generated using known dilutions of a DNA standard template such as linearized plasmid DNA containing the sequence of interest.

1. Prepare eight serial tenfold dilutions of Q-PCR Standard Stock (2×10^{11} copies/mL of ds linearized plasmid DNA) in standard diluent to obtain Standards S1 (2×10^{10} copies/mL) through S8 (2×10^{3} copies/mL) and standard curve with the range 10–10^{8} copies per Q-PCR reaction. Keep standards on ice.
2. Thaw Test Article and Reference vector and keep on ice.
3. Set up DNAse digest reactions for Test Article and Reference vector by mixing 10 μL of vector with 35 μL of 1× DNase buffer and 5 units of DNase I.

4. Set up DNase control by mixing 10 μL of Standard S1 (2×10^8 copies of plasmid) with 35 μL of 1× DNase buffer and 5 units of DNase I.
5. Mix by tapping and incubate at RT for 15 min.
6. Add 50 μL of 5 mM EDTA pH 8.0 to each digest tube and heat inactivate DNase for 10 min at 65°C. Use centrifuge tubes to collect condensate. Keep the tubes on ice.
7. Prepare three independent 10-, 30-, and 100-fold dilutions for each vector sample in Q-PCR dilution buffer (see Notes 5 and 6).
8. Perform additional 100-fold dilution in Q-PCR dilution buffer for each vector sample. Keep the tubes on ice.
9. Prepare 100-fold dilution for DNase control sample. Keep the tube on ice.
10. At this point you should have eight standards and seven samples ready for loading into Q-PCR reactions.
11. Thaw TaqMan probe and primers and prepare Q-PCR master mix according to the table below (see Note 6). Volume of Q-PCR Master mix should be sufficient to set up triplicate reactions for each of the eight standards, three Test Article dilutions, three Reference vector dilutions, DNase control, and No Template Control (NTC)

Reagents	Volume per reaction (μL)	Final concentration
Universal master mix (2×)	12.5	1×
Forward primer (50 μM)	0.5	1 μM
Reverse primer (50 μM)	0.5	1 μM
TaqMan probe (100 μM)	0.05	0.2 μM
Nuclease-free water	6.45	NA

12. Place 96-well optical reaction plate into plate rack and aliquot 20 μL Q-PCR Master Mix into each well.
13. Add 5 μL per well of each standard in triplicates.
14. Add 5 μL per well of Test Article, Reference vector, and DNase control samples in triplicates.
15. Add 5 μL Q-PCR dilution buffer as NTC in triplicates.
16. Seal the plate with optical adhesive cover, spin at 2,000 ×*g* for 2 min, and insert into ABI 7500.

Run 7500 System software Absolute Quantification (Standard Curve) protocol for "96-well clear plate" with the following cycler conditions (sample volume: 25 μL; run mode: 7500; data collection: Stage 3 step 2; standard thermo profile (profile may be modified if required); auto increment: 0 for all stages; ramp rate: 100% for all stages).

17. The software calculates the copy number of target sequence in each well (Qty, Column 7 in SDS data report) based on the Slope and Intercept of the Standard curve (C_t value vs. LOG (copy number of each standard)). The correlation coefficient (Pearson R^2) of the Standards reflects the average C_t value for each standard vs. the LOG (copy number of each standard). $\text{Qty} = 10^{(C_t - \text{Intercept})/\text{Slope}}$, where Qty is the quantity of plasmid genome equivalent in 5 μL of diluted sample. The SDS software also calculates mean quantity and standard deviation (SD) for triplicate reactions.
18. To calculate the Test Article and Reference vector titers (vg/mL) multiply the mean quantity (vg/5 μL, Column 8) by 4×10^5 (total dilution of vector during sample preparation, see explanation below) and by initial post-DNase dilution factor (10×, 30×, or 100×).

    Each vector has been diluted prior to the Q-PCR reaction:

Assay step	Dilution
DNase digest	Fivefold
EDTA addition	Twofold
Initial post-DNAse dilution	10-, 30-, or 100-fold
PrePCR dilution	100-fold
Conversion factors used in calculations	
1 plasmid genome equivalent to 2 vg	2× (see Note 8)
Volume conversion from 5 μL to 1 mL	200×
Total	$4 \times 10^5 \times$ (10 or 30 or 100)

19. Calculate the Q-PCR reaction efficiency using the formula given below:

$$\text{Amplification efficiency (\%)} = [10^{(-1/\text{slope})} - 1] \times 100$$

20. Determine the efficiency of the DNase in reaction by calculating the amount of undigested DNA as percent of DNA loaded in DNase control reaction (10^5 copies) as follows:

$$\text{Undigested DNA (\%)} = 100 \times \text{Mean quantity} / 10^5 \text{copies}$$

For our laboratory, a valid assay should achieve the following acceptance criteria: (1) R^2 of standard curve ≥0.985; (2) reaction efficiency 100 ± 10%; (3) mean C_t value in NTC higher than C_t value of the lowest standard or undetectable; (4) undigested DNA from the DNase control ≤1%; (5) calculated

Reference Vector titer within 25% of the established value; (6) relative standard deviation (RSD) of mean titer calculated for three dilutions ≤25%. If assay parameters do not meet these acceptance criteria, the assay is invalid and is repeated.

3.4. Vector Infectivity by Limiting Dilution with Q-PCR Readout

This assay evaluates specific infectivity of AAV2 vectors in vitro measured as a ratio of vector genomes per infectious unit (vg/IU), also referred to as a "particle to infectivity ratio." HeLa cells expressing AAV2 rep and cap genes (C12 cells) (27) seeded on a 96-well plate are infected with multiple log dilutions of Test Article in the presence of adenovirus. Under these conditions infectious vector particles enter the cell and vector genomes are replicated. Ten replicate infections are performed for each vector dilution. An adenovirus only control is included as a negative control to define background replication. After 48 h incubation, cells are lysed and transgene is quantified in each well by Q-PCR. Wells with C_t values below the negative control threshold value (i.e., indicating vector genome replication) are scored as positive, and the pattern of positive wells as a function of Test Article dilution are used to determine the infectivity titer using the $TCID_{50}$ method (28).

1. Seed C12 cells into 96-well plates: 1×10^4 cells, 100 μL medium per well, 80 wells per plate.
2. The next day, prepare 12 serial tenfold dilutions of Test Article.
3. Aspirate the medium and add 50 μL of culture medium containing 2.5×10^5 vg of adenovirus to each of the ten wells in rows A through H (80 wells).
4. Ad control: Add 50 μL of medium to each of the ten wells in row H.
5. Use 50 μL of each Test Article dilution 10^{-6} to 10^{-12} to transduce ten wells in each row of the plate (ten replicate wells per row, rows A (dilution 10^{-6}) through G (dilution 10^{-12})).
6. Incubate the cells at 37°C, 5% CO_2 for 48 h and lyse by adding 10 μL of Lysis buffer (0.1 M Tris–HCl pH 8.0 1% SDS) per well. Mix by pipetting five to ten times. Make sure to avoid cross-contamination between wells.
7. Seal the plate with adhesive cover, incubate at 95°C for 15 min and then centrifuge for 1 min, 1,000 × *g*.
8. Dilute each sample 100-fold using TE as diluent and a new 96-well plate.
9. Analyze diluted samples by real-time Q-PCR.
10. Prepare 2 mL of Q-PCR Master mix (1,250 μL TaqMan Universal PCR Master Mix, 655 μL H_2O, 45 μL target-specific

forward primer (50 μM), 45 μL target-specific reverse primer (50 μM), 5 μL target-specific probe (100 μM)). Vortex.

11. Add 20 μL of Master mix into each of 96 wells on a Q-PCR plate.
12. Add 5 μL of H_2O in all wells in column 11 (negative control).
13. Add 5 μL of diluted sample into each of the ten wells in rows A through H according to the schematic below. Start with the Ad only samples in row H, ascending toward row A.

	1	2	3	4	5	6	7	8	9	10	11	12
A	Dilution 10^{-6} →										H_2O	
B	Dilution 10^{-7} →										H_2O	
C	Dilution 10^{-8} →										H_2O	
D	Dilution 10^{-9} →										H_2O	
E	Dilution 10^{-10} →										H_2O	
F	Dilution 10^{-11} →										H_2O	
G	Dilution 10^{-12} →										H_2O	
H	Ad only →										H_2O	

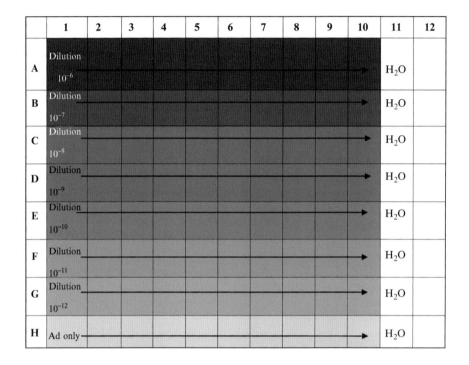

14. Seal the plate with an optical adhesive cover, centrifuge for 2 min at 1,000 × *g*, then place into ABI 7500.
15. Select the appropriate probe option and run absolute quantification Q-PCR program.
16. Data obtained will be expressed as C_t value for each individual reaction. Using these data calculate the mean and SD of the eight negative control (Water) replicates; mean and SD of the ten "Ad only" replicates, and threshold value for scoring positive and negative sample wells. Threshold is determined by subtracting 3× standard deviation (SD) of

"Ad only" from "Ad only" mean value. All wells having a C_t value lower than the threshold value are scored as positive.

17. Determine the "positive to total" replicate ratio for each Test Article dilution. The positive ratio is the quotient of the number of positive replicates divided by the total number of replicates.
18. Sample results are shown in the following schematic:

Log dilution:	6	7	8	9	10	11	12	Ad only	Negative
Ct value:	24.03	22.07	29.6	33.4	35.99	38.61	37.17	37.4	36.7
	24.94	20.65	30.53	33.42	36.44	37.12	37.2	36.93	38.34
	25.55	20.36	30.45	32.67	36.82	37.08	36.96	37.41	37.14
	24.14	21.41	31.22	33.39	36.92	37.9	37.01	37.31	39.83
	24.79	22.23	30.14	33.05	36.84	38.01	37.15	37.18	ND
	24.4	22.76	30.13	33.71	36.04	36.45	36.89	36.91	37.69
	23.94	22.92	29.99	33.1	37.02	37.45	36.95	37.37	38.98
	24.35	23.11	30.6	33.16	36.18	37.39	37.4	37.27	39.11
	24.24	23.34	29.98	33.58	36.61	37.21	37.25	36.94	
	22.33	27.12	30.42	33.39	36.16	37.82	37.01	37.73	
Positive ratio:	1	1	1	1	0.5	0.1	0		
				Ad mean		37.25	Negative mean		38.26
				SD		0.26	SD		1.13
Positive wells (C_t < threshold) are those < 36.47				Threshold C_t value		36.47			

19. Calculate the infectious titer of the Test Article as Infectious Units (IU) per mL.

$$\text{Infectious titer}\,[\text{IU} / \text{mL}] = 10^{[1+S-0.5]} / V[\text{mL}],$$

where S is the logarithm of the initial dilution plus the sum of ratios of infectious-positive wells per total wells at each subsequent dilution and V is the volume of the diluted Test Article in mL.

For this example, $S = 5 + 1 + 1 + 1 + 1 + 0.5 + 0.1 = 9.6$

Therefore, infectious titer $= 10^{(1+9.6-0.5)}/0.05\ \text{mL} = 10^{10.1}/0.05\ \text{mL} = 2.5 \times 10^{11}$ (IU/mL)

20. Calculate the specific infectivity as a ratio of vg to IU.

$$\text{Specific infectivity of the Test Article}\,[\text{vg} / \text{IU}] = \frac{\text{Vector genome titer}\,[\text{vg} / \text{mL}]}{\text{Infectious titer}\,[\text{IU} / \text{mL}]}$$

3.5. Vector Potency by Transduction and Transgene ELISA

The use of a cell culture-based method to measure transduction titer is useful as one measure of functional activity of AAV vectors. Varying doses of an AAV vector are added to target cells in culture, and subsequent expression of the transgene encoded protein is measured by ELISA. Requirements for the assay are (1) use of a target cell that is adequately susceptible to transduction by the vector; and (2) establishing a reliable method to measure transgene product resulting from target cell transduction. The following example involves quantification of human coagulation factor IX (hFIX) following transduction of a human liver cells line (HepG2) with recombinant AAV2 encoding hFIX expressed by a liver-specific promoter (see Notes 9 and 10).

1. One day prior to transfection, seed HepG2 cells in 24-well flasks at 3×10^5 cells per well.
2. On the day of transfection prepare three dilutions of Test Article and Reference vector to obtain vector concentrations of 1.2×10^9, 2.4×10^9, and 4.8×10^9 vg/mL in DMEM/10% FBS/Pen-Strep.
3. Aspirate the medium from HepG2 cells.
4. Add 250 μL of each dilution of Test Article and Reference Vector per well in triplicates according to schematic shown below.
5. Use 250 mL of DMEM/10% FBS/Pen/Strep medium for negative control wells and remaining three wells.
6. Recommended transduction scheme for 24-well plate (see Note 9):

MOI 1000 Test Article 3e8 vg Replicate 1	MOI 1000 Test Article 3e8 vg Replicate 2	MOI 1000 Test Article 3e8 vg Replicate 3	MOI 1000 Reference 3e8 vg Replicate 1	MOI 1000 Reference 3e8 vg Replicate 2	MOI 1000 Reference 3e8 vg Replicate 3
MOI 2000 Test Article 6e8 vg Replicate 1	MOI 2000 Test Article 6e8 vg Replicate 2	MOI 2000 Test Article 6e8 vg Replicate 3	MOI 2000 Reference 6e8 vg Replicate 1	MOI 2000 Reference 6e8 vg Replicate 2	MOI 2000 Reference 6e8 vg Replicate 3
MOI 4000 Test Article 1.2e9 vg Replicate 1	MOI 4000 Test Article 1.2e9 vg Replicate 2	MOI 4000 Test Article 1.2e9 vg Replicate 3	MOI 4000 Reference 1.2e9 vg Replicate 1	MOI 4000 Reference 1.2e9 vg Replicate 2	MOI 4000 Reference 1.2e9 vg Replicate 3
Negative control Replicate 1	Negative control Replicate 2	Negative control Replicate 3	These three wells can be used for cell count verification before transduction		

7. Incubate the cells in 37°C, 5% CO_2 for 12–18 h.
8. After incubation, aspirate the medium from cells and add 1 mL of prewarmed (37°C) DMEM/10% FBS/Pen/Strep to each well.

9. Incubate the transduced cells in 37°C, 5% CO_2 incubator for 36–48 h.
10. Remove the plate from the incubator; gently rock the plate to mix the medium; and then transfer 500 μL medium from each transduced well into correspondingly labeled microcentrifuge tubes. Samples should be stored on ice prior to analysis by ELISA if performed on the same day, or at -80°C for extended storage.
11. Coat ELISA plate with an appropriate primary antibody dilution and incubate at 2–8°C for 12–18 h.
12. Remove primary antibody solution from wells by tapping the plate onto the paper towel.
13. Wash three times with 200 μL of Wash buffer per well.
14. Remove the wash and block the plate with freshly prepared blocking solution (ELISA wash buffer containing 5% nonfat milk). Add 200 μL of blocking solution per well, seal the plate, and incubate at RT for 2 h or at 2–8°C for 4–6 h.
15. Prepare fresh hFIX protein standard.
    (a) Dissolve 1 vial of TriniCHECK Level 1 in 1 mL ELISA dilution buffer to prepare stock with the concentration of 5 μg HFIX/mL. Swirl gently and allow to hydrate for 20 min at RT. Reconstituted standard is stable for 24 h on ice or at 2–8°C.
    (b) Dilute 5 μg/mL stock in ELISA dilution buffer to prepare standard curve – seven HFIX standards with the concentration range 3–200 ng/mL.
16. Retrieve medium samples from HepG2 transduction. Thaw and keep on ice.
17. Wash the plate three times with 200 μL of Wash buffer per well.
18. Load 50 μL ELISA dilution buffer (BLANK) in wells 1H, 5H, and 9H.
19. Load the lowest standard first into wells 1G, 5G, and 9G.
20. Load 50 μL per well of each transduction sample in triplicates. Start from NC, bottom row, and continue loading from the bottom row upward (i.e. from the lowest to the highest MOI).
21. Remove the bubbles, seal the plate, and incubate at RT for 2 h or at 4°C for 12–18 h if needed.
22. Prepare 1:2,200 dilution of goat anti-hFIX-HRP antibody in ELISA dilution buffer.
23. Wash the plate three times with 200 μL of Wash buffer per well.
24. Add 100 μL of diluted goat anti-hFIX-HRP antibody to each well. Seal the plate and incubate for 2 h.

25. Wash three times with 200 μL of Wash buffer per well.
26. Add 11 mL ABTS solution into a 50-mL reservoir.
27. With a multichannel pipette, add 100 μL ABTS solution into each well. It is recommended to add ABTS solution starting from the lowest expected HFIX concentration, and moving toward the highest HFIX concentration. This reduces the error in color development caused by the time needed to add ABTS solution. Work carefully but quickly.
28. Mix by gently tapping the side of the plate. Avoid spilling.
29. Set the timer on count up. Read at 3 and 5 min using SoftMaxPro software "Basic endpoint" protocol with the following setting selection: (1) endpoint reading; (2) 405 nm; (3) automix before the first read.

3.6. Vector Purity by SDS-PAGE/Silver Staining

1. Thaw Test Article and Reference Vector aliquots and keep on ice.
2. Adjust Test Article and Reference Vector concentration to 5×10^{12} vg/mL with DPBS.
3. Aliquot 16 μL of the diluted vector into a fresh microcentrifuge tube.
4. Add 10 μL Milli-Q water, 4 μL 10× sample reducing agent and 10 μL 4× LDS sample buffer. Mix.
5. Incubate samples in a heat block at 95°C for 5 min, microcentrifuge at $10{,}000 \times g$ for 1 min, and set on ice while preparing the gel.
6. Load 20 μL of each sample in duplicate (see Notes 11 and 12). Load molecular weight markers.
7. Run the gel at constant V (~200 V) until the blue dye from the loading buffer just reaches the bottom of the cassette (~40 min). Do not let the dye front run off.
8. Fix gel in 200 mL of fixing solution for ~10 min while shaking at ~60 rpm.
9. Decant fixing solution and sensitize the gel by incubating in 100 mL sensitizing solution for ~30 min while shaking at ~60 rpm.
10. Decant solution. Repeat sensitizing step for ~30 min with fresh 100 mL sensitizing solution.
11. Decant sensitizing solution and rinse gel with 200 mL Milli-Q water for ~10 min while shaking at ~60 rpm. Repeat rinsing step twice.
12. Decant Milli-Q water and stain the gel in 100 mL staining solution for ~15 min while shaking at ~60 rpm.
13. Decant staining solution and rinse gel with 200 mL Milli-Q water for ~5 min while shaking at ~60 rpm. Repeat rinsing step twice.

14. Decant water and develop the gel by immersing into 100 mL developing solution for 3–15 min while shaking at ~60 rpm.
15. Observe gel staining. If desired staining intensity is reached, stop color development by adding 5 mL Stopper solution. Incubate for ~10 min while shaking at ~60 rpm.
16. Decant stopping solution and wash gel with 200 mL Milli-Q water for ~5 min while shaking at ~60 rpm. Repeat washing step twice.
17. Decant the water, add 100 mL Gel-Drying Solution and shake at ~60 rpm for 15–20 min.
18. Remove two sheets of cellophane from the package and soak (one at a time) in Gel-Drying Solution for 15–20 s. Do not soak more than 2 min.
19. Place the gel between two pieces of cellophane and place into DryEase Gel Drying Frame.
20. Align the other frame so that its corner pins fit into the appropriate holes on the bottom frame. Secure the four sides of the frame with plastic clamps.
21. Dry gel for 12–36 h on the bench.
22. When the cellophane is dry, remove the gel/cellophane sandwich, from the frame, trim off excess cellophane.
23. Press dried gel between the pages of a book under light pressure for ~48 h. Gel will remain flat for scanning or display.

3.7. Vector Purity by Optical Density

UV absorbance of denatured AAV vector provides a simple, rapid, and direct method for assessing purity of AAV vectors that can complement SDS-PAGE and residual specific nucleic acid measurements. AAV particles are composed of a defined amount of protein and nucleic acids. For highly purified vector, concentration can be measured based on the known extinction coefficients for AAV capsid protein and DNA. The extinction coefficients for the protein component of an AAV2 vector particle are $3.72 \times 10^6\ M^{-1}\,cm^{-1}$ at 260 nm and $6.61 \times 10^6\ M^{-1}\,cm^{-1}$ at 280 nm (see Note 13). The extinction coefficients for the DNA component of an AAV vector particle, which varies depending on the base composition, are approximately 20 $g^{-1}\ cm^{-1} \times MW_{DNA}$ at 260 nm and 11 $g^{-1}\ cm^{-1} \times MW_{DNA}$ at 280 nm. The presence of extra proteins, nucleic acids, and other UV absorbing impurities will augment optical density to a degree proportional to the amounts of such impurities present.

1. Prepare Test and Reference Vectors in microfuge tubes by adding 0.5 μL of 20% stock SDS per 100 μL vector solution, and heat at 75°C for 10 min. Microcentrifuge briefly to recover condensate (see Note 14).
2. Switch on the spectrophotometer, and complete diagnostic tests, as appropriate.

3. Turn on the Visible (Tungsten) and UV (D2) lamps.
4. Open the sample port cover of the DU800 unit, and place a clean quartz cuvette in the cuvette holder. It is important that the cuvette be seated properly.
5. Using the pipetter and clean pipette tips, and leaving the cuvette in place, rinse the cuvette three times with PBS180/0.001% PF68 excipient. For each rinse, add ~100 μL, repipette the volume at least three times, and then withdraw the fluid completely.
6. Add 100 μL of Diluent to the cuvette, and close the sample port cover. Press "blank," and wait while the unit automatically calibrates to an optical density of zero.
7. Open the sample port cover and withdraw the Diluent from the cuvette. Add ~100 μL of Diluent to the cuvette, and close the sample port cover. Measure OD_{260} and OD_{280}.
8. Open the sample port cover and completely withdraw the fluid from the cuvette.
9. Add 100 μL of Reference Vector (Rep r1) to the cuvette, and close the sample port cover. Measure OD_{260} and OD_{280}.
10. Repeat measurement with another ~100 μL of the Reference Vector (Ref r2).
11. Repeat measurement with a third ~100 μL aliquot of the Reference Vector (Ref r3).
12. Open the sample port cover and withdraw the fluid from the cuvette. Using the pipetter and clean pipette tips, and leaving the cuvette in place, rinse the cuvette three times with excipient, and then withdraw the fluid carefully using the pipette.
13. For concentrated samples, the Test Article is diluted with Diluent to achieve OD values that are in the range 0.05–0.5 to ensure optimal accuracy and consistency of results.
14. Add ~100 μL of Test Article (TA r1) to the cuvette, and close the sample port cover. Measure Test Article OD at wavelengths 260 and 280 nm.
15. Open the sample port cover and withdraw the fluid from the cuvette. Repeat measurement with a second ~100 μL of the Test Article (TA r2).
16. Repeat measurement with a third ~100 μL aliquot of the Test Article (TA r3).
17. After completing the OD measurements of samples, the cuvette must be rinsed thoroughly with DI water and stored appropriately.
18. Calculate the mean and SD for the Reference and Test Articles. Normalize data to calculate OD_{260} and OD_{280} units per 10^{13} vg, and the ratio of OD_{260}:OD_{280}. Compare the

experiment values with expected/historic values as a measure of vector purity.

3.8. Residual Mammalian DNA by Real-Time Q-PCR

1. Prepare human genomic DNA standard by digesting 250 μg with *Eco*RI in 200 μL reaction.
2. Perform two phenol extractions using 200 μL phenol–chloroform–isoamylalcohol (25:24:1) and vortex tube for 1 min.
3. Precipitate DNA with ethanol and dissolve DNA pellet in 50 μL TE. Aliquot and store at −20°C.
4. Calculate the DNA contents using standard OD260/280 method.
5. Dilute genomic DNA in nuclease-free water to prepare standard S1 with the concentration of 20 ng/μL. Prepare at least 600 μL. Prepare seven serial fourfold dilutions of standard S1 in nuclease-free water to obtain standard curve S1–S8 the with range 10^5 pg/5 μL (S1) to 1.28 pg/5 μL (S8).
6. Thaw aliquots of Test and Reference Vectors, and store on ice.
7. For Test and Reference Vectors, set up three independent dilutions in Q-PCR dilution buffer, to obtain a 10-, 30-, and 100-fold dilutions, by adding 10 μL aliquots into 90, 290, and 990 μL Q-PCR dilution buffer. Vortex, spin down briefly.
8. Thaw genomic DNA standards, vortex briefly, spin down, and keep at room temperature. Thaw 18S rRNA assay reagent, vortex briefly, and keep on ice.
9. Prepare Q-PCR reaction master mix by mixing per well (PCR reaction): 12.5 μL TaqMan Universal Master Mix, 1.25 μL 18S rRNA Assay Reagent (20×), and 6.25 μL H_2O. Prepare enough for triplicate reactions for three Test Article dilutions, three reference vector dilutions, three NTCs (No Template Controls) + triplicate reactions for each standard.
10. Place 96-well optical reaction plate into plate rack, and aliquot 20 μL Q-PCR reaction master mix into each well used.
11. Using a P20 pipette, add 5 μL per well of each standard in triplicates.
12. Add 5 μL per well of each Test Article and Reference Vector.
13. Add 5 μl Q-PCR dilution buffer as NTC in triplicates.
14. Seal the plate with optical adhesive cover and centrifuge at 2,000 × *g* for 2 min.
15. Run PCR using the following Thermo profile: Stage 1: 1 Rep (50°C, 2:00 min); Stage 2: 1 Rep (95°C, 10:00 min); Stage 3: 40 Reps (95°C, 0:15 min; 60°C, 1:00 min). Auto Increment: 0 for all stages. Ramp rate: 100% for all stages.
16. The SDS software calculates quantity per well based on the slope and intercept of the standard curve (C_t vs. LOG (copy

number of each standard). Quantity $= 10^{(C_t - \text{Intercept})/\text{Slope}}$. The correlation coefficient (Pearson R^2) of the standards reflects the average C_t value for each standard vs. the LOG (copy number of each standard)).

17. To obtain the Host Cell DNA concentration/mL in the Test Article and Reference Vector samples, multiply the copy number per well by 200 and then by the dilution factor (10, 30, 100).
18. Determine the average sample concentration from the three different dilutions and determine the standard deviation.
19. Calculate the Host Cell DNA concentration per 10^9 vg of Test Article and Reference Vector.

3.9. Residual Plasmid DNA by Real-Time Q-PCR

The TaqMan® real-time Q-PCR procedure described herein requires a set of target-specific Q-PCR reagents, such as primers and probe, to detect specific sequences in plasmids (usually antibiotic resistance gene) used for vector biosynthesis. If sequence common for all plasmids used in vector manufacture can be identified, total residual plasmid can be determined in a single Q-PCR assay. If a single common target sequence is not available, for example, when Amp^R and Kan^R plasmids are used in the manufacture of a specific vector and plasmid backbones do not contain identical sequences, total residual plasmid should be calculated as a sum of residual Amp^R and residual Kan^R DNA each determined in a separate assay.

1. Prepare three independent dilutions of the Test Article and Reference vector in Q-PCR dilution buffer.
2. Prepare plasmid DNA standards containing the target sequence of interest, e.g., Amp^R, as described in Section 3.2., then prepare standard curve dilutions as described in Section 3.3., step 1. Next follow Section 3.3, steps 11 through 16, omitting the DNAse control.
3. The SDS software calculates copy number per well based on the slope and intercept of the standard curve (C_t vs. LOG (copy number of each standard)).

   Unknown $= 10^{(C_t - \text{Intercept})/\text{Slope}}$. The correlation coefficient (Pearson R^2) of the standards reflects the average C_t value for each standard vs. the LOG (copy number of each standard).
4. Calculate amplicon concentration in each of the three Test Article dilutions, each of the three Reference vector dilutions, and mean concentrations, respectively:

$$\text{Amp [copies / mL]} = \text{Mean [copies / well or } 5\,\mu\text{L]} \times 200 \times \text{dilution factor}$$

5. Conversion of copy number concentration to a mass (e.g., pg/mL) may be performed using assumption about the average size of residual plasmid DNA copies (see Note 15).

4. Notes

1. The probe is a ^{32}P-labeled fragment of DNA corresponding to AAV Cap. Details for the preparation of this probe will vary with DNA source and AAV serotype. For AAV2 Cap described herein, we obtained a 2,742-bp XbaI/HindIII fragment. The fragment was purified by agarose gel electrophoresis, and then subjected to random primer labeling using ^{32}P-α-dCTP (26).
2. The sensitivity of the wtAAV assay described here is dependent on the efficiency of infection of target cells. Wild-type AAV2 infects HEK293 cultured cells efficiently, therefore high sensitivity can be achieved (approximately one wtAAV particle per 10^9 rAAV particles can be detected). In contrast, many other AAV serotypes do not efficiently infect cells in culture, resulting in correspondingly lower sensitivity achievable in this type of assay.
3. The wtAAV concentration in the spike control can be reduced to increase the sensitivity of the assay as long as the lower spike concentration is still detectable.
4. It is important to use linearized plasmid for the preparation of standard. Use of supercoiled plasmid DNA results in delayed amplification (higher C_t values) of the standards possibly due to lower accessibility of template for primer/probe binding. Therefore, use of supercoiled plasmid DNA will result in a higher apparent titer of the Test Article, approximately two to threefold higher than the more accurate value obtained using linearized plasmid DNA for the standard curve. The use of a linearized plasmid DNA template for Q-PCR standards results in primer/probe binding with an efficiency similar to the actual Test Article.
5. It is critical to include nonionic surfactant in Q-PCR dilutions. Highly purified recombinant AAV corresponds to a dilute protein solution, for example, vector preparation with the concentration of 10^{13} vg/mL corresponds to ~70 μg/mL, and vector dilution at 10^9 vg/mL concentration, range required for various QC assays, corresponds to a 7 ng/mL. Vectors in such dilute solutions are readily susceptible to loss due to irreversible binding to a variety of surfaces, such as microfuge tube walls, pipette tips, syringe, and filter surfaces (29). The inclusion of a nonionic surfactant (e.g., 0.001% Pluronic F68) in solutions used for storage and dilution of AAV vectors is necessary to prevent nonspecific loss, and can dramatically improve QC assay accuracy and precision.

6. Protocol and standard curve described herein should work well for vectors with titer in the range 10^{11}–10^{13} vg/mL. Dilution scheme may be adjusted for vector preparations with higher or lower titer. In initial experiments for Test Articles with unknown titer, serial 5× or 10× vector dilutions may be used to establish appropriate dilution ranges for the standard curve range recommended herein.
7. Optimization of primer/probe concentrations in Q-PCR reaction may be required for specific sets of primers and probes.
8. Titer calculation for self-complementary vectors should take into account that one plasmid genome copy is equivalent to one vector genome copy.
9. Transduction efficiency depends on vector's ability to infect cells, therefore may vary for different AAV serotypes. MOIs should be empirically determined for each vector.
10. Semiquantitative assessment of transgene expression can also be performed by Western blot. Transgene expression as a function of vector dose may be useful as a measure of transduction/expression for nonsecreted/membrane-associated proteins and for comparison of different vector.
11. Always wear gloves when performing silver staining. Touching the prestained gel with ungloved hands will stain fingerprints and ruin the gel.
12. The loading recommended for SDS-PAGE using silver stain for protein detection is in the range 5×10^{10} to 10^{11} vg per lane on a minigel. For purified vectors free of empty capsids, this corresponds to approximately 300–600 ng vector protein per lane, enough to provide strong virus protein (VP) bands by silver staining and detection of protein impurities corresponding to >2% of vector. Lower vg per lane may be appropriate if vector preparations contain high amounts of impurities.
13. Spectrophotometric extinction coefficients AAV serotypes other than AAV2 will differ somewhat from those reported for AAV2 because of variation in the number of aromatic amino acids. Extinction coefficients can be estimated using algorithms as given in www.basic.northwestern.edu/biotools/proteincalc.html.
14. Denaturation of the AAV particle using SDS ensures that the chromophores are fully exposed, and prevents Rayleigh light scattering that occurs for intact AAV particles.
15. This procedure provides the copy number of the PCR amplicon used to measure residual plasmid DNA (e.g., Amp^R), but does not directly provide a mass of this impurity. Additional characterization of residual plasmid DNA sequences can be performed, including assessment of nuclease sensitivity

(to indicate whether the impurity is nuclease sensitive (e.g., accessible, soluble) or resistant (e.g., AAV encapsidated)), and Southern blot analysis to estimate the size of residual plasmid DNA fragments. The vector purification methods used in our laboratory includes an efficient nuclease treatment step to remove accessible DNA, and essentially all residual plasmid DNA is nuclease resistant. For such AAV-encapsidated residual DNA species, one approach to convert copy number to mass is to assume that each copy of DNA has a length corresponding to the packaging capacity of AAV, i.e., ~4,700 bases. Using this "worst case" assumption, an estimate of the mass of the residual DNA impurity in pg/mL is obtained by multiplying the copy number per mL times the mass of a 4,700-base polynucleotide (~3 pg/10^6 copies).

Acknowledgments

We acknowledge Katherine High, Bernd Hauck, Jennifer Wellman, and Guang Qu for helpful scientific discussions, and Lisa Africa and Xingge Lui for expert technical support. We also acknowledge The Children's Hospital of Philadelphia Center for Cellular and Molecular Therapeutics, and the NIH National Heart, Lung and Blood Foundation Gene Therapy Resource Program (HHSN268200748203C), for support.

References

1. Carter, B.J. (2005). Adeno-associated virus vectors in clinical trials. *Hum Gene Ther.* **16**, 541–550.
2. Warrington, K.H. and Herzog, R.W. (2006). Treatment of human disease by adeno-associated viral gene transfer. *Hum Genet.* **119**, 571–603.
3. Fiandaca, M., Forsayeth, J., and Bankiewicz, K. (2008). Current status of gene therapy trials for Parkinson's disease. *Exp Neurol.* **209**, 51–57.
4. Christine, W.C., Starr, P., Larson, P., Eberling, J.L., Jagust, W.J., Hawkins, R., *et al.* (2009). Safety and tolerability of putaminal AADC gene therapy for Parkinson disease. *Neurology* **73**, 1662–1669.
5. Maguire, A.M., High, K.A., Auricchio, A., Wright, J.F., Pierce, E.A., Testa, F., *et al.* (2009). Age-dependent effects of RPE65 gene therapy for Leber's congenital amaurosis: a phase 1 dose-escalation trial. *Lancet* **374**, 1597–1605.
6. 21CFR Parts 120 and 211 – Current Good Manufacturing Practice in manufacturing, processing, packing or holding of drugs; General and current Good Manufacturing Practice for finished pharmaceuticals. http://www.access.gpo.gov/nara/cfr/index.html
7. US Dept Health Human Services, Food and Drug Administration, Center for Biologics Evaluation and Research. Guidance for FDA and Sponsors: Content and review of Chemistry, Manufacturing, and Control (CMC) information for human gene therapy Investigational New Drug applications (INDs). November, 2004. http://www.fda.gov/cber/guidelines.htm
8. US Dept Health Human Services, Food and Drug Administration, Center for Drug Evaluation and Research, Center for Biologics Evaluation and Research. Guidance for Industry: INDs – Approaches to complying with CGMP during Phase 1. January 2006. http://www.fda.gov/cber/guidelines.htm
9. Gombold, J., Peden, K., Gavin, D., Wei, Z., Baradaran, K., Mire-Sluis, A., and Schenerman, M..(2006). Lot release and characterization testing of live-virus-based vaccines and gene

therapy products, Part 1: Factors influencing assay choices. *BioProcess Int.* **4**, 46–56.

10. Muzyczka, N. (1992). Use of adeno-associated virus as a general transduction vector for mammalian cells. *Curr. Top. Microbiol. Immunol.* **158**, 97–129.
11. Allen, J.M., Debelak, D.J., Reynolds, T.C., and Miller, A.D. (1997). Identification and elimination of replication-competent adeno-associated virus (AAV) that can arise by non-homologous recombination during AAV vector production. *J. Virol.* **71**, 6816–6822.
12. Samulski, R.J., Shang, L.-S., and Shenk, T. (1989). Helper-free stocks of recombinant adeno-associated viruses: normal integration does not require viral gene expression. *J. Virol.* **61**, 3096–3101.
13. Grimm, D., Kern, A., Rittner, K., and Kleinschmidt, J.A. (1998). Novel tools for production and purification of recombinant adenoassociated virus vectors. *Hum. Gene Ther.* **9**, 2745–2760.
14. Clark, K.R., Liu, X., McGrath, J.P., and Johnson, P.R. (1999). Highly purified recombinant adeno-associated virus vectors are biologically active and free of detectable helper and wild-type virus. *Hum. Gene Ther.* **10**, 1031–1039.
15. US Department of Health and Human Services, Food and Drug Administration, Center for Biologics Evaluation and Research, Draft Guidance for Industry: Potency Tests for Cellular and Gene Therapy Products. October 2008. http://www.fda.gov/cber/guidelines.htm
16. Grimm, D., Kern, A., Pawlita, M., Ferrari, F.K., Samulski, R.J., and Kleinschmidt, J.A. (1999). Titration of AAV-2 particles via a novel capsid ELISA: Packaging of genomes can limit production of recombinant AAV-2. *Gene Ther.* **5**, 1322–1330.
17. Sommer, J.M., Smith, P.H., Parthasarathy, S., Isaacs, J., Vijay, S., Kieran, J., *et al.* (2003). Quantification of adeno-associated virus vectors and empty capsids by optical density measurement. *Mol. Ther.* **7**, 122–128.
18. Chadeuf, G., Ciron, C., Moullier, P., and Salvetti, A (2005). Evidence for encapsidation of prokaryotic sequences during recombinant adeno-associated virus production and their in vivo persistence after vector delivery. *Mol. Ther.* **12**, 744–753.
19. Report from the CHMP gene therapy expert group meeting. European Medicines Agency 2005; EMEA/CHMP/183989/2004. www.emea.europa.eu/pdfs/human/genetherapy/18398904en.pdf
20. Hauck, B., Murphy, S., Smith, P.H., Qu, G., Liu, X., Zelenaia, O., *et al.* (2009). Undetectable transcription of cap in a clinical AAV vector: Implications for preformed capsid in immune responses. *Mol. Ther.* **17**, 144–152.
21. Smith, P.H., Wright, J.F., Qu, G., Patarroyo-White, S., Parker, A., and Sommer, J.M. (2003). Packaging of host cell and plasmid DNA into recombinant adeno-associated virus vectors produced by triple transfection. *Mol. Ther.* **7**, S348.
22. EMEA Committee for Medicinal Products for Human Use (CHMP). Questions and Answers on Gene therapy. EMA/CHMP/GTWP/212377/2008. http://www.ema.europa.eu/pdfs/human/genetherapy/21237708en.pdf
23. US Dept Health Human Services, Food and Drug Administration, Center for Drug Evaluation and Research, Center for Biologics Evaluation and Research, Center for Veterinary Medicine. Guidance for Industry: Comparability protocols – Protein drug products and biological products – Chemistry, Manufacturing, and Controls information. September 2003. http://www.fda.gov/cber/guidelines.htm
24. US Dept Health Human Services, Food and Drug Administration, Center for Drug Evaluation and Research, Center for Biologics Evaluation and Research. Guidance for Industry: Analytical procedures and methods validation. Chemistry, manufacturing and controls documentation. August 2000. http://www.fda.gov/cber/guidelines.htm
25. US Dept Health Human Services, Food and Drug Administration, Center for Drug Evaluation and Research, Center for Biologics Evaluation and Research. Guidance for Industry: Q1A (R2) Stability testing of new drug substances and products. November 2003. http://www.fda.gov/cber/guidelines.htm
26. Sambrook, J., and Russell, D.W. (2001). Molecular cloning: A Laboratory Manual. Cold Spring Harbor Laboratory Press, Cold Spring Harbor, New York.
27. Clark, K.R., Voulgaropoulou, F., Fraley, D.M., and Johnson, P.R. (1995). Cell lines for the production of recombinant adeno-associated virus. *Hum. Gene Ther.* **6**, 1329–1341.
28. Zhen, Z., Espinoza, Y., Bleu, T., Sommer, J.M., and Wright, J.F. (2004). Infectious titer assay for adeno-associated virus vectors with sensitivity sufficient to detect single infectious events. *Hum. Gene Ther.* **15**, 709–715.
29. Simpson, R.J. (2009). Stabilization of Proteins for Storage. *In* Basic Methods in Protein Purification and Analysis. Eds. Simpson R.J., Adams, P.D., and Golemis, E.A. Cold Spring Harbor Laboratory Press, Cold Spring Harbor, New York, pp 385–400.

Chapter 12

Baculoviruses Mediate Efficient Gene Expression in a Wide Range of Vertebrate Cells

Kari J. Airenne, Kaisa-Emilia Makkonen, Anssi J. Mähönen, and Seppo Ylä-Herttuala

Abstract

Baculovirus expression vector system (BEVS) is well known as a feasible and safe technology to produce recombinant (re-)proteins in a eukaryotic milieu of insect cells. However, its proven power in gene delivery and gene therapy is still poorly recognized. The basis of BEVS lies in large enveloped DNA viruses derived from insects, the prototype virus being *Autographa californica* multiple nucleopolyhedrovirus (AcMNPV). Infection of insect cell culture with a virus encoding a desired transgene under powerful baculovirus promoter leads to re-protein production in high quantities. Although the replication of AcMNPV is highly insect specific in nature, it can penetrate and transduce a wide range of cells of other origin. Efficient transduction requires only virus arming with an expression cassette active in the cells under investigation. The inherent safety, ease and speed of virus generation in high quantities, low cytotoxicity and extreme transgene capacity and tropism provides many advantages for gene delivery over the other viral vectors typically derived from human pathogens.

Key words: Baculovirus, Viral vector, Insect cells

1. Introduction

Baculoviruses are safe, fast, and cost-effective tools to express the desired genes in a wide range of cells (1). This has made the baculovirus expression vector system (BEVS) a widely used standard, especially when complex recombinant proteins are in focus (2). Originally derived from larvae, baculoviruses infect insect cells well (3). However, the most widely used and studied

Otto-Wilhelm Merten and Mohamed Al-Rubeai (eds.), *Viral Vectors for Gene Therapy: Methods and Protocols*, Methods in Molecular Biology, vol. 737, DOI 10.1007/978-1-61779-095-9_12, © Springer Science+Business Media, LLC 2011

baculovirus, *Autographa californica*, a multiple polyhedrovirus, is able to penetrate and deliver genes also into nontarget cells (2, 4). Numerous cells including primary, progenitor, and stem cells from various origins (human, monkey, pig, rabbit, rat, feline, mouse, fish, avian, frog, etc.) have been shown to be permissive for a baculovirus-mediated gene delivery if the virus is just equipped with an appropriate expression cassette, which is functional in the target cell (1, 5–8). However, in the nontarget cells, AcMNPV is incapable for replication or proper viral gene expression (4, 9). It is thus gentle (nontoxic) for the vertebrate cells and harmless for a human, and can be handled in level 1 biosafety facilities (10, 11). In addition to safety, baculoviruses provide several advantages over the other gene delivery systems. Working is easy and does not require any special skills or equipments. Transgene capacity is in practice unlimited (at least 100 kbp) and multiple high-titer viruses can be generated and titered in a short time frame. In addition, insect cells are easy to cultivate in a serum-free medium on plates, or in suspension, and the produced viruses are easy to store at 4°C. Furthermore, virus and protein production is amenable to automation and AcMNPV can transduce dividing and dormant cells. Finally, the BEVS is recognized by the regulatory agencies (EMEA, FDA). So, it is not surprising that the popularity of baculoviruses is still increasing and they are nowadays applied and studied for many purposes such as drug screening (12), viral vector production (13, 14), eukaryotic surface displaying (15), gene therapy (16, 17), and vaccination (18, 19) in addition to traditional use as biopesticides (3, 20) and recombinant protein production tools (2).

Baculoviruses are large (30–60 × 250–300 nm), rod-shaped double-stranded DNA (80–180 Kbp) viruses that infect and replicate only in arthropod hosts, mostly in Lepidopteran larvae (21). Baculoviruses are highly species specific and well known (22). They have been studied for decades as a biopesticides, and since early 1980s, as tools to produce recombinant proteins in eukaryotic milieu of insect cells (1). Baculoviruses exist in two natural forms. The budded virus (BV) spreads the infection within the host and the occlusion derived virus (ODV) between the hosts. The two different viruses have identical capsids, but differ in their envelope which is derived either from the cell (BV) or nuclear membrane (ODV) (23). The BV form is usually used for biotech purposes to infect the desired cells. Commercially available baculovirus vectors typically lack polyhedrin gene, which is dispensable for virus propagation in a laboratory (24).

The type species of *Baculoviridae*, AcMNPV, is unable to replicate in cells other than those of insect origin. Its genome (134 Kbp) has been sequenced and predicted to encode 156 proteins (25). AcMNPV enters cells via a low pH-dependent endocytosis (1, 26–28). The major envelope glycoprotein, gp64, is necessary and sufficient for the pH dependent membrane fusion during virus entry allowing virion attachment on the cell surface and escape of the viral capsids from the endosomes (29). Gp64 is also needed for the efficient virion budding from the insect cells (30). The receptor(s) for AcMNPV is (are) not known, but wide range of target cells suggest that the molecule(s) is (are) a common cell surface molecule such as integrin, phospholipid, or heparan sulfate proteoglycan. For more detailed information of all aspects of baculoviruses, see "The Baculoviruses" edited by Miller (22). The truly beneficial nature of AcMNPV as a safe and efficient vector has raised recently a lot of interest in using these viruses for a universal gene delivery *in vitro* and *in vivo*, the ultimate goal being baculovirus-mediated gene therapy (1, 7, 31). This chapter describes useful protocols to generate, amplify, concentrate and purify baculoviruses to infect or transduce insect and vertebrate cells, respectively.

2. Materials

2.1. Insect Cell Culture

1. *Spodoptera frugiperda 9* (Sf9) insect cells (ATCC-CRL-1711 or Invitrogen), stored in liquid nitrogen.
2. Serum-free insect cell growth media e.g. Insect Xpress (BioWhittaker) or Sf-900 II SFM (Invitrogen).
3. T-75 (75 cm^2) cell culture flask (Sarstedt).
4. Trypan blue 0.4% (Sigma). This substance is toxic. Avoid exposure and pay attention to safety precautions when handling product.

2.2. Baculovirus Generation

1. Clone the gene of interest into a donor plasmid compatible with the transposon-based baculovirus generation. Suitable vectors are commercially available from Invitrogen (Bac-to-Bac system) and by request from authors of this topic. We provide by request, the pBVboost-series vectors (16, 32, 33), which are improved derivatives of the pFastBac vector (Bac-to-Bac system).
2. Electrocompetent DH10Bac (Invitrogen) or DH10BacΔTn7 *Escherichia coli* cells (32) by a request from the author's laboratory.
3. LB-medium: 10 g Bacto-tryptone, 5 g Bacto-yeast extract, and 10 g NaCl. Add H_2O to 1 L and adjust pH to 7.5 with 1 M NaOH.

4. LB-plates: 100 ml of 5× LB-medium and 7.5 g Bacto-agar. Add H_2O to 500 ml, autoclave and let the temperature to settle to 53°C in water bath. Add appropriate antibiotics and cast the plates (about 15 plates per 100 ml).
5. LB_{tg}-plates: LB with 10 μg/ml tetracycline and 7 μg/ml gentamicin.
6. LB_{stg}-plates: LB_{tg} with 10% sucrose.
7. S.O.C.: 2 g Bacto-tryptone, 0.5 g yeast-extract, 1 ml 1 M NaCl, 0.25 ml 1 M KCl, and 1 ml 2 M Mg-stock solution (1 M $MgSO_4$, 1 M $MgCl_2$). Add water to 1 L.
8. Boost solution 1: 15 mM Tris–HCl, pH 8, 10 mM EDTA, 100 μg/ml RNAse A. Filter sterilize and store at 4°C.
9. Boost solution 2: 0.2 M NaOH, 1% (W/V) SDS. Filter sterilize and store at room temperature.
10. Boost solution 3: 3 M KCH_3COO, pH 5.5. Autoclave and keep at room temperature.
11. TE-buffer: 10 mM Tris–HCl, 1 mM EDTA, pH 8.

2.3. Virus Characterization by Triple PCR

1. 100% dimethyl sulfoxide, DMSO.
2. 10× Dynazyme-buffer (Finnzymes).
3. 10 mM dNTP-mix (Finnzymes).
4. pUC/M13 Forward primer (Invitrogen) (5′ GTT TTC CCA GTC ACG AC 3′), 10 pmol/μl, pUC/M13 Reverse primer (Invitrogen) (5′ CAG GAA ACA GCT ATG AC 3′), 10 pmol/μl and F835R primer, 10 pmol/μl (5′ TGG GAG GGG TGG AAA TGG AG 3′).
5. Isolated bacmid-DNA as a template (see Subheading 3.4).
6. Dynazyme II DNA Polymerase (Invitrogen), store at –20°C.
7. PCR-grade H_2O, also used as a negative control.
8. LE agarose (SeaKem).
9. 0.5× Tris–Borate–EDTA-buffer (TBE), (44.5 M Tris–Borate, 1 mM EDTA, pH 8.3).
10. Sybr Safe (Invitrogen) for a DNA staining.
11. Gene Ruler DNA ladder mix (MBI Fermentas).

2.4. Virus Titering

1. Sf9 insect cells
2. Serum-free insect cell growth media e.g. Insect Xpress (BioWhittaker) or Sf-900 II SFM (Invitrogen).
3. Trypan blue 0.4% (Sigma). This substance is toxic. Avoid exposure and pay attention to all necessary safety precautions when handling the product.
4. The virus of interest.

2.5. Virus Concentration and Purification

1. Sf9 cells (ATCC-CRL-1711 or Invitrogen).
2. Serum-free insect cell growth media, e.g. Insect Xpress (BioWhittaker) or Sf-900 II SFM (Invitrogen).
3. High titer virus of interest.
4. Dulbecco's Phosphate Buffered Saline (DPBS, 1×, w/o calcium and magnesium) (BioWhittaker).
5. Saccharose (MP Biomedicals) for preparation of 20, 25, and 50% w/v saccharose solutions in DPBS. Filter-sterilize through a 0.22-μm filter. Store at 4°C.

2.6. SDS-Polyacrylamide Gel Electrophoresis and Immunoblotting

1. 1.5 M Tris–HCl, pH 8.8 for 10% running gel. Store at 4°C.
2. 0.5 M Tris–HCl, pH 6.8 for 4% stacking gel. Store at 4°C.
3. 10% SDS. Store at room temperature.
4. 30% Acrylamide/Bis solution (37.5:1) (Biorad) (This solution is toxic. Avoid exposure and pay attention to all necessary safety precautions when handling the product). Store at 4°C.
5. 10% w/v Ammonium persulfate (JTBaker). Prepare 10% solution in water, aliquot in amounts suitable for single use and store at −20°C.
6. TEMED (*N*,*N*,*N*′,*N*′-tetramethyl ethylenediamine; Sigma). The substance is considered corrosive. Avoid exposure and especially inhalation. Pay attention to all necessary safety precautions when handling the product.
7. 4× sample buffer (0.125 M Tris–HCl, pH 6.8, 4% SDS, 20% Glycerol, 0.004% bromophenol blue, 10% 2-mercaptoethanol) for sample reduction. Store at 4°C.
8. Prestained molecular weight marker standard such as the SeeBlue Plus2 (Invitrogen).
9. SDS-running buffer (0.192 M glycine, 25 mM Tris–HCl, 0.1% SDS, pH 8.8). Store at 4°C.
10. Nitrocellulose membrane: Trans-Blot Transfer Medium (Biorad) and 3MM Chr Chromatography paper (Whatman).
11. Transfer buffer: 0.025 M Tris base, 0.04 M Glycine, pH 8.3 +20% v/v methanol. Store at 4°C.
12. Tris-buffered saline with Tween (TBS-T) (5 M NaCl, 0.2 M Tris–HCl +0.1% Tween-20 (Sigma)).
13. Blocking buffer: 5% w/v nonfat dry milk in TBS-T.
14. Primary gp64-antibody (anti-baculovirus Envelope gp64 Clone AcV5, Bioscience or Sigma), diluted in blocking buffer.
15. Secondary antibody (Goat anti-mouse IgG (H+L)-HRP Conjugate, BioRad) diluted in blocking buffer.
16. APA-buffer (0.1 M $NaHCO_3$, 1 mM $MgCl_2 \cdot 6H_2O$, pH 9.8).
17. NBT/BCIP Stock solution (Roche).

2.7. Transduction of Vertebrate Cells

1. Concentrated and purified high-titer baculovirus.
2. Exponentially growing healthy cultures of mammalian cells [e.g. human liver cell line; HepG2 (ATCC: CRL-11997)].
3. Cell culture growth medium [complete, serum-free, and selective (optional)]. RPMI-1640 (BioWhittaker; Cambrex) supplemented with 10% fetal bovine serum (FBS, HyClone) and antibiotics (100 U/ml of penicillin and 100 μg/ml of streptomycin; Gibco/BRL) is highly recommended (34). HepG2 cells need 2 mM l-glutamine (l-glutamax, Gibco/BRL), 0.1 mM nonessential amino acids (Gibco/BRL), and 1.0 mM Na-pyruvate (Gibco/BRL)).
4. Dulbecco's phosphate buffered saline (DPBS, Sigma): 2.7 mM KCl, 1.5 mM KH_2PO_4, 136.9 mM NaCl, and 8.9 mM $Na_2HPO_4{\cdot}7H_2O$.
5. Trypsin–EDTA (Sigma).
6. Trypan Blue 0.4% (Sigma). This substance is toxic. Avoid exposure and pay attention to all necessary safety precautions when handling product.
7. Sodium butyrate (Sigma) (Optional). 1 M sodium butyrate solution: Dissolve sodium butyrate at a concentration of 148 mg/ml in 1× PBS and sterilize the solution by passing it through a 0.22-μm filter.
8. Tissue culture dishes (6-well plate) (CellStar, Greiner Bio-One).

3. Methods

The most commonly used cell line for AcMNPV propagation is Sf-9 (35), a derivative of the *S. frugiperda* cell line Sf21. These cells are derived from ovarian tissue of the fall army-worm and grow well in adherent and suspension cultures also in a serum-free medium. Another popular insect cell line, especially for production of secreted proteins, is BTI-TN-5B1-4 (also known as Tn5 or High-Five™) derived from *Trichoplusia ni* egg cell homogenate. However, these cells tend to form aggregates in suspension culture (36) and have recently shown to suffer from the latent infection by a novel nodavirus, TNCL (=Tn5 cell line virus) (37). If complex glycosylation (terminal galactose and/or sialic acid residues) of the recombinant product is needed, engineered "sweet" insect cell lines (sfSWT) exits for this purpose (2, 38). It is important to work with healthy cells of known origin and to avoid clonal selection in cell passaging (39).

In order to enable the expression of the recombinant protein in insect cells, the gene for the desired protein is usually placed under a strong polyhedrin promoter (*polh*) of AcMNPV (24). However, the *polh* promoter is activated very late in the infection at a point when the host cell machinery for post-translational modifications is no longer working properly. This may be a problem, for example, if the biological activity of the protein depends on proper glycosylation (2). Other promoters, which will activate earlier than the *polh* such as the promoter for the *p10* gene (p10), the major capsid protein gene (*vp39*), the basic 6.9 kDa protein gene (*cor*), and the viral *ie1* gene (*ie1*) (24, 40) may then help along with the Sweet cells as described above. For transduction of vertebrate cells, a promoter active in the target cells, such as CMV (cytomegalovirus), RSV (Rous sarcoma virus), or CAG (chicken β-actin promoter) must be used since the *polh* is inactive in these cells (41).

The methods to generate baculoviruses have improved much since the original AcMNPV generation protocol by homologous recombination in early 1980s (for recent reviews see Refs. 1 and 42). Many commercially available vectors and kits are now available. The most popular of them is still the transposon-based system known as the Bac-To-Bac system (Invitrogen), which uses a site-specific transposition with Tn7 to insert foreign genes into a bacmid DNA (baculovirus genome) propagated in *E. coli* cells (32, 43). After cloning the desired expression construct into a donor plasmid, the recombinant baculoviral genome(s) (bacmid) is prepared simply by transforming DH10BacΔTn7 *E. coli* cells with the donor vector(s). The recombinant bacmid containing *E. coli* clones are selected by color (lacZ), and the bacmid purified from a single white colony is used to transfect insect cells. This method is fast and handy since it eliminates the need for time-consuming plaque purification. Pure recombinant viruses can be prepared within 7–10 days. However, the original system suffers from the poor selection scheme. In addition, the identification of recombinant baculovirus genomes in *E. coli* is hampered by colony sectoring and multiple colony morphologies (44). To avoid the caveats (low frequency generation of the true recombinant bacmids), we have improved the original system while retaining a simple, rapid, and convenient virus production (32). The method is based on the modified donor vector (pBVboost) and an improved selection scheme of the baculovirus genomes (bacmids) in *E. coli* with a mutated *SacB* gene. Multiple recombinant bacmids can be generated at a frequency of ~10^7 re-bacmids per microgram of donor vector without background in 5–6 days. The BVboost system supports also efficient setups for a high-throughput screening and gene expression purposes (33). A tetra-promoter variant of the pBVboost, pBVboostFG, enables facile all-in-one strategy for gene expression in mammalian, bacterial and insect cells (31).

3.1. Growth and Maintenance of Insect Cells in Adherent and Suspension Culture

1. Prepare a 15-ml centrifuge tube filled with 9 ml of prewarmed (28°C) insect cell medium.
2. Thaw a frozen vial of low passage cells (10,000,000 cells/ml) as quickly as possible in a 28°C water bath.
3. Move thawed cells to a centrifuge tube, which contains the prewarmed medium and centrifuge at 120×*g* for 7 min at room temperature.
4. Remove the supernatant and resuspend the cells into 15 ml of fresh and warm medium (28°C). Transfer the cells to a T75-cell culture flask and grow at 28°C until confluent.
5. Gently detach the cells with a cell scraper or by pipetting. Add 10 ml of prewarmed (28°C) medium on the cells and transfer the suspension into a new 50-ml centrifuge tube.
6. Determine the viability of the cells with Trypan Blue. In practice, take 10 μl of cells, 40 μl of medium, and add 50 μl of Trypan Blue. Mix, insert the stained cells into a cell count chamber and count the fields twice.
7. Dilute the cell suspension to 500,000 cells/ml in a volume of 25 ml in a 250-ml Erlenmeyer bottle. For a volume of 25 ml suspension culture, approximately 12.5×10^6 cells are thus needed. Grow the cells in a shaker at 110 rpm, 28°C until the cell density reaches approximately 2.0×10^6 cells/ml (see Note 1).
8. After achieving the appropriate density, count the cells and subculture 500,000 cells/ml to 25 ml of fresh medium. Repeat the step three times.
9. After the initial adaptation, the cells can be maintained in a suspension culture by subculturing to density of 200,000–300,000 cells/ml when the viable cell count reaches density of $2.0–4.0 \times 10^6$ cells/ml. After 40 passages or 3 months of passaging, the cells need to be replaced by healthy low passage cells (see Note 2).

3.2. Generation of Recombinant Baculovirus Genomes (Bacmids) via Transposition in E. coli

1. This protocol is a derivative of the original transposition-based method by Luckow et al. (43) (commercially available as Bac-to-Bac system) and assumes the use of Bio-Rad MicroPulser for electro-transformation. To achieve the most convenient and background-free bacmid generation, BVboost system vectors and DH10BacΔTn7 *E. coli* cells are recommended (32). They are available from the author's lab.
2. Prepare LB_{tg}- or LB_{stg} (BVboost based vectors)- agar plates containing 100 μg/ml Bluo-gal and 100 mM IPTG (see Note 3).
3. Thaw electrocompetent DH10Bac cells on ice and move 40 μl of the thawed cells into prechilled Eppendorf tube (see Note 4).

4. Add 1 ng of the donor plasmid in 5 μl of TE and gently mix the DNA with the cells.
5. Transfer the mixture into a cooled cuvette and carry out the electroporation according to instructions of the electroporator. For the Bio-Rad MicroPulser, use 1.8 kV setting for the Ec1.
6. Add 1 ml S.O.C. to the mixture.
7. Transfer the mixture into a 2-ml Eppendorf tube and incubate in a shaker at 250 rpm for 4 h at 37°C.
8. Serially dilute the mixture using S.O.C. to 10^{-1}, 10^{-2}, and 10^{-3} and spread 100 μl of original mixture and each dilution on the LB_{tg} or LB_{stg}-plates.
9. Incubate for at least 24 h at 37°C. All the appeared colonies represent recombinant baculoviral genomes if boost-series vectors and DH10BacΔTn7 *E. coli* are used. In the other cases, wait 48 h and verify the white color colony phenotype (indicating the true re-bacmid) further by streaking and cultivating potential white colonies on fresh plates at least 24 h at 37°C (see Note 5).

3.3. Isolation of a Recombinant Virus Genome (Bacmid)

1. Pick white colonies (2–10 from Subheading 3.2, step 9) and inoculate into 5 ml of LB medium supplemented with 10 μg/ml tetracycline and 7 μg/ml gentamicin. Use 50-ml Nunc tubes. Grow the bacteria in a shaker (250 rpm) at 37°C overnight.
2. Transfer 1 ml of the culture into a 1.5-ml tube and centrifuge at 14,000 × *g* for 1 min.
3. Remove the supernatant and resuspend the cells into 300 μl of the boost-solution 1 by gently mixing with pipette.
4. Add 300 μl of the boost-solution 2 and mix by inverting the tube few times. Incubate at room temperature for 5 min.
5. Add slowly 300 μl of the boost-solution 3 by mixing gently during addition. A thick white precipitate (*E. coli* genomic DNA and proteins) is formed. Place the mixture on ice for 5 min.
6. Centrifuge at 14,000 × *g* for 10 min. While the tube is in a centrifuge, add 0.8 ml of absolute isopropanol into a new 2-ml tube.
7. Gently transfer the supernatant into the isopropanol containing tube. Avoid the white precipitate. Mix gently by inverting the tube several times and hold on ice for 5 min.
8. Centrifuge the tube at 14,000 × *g* for 15 min at 4°C.
9. Remove the supernatant and add 500 μl of 70% ethanol. Invert the tube for several times to wash the pellet and centrifuge at 14,000 × *g* for 5 min at 16°C.
10. Remove the supernatant and air-dry the pellet for 5–10 min.
11. Dissolve the DNA into 40 μl of TE-buffer by gently tapping the tube (see Note 6).

3.4. Verification of a Recombinant Bacmid (A Virus Genome) by PCR (Optional)

1. To verify the transposition of the insert into the bacmid by PCR, purified bacmid DNA from the Subheading 3.3 (step 11) is used as a template.
2. Prepare the following PCR-mixture per purified bacmid and put the tube on ice (see Note 7):

   5 μl 100% DMSO

   5 μl 10× Dynazyme-buffer

   1 μl 10 mM dNTP-mix

   2 μl F835R primer

   2 μl pUC/M13 Forward primer

   2 μl pUC/M13 Reverse primer

   3 μl bacmid-DNA

   1 μl Dynazyme II DNA Polymerase

   29 μl PCR-grade H_2O
3. As a negative control, use bacmid DNA purified from a blue colony. Optionally, inoculate a blue colony directly from the culture plate into PCR-mix in which the 3 μl of template Bacmid has been replaced with 3 μl of H_2O.
4. The program for PCR reaction is:
   (a) 4 min at 95°C
   (b) 30 s at 95°C
   (c) 30 s at 55°C
   (d) 30 s at 72°C
   (e) Repeat steps 2–4 for 25 times
   (f) 10 min at 72°C
5. Run the amplified DNA in a 1% agarose gel electrophoresis using 0.5× TBE buffer. The DNA is amplified by using a combination of the pUC/M13 and F835 forward and pUC/M13 reverse primers (Fig. 1). If the insert is not integrated into a bacmid, a 325-bp sized band is amplified in the PCR. On the other hand, the successful transposition is visualized by a 200-bp sized fragment in the gel. In the latter case, a band corresponding to size of the transposon plus the transgene (in the case of the pFastBac1 vector, 2,300 bp plus the insert) may also be visible (large inserts may not amplify well). For a direct verification of the recombinant bacmid, replace one of the pUC/M13 primers with an insert specific primer and optimize the PCR conditions.

3.5. Virus Generation

1. Cultivate insect cells in suspension as described in the Subheading 3.1 and seed 1.5×10^6 Sf9 cells into 35-mm wells of a 6-well plate in 2 ml of serum-free medium (see Note 8).

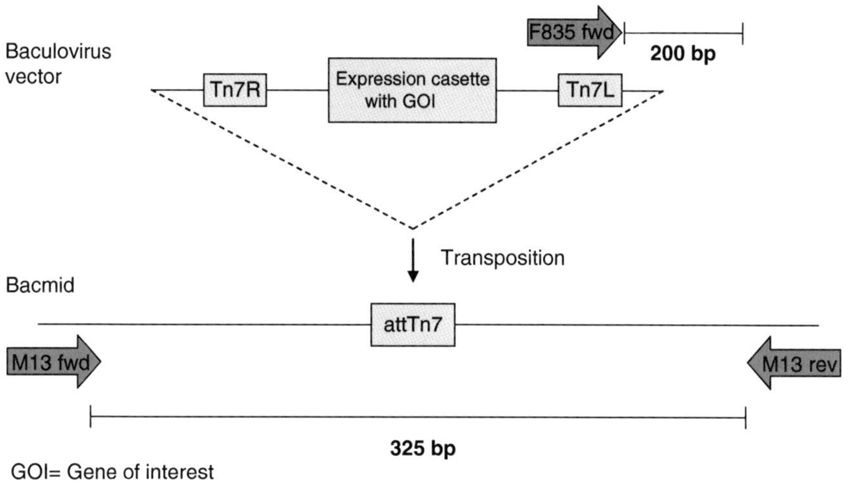

Fig. 1. A schematic presentation of the PCR analysis. The bacmid contains M13 Forward and M13 Reverse priming sites flanking the mini-attTn7 site within the lacZ α-complementation region. The Tn7L-site contains a priming site for F835 primer. A 200-bp sized fragment will be amplified only from the recombinant bacmids. Parental bacmid background creates a 325-bp fragment. Depending on the size of the cloned insert, it may be visible from the recombinant bacmid.

2. Allow the cells to attach for 1 h at 28°C.
3. In the meantime, dilute in an Eppendorf tube 5 μl of bacmid DNA (from Subheading 3.3, step 11) into 100 μl of insect cell medium (solution A), and for solution B, 6 μl of thoroughly mixed Cellfectin® transfection reagent (Invitrogen) into 100 μl of insect cell medium.
4. Combine solutions A and B by mixing gently and incubate for 20–45 min at room temperature.
5. Add 0.8 ml of insect cell medium to tubes from the step 4 and mix gently. Remove media from the wells and overlay 1 ml of diluted lipid–DNA complexes onto cells to transfect them.
6. Incubate the plate for 5 h at 28°C.
7. Remove the transfection solution and add 2 ml of insect cell medium.
8. Incubate for 72–96 h at 28°C.

3.6. Virus Harvest and Storage

1. Harvest the supernatant to a sterile tube. Centrifuge at 500 × *g* for 5 min to clarify the cell debris. Transfer the cleared primary virus supernatant to a fresh tube (see Note 9).
2. Primary virus stock tube can be stored, protected from the light, wrapped in folio for 6 months at 4°C. For a long-term storage, store at –70°C freezer (see Note 10).
3. Determine the viral titer by the end-point dilution procedures as described in the next section (optional).

3.7. Virus Titration by End-Point Dilution Method

1. To perform the end point titration, use only healthy Sf-9 cells, which are in an exponential crowing phase (approximately when the cells reach the density of 10^6 cells/ml after subculturing).
2. Count the cells and divide 2×10^4 cells to a 96-well plate containing 100 μl of medium/well. Perform the titration parallel at least in two plates.
3. Incubate the plates at minimum for 1 h at 27–28°C. This allows the cells to attach to the bottom of the wells.
4. Incubate the virus for 30 min at 27–28°C and perform the virus dilutions to a prewarmed (28°C) insect cell medium as shown below.
5. Start the dilution from 10^{-1} (10 μl of virus stock+90 μl medium) and then proceed to 10^{-3} dilution, which serves as a positive control. As a negative control, use cells and medium without the virus.

   1:100: 30 μl (10^{-1})+2,970 μl medium = 10^{-3}
   1:100: 30 μl (10^{-3})+2,970 μl medium = 10^{-5}
   1:10: 300 μl (10^{-5})+2,700 μl medium = 10^{-6}
   1:10: 300 μl (10^{-6})+2,700 μl medium = 10^{-7}
   1:10: 300 μl (10^{-7})+2,700 μl medium = 10^{-8}
   1:10: 300 μl (10^{-8})+2,700 μl medium = 10^{-9}
   1:10: 300 μl (10^{-9})+2,700 μl medium = 10^{-10}
   1:10: 300 μl (10^{-10})+2,700 μl medium = 10^{-11}
   1:10: 300 μl (10^{-11})+2,700 μl medium = 10^{-12}

6. After the cells have attached to the bottom of the wells, remove the medium and add the virus dilutions (100 μl/well).
7. Incubate the plates for 7 days at 27–28°C.
8. After the incubation, prepare new 96-well plates with 2.5×10^4 cells and 80 μl of medium per well. Incubate the plates at 27–28°C for a minimum of 1 h.
9. Add 20 μl of medium from the step 8 to the new plates and incubate for 3–4 days at 27–28°C.
10. Analyze the wells for infected cells using a microscope. The well is counted as infected if it has at least one infected cell. Compare the virus samples to positive and negative controls to avoid false positives. Baculovirus infection is lytic for insect cells and the infected cells are swollen and/or disintegrated (see Note 11).
11. The titer is calculated by estimating the virus dilution that would infect 50% of the cultures (the end-point dilution

method) (24). Fill the table below with the data and follow the example [modified from O'Reilly et al. (24)] to calculate the titer (see Note 12):

End-point dilution results

Dilution (10^{-X})		7	8	9
Positive rate/10 wells		10/10	4/10	0/10
Positive well number		10	4	0
Negative well number		0	6	10
Total positive wells		14	4	0
Total negative wells		0	6	16
Total positive rate		14/14	4/10	016
Positive well %		100	40	0
PD	0.83333			
TCID50	1.46780E-08			
TCID50/ml	6.81292E+08			
PFU/ml	4.70092E+08			

The dilution that would have given 50% infection lies between 10^{-7} and 10^{-8} in this example. This information is needed to calculate the proportionate distance (PD), which allows calculation of the titer:

PD = (*A*–50)/(*A*–*B*), where the *A* is the % response above 50% and *B* is the response below 50%.

In this example: PD = (100–50)/(100–40) = 0.83333.

The dose that would have given a 50% infection, the $TCID_{50}$, is then calculated using the formula:

Log $TCID_{50}$ = log of the dilution giving a response greater than 50% – the PD of that respond.

Thus, log $TCID_{50}$ = –7–0.8333 = –7.8333.

Therefore: $TCID_{50} = 10^{-7.8333}$

$$= 1.46780 \times 10^{-8}$$

The titer of the virus (pfu/ml) can then be calculated:

$$1/\ TCID_{50} = 1/1.46780 \times 10^{-8}$$

$$= 6.81292 \times 10^{7}$$

$TCID_{50}$/ml = $6.81292 \times 10^{7}/0.1$ (0.1 refers to 100 μl of virus used to infect the wells).

$$= 6.81292 \times 10^{8}$$

This can be converted to pfu/ml by using the relationship pfu = $TCID_{50} \times 0.69$.

So the titer of the example virus is:

$$6.81292 \times 10^{8} \times 0.69 = 4.70092 \times 10^{8}\ \text{pfu/ml}.$$

3.8. Virus Amplification for the Secondary Virus Stock

1. Infect suspension culture (as described in the Subheading 3.1) using the primary virus stock from the Subheading 3.6 using low MOI of 0.01–0.1 for 96 h (see Note 13).
2. Centrifugate the culture at 1,000 ×*g* for 5 min to harvest the secondary virus stock.
3. Transfer the clarified supernatant containing the viruses to a fresh tube and store protected from light at 4°C.
4. Determine the titer of secondary virus stock as described in Subheading 3.7 (see Note 14).

3.9. Preparation, Concentration, and Purification of a High-Titer Virus

1. Set up a suspension cell culture of 500 ml in which there are 250,000–300,000 cells/ml. Cultivate by shaking at 110 rpm for approximately 3 days at 28°C or until the cell density has reached 1,000,000 cells/ml.
2. Infect the cells with low MOI (<0.1) and let the infection proceed for 4 days (see Note 15).
3. Remove the cells and cell debris by centrifugation at 5,000 × g for 20 min at 4°C using 750-ml bottles.
4. Take the supernatant in to a clean 750-ml centrifugation bottle having 75 ml of cold 25% saccharose solution at the bottom. Centrifuge at 5,000 ×*g* for 20–24 h at 4°C.
5. Decant the supernatant and remove all drops of liquid from the bottle. Suspend the virus pellet to 25 ml of DPBS at room temperature.
6. The ultracentrifugation is assumed to be performed with the Beckman Counter Optima L-90-K with SW28 rotor and 35-ml buckets. Fill two ultracentrifugation tubes on ice with 6 ml of cold 50% saccharose solution and overlay the layers carefully with 5 ml of cold 20% saccharose solution (see Note 16).
7. Add carefully the virus suspension from the step 5 to the top of the gradient tubes.
8. Fill the tubes with DPBS so that the total volume will be 32 ml/tube.
9. Ultracentrifuge at 82705 × *g* for 1 h at 4°C.
10. Collect the virus bands, locating between the gradients, with an 18-G (BD) needle syringe. Transfer the viruses on ice to new ultracentrifugation tubes filled with 6 ml of cold DPBS.
11. Fill the tubes with cold DPBS to 32 ml.
12. Ultracentrifuge at 82705 × *g* for 1 h at 4°C.
13. Remove all the supernatant and add 400–500 μl of cold DPBS on top of the pellets.

14. Store overnight at 4°C (see Note 17).
15. Suspend the virus pellets gently by pipetting to the surrounding solution and combine them (see Note 18).
16. Pack the virus suspension to a Pierce Dialysis cassette (10K) and dialyze in a sterile decanter class three times (first 2× 3 h and then once overnight) at 4°C against 500 ml DPBS.
17. Collect the virus and store at 4°C in the dark.
18. Determine the virus titer (Subheading 3.7) and characterize the concentrated and purified virus stock by immunoblotting as described below.

3.10. Virus Characterization by Immunoblotting

1. These instructions assume the use of Biorad's Mini Protean electrophoresis system. However, the protocol is adaptable for other equipments too.
2. Prepare a 1.5-mm thick 10% SDS-PAGE gel using clean glass plates. For a one running gel, mix 3.0 ml of H_2O, 1.875 ml of 1.5 M Tris–HCl, 75 μl of 10% SDS, 2.5 ml of 30% AA-Bis, and 37.5 μl of 10% APS. Just before pouring the solution between the class plates, add 3.75 μl of TEMED. Leave some space for the stacking gel and overlay the gel carefully with H_2O.
3. Allow the gel to polymerize for half an hour.
4. Prepare the 4% stacking gel by mixing 3.05 ml of H_2O, 1.25 ml of 0.5 M Tris–HCl, 50 μl of 10% SDS, 0.65 ml of 30% AA-Bis, and 25 μl of 10% APS.
5. Discard the water from the top of the running gel carefully.
6. Add 5 μl of TEMED to the stacking gel mix just before pouring the stacking gel mix onto the top of the running gel.
7. Insert the comb and allow the gel to polymerize for 45 min.
8. Set up the running chamber by filling it with the Running buffer.
9. Remove the comb and wash the wells with the Running buffer.
10. Prepare the samples by mixing the virus with the sample buffer (see Note 19).
11. Boil the samples for 5 min at 95°C.
12. Chill the samples on ice, spin shortly and load on the gel along with the molecular weight markers.
13. Connect the chamber to a power supply (BioRad HC) and run the samples first for 10 min with 100 V (to get the

samples through stacking gel), and then for approximately 1 h more with 180 V.

14. After the run, unassemble the chamber, cut away the stacking gel, and incubate the gel, fiber pads, nitrocellulose membrane, and Whatman 3MM chromatography papers (cut in the size of gels) in the cold (4°C) Transfer-buffer for 15 min.
15. Assemble the blotting system according to manufacturer's instructions. If Biorad's mini trans blot cell-system is used, place first a fiber pad on the black side of the blotting cassette, then a chromatography paper, followed by the gel and the membrane. Next, a chromatography paper again and then the second fiber pad (see Note 20).
16. Close the cassette and assemble the blotting apparatus so that the black face of the blotting cassette face to the black side in the cassette assembler.
17. Fill the blotting apparatus with the Transfer buffer (cooled during running by ice block) and transfer the proteins from the gel onto the membrane for 1 h at 100 V. Use magnetic stirring in the buffer chamber.
18. After the transfer, unassemble the blotting cassettes, mark the membranes, and wash them once with TBS for 5 min on a rocking platform (see Note 21). Block the membrane for 1 h at room temperature or optionally overnight at 4°C in a blocking buffer.
19. Wash the membrane for 5 min in TBST.
20. Dilute the primary gp64-antibody 1:1,000 into blocking buffer and incubate for 1 h at room temperature.
21. Wash the membrane for 4× 5 min with the TBST.
22. Dilute the secondary antibody (typically 1:2,000) to blocking buffer and incubate for 1 h at room temperature.
23. Wash the membrane for 4× 5 min with the TBST.
24. Incubate the membrane for 5 min in the APA-buffer.
25. Add 80 μl of NBT/BCIP Stock solution to 5 ml of APA-buffer to stain the membrane.
26. Add the staining solution on the membrane and incubate at room temperature until the color develops.
27. Stop the color reaction by washing the membrane with several times with H_2O (see Note 22).
28. Dry the membrane on a chromatography (Whatman) paper.

3.11. Transduction of Vertebrate Cells

This protocol is designed for cells grown in 6-well plate. If multiwell plates or dishes of different diameter are used, scale the cell density and reagent volumes accordingly. Use only subconfluent healthy cells.

1. Remove the medium, rinse the cells once with 5–10 ml of sterile PBS and trypsinize them (1 ml of trypsin for 90-mm culture dish).
2. Transfer the cell suspension to a 15-ml centrifuge tube containing 4 ml of fresh growth medium (see Note 23).
3. Centrifugate at 47–92 × *g* (500–700 rpm in a Heraeus, Megafuge 1.0) for 5–7 min at room temperature.
4. Remove the medium and suspend the cell pellet in 1–4 ml of fresh medium.
5. Count the cell density and replate them on a 6-well plate at a density of about $1–2 \times 10^5$ cells (see Note 24).
6. Add 3 ml of growth medium per well and incubate the cultures for 18–20 h at +37°C in a humidified incubator with 5% CO_2.
7. Replace the culture medium with 1 ml of prewarmed medium.
8. Add desired MOI of virus and incubate the cells for at least 2 h at +37°C with 5–10% CO_2 (see Notes 25–32).
9. Remove the virus containing medium by aspiration.
10. Add 3 ml of prewarmed culture medium to wells and return the plate to the incubator (see Note 33).
11. Examine the cells 24–72 h after the transduction. Figure 2 shows an example of transduced cells.

Fig. 2. Fluorescence microscopy pictures of the baculovirus-mediated EGFP expression in a human liver (HepG2), rabbit smooth muscle (RaaSMC), and human kidney (HEK293T) cells. Cells were transduced with Ba-CAG-EGFP/WPRE virus (34) in the optimized cell culture medium (RPMI 1640). Original magnification, 100×.

4. Notes

1. This density can normally be reached within 5–7 days.
2. The cell density has an effect on the growth rate. Subculturing to densities below 200,000 should be avoided and can be detrimental to the cells.
3. Bluo-gal and IPTG are optional for the BVboost system vectors. X-gal (60 μg/ml) can be used instead of the Bluo-gal, but it will result in a lower color intensity.
4. To achieve 100% re-bacmid generation rate, use Boost series vectors and DH10BacΔTn7 *E. coli* cells (32). The cryptic *att*Tn7-site is blocked in the DH10BacΔTn7 cells, which improves markedly the transposition efficacy between the donor vector and a bacmid.
5. Even if Bluo-gal is used for the color selection, the true white colonies may be hard to separate from the faint blue colonies representing the parental bacmid background. Therefore, at least 48-h incubation and additional streaking of potential colonies is needed if the Bac-to-Bac system cells (DH10Bac) or vectors (like pFastBac or its derivatives) are used.
6. The pellet can also be left to dissolve overnight at 4°C. Bacmid DNA can be stored at −20°C. Avoid repeated freeze/thaw cycles.
7. If needed, multiply the volumes by the amount of samples and prepare a PCR master mix.
8. Use only healthy cells in the mid-log growth phase ($1–2 \times 10^6$ cell/ml) with viability over 97%.
9. Cell pellet can be used for the verification of virus generation (see Subheading 3.10). Cell pellets can also be used to analyze recombinant protein production if the transgene is under an insect cell active promoter.
10. If the virus was generated in a serum-free medium, addition of 5–10% FBS to virus stock may help to maximize its stability. However, the stocks can usually be maintained at dark even without FBS at least half a year at 4°C.
11. A fluorescent protein marker in a donor vector or a Bacmid genome makes the titration faster and easier (33, 45).
12. Yes, it is a complicated equation and calculation. An automatic Windows Excel-based data-sheet is available from the author's laboratory to ease the titer calculation.
13. It is essential to use low MOI in this step to avoid the accumulation of defective interfering particles (24). For preparing 30 ml of secondary virus stock, 60 μl of the primary stock should lead to desired MOI and virus amplification without a

fear of overinfection. Optionally, determine the primary viral titer before infection as described in the Subheading 3.7.

14. At this point the titer is typically ~10^8 pfu/ml for a good quality virus.
15. For a routine use, infection with 1:1,000 dilution of secondary virus stock (i.e. 0.5 or 1–500 or 1,000 ml culture, respectively) is usually fine. However, this may easily lead to actual MOIs over 0.1. To avoid possible virus degeneration by using higher MOI than 0.1, titer the virus before infection as described in the Subheading 3.7. Often low MOI of 0.1–1 is also beneficial in protein production. Less virus stock is also needed per batch and infection can be initiated at lower cell density (~1×10^6 cell/ml). As opposite to the high MOI infection (MOI ≥5), the low MOI infection allows the cells to divide several times before they are infected by the viruses produced by the cells first infected in the culture. The disadvantage of the low MOI infection is that some, especially protease sensitive, proteins may be vulnerable for unsynchronized and extended cultivation time.
16. Avoid air bubbles in the gradient.
17. Incubation overnight at 4°C makes the virus easier to suspend with DPBS.
18. The virus is fragile. Do not vortex or use a pipet with a small hole.
19. The sample volume depends on the gel thickness and the used comb. In the present protocol, the maximum sample volume in a 1.5-mm gel, poured with the 10-well comb, is 40 μl.
20. Make sure that no air bubbles are trapped between the gel and the membrane. Mark the gels (and after blotting the membranes) by cutting their corners.
21. All the washes and antibody incubations are performed on the rocking platform.
22. The color development will not stop immediately. Take this into account to avoid overstaining.
23. Serum does not harm the transduction efficacy. The RMPI 1640 is highly recommended as a suitable culture medium (34).
24. The exact cell density depends on the size and growth-rate of the cells and must be determined empirically. The best transduction efficiency will typically be achieved when cells are in 50–75% confluency.
25. Handle the fragile baculoviruses gently (no vortexing or vigorous pipetting).
26. Transduction efficacy may increase if viruses are left onto cells for an extended period of time or until examination. Extended transduction time will not harm the cells.

27. Due to the low cytotoxicity of baculoviruses, cells can also be retransduced (supertransduced) several times to prolong and enhance the transduction efficacy (46).
28. Instead of concentrated virus stock in PBS, secondary (high-titer) virus stock may be used diluted at least 1:1 to vertebrate culture medium. Incubate cells with virus at +37°C at least for 2 h (8, 47).
29. Lower incubation temperature (e.g. +25°C) may enhance transduction efficacy (47).
30. Microtubule depolymerizing agents augment baculovirus-mediated gene delivery (48).
31. Recombinant baculoviruses having vesicular stomatitis virus glycoprotein (VSVG) or other envelope modification may increase transduction efficacy (15, 49–53). Note, however, that VSV-G may be toxic to the cells.
32. Equip the expression cassette with Woodschock post-transcriptional element (WPRE) to improve results in most of the cells (34). The choice of promoter may also make a difference. Chicken beta-actin promoter has been shown to drive better transgene expression than CMV in some cells (54, 55). Cell-type specific gene expression has been achieved with tissue-specific promoters (56–59).
33. Addition of sodium butyrate directly to the growth medium to a final concentration of 2.5–5 mM facilitates the transduction efficiency of baculovirus (5, 16).

Acknowledgments

We would like to thank Ms. Tarja Taskinen for technical assistance and valuable comments on the manuscript. This work was supported by grant from the European Union (LHSB-CT-2006-037541).

References

1. Airenne, K. J., Mähönen, A. J., Laitinen, O. H., and Ylä-Herttuala.S. (2009) Baculovirus-mediated gene transfer: An emerging universal concept, in *Gene and cell therapy: Therapeutic mechanisms and strategies*, (Templeton, N. S. ed.), CRC Press, Boca Raton, pp. 263–307.
2. Kost, T. A., Condreay, J. P., and Jarvis, D. L. (2005) Baculovirus as versatile vectors for protein expression in insect and mammalian cells. *Nat. Biotechnol.* **23**, 567–575.
3. Summers, M. D. (2006) Milestones leading to the genetic engineering of baculoviruses as expression vector systems and viral pesticides. *Adv. Virus Res.* **68**, 3–73.
4. Volkman, L. E. and Goldsmith, P. A. (1983) In vitro Survey of *Autographa californica* Nuclear Polyhedrosis Virus Interaction with Nontarget Vertebrate Host Cells. *Applied and Environmental Microbiology* **45**, 1085–1093.
5. Condreay, J. P., Witherspoon, S. M., Clay, W. C., and Kost, T. A. (1999) Transient and

stable gene expression in mammalian cells transduced with a recombinant baculovirus vector. *Proc. Natl. Acad. Sci. U S A* **96**, 127–132.

6. Song, S. U., Shin, S. H., Kim, S. K., Choi, G. S., Kim, W. C., Lee, M. H., Kim, S. J., Kim, I. H., Choi, M. S., Hong, Y. J., and Lee, K. H. (2003) Effective transduction of osteogenic sarcoma cells by a baculovirus vector. *J. Gen. Virol.* **84**, 697–703.
7. Hu, Y. C. (2008) Baculoviral vectors for gene delivery: a review. *Curr. Gene Ther.* **8**, 54-65.
8. Cheng, T., Xu, C. Y., Wang, Y. B., Chen, M., Wu, T., Zhang, J., and Xia, N. S. (2004) A rapid and efficient method to express target genes in mammalian cells by baculovirus. *World J. Gastroenterol.* **10**, 1612–1618.
9. Granados, R. R. (1978) Replication phenomena of insect viruses *in vivo* and *in vitro*., in *Safety Aspects of Baculoviruses as Biological Insecticides*, (Miltenburger, H. G. ed.), Bundesministerium für Forschung und Technologie, Bonn, pp. 163–184.
10. Kost, T. A. and Condreay, J. P. (2002) Innovations-Biotechnology: Baculovirus vectors as gene transfer vectors for mammlian cells: Biosafety considerations. *J. Am. Biol. Safety Ass.* 7, 167–169.
11. Burges, H. D., Croizier G., and Huger, J. (1980) A review of safety tests on baculoviruses. *Entomaphaga* **25**, 329–340.
12. Kost, T. A., Condreay, J. P., Ames, R. S., Rees, S., and Romanos, M. A. (2007) Implementation of BacMam virus gene delivery technology in a drug discovery setting. *Drug Discov. Today* **12**, 396–403.
13. Lesch, H. P., Turpeinen, S., Niskanen, E. A., Mahonen, A. J., Airenne, K. J., and Yla-Herttuala, S. (2008) Generation of lentivirus vectors using recombinant baculoviruses. *Gene Ther.* **15**, 1280–1286.
14. Smith, R. H., Levy, J. R., and Kotin, R. M. (2009) A Simplified Baculovirus-AAV Expression Vector System Coupled With One-step Affinity Purification Yields High-titer rAAV Stocks From Insect Cells. *Mol. Ther.* **17**, 1888–1896.
15. Oker-Blom, C., Airenne, K. J., and Grabherr, R. (2003) Baculovirus display strategies: Emerging tools for eukaryotic libraries and gene delivery. *Brief. Funct. Genomic. Proteomic.* **2**, 244–253.
16. Airenne, K. J., Hiltunen, M. O., Turunen, M. P., Turunen, A. M., Laitinen, O. H., Kulomaa, M. S., and Yla-Herttuala, S. (2000) Baculovirus-mediated periadventitial gene transfer to rabbit carotid artery. *Gene Ther.* **7**, 1499–1504.
17. Wang, C. Y., Li, F., Yang, Y., Guo, H. Y., Wu, C. X., and Wang, S. (2006) Recombinant baculovirus containing the diphtheria toxin A gene for malignant glioma therapy. *Cancer Res.* **66**, 5798–5806.
18. Tani, H., Abe, T., Matsunaga, T. M., Moriishi, K., and Matsuura, Y. (2008) Baculovirus vector for gene delivery and vaccine development. *Future Virol.* **3**, 35–43.
19. Hu, Y. C., Yao, K., and Wu, T. Y. (2008) Baculovirus as an expression and/or delivery vehicle for vaccine antigens. *Expert. Rev. Vaccines.* 7, 363–371.
20. Black, B. C., Brennan, L. A., Dierks, P. M., and Gard, I. E. (1997) Commercialization of Baculoviral Insecticides, in *The Baculoviruses*, (Miller, L. K. ed.), Plenum Press, New York, pp. 341–387.
21. Theilmann, D. A., Blissard, G. W., Bonning, B., Jehle, J., O'Reilly, D. R., Rohrmann, G. F., Thiem, S., and Vlak, J. M. (2005) Family Baculoviridae, in *Virus Taxonomy: Eight Report of the International Committee on Taxonomy of Viruses*, (Fauquet, C. M., Mayo, M. A., Maniloff, J., Desselberger, U. and Ball, L. A., eds.), Elsevier, London, pp. 177–185.
22. Miller, L.K. (1997) *The Baculoviruses.* Plenum Press, New York.
23. Funk, C. J., Braunagel, S. C., and Rohrmann, G. F. (1997) Baculovirus Structure, in *The Baculoviruses*, (Miller, L. K. ed.), Plenum Press, New York, pp. 7–32.
24. O'Reilly, D.R., Miller, L. K., and Luckov, V. A. (1994) *Baculovirus expression vectors. A laboratory manual.* Oxford University Press, New York.
25. Ayres, M. D., Howard, S. C., Kuzio, J., Lopez-Ferber, M., and Possee, R. D. (1994) The complete DNA sequence of Autographa californica nuclear polyhedrosis virus. *Virology* **202**, 586–605.
26. Laakkonen, J. P., Makela, A. R., Kakkonen, E., Turkki, P., Kukkonen, S., Peranen, J., Yla-Herttuala, S., Airenne, K. J., Oker-Blom, C., Vihinen-Ranta, M., and Marjomaki, V. (2009) Clathrin-independent entry of baculovirus triggers uptake of E. coli in non-phagocytic human cells. *PLoS. ONE.* **4**, e5093.
27. Blissard, G. W. (1996) Baculovirus–insect cell interactions. *Cytotechnology* **20**, 73–93.
28. Wang, P., Hammer, D. A., and Granados, R. R. (1997) Binding and fusion of Autographa californica nucleopolyhedrovirus to cultured insect cells. *J. Gen. Virol.* **78**, 3081–3089.
29. Blissard, G. W. and Wenz, J. R. (1992) Baculovirus gp64 envelope glycoprotein is sufficient to mediate pH-dependent membrane fusion. *J. Virol.* **66**, 6829–6835.

30. Oomens, A. G. and Blissard, G. W. (1999) Requirement for GP64 to drive efficient budding of Autographa californica multicapsid nucleopolyhedrovirus. *Virology* **254**, 297–314.
31. Laitinen, O. H., Airenne, K. J., Hytonen, V. P., Peltomaa, E., Mahonen, A. J., Wirth, T., Lind, M. M., Makela, K. A., Toivanen, P. I., Schenkwein, D., Heikura, T., Nordlund, H. R., Kulomaa, M. S., and Yla-Herttuala, S. (2005) A multipurpose vector system for the screening of libraries in bacteria, insect and mammalian cells and expression in vivo. *Nucleic Acids Res.* **33**, e42.
32. Airenne, K. J., Peltomaa, E., Hytonen, V. P., Laitinen, O. H., and Yla-Herttuala, S. (2003) Improved generation of recombinant baculovirus genomes in Escherichia coli. *Nucleic Acids Res.* **31**, e101.
33. Karkkainen, H. R., Lesch, H. P., Maatta, A. I., Toivanen, P. I., Mahonen, A. J., Roschier, M. M., Airenne, K. J., Laitinen, O. H., and Yla-Herttuala, S. (2009) A 96-well format for a high-throughput baculovirus generation, fast titering and recombinant protein production in insect and mammalian cells. *BMC. Res. Notes* **2**, 63.
34. Mahonen, A. J., Airenne, K. J., Purola, S., Peltomaa, E., Kaikkonen, M. U., Riekkinen, M. S., Heikura, T., Kinnunen, K., Roschier, M. M., Wirth, T., and Yla-Herttuala, S. (2007) Post-transcriptional regulatory element boosts baculovirus-mediated gene expression in vertebrate cells. *J. Biotechnol.* **131**, 1–8.
35. Vaughn, J. L., Goodwin, R. H., Tompkins, G. J., and McCawley, P. (1977) The establishment of two cell lines from the insect Spodoptera frugiperda (Lepidoptera; Noctuidae). *In Vitro* **13**, 213–217.
36. Wickham, T. J., Nemerow, G. R., Wood, H. A., and Shuler, M. L. (1995) Comparison of different cell lines for the production of recombinant baculovirus proteins, in *Baculovirus expression protocols*, (Richardson. C.D. ed.), Humana Press, Totowa, pp. 385–395.
37. Li, T. C., Scotti, P. D., Miyamura, T., and Takeda, N. (2007) Latent infection of a new alphanodavirus in an insect cell line. *J. Virol.* **81**, 10890–10896.
38. Harrison, R. L. and Jarvis, D. L. (2006) Protein N-glycosylation in the baculovirus-insect cell expression system and engineering of insect cells to produce "mammalianized" recombinant glycoproteins. *Adv. Virus Res.* **68**, 159–191.
39. Hughes, P., Marshall, D., Reid, Y., Parkes, H., and Gelber, C. (2007) The costs of using unauthenticated, over-passaged cell lines: how much more data do we need? *Biotechniques* **43**, 575, 577–2.
40. Miller, L. K. (1993) Baculoviruses: high-level expression in insect cells. *Curr. Opin. Genet. Dev.* **3**, 97–101.
41. Stanbridge, L. J., Dussupt, V., and Maitland, N. J. (2003) Baculoviruses as Vectors for Gene Therapy against Human Prostate Cancer. *J. Biomed. Biotechnol.* **2003**, 79–91.
42. Hitchman, R. B., Possee, R. D., and King, L. A. (2009) Baculovirus expression systems for recombinant protein production in insect cells. *Recent Pat Biotechnol.* **3**, 46–54.
43. Luckow, V. A., Lee, S. C., Barry, G. F., and Olins, P. O. (1993) Efficient generation of infectious recombinant baculoviruses by site-specific transposon-mediated insertion of foreign genes into a baculovirus genome propagated in Escherichia coli. *J. Virol.* **67**, 4566–4579.
44. Leusch, M. S., Lee, S. C., and Olins, P. O. (1995) A novel host-vector system for direct selection of recombinant baculoviruses (bacmids) in Escherichia coli. *Gene* **160**, 191–194.
45. Cha, H. J., Gotoh, T., and Bentley, W. E. (1997) Simplification of titer determination for recombinant baculovirus by green fluorescent protein marker. *Biotechniques* **23**, 782–4, 786.
46. Hu, Y. C., Tsai, C. T., Chang, Y. J., and Huang, J. H. (2003) Enhancement and prolongation of baculovirus-mediated expression in mammalian cells: focuses on strategic infection and feeding. *Biotechnol. Prog.* **19**, 373–379.
47. Hsu, C. S., Ho, Y. C., Wang, K. C., and Hu, Y. C. (2004) Investigation of optimal transduction conditions for baculovirus-mediated gene delivery into mammalian cells. *Biotechnol. Bioeng.* **88**, 42–51.
48. Salminen, M., Airenne, K. J., Rinnankoski, R., Reimari, J., Valilehto, O., Rinne, J., Suikkanen, S., Kukkonen, S., Yla-Herttuala, S., Kulomaa, M. S., and Vihinen-Ranta, M. (2005) Improvement in nuclear entry and transgene expression of baculoviruses by disintegration of microtubules in human hepatocytes. *J. Virol.* 79, 2720–2728.
49. Barsoum, J., Brown, R., McKee, M., and Boyce, F. M. (1997) Efficient transduction of mammalian cells by a recombinant baculovirus having the vesicular stomatitis virus G glycoprotein. *Hum. Gene Ther.* **8**, 2011–2018.
50. Raty, J. K., Airenne, K. J., Marttila, A. T., Marjomaki, V., Hytonen, V. P., Lehtolainen, P., Laitinen, O. H., Mahonen, A. J., Kulomaa, M. S., and Yla-Herttuala, S. (2004) Enhanced gene delivery by avidin-displaying baculovirus. *Mol. Ther.* **9**, 282–291.

51. Tani, H., Limn, C. K., Yap, C. C., Onishi, M., Nozaki, M., Nishimune, Y., Okahashi, N., Kitagawa, Y., Watanabe, R., Mochizuki, R., Moriishi, K., and Matsuura, Y. (2003) In vitro and in vivo gene delivery by recombinant baculoviruses. *J. Virol.* 77, 9799–9808.
52. Kitagawa, Y., Tani, H., Limn, C. K., Matsunaga, T. M., Moriishi, K., and Matsuura, Y. (2005) Ligand-directed gene targeting to mammalian cells by pseudotype baculoviruses. *J. Virol.* **79**, 3639–3652.
53. Ojala, K., Mottershead, D. G., Suokko, A., and Oker-Blom, C. (2001) Specific binding of baculoviruses displaying gp64 fusion proteins to mammalian cells. *Biochem. Biophys. Res Commun.* **284**, 777–784.
54. Shoji, I., Aizaki, H., Tani, H., Ishii, K., Chiba, T., Saito, I., Miyamura, T., and Matsuura, Y. (1997) Efficient gene transfer into various mammalian cells, including non-hepatic cells, by baculovirus vectors. *J. Gen. Virol.* 78, 2657–2664.
55. Spenger, A., Ernst, W., Condreay, J. P., Kost, T. A., and Grabherr, R. (2004) Influence of promoter choice and trichostatin A treatment on expression of baculovirus delivered genes in mammalian cells. *Protein Expr. Purif.* **38**, 17–23.
56. Li, Y., Wang, X., Guo, H., and Wang, S. (2004) Axonal transport of recombinant baculovirus vectors. *Mol. Ther.* **10**, 1121–1129.
57. Li, Y., Yang, Y., and Wang, S. (2005) Neuronal gene transfer by baculovirus-derived vectors accommodating a neurone-specific promoter. *Exp. Physiol* **90**, 39–44.
58. McCormick, C. J., Challinor, L., Macdonald, A., Rowlands, D. J., and Harris, M. (2004) Introduction of replication-competent hepatitis C virus transcripts using a tetracycline-regulable baculovirus delivery system. *J. Gen. Virol.* **85**, 429–439.
59. Park, S. W., Lee, H. K., Kim, T. G., Yoon, S. K., and Paik, S. Y. (2001) Hepatocyte-specific gene expression by baculovirus pseudotyped with vesicular stomatitis virus envelope glycoprotein. *Biochem. Biophys. Res. Commun.* **289**, 444–450.

Chapter 13

Herpes Simplex Virus Type 1-Derived Recombinant and Amplicon Vectors

Cornel Fraefel, Peggy Marconi, and Alberto L. Epstein

Abstract

Herpes simplex virus type 1 (HSV-1) is a human pathogen whose lifestyle is based on a long-term dual interaction with the infected host, being able to establish both lytic and latent infections. The virus genome is a 153 kbp double-stranded DNA molecule encoding more than 80 genes. The interest of HSV-1 as gene transfer vector stems from its ability to infect many different cell types, both quiescent and proliferating cells, the very high packaging capacity of the virus capsid, the outstanding neurotropic adaptations that this virus has evolved, and the fact that it never integrates into the cellular chromosomes, thus avoiding the risk of insertional mutagenesis. Two types of vectors can be derived from HSV-1, recombinant vectors and amplicon vectors, and different methodologies have been developed to prepare large stocks of each type of vector. This chapter summarizes (1) the two approaches most commonly used to prepare recombinant vectors through homologous recombination, either in eukaryotic cells or in bacteria, and (2) the two methodologies currently used to generate helper-free amplicon vectors, either using a bacterial artificial chromosome (BAC)-based approach or a Cre/loxP site-specific recombination strategy.

Key words: HSV-1, Recombinant vectors, Amplicon vectors

1. Introduction: HSV-1 and Its Derived Vectors

HSV-1. Herpes simplex virus type 1 (HSV-1) is a major human pathogen whose lifestyle is based on a long-term dual interaction with the infected host. After initial infection and lytic multiplication at the body periphery, generally at oral or genital epithelial cells, the virus enters the sensory neurons that innervate the infected epithelia and, following retrograde transport of the capsids to the cell bodies, establishes a lifelong latent infection in sensory ganglia. Periodic reactivation from latency usually leads to the return of the virus to epithelial cells, where it produces secondary lytic infections (recurrences) resulting in mild illness symptoms, such as cold sores. Often, infectious virus can be detected in the

Otto-Wilhelm Merten and Mohamed Al-Rubeai (eds.), *Viral Vectors for Gene Therapy: Methods and Protocols*, Methods in Molecular Biology, vol. 737, DOI 10.1007/978-1-61779-095-9_13,

saliva of people presenting no clinical symptoms of disease. In rare cases, HSV-1 can spread centripetally into the central nervous system, to cause devastating encephalitis. For a comprehensive review on HSV-1 lytic and latent cycles, see ref. (1).

The HSV-1 particle is made up of four concentric layers. The DNA virus genome is enclosed into an icosahedric capsid, which is surrounded by the tegument, a rather unstructured layer containing some 20 virus-encoded proteins. The tegument is delimited by the envelope, which is a lipid membrane of cellular origin, containing a dozen virus-encoded glycoproteins (see Fig. 1). HSV-1 enters epithelial cells and neurons by fusion of the virus envelope with the plasma or endosomal membranes, and the virus capsids are transported to the nuclear pores through association with microtubules, from where the viral DNA is released into the nucleus (2). During lytic infection, the virus 153 kbp double-stranded DNA genome expresses more than 80 genes that are temporarily regulated in a cascade fashion, giving rise to three phases of gene expression. The expression cascade, which is regulated mainly at the transcriptional level, begins with the expression of the immediate-early (IE) genes. Five viral IE genes are expressed first, and four of these encode regulatory proteins (ICP0, ICP4, ICP22, and ICP27) that are responsible for controlling viral gene expression during subsequent early and late phases of the replication cycle and for inducing shutoff of cellular protein synthesis (3). Transcription of IE genes occurs in the absence of de novo viral protein synthesis and is highly stimulated by a virion protein known as VP16, which is a powerful transcription factor that, in conjunction with cellular proteins, acts on DNA motifs present only in the IE regulatory regions to

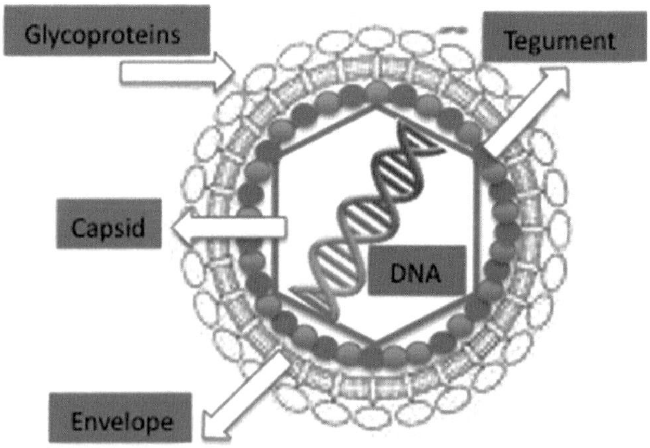

Fig. 1. The HSV-1 particle is composed of four concentric layers: the lipid envelope carrying 12 virus-encoded glycoproteins, the tegument layer, which is composed of some 20 different virus-encoded protein species, then the icosahedric capsid, and inside the capsid, the DNA molecule representing the virus genome.

upregulate expression (4). The early gene products that are synthesized next include enzymes that, like thymidine kinase and ribonucleotide reductase, act to increase the pool of deoxynucleotides of the infected cells, and several replication proteins that are directly involved in viral DNA synthesis. The last functions to be expressed are the late genes, which encode proteins involved in the packaging of virus DNA, as well as the structural proteins involved in the assembly of the virion particle, including the capsid, the tegument, and the glycoproteins (1). Some of these structural proteins, like the tegumentary VP16, play major regulatory roles in the next infectious cycle. Capsids are assembled in the nucleus but the tegument and the envelope are acquired in the cytoplasm, most probably by budding into endosomes, and are released by exocytosis at cell membranes (5).

During latency in sensory neurons, the viral genome remains as a circular chromatinized episome (6) within the cell nuclei and undergoes dramatic changes resulting in an almost complete silencing of transcription. Only one region of the viral genome, known as the LAT locus, is actively transcribed during latency, due to the presence of a latency-associated promoter (LAP) that remains active during this phase of the infection, resulting in the synthesis of nonmessenger RNA molecules of unknown function (the latency-associated transcripts, or LATs), which accumulates in the nucleus of the latently infected neurons (7). Very recently, the LAT locus has been shown to express miRNA molecules that can downregulate expression of lytic viral genes (8). The latent virus genome can reactivate in response to a wide variety of stimuli that allow it to enter the lytic phase of the HSV-1 life cycle. For a more specific review on HSV-1 latency, see ref. (9).

HSV-1-based vectors. HSV-1-based vectors have the capacity to deliver up to 150 kbp of foreign DNA to the nucleus of most proliferating and quiescent mammalian cells, making this family of vectors a very interesting tool for gene transfer and gene therapy. The uniqueness of HSV-1-based vectors stems from several properties of HSV-1: (1) the very large capacity of the virus particle, (2) the virus can infect many different cell types, both quiescent and proliferating mammalian cells, (3) the virus DNA will not integrate into host chromosomes, thus reducing the risk of insertional mutagenesis, (4) the complexity of the virus genome, which contains approximately 40 genes that are not essential for virus replication and can therefore be deleted without disturbing virus production in cultured cells and (5) the capacity of HSV-1 to infect the nervous system, including the ability to trans-synaptically spread from neuron to neuron in both anterograde and retrograde directions and the capacity to establish latent infections in neurons.

Three different types of vectors can be derived from HSV-1, which attempt to exploit one or more of these properties.

Recombinant attenuated viruses are replication-competent HSV-1 vectors carrying mutations that restrict spread and lytic viral replication to cancer cells without causing major toxicity to the healthy tissues. They are used mainly as oncolytic viruses (10). Defective, replication-incompetent nonpathogenic recombinant HSV-1 vectors lack one or more essential replication genes but retain many advantageous features of wild-type HSV-1, particularly the ability to express transgenes after having established latent infections in central and peripheral neurons (11). Amplicon vectors are defective, helper-dependent vectors that carry no viral genes and take advantage of the large carrier capacity of the virus particle to deliver long transgenic sequences (12, 13). In all three cases, the vector particles are basically identical to that of wild-type HSV-1, which are complex particles made up of some 40 different virus-encoded structural proteins. For a recent comprehensive review on HSV-1-based vectors, see ref. (14). From the methodological point of view, we can, however, consider that there are only two types of vectors, the recombinant vectors, which can be defective or attenuated, and the amplicon vectors, which are always defective. In the following sections, we will therefore present protocols to generate and prepare each of these two types of HSV-1-based vectors.

Recombinant HSV-1 vectors. Replication-defective recombinants are viral vectors where essential genes for viral replication are either mutated or deleted. Therefore, these mutants cannot grow except in transformed cell lines, where they are complemented in trans. To date, several replication-defective HSV-1 vectors have been constructed in which the immediate-early (IE) genes, expressing regulatory infected cell proteins (ICP) 0, 4, 22, 27, have been deleted in various combinations (15–18). ICP4 and ICP27 are essential for replication and the deletion of one or both of these genes requires adequate complementing cell lines, such as the Vero-7b cell line (19), capable of providing in trans the proteins encoded by deleted viral genes.

Attenuated HSV-1 recombinant vectors carry deletions in some nonessential genes, resulting in viruses that retain the ability to replicate in vitro, but are compromised in vivo, in a context-dependent manner (20, 21). Among the limitations to the use of HSV-1 is the fact that wild-type virus is highly pathogenic and entry in the brain can cause fatal encephalitis. Toxic and/or pathogenic properties of the virus must, therefore, be disabled prior its use as a gene delivery vector. Several genes involved in HSV-1 replication, virulence and immune evasion, nonessential for viral life cycle in vivo, have been identified. These genes are usually involved in multiple interactions with cellular proteins, which optimize the ability of the virus to grow within cells. Understanding such interactions has permitted the deletion/modification of these genes, alone or in combination, to create virus mutants with a reduced ability to replicate in normal quiescent cells, but that can replicate in tumor or dividing cells (22).

In spite of their fundamental biological differences, defective and attenuated HSV-1 recombinant vectors are constructed and prepared using very similar methodologies. The only significant practical difference is the need to use complementing cell lines to produce the defective recombinants. Classically, recombinant HSV-1 vectors were constructed by homologous recombination in eukaryotic cells, by cotransfecting the virus genome and a plasmid carrying the desired transgene surrounded by virus sequences to favor recombination. A more recent approach uses an HSV-1 genome cloned as a bacterial artificial chromosome and the desired transgenes are introduced into the virus genome by homologous recombination in bacteria. We will describe here both approaches.

1.1. Construction of Recombinant HSV-1 Vectors by Homologous Recombination in Eukaryotic Cells

Alterations of the HSV-1 genome in eukaryotic cells can be achieved in a number of ways. These usually require a process in which portions of the virus genome, which have been cloned into plasmid vectors, are first modified in vitro; then the modified sequences are introduced into the virus genome and recombinant viruses are selected. Several methods have been described to insert DNA sequences into the viral DNA. Recombination into specific sites within the viral genome has been achieved in vitro using a site-specific recombination system derived from phage P1 (23–25). It is possible, however, to significantly enhance the frequency of recombination using a two-step method through homologous recombination in cultured cells (26). The first step is the insertion of a reporter gene cassette flanked by *Pac*I or *Pme*I restriction enzymes sites, not otherwise found in the viral genome. Green fluorescent protein (GFP) and LacZ are two convenient marker genes that allow easy selection of the mutated virus genome (Fig. 2, left and center panels). The second step is the substitution of the reporter gene with a second foreign DNA, carrying the transgene of interest, by digestion of the vector DNA with *Pac*I or *Pme*I to remove the reporter gene and subsequent repair of the vector

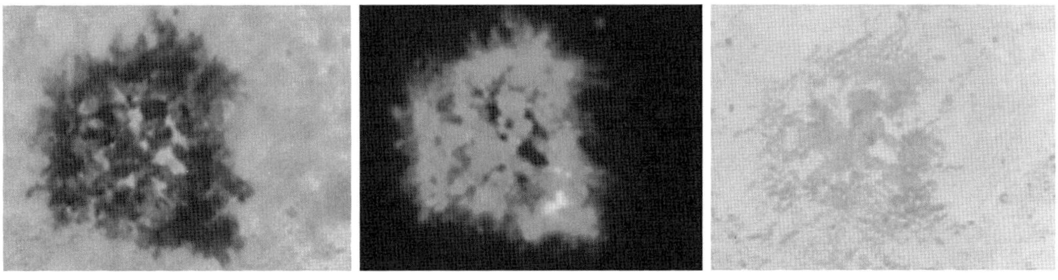

Fig. 2. Plaque phenotype of a recombinant HSV-1 expressing both β-galactoside (*left panel*) and GFP (*middle panel*) reporter genes. Compare with the "clear plaque" phenotype (*right panel*).

genome by homologous recombination with a transgene expression plasmid. Potential recombinants, identified by a "clear plaque" phenotype (not expressing GFP or LacZ, as in the right panel of Fig. 2), arose at high frequency (80–100%). For details on the construction of recombinant vectors by homologous recombination in eukaryotic cells, refer to Subheadings 2.1 and 3.1.

1.2. Construction of Recombinant HSV-1 Vectors by Homologous Recombination in Bacteria: ET Recombination and galK-Positive/Negative Selection

The cloning of large DNA virus genomes, such as that of HSV-1 (27, 28) as bacterial artificial chromosomes (BAC), has facilitated the easy construction of recombinant viruses by homologous recombination in *Escherichia coli*. The protocol below describes the construction of recombinant HSV-1 by using the λ prophage homologous recombination system (red α and red β genes) and *galK* selection. The *galK* selection method is a two-step system: In the first step, a *galK* cassette, flanked by at least 50 nucleotides of homology to specified positions on the HSV-1 BAC DNA, is inserted via homologous recombination into the BAC (*galK*-positive selection). In the second step, the *galK* cassette is replaced by homologous recombination with an oligonucleotide or PCR product that contains appropriate homology arms and selection against *galK*. This method allows constructing a recombinant HSV-1 within 2–3 weeks. For details on the construction of recombinant HSV-1 vectors by homologous recombination in bacteria, refer to Subheadings 2.2 and 3.2.

HSV-1-based amplicon vectors. HSV-1-based amplicon vectors carry no viral genes, they are therefore replication defective and depend on helper functions for production. Helper functions can be provided either by replication competent, but packaging-defective HSV-1 genomes cloned as set of cosmids (29) or bacterial artificial chromosome (BAC) (30). Following transfection into mammalian cells, sets of cosmids that overlap and represent the entire HSV-1 genome can form circular replication-competent viral genomes via homologous recombination. These reconstituted viral genomes give rise to infectious virus progeny. Similarly, BACs that contain the entire HSV-1 genome also produce infectious virus progeny in transfected cells. If the viral DNA packaging/cleavage (*pac*) signals are deleted from the HSV-1 cosmids or HSV-1 BACs, reconstituted virus genomes are packaging defective; however, in the absence of the *pac* signals, these genomes can still provide all helper functions required for the replication and packaging of cotransfected amplicon DNA. The resulting amplicon vector stocks are essentially free of helper virus contamination.

Alternatively, helper-free amplicon vector stocks can be prepared using a helper system based on the deletion of the *pac* signals of the helper virus genome by Cre/loxP-based site-specific recombination, in order to inhibit its cleavage/encapsidation in the cells that are producing the amplicons (31). This helper virus, named HSV-1-LaLΔJ helper, carries a unique and ectopic *pac* signal, flanked by two loxP sites in parallel orientation. This is therefore a Cre-sensitive virus that cannot be packaged in Cre-expressing cells due to deletion of the floxed packaging signal. Nevertheless, some helper genomes can escape action of the Cre recombinase, allowing the production of some contaminant helper particles. For this reason, the two genes surrounding the cleavage/packaging signal, respectively, encoding a virulence factor known as ICP34.5 and the essential protein ICP4, were further deleted from the helper virus genome. Although the amplicon stocks prepared with this helper virus (in a complementing cell line expressing both Cre and ICP4 proteins) still can contain a small amount of contaminating helper particles, this helper is replication incompetent and cannot spread upon infection of target cells or tissues. Use of the HSV-1 LaLΔJ helper virus generally results in the production of large stocks of amplicon vectors only barely contaminated (0.05–0.5%) with defective, nonpathogenic helper particles. In the next two sections, we will describe methods to prepare helper-free amplicon vectors using (1) the HSV-1 genome cloned as a bacterial artificial chromosome, and (2) the Cre/loxP site-specific dependent system.

1.3. Packaging of HSV-1 Amplicon Vectors Using a Replication-Competent, Packaging-Defective HSV-1 Genome Cloned as a Bacterial Artificial Chromosome

Subheadings 2.3 and 3.3 describe the preparation of (1) large amounts of HSV-1 BAC DNA, (2) cotransfection of amplicon DNA and HSV-1 BAC DNA into VERO 2-2 cells by cationic liposome-mediated transfection, and (3) concentration/purification and titration of packaged amplicon vector stocks.

1.4. Packaging of Amplicon Vectors Using a Replication Incompetent, Cre/loxP-Sensitive Helper Virus

Subheadings 2.4 and 3.4 describe the technology required (1) to prepare, purify, and titrate the HSV-1-LaLΔJ helper virus and (2) to prepare, purify, and titrate amplicon stocks that have been produced using this virus as helper system.

2. Materials

2.1. Construction of Recombinant HSV-1 Vectors by Homologous Recombination in Eukaryotic Cells

1. T-175 cm^2 and T-75 cm^2 flasks, blue filter cup. Nunc A/S, Denmark.
2. 24-, 48-, 96-well cell culture cluster, flat bottom with lid, polystyrene tissue culture treated. Corning Incorporated.
3. 60 mm × 15 mm dish polystyrene. Corning Incorporated.

4. 10 ml tubes (PST test tubes with two-position closure cap, individually wrapped. Artiglass).
5. ART pipette tips for p20, p200, p1000 with filter.
6. Tips for viral DNA with filter.
7. 2, 5, 10, 25 ml disposable serological pipettes. Corning Incorporated.
8. 15 and 50 ml centrifuge screw cap tube, polypropylene. Corning Incorporated.
9. 1.5 ml Eppendorf safe-lock tube.
10. Cell scrapers 18 cm handle/1.8 cm blade (BD Falcon, ref.# 353085) and 25 cm handle/1.8 cm blade (BD Falcon, ref.# 353086).
11. 50 ml Reagent Reservoir, polystyrene. Costar, ref.# 4870.
12. Stericup, vacuum disposable filtration system, 0.22 and 0.45 μm membrane. Millipore.
13. Phenol:Chloroform:Isoamyl alcohol (25:24:1). Ambion, Applied Biosystems.
14. Chloroform: Isoamyl alcohol (24:1). Ambion, Applied Biosystems.
15. Geneticin G418. Roche ref.# 1464990 (5 g).
16. Proteinase K (20 mg/ml). Euroclone EMR02201.
17. Trypsin: 0.25% trypsin/0.02% EDTA. Life Technologies.
18. Sterile Glycerol. Sigma.
19. TE: 10 mM Tris–HCL, pH 8.0, 1 mM EDTA.
20. Vero (African green monkey kidney, ATCC) cells. Other complementing cell lines (such as VERO-7b cells (19)) are required to propagate HSV-1 essential gene deletion viruses.
21. Cell culture medium: Dulbecco's Modified Eagle's Medium (DMEM) high glucose (Life Technologies) supplemented with 10% fetal bovine serum (FBS, HyClone) and 2 mM glutamine, 100 units/ml penicillin, 100 μg/ml streptomycin.
22. 0.5 M EDTA Stock Solution: Dissolve 16.81 g of EDTA in 90 ml distilled water. Adjust pH to 8.0 with NaOH. Adjust volume to 100 ml with water. Store at room temperature.
23. 1× PBS (Phosphate-Buffered Saline): Dissolve 8 g of NaCl, 0.2 g of KCl, 1.44 g of Na_2HPO_4, 0.24 g of KH_2PO_4 in 800 ml distilled H_2O. Adjust pH to 7.4 and bring the volume to 1 L with additional distilled H_2O. Sterilize by autoclaving. Store at room temperature.
24. 1× TBS (Tris-Buffered Saline): Dissolve 8 g of NaCl, 0.2 g of KCl, 3 g of Tris base in 800 ml of distilled H_2O. Adjust the pH to 7.4 with HCl and bring the volume to 1 L with

additional distilled H_2O. Sterilize by autoclaving. Store at room temperature.

25. Lysis buffer: 10 mM Tris–HCL, pH 8.0, 10 mM EDTA, 0.6% SDS. The proteinase K 0.25 mg/ml is added at the moment of the use.
26. 10% sodium dodecyl sulfate solution: 10% (w/v) in distilled H_2O. Sterilize by passage through a 0.22 μm Stericup filter. Store at room temperature. Note: precipitation of SDS is not unusual, warm gently to redissolve.
27. 2× HBS: Dissolve 10 g HEPES (0.021 M), 0.25 g $Na_2HPO_4{\cdot}2H_2O$ * (0.702 mM), 0.74 g KCl (5 mM), 2 g glucose ** (5.6 mM). Adjust pH to 7.05 with NaOH 5 M and bring the volume to 2 L with additional distilled H_2O. Accurate pH of this solution is critical. Sterilize by passage through a 0.22 μm Stericup filter. Store at 4°C.* (0.188 g if it is $Na_2HPO_4{\cdot}7H_2O$, 0.251 g if it is $Na_2HPO_4{\cdot}12H_2O$, 0.09966 g if it is anhydrous Na_2HPO_4); ** (1.11 g if it is glucose monohydrate).
28. $CaCl_2$ (2 M): Dissolve 29.4 g $CaCl_2$ in 70 ml H_2O. Adjust the volume to 100 ml with additional distilled H_2O. Sterilize by passage through a 0.2 μm filter. Store in aliquots at −20°C.
29. 1.5% Methylcellulose overlay: add 1.5 g of methylcellulose to 100 ml PBS, pH 7.5 in a sterile bottle containing a stir bar. Autoclave the bottle on liquid cycle for 45 min. After the solution cools, add 350 ml of DMEM high glucose supplemented with 2 mM glutamine, 100 units/ml penicillin, 100 μg/ml streptomycin. Mix well, place the bottle on a stir plate at 4°C overnight or until the methylcellulose is completely dissolved. Once the methylcellulose has been solubilized, add 50 ml of fetal bovine serum.
30. 1% crystal violet in solution (50:50 methanol:dH_2O v/v).

Materials and solutions for viral stock preparation and Optiprep gradient

1. T150–175 cm^2 tissue culture flasks.
2. 15 and 50 ml centrifuge screw cap tube. Corning Incorporated.
3. 50 ml tubes Nalgene Centrifuge Oak Ridge copolymer.
4. OptiSeal polyallomer centrifuge tubes and plugs 5/8 × 2¾ in. 11.2 ml capacity.
5. Needles: 18 G 1½ in.
6. Syringes: 10 cm^3.
7. Sonicator: Sonicator Ultrasonic Processor. Misonix, NY.
8. 21/2 in. Cup Horn for sonicator. Misonix, NY.
9. Centrifuge Beckman Avanti J25.

10. Ultracentrifuge Beckman Coulter Optima LE-80K.
11. Rotor JA-20.
12. Rotor Vti65.1 Beckman (ref.# 362181).

2.1.2. Solutions

Iodixanol (Optiprep™, Axis-Shield (Norway), ref.# 1030061).

Solution B: 2.8 ml 5 M NaCl, 6 ml HEPES 1 M, pH 7.3, 1.2 ml EDTA 0.5 M, pH 8.0. Bring to 100 ml and sterile filtrate. Keep at 4°C.

Solution C: 2.8 ml 5 M NaCl, 1 ml HEPES 1 M, pH 7.3, 200 μl EDTA 0.5 M, pH 8.0. Bring to 100 ml and sterile filtrate. Keep at 4°C.

Solution D: 5 volumes of Optiprep + 1 volume of solution B (5:1) Prepare fresh before use (e.g., 10 ml Optiprep + 2 ml B is enough for one Beckman, Optiseal tube (ref.# 362181) and one balancing tube).

Solution E: Virus (0.5 ml) + solution C (4.3 ml); total of 4.8 ml/ tube for Beckman, Optiseal tube. Sonicate the virus to break up the clumps before adding it to solution C. Use total volumes of Solution C for balancing tubes. Use no more than virus obtained from three T175 tissue culture flasks for each OptiSeal polyallomer centrifuge tube. Make just before use.

Solution F: Top-up solution; 1.27 ml solution C + 1 ml solution D. This equals a final concentration of 22% Optiprep that the gradient will have. Prepare fresh before use.

2.2. Construction of Recombinant HSV-1 Vectors by Homologous Recombination in Bacteria

2.2.1. Generation of a Targeting DNA Fragment by PCR Amplification

1. Plasmid pgalK (32).
2. Oligonucleotide primers (Microsynth AG, Balgach, Switzerland).
3. QIA Quick PCR purification kit (Qiagen, ref.# 28106).
4. Restriction endonuclease *Dpn*I (New England Biolabs).
5. Electrophoresis grade agarose (Lonza, Rockland, USA).
6. TAE buffer 25×: Dissolve 121 g Tris base and 16.8 g EDTA in 970 ml H_2O, add 30 ml glacial acetic acid.
7. MinElute Gel Extraction Kit (Qiagen, ref.# 28604).

2.2.2. Preparation of Electrocompetent E. coli, Electroporation, and Selection of galK-Positive and galK-Negative Bacteria

1. *E. coli* SW102 (32) (see Note 1).
2. HSV-1 BAC; e.g., YE102bac (28).
3. Gene pulser cuvettes, 0.1 cm (BioRad).
4. M9 salts 1×: Dissolve 6 g Na_2HPO_4, 3 g KH_2PO_4, 1 g NH_4Cl, 0.5 g NaCl in 1 L H_2O; autoclave (121°C, 15 min, store at room temperature).
5. M63 minimal plates containing galactose, biotin, leucine, and chloramphenicol. Dissolve 7.5 g agar in 400 ml H_2O; autoclave.

Cool down to 50°C and add 100 ml 5× M63 medium, 0.5 ml 1 M $MgSO_4$ solution, 5 ml 20% galactose (autoclaved stock solution), 2.5 ml biotin (0.2 mg/ml stock solution, autoclaved), 2.25 ml leucine (10 mg/ml stock solution, autoclaved), and 500 μl chloramphenicol (12.5 mg/ml).

6. M63 medium 5×: Dissolve 10 g $(NH_4)_2SO_4$, 68 g KH_2PO_4, 2.5 mg $FeSO_4·7H_2O$ in 800 ml H_2O, adjust pH to 7.0 with KOH, add H_2O to 1 L, autoclave.
7. McConkey indicator plates containing chloramphenicol: Dissolve 25 g McConkey agar in 500 ml H_2O, autoclave. Cool down to 50°C and add 5 ml of 20% galactose (autoclaved stock solution in H_2O) and 500 μl chloramphenicol (12.5 mg/ml).
8. M63 minimal plates containing glycerol, 2-deoxy-galactose (DOG), biotin, leucine, and chloramphenicol. (Preparation: Dissolve 1.5 g agar in 70 ml H_2O; autoclave. Cool down to 50°C and add 10 ml 2% DOG (0.2 g in 10 ml H_2O), autoclave. Cool down to 50°C and add 20 ml 5× M63 medium, 0.1 ml 1 M $MgSO_4$ solution, 1 ml 20% galactose (autoclaved stock solution), 0.5 ml biotin (0.2 mg/ml stock solution, autoclaved), 0.45 ml leucine (10 mg/ml stock solution, autoclaved), and 100 μl chloramphenicol (12.5 mg/ml).

2.2.3. Isolation and Analysis of BAC DNA from E. coli (Miniprep Protocol)

1. Solution P1: 50 mM Tris–HCl pH 8.0, 10 mM EDTA, 100 μg/ml RNAseA. Filter sterilize, store at 4°C.
2. Solution P2: 0.2 N NaOH. Filter sterilize, store at room temperature.
3. Solution P3: 3 M KOAc, pH 5.5. Autoclave, store at 4°C.
4. Isopropanol.
5. Selected restriction endonucleases (New England Biolabs).

2.2.4. Transfection of Mammalian Cells with BAC DNA and Reconstitution of Recombinant HSV-1

1. Vero cells (African green monkey cells, ATCC).
2. Dulbecco's modified Eagle medium (DMEM; Life Technologies, ref.# 31885) with 10% fetal bovine serum (FBS).
3. 0.25% trypsin/0.02% EDTA (Life Technologies, ref.# 51985).
4. Opti-MEM I reduced serum medium (Life Technologies, ref.# 25300).
5. Plasmid p116 (Dr. K. Tobler, University of Zurich, Zurich, Switzerland) (see Note 2).
6. LipofectAMINE reagent (Life Technologies, ref.# 11514-015).
7. Phusion DNA Polymerase (Finnzymes, Espoo, Finland; ref.# F-530L).

8. DMSO (Finnzymes, Espoo, Finland; ref.# F-515).
9. 5× GC buffer (Finnzymes, Espoo, Finland; ref.# F-519).
10. Solution P1: 50 mM Tris–HCl, pH 8.0, 10 mM EDTA, 100 μg/ml RNase A (filter sterilize, store at 4°C).
11. Solution P2: 0.2 N NaOH, 1% SDS (filter sterilize, store at room temperature).
12. Solution P3: 3 M KOAc, pH 5.5 (filter sterilize, store at 4°C).

2.3. Packaging of HSV-1 Amplicon Vectors Using a Replication-Competent, Packaging-Defective HSV-1 Genome Cloned as BAC

2.3.1. Preparation of HSV-1 BAC DNA

1. *E. coli* clones of HSV-1 BAC fHSVΔpacΔ27ΔKn and plasmid pEBHICP27 (30).
2. LB medium containing 12.5 μg/ml chloramphenicol.
3. Plasmid Maxi Kit (Qiagen, ref.# 12163), which includes Qiagen tip 500 columns and buffers P1, P2, P3, QBT, QC, QGT, and QF.
4. TE buffer pH 7.4.
5. Restriction endonucleases *Hin*dIII and *Kpn*I (New England Biolabs).
6. TAE electrophoresis buffer (10×): 24.2 g Tris base, 5.71 ml glacial acetic acid, 3.72 g $Na_2EDTA \cdot 2H_2O$, H_2O to 1 L. Store at room temperature.
7. Graduated snap-cap tubes 17×100 mm (e.g., Falcon ref.# 2059), sterile.
8. Sorvall GSA and SS-34 rotors.
9. 120 mm diameter folded filters (Schleicher and Schüll, Dassel, Germany).
10. Ultra-Clear Centrifuge tubes 13×51 mm (Beckman, Munich, Germany).
11. TV 865 ultracentrifuge rotor (Sorvall).
12. 1 ml disposable syringes.
13. 21-gauge and 36-gauge hypodermic needles.
14. UV-lamp (366 nm).
15. Dialysis cassettes, Slide-A-Lizer 10 K (10,000 MWCO; Pierce, Rockford, USA).
16. UV spectrophotometer (Ultrospec 3000, Pharmacia).

2.3.2. Preparation of HSV-1 Amplicon Vector Stocks

1. Vero 2-2 cells (33).
2. An amplicon plasmid (see Note 3).
3. Dulbecco's modified Eagle medium (DMEM; Life Technologies) with 10% and 6% fetal bovine serum.
4. G418 (Geneticin; Life Technologies).
5. 0.25% trypsin/0.02% EDTA (Life Technologies, ref.# 51985).

6. Opti-MEM I reduced serum medium (Life Technologies, ref.# 25300).
7. HSV-1 BAC fHSVΔpacΔ27ΔKn and pEBHICP27 plasmid DNA (30) (Dr. Y. Saeki, Ohio State University, Columbus, OH, USA: saeki.6@osu.edu).
8. HSV-1 amplicon DNA (maxiprep DNA isolated from *E. coli*).
9. LipofectAMINE reagent (Life Technologies, ref.# 11514-015).
10. 10, 30, and 60% (w/v) sucrose in PBS.
11. Phosphate-buffered saline (PBS).
12. 75 cm^2 tissue culture flasks.
13. Humidified 37°C, 5% CO_2 incubator.
14. 60 mm diameter tissue culture dishes.
15. 15 ml conical centrifuge tubes.
16. Dry ice/ethanol bath.
17. Probe sonicator.
18. 0.45 μm syringe-tip filters (Sarstedt polyethersulfone membrane filters).
19. 20 ml disposable syringes.
20. 30 ml centrifuge tubes (Beckman Ultra-Clear 25 × 89 mm and 14 × 95 mm).
21. Sorvall SS-34 rotor.
22. Fiber-optic illuminator.
23. Ultracentrifuge (Sorvall) with Beckman SW28 and SW40 rotors.

2.3.3. Harvesting, Purification, and Titration of HSV-1 Amplicon Vectors

1. Vero cells (clone 76; ECACC, ref.# 85020205); BHK cells (clone 21; ECACC, ref.# 85011433); 293 cells (ATCC, ref.# 1573).
2. DMEM (e.g., Life Technologies, ref.# 31885) supplemented with 10% and 2% FBS.
3. 4% (w/v) paraformaldehyde solution.
4. X-gal staining solution: 20 mM $K_3Fe(CN)_6$, 20 mM $K_4Fe(CN)_6 \cdot 3H_2O$, 2 mM $MgCl_2$ in PBS pH 7.5. Filter sterilize and store up to 1 year at 4°C. Before use, equilibrate solution to 37°C and add 20 μl/ml of 50 mg/ml of 5-bromo-4-chloro-3-indolyl-β-d-galactopyranoside (X-gal) in DMSO. Store X-gal solution in 1 ml aliquots up to several years at −20°C in the dark.
5. GST solution: 2% (v/v) goat serum and 0.2% (v/v) Triton X-100 in PBS. Store up to 1 month at 4°C.
6. Primary and secondary antibodies specific for detection of the transgene product.

7. 24-well tissue culture plates.
8. Humidified 37°C, 5% CO_2 incubator.
9. Inverted fluorescence microscope.
10. Inverted light microscope.

2.4. Packaging of Amplicon Vectors Using a Replication-Incompetent Cre/loxP-Sensitive Helper Virus

1. Vero (African green monkey cells, ATCC).
2. Vero-7b cell line (19).
3. Gli36 cell line (34).
4. TE-Cre-Grina cell line (31).
5. Growth medium: DMEM (Invitrogen, Paisley, UK) supplemented with 10% fetal bovine serum (FBS, Invitrogen), 100 units/ml penicillin and 100 μg/ml streptomycin, both from Invitrogen. All cell lines are maintained at 37°C in humidified incubators containing 5% CO_2.
6. Maintenance medium: medium 199 (Lonza) supplemented with 1% FBS.
7. Phosphate-buffered saline (PBS).
8. Geneticin (G418) (Cayla, Invivogen).
9. Opti-MEM (Ultra-MEM, Lonza).
10. LipofectAmine Plus reagent (Invitrogen).
11. Polystyrene roller bottles (VWR, ref.# 734-0008).

3. Methods

3.1. Construction of Recombinant HSV-1 Vectors by Homologous Recombination in Eukaryotic Cells

3.1.1. Viral DNA Preparation for Transfection

Engineering a new recombinant virus requires purified infectious viral DNA along with a plasmid containing the specific sequences of interest flanked by sufficient amounts of viral sequences (at least 500–1,000 bp of flanking HSV-1 sequences) homologous to the targeted gene locus within the HSV-1 genome. The quality of the above reagents will determine the frequency and the efficacy of generating the desired recombinant virus. The quality of the viral DNA can be evaluated by its capability to produce plaques following transfection of 1–5 μg of viral DNA, depending on the deletions present in the HSV-1 genome. The protocol we use is as follows:

1. Infect a T75 cm^2 (8×10^6 cells/flask) or a T175 cm^2 (24×10^6 cells/flask) subconfluent-confluent monolayer of cells, either Vero cells or a transcomplementing cell line such as Vero-7b (see Note 4) at multiplicity of infection (MOI) of 1–3 PFU/cell with the parental HSV-1 virus strain that will be genetically modified.

2. Allow infection to proceed for ~18–24 h depending on the cells and the virus strain used.
3. Wait until the infection is completed, all the cells should be rounded-up and still adhere to the flask.
4. Remove the cells by tapping the flask or use a cell scraper to dislodge them.
5. Pellet the cells for 5–10 min at 1,204 × *g* in a 15 ml or 50 ml centrifuge screw cap tube.
6. Wash the cells one time with 5 ml PBS pH 7.5.
7. Lyse the cells with 2–3 ml of Lysis buffer (see Subheading 2.1).
8. Incubate the tube at 37°C overnight (ON) in an orbital shaker.
9. Extract DNA two times with Phenol:Chloroform:Isoamyl alcohol (25:24:1). It is important to invert the tube enough to achieve proper mixing of the phases but being careful not to be too vigorous. When removing the aqueous phase to a new tube, go as close to the interface as possible. The DNA present at the interface is extremely viscous.
10. Extract two times with Chloroform:Isoamylic alcohol (24:1).
11. Remove the aqueous phase to a new 15 ml centrifuge screw cap tube, going again as close to the interface as possible.
12. Add two volumes of cold isopropanol or cold ethanol. Mix well.
13. The DNA can be spooled on a heat-sealed glass Pasteur pipette or the mixture can be stored ON at minus 20°C. If spooling the DNA, remove the isopropanol or ethanol by capillarity and transfer the pipette with the DNA into a new tube with distilled H_2O or in TE buffer (see Subheading 2.1) (see Note 5).

3.1.2. Cotransfection of Plasmid and Viral DNA to Generate a Recombinant Virus

Once the plasmid has been constructed (see Note 6) and the viral DNA has been prepared, it is possible to transfer the transgene from the plasmid to the HSV-1 genome, using the active recombination machinery of the virus, by transfecting together both the linearized plasmid and purified viral DNA into permissive cells such as Vero cells or complementing cell lines such as Vero-7b, if the viral genome is deleted in essential genes (see Note 7) using calcium phosphate method (see Note 8). The protocol we use is as follows:

1. The day before transfection, seed $8–9 \times 10^5$ Vero cells (in DMEM + 10%FBS + pen/strep) for replication-competent viral DNA, or the pertinent complementing cell line for replication-defective viral DNA, into 60 mm plates. Adjust the final volume to 3 ml of growth medium.

2. The next morning, cells should have reached 70–80% confluence. Set up transfections as follows:
3. For each transfection event, add 500 μl of HBS (pH 7.05) in a sterile 10 ml polystyrene tube.
4. To each tube add viral DNA (about 1–5 μg) using tips with wide ends, and then linearize plasmid DNA (the amount of plasmid DNA should be equal to about 10× to 50× equivalents of viral DNA). Mix contents of each tube.
5. Add 30 μl of 2 M $CaCl_2$ to each tube, drop by drop, mix each tube immediately after adding 2 M $CaCl_2$ by blowing air into the tubes (use 2 ml pipettes) to facilitate the precipitate formation avoiding large clumps.
6. Incubate for 10–20 min at room temperature to allow the precipitate to form.
7. In the meantime, aspirate the medium from the plates and rinse the cells three times with 3 ml HBS.
8. Pipet transfection mixture up and down from the tube and add it carefully drop by drop to cell monolayer and gently mix by moving the dish, place the plates at 37°C in the CO_2 incubator for 15–20 min.
9. Carefully overlay 3 ml DMEM + 10% FBS-completed medium (see Subheading 2.1) and incubate the transfected cells at 37°C for 4–6 h in 5% CO_2 incubator.
10. 4–6 h later, aspirate the medium from the plates and wash once with 3 ml DMEM + 10% FBS-completed medium.
11. Gently add 3 ml 20% glycerol (in DMEM + 10% FBS-completed medium). Leave exactly for 3 min on the monolayer.
12. Carefully remove all the glycerol solution by aspiration and wash monolayer four times with DMEM + 10% FBS-completed medium.
13. Overlay monolayer with 3 ml DMEM + 10% FBS-completed medium and incubate at 37° in 5% CO_2.
14. Observe the plates under the microscope for the production of HSV-1 cytopathic effect (CPE) indicating the presence of infectious foci. This usually takes 3–5 days depending on the virus and the cell type used to propagate the recombinants.
15. When the viral plaques are open harvest the monolayer with the medium and isolate the virus from the cell pellet by three cycles of freeze–thaw and sonication followed by centrifugation at 771 × *g* for 5–10 min to eliminate the cellular debris. Store the unpurified virus at −80°C until doing the limiting dilution to isolate the recombinant virus from the parental virus.

3.1.3. Limiting Dilution to Isolate and Purify the Recombinant Virus

The advantage of doing limiting dilution is to avoid contamination of possible recombinants with parental virus by using the standard plaque isolation technique in which the single plaques are picked following methylcellulose or agarose overlay, since it is difficult to find well-isolated plaques on the plate. The stock of virus obtained after cotransfection should be used to isolate the recombinant. If recombination led to a deletion of an essential gene, the screen of the recombinant can be confirmed by plating the virus on both the complementing cell line (e.g., Vero-7b derived cell line that expresses the HSV-1 immediate-early genes ICP4 and ICP27 required for virus replication) and on noncomplementing cell line (e.g., Vero cells). Selection of a nonessential gene cannot be easily selected and it is for this reason that usually a marker gene that can be substituted in a second time by the desired gene is introduced in a first step. To isolate the recombinant virus it is necessary to go through not less than three rounds of limiting dilution, as follows:

1. Titer the virus stock obtained from the cotransfection event, as described in Subheading 3.1.5.
2. Detach the cell monolayer with trypsin, count cells, and transfer 2–3 × 10^6 cells in a final volume of 2.0 ml in DMEM + 10% FBS-completed medium to 15 ml centrifuge screw cap tube.
3. Add 20–30 PFU of virus stock to the cells. Rock at 37°C for 1 h to adsorb the virus (see Note 9).
4. Add 8.0 ml DMEM + 10% FBS-completed medium to the 2.0 ml of the infected cell to reach a final volume of 10 ml. Using a 50 ml Reagent Reservoir and a multichannel pipette, dispense 0.1 ml into each well of a 96-well plate. Incubate at 37°C and wait for appearance of plaques.
5. Identify the wells containing single plaques and mark them. Carefully inspect the edges of the wells under high power to ensure that no additional plaques are present.
6. Once the plaques have started opening up (normally 2–3 days after infection, depending on the recombinants and the cells), freeze the plate at −80°C, and thaw at 37°C in an incubator (see Note 10).
7. Using a p200 Pipetman, scrape the cells from the bottom of each well identified to contain a single plaque and pipet the entire contents of the well into an Eppendorf tube and freeze at −80°C.
8. Repeat steps 1–7 two more times (second and third limiting dilutions).
9. Plate out 2–3 × 10^6 cells into a 96-well plate again (cells in a final volume of 10 ml complete medium, 0.1 ml/well).
10. The next day, inoculate each well with 1/10 of the preliminary stock for each single virus in ~50 μl of medium.

Adsorb virus as usual (at 37°C for 1 h, rocking every 15 min) and then add other ~50 μl of medium to each well. It is useful at this point to infect positive and negative control viruses into some of the wells. Incubate at 37°C and wait for the appearance of the plaques. If the recombinant virus is a defective one, it is better to infect 24–48-well plates with at least half of the volume of each single virus to get sufficient viral DNA to perform a Southern blot.

11. Wait for all the cells in each well to round up. Aspirate the medium and add 200 μl lysis buffer to the cells in preparation for Southern blot analysis of viral DNA.
12. At this point, it is useful to characterize the recombinant viral DNA by Southern blot hybridization analyses using probes specific for the inserted sequences (see Notes 11 and 12).

3.1.4. Preparation of High Titer Replication-Defective Recombinant Viral Stock

After the recombinant has been isolated, following three rounds of limiting dilution, prepare a midi stock of the desired virus from a monolayer of cells in 2–3 × 150–175 cm^2 tissue culture flasks and obtain the titer of the stock for preparation of the large stock (see Subheading 3.1.5). The following procedure is used to prepare a virus stock of replication-defective recombinant virus and it can be scaled up or down depending on specific needs:

1. Seed 24 × 150–175 cm^2 flasks of complementing cells with 10 × 10^6 cells for each flask, to get confluent monolayers on the next day, in 20 ml of DMEM + 10% FBS-completed medium. Incubate at 37°C in 5% of CO_2.
2. The next day wash the cells with TBS twice, add enough trypsin to cover the monolayer and detach the cells. Collect the cells in 50 ml Corning screw cap tubes and pellet at 1,204 × *g* for 10 min. Discard supernatants and resuspend cell pellets in a small volume, combine all cell pellets in one 50 ml Corning screw cap tube, re-pellet and resuspend in a final volume of no more the 20 ml. Infect the cells in suspension with the recombinant virus at MOI of 0.05–0.08 PFU/cell. Gently rock the tube for 1 h at 37°C (see Note 13).
3. Plate back infected cells into the 150–175 cm^2 flasks and add growth medium until 20 ml.
4. Incubate until all the cells are infected (24–36 h post infection).
5. Scrape out infected cells into their own medium and pipet all into 50 ml Corning screw cap tubes.
6. Pellet at 1,204 × *g* for 15 min at 4°C. Decant supernatant and save it in ice.
7. Resuspend cell pellets in a small volume of supernatant, combine, and re-pellet.

8. Resuspend the final pellet in 2–3 ml of 15 ml Corning screw cap tube. If you stop at this point put everything at −80°C.
9. Freeze–thaw the cell pellet three times; vortex every time you thaw. After final thaw, sonicate three times for 10–15 s with 10 s of pause in ice after each sonication. The viral suspension you obtain should be homogeneous.
10. Pellet the cell debris at 1,734 × *g* for 15 min at 4°C.
11. Transfer the supernatant derived from the pellet into 50 ml Oak Ridge polypropylene tube along with original viral supernatant. Spin down at 48,384 × *g* for 30 min at 4°C in JA-20 rotor.
12. Decant and discard supernatant. Carefully remove remaining supernatant and resuspend pellet by vortexing or pipetting in less than 1 ml growth medium. You can spin the tube to break bubbles.
13. Do not freeze the virus but directly purify it on gradients (see Subheading 3.1.6). After purification, resuspend the virus in PBS, aliquot it in small volumes and freeze at −80°C.

3.1.5. Titration of Virus Stock

The following procedure can be used to obtain the titer of virus stocks of any size.

1. One day prior titration of the virus stock prepare six-well tissue culture plates with 0.5×10^6 cells (e.g., Vero cells if no essential viral gene has been deleted, or corresponding complementing cells if the recombinant virus contains deletions in any essential gene) to titer the virus. The critical point is that the day of titration the cell monolayer should be confluent.
2. Prepare a series of tenfold dilutions (10^{-2} to 10^{-10}) of the virus stock in Eppendorf tubes with 1 ml DMEM + 10% FBS-completed medium.
3. Add 100 μl of each dilution to confluent cells in a single well of a six-well culture plate.
4. Allow the virus to infect the cells for a period of 1 h at 37°C in a CO_2 incubator. Rock the plates every 15 min to distribute the inoculum to all cells in the monolayer.
5. Aspirate off the virus inoculum, overlay the monolayer with 3 ml of 1.5% methylcellulose in DMEM + 10% FBS-completed medium.
6. Reincubate the plates for 3–5 days until well-defined plaques appear.
7. Aspirate off the methylcellulose and stain with 2 ml of 1% crystal violet solution (50:50 methanol:dH_2O v/v) for 10–20 min. The stain fixes the cells and the virus.

8. Count the number of plaques per well, determine the average for each dilution (if it is in duplicate), and multiply by a factor of 10 to get the number of plaque-forming units/ml (PFU/ml) for each dilution. Multiply this number by 10 to the power of the dilution to achieve the titer in PFU/ml (see Note 14).

3.1.6. Purification of Recombinant HSV-1 Stock

In order to purify the virus from cell debris or proteins and to prepare sufficient virus stock at high titer for use in preclinical experiments, the virus pellet can be resuspended in PBS and purified through sucrose, dextrane, or iodixanol gradients. The following protocol is based on the purification of the virus with iodixanol gradient (Optiprep™ Axis-Shield, ref.# 1030061).

Gradient preparation and run:

Iodixanol gradient is self forming. Prepare solutions B–F (see Subheading 2.1.2). It is better to keep all solutions cold in ice and to precool the ultracentrifuge rotor.

1. Prepare OptiSeal polyallomer centrifuge tubes for run in a rotor Vti65.1 by using a pipette or a syringe, pipet 4.4 ml solution D into each tube.
2. Solution E can be mixed before adding it into the tube containing the solution D. Sonicate the virus to break up the clumps before adding it to solution C to obtain the solution E.
3. Slowly, add 4.8 ml solution E into each tube. Be careful to avoid clogging of neck and bubble formation. To remove bubbles, use a syringe.
4. Top up the tubes with solution F (about 1.5 ml for each tube).
5. Leave just a small air bubble in neck of tube and close up with the cap.
6. Weight the tubes to be sure that they are exactly balanced; if they are not, equilibrate them by adding solution F.
7. Dry outside the tubes if necessary and place them into the rotor. Place plugs and caps and close by using 120 in.-lbs torque value.
8. Place rotor in centrifuge, close door, turn on vacuum, enter run specifications in ultracentrifuge: speed 296,516 × *g*, time from 4 to 15 h (depending if you want to collect the virus the same day or the following day), at 4°C, acceleration max, no brake during deceleration.
9. Start run. During the run, check if centrifuge attained full speed.
10. At the end of run turn off vacuum, remove rotor, and put carefully the tubes on ice. You will see a band in them about in the middle of the gradient.

3.1.7. Collection of Virus Particles

1. Collect the band by puncturing the side of the tube with a needle and syringe 2–3 mm under the band. Be careful not to aspire too much volume to avoid the collection of debris.
2. Place the collected virus into 50 ml tubes Nalgene Centrifuge Oak Ridge and dilute with PBS until fill up the tube to remove residual iodixanol solution.
3. Centrifuge the tubes in Beckman centrifuge, 48,384×*g* for 30 min at 4°C in JA-20 rotor, to concentrate the virus. If you have performed your gradient in more than three ultracentrifuge tubes, place the collected virus into two JA-20 oak ridge tubes and dilute with PBS and centrifuge them.
4. Discard supernatant and resuspend the pellet in about 1–2 ml of PBS. If the pellet is too clumped leave it to dissolve ON in ice in cold room.
5. Carefully transfer the virus in a 10 ml tube, sonicate it to break up clumps (two to three times, 5–10 s each time, with 10 s pause in ice between sonications). The viral suspension should be homogeneous before to aliquot in Eppendorf tubes.
6. Aliquot the virus in small volumes.
7. Store the aliquots at −80°C (see Notes 15 and 16).

3.2. Construction of Recombinant HSV-1 Vectors by Homologous Recombination in Bacteria

3.2.1. Generation of a galK+ Targeting DNA Fragment by PCR Amplification

1. Design primers with 50 bp homology to either side of the targeting sequence on the HSV-1 genome, followed by sequences (underlined below) that bind to the *galK* cassette in plasmid pgalK, which serves as the template for the PCR reaction. Forward primer: 5′ 50-nucleotide homology arm-<u>CGTGTTGACAATTAATCATCGGCA3′</u>. Reverse primer: 5′ 50-nucleotide homology arm-<u>TCAGCACTGTCCTGCTCCTT</u> 3′.
2. Perform PCR amplification as follows: 10 μM of each primer, 10 μl 5× GC buffer, 3 μl DMSO, 10 μM dNTPs, 10 ng pgalK template DNA, 0.5 μl Phusion DNA Polymerase, H_2O to 50 μl; 94°C for 15 s, 60°C for 30 s, and 72°C for 1 min, for 30 cycles.
3. Purify the PCR reaction by using the QIA Quick PCR purification kit (Qiagen).
4. Add *Dpn*I (10 U, New England Biolabs) and incubate for 2 h at 37°C (see Note 17).
5. Separate the fragments by agarose gel electrophoresis (1% agarose in 1× TAE), and excise the band containing the *galK+* targeting DNA fragment. Purify the DNA by using the MinElute Gel Extraction Kit (Qiagen), precipitate and wash with ethanol, resuspend in 30 μl H_2O.

6. To determine the DNA concentration, measure the absorbance at 260 nm (A_{260}) and 280 nm (A_{280}) using a UV spectrophotometer (see Note 18).

3.2.2. Preparation of Electrocompetent E. coli SW102

1. Inoculate 5 ml of LB medium with *E. coli* SW102. Incubate over night at 32°C on a shaker.
2. The next day, prepare the following: two water baths, one at 32°C, the other at 42°C; an ice/water slurry; 50 ml ice cold H_2O; a pre-cooled centrifuge (0°C).
3. Inoculate 25 ml of LB medium in a 50 ml Erlenmeyer flask with 500 μl of the overnight culture. Incubate at 32°C in a shaking water bath until the OD_{600} reaches approximately 0.6.
4. Transfer 10 ml of the culture to another 50 ml Erlenmeyer flask and incubate for exactly 15 min at 42°C.
5. Cool the culture in an ice/water slurry, transfer to two 15 ml Falcon tubes, and centrifuge for 5 min at 5,500 × *g* and 0°C.
6. Remove all supernatant and resuspend the pellet in 1 ml of ice cold H_2O by gently swirling the tube in the ice/water slurry.
7. Add 9 ml of ice cold H_2O and pellet again (5 min, 5,500 × *g*, 0°C). Repeat step 6.
8. Remove all supernatant by inverting the tubes on a paper towel. Resuspend the pellet in the remaining liquid (approximately 50 μl) and keep on ice until used for electroporation.

3.2.3. Electroporation of HSV-1 BAC DNA into E. coli SW102

1. Mix 25 μl of electrocompetent bacteria with 2 μl of HSV-1 BAC DNA (e.g., YE102bac) (28). Transfer the suspension into a 0.1 cm cuvette (BioRad) and electroporate at 25 μF, 1.75 kV, and 200 ohm.
2. Recover the bacteria with 1 ml LB medium, and incubate the culture for 1 h at 32°C on a shaker. Plate 1:10, 1:100, and 1:1,000 dilutions (in LB medium) onto LB agar plates containing the appropriate antibiotic (e.g., 12.5 μg/ml of chloramphenicol for YE102bac).

3.2.4. Electroporation of the galK-Targeting DNA into E. coli SW102 Containing the HSV-1 BAC and Screening for galK-Positive Clones (galK-Positive Selection)

1. Prepare electrocompetent *E. coli* SW102 containing the HSV-1 BAC as described in Subheading 3.2.2, except that in steps 1 and 3, the LB medium should contain the appropriate antibiotic (e.g., 12.5 μg/ml of chloramphenicol for YE102bac).
2. Mix 25 μl of electrocompetent bacteria with 2 μl of the PCR product from Subheading 3.2.1, step 5 (approximately 30 ng). Transfer suspension into a 0.1 cm cuvette (BioRad) and electroporate at 25 μF, 1.75 kV, and 200 ohm.
3. Recover the bacteria with 1 ml LB medium and transfer into a 15 ml Falcon tube. Add another 9 ml of LB medium and incubate culture for 1 h at 32°C on a shaker.

4. Wash the bacteria twice in 1× M9 salts as follows: Transfer 1 ml of the culture into an Eppendorf tube and pellet for 15 s at 17,900 ×*g*. Remove the supernatant with a pipette. Resuspend the pellet in 1 ml 1× M9 salts and centrifuge again. Repeat this washing step once more (see Note 19). Then resuspend the pellet in 400 μl of 1× M9 salts and plate serial dilutions in 1× M9 salts (1:10, 1:100, 1:1,000) onto M63 minimal medium plates containing galactose and the appropriate antibiotic (e.g., 12.5 μg/ml of chloramphenicol for YE102bac). Incubate for 3 days at 32°C.
5. Streak several colonies onto McConkey indicator plates containing galactose and the appropriate antibiotic (e.g., 12.5 μg/ml of chloramphenicol for YE102bac). Red colonies will indicate *galK*+ bacteria. The *galK*-positive selection step is normally very efficient and it is not necessary to further analyze the clones. However, to confirm the correct insertion of the *galK* cassette, HSV-1 BAC DNA can be prepared and analyzed as described in Subheading 3.2.6.

3.2.5. Electroporation of the Targeting DNA into galK-Positive E. coli SW102 Containing the HSV-1 BAC and Screening for galK-Negative Clones (galK-Negative Selection)

1. Pick single bright red colonies (*galK*+) from Subheading 3.2.4, step 5, and inoculate 5 ml LB medium containing the appropriate antibiotic (e.g., 12.5 μg/ml of chloramphenicol for YE102bac).
2. Prepare electrocompetent bacteria as described in Subheading 3.2.2, except that in steps 1 and 3, the LB medium should contain the appropriate antibiotic (e.g., 12.5 μg/ml of chloramphenicol for YE102bac).
3. For insertions of foreign DNA, prepare a linear targeting DNA by PCR amplification as described in Subheading 3.2.1. The forward and reverse primers contain the same 50 nucleotides of targeting sequence at the 5′ end as the primers designed in Subheading 3.2.1 step 1, followed by sequences that bind to the 3′ and 5′ ends of the DNA fragment to be inserted. For HSV-1 gene deletions, design an oligonucleotide that contains the 50 nucleotides of 5′ targeting sequence followed by the 50 nucleotides of 3′ targeting sequence of Subheading 3.2.1, step 1. For point mutations, the altered nucleotide(s) can be inserted into an oligonucleotide as described above, in the center of the 50 nucleotides 5′ and 3′ targeting sequences. It is not necessary to use double-stranded oligonucleotides, although the efficiency of a double-stranded DNA is somewhat higher.
4. Mix 25 μl of electrocompetent bacteria from step 2 with 2 μl of the targeting DNA from Subheading 3.2.5, step 3 (100–200 ng). Transfer suspension into a 0.1 cm cuvette (BioRad) and electroporate at 25 μF, 1.75 kV, and 200 ohm.

5. Recover the bacteria with 10 ml LB medium and transfer into a 50 ml Erlenmeyer flask; incubate culture for 4.5 h at 32°C on a shaker.
6. Wash and dilute the bacteria as in Subheading 3.2.4, step 4. Then plate bacteria onto M63 minimal medium plates containing glycerol, leucine, biotin, 2-deoxy-galactose (DOG), and the appropriate antibiotic (e.g., 12.5 μg/ml of chloramphenicol for YE102bac). Incubate for 3 days at 32°C (see Note 20).

3.2.6. Isolation and Characterization of HSV-BAC DNA from Small Bacterial Cultures

1. Inoculate single bacterial colonies into 5 ml LB medium containing the appropriate antibiotic in 15 ml Falcon tubes. Incubate over night at 32°C on a shaker.
2. Centrifuge tubes for 10 min at 2,000 × *g* and 4°C. Discard supernatant, resuspend pellet in 300 μl of solution P1, and transfer suspension into an Eppendorf tube.
3. Add 300 μl of solution P2, mix gently, and incubate for 5 min at room temperature.
4. Slowly add 300 μl of solution P3, mix gently, and incubate on ice for at least 5 min.
5. Centrifuge for 10 min at 10,600 × *g* and 4°C. Transfer supernatant into a new Eppendorf tube that contains 800 μl of isopropanol. Mix by inverting the tube several times and incubate on ice for at least 5 min.
6. Centrifuge for 15 min at 10,600 × *g* and 4°C. Remove supernatant and wash the pellet with 500 μl of 70% ethanol. Invert tube several times and then centrifuge again for 5 min at 10,600 × *g* and 4°C.
7. Aspirate supernatant and air-dry pellet at room temperature. Then, resuspend the DNA in 40 μl of TE buffer in a 37°C water bath. Use 10 μl of the DNA for restriction endonuclease analysis and agarose gel electrophoresis (0.5% agarose in TAE) to confirm the mutation in the HSV-1 DNA.

3.2.7. Transfection of Mammalian Cells with BAC DNA and Reconstitution of Recombinant HSV-1

1. Maintain Vero cells in DMEM + 10% FBS. Propagate the culture twice a week by splitting 1/5 in fresh medium (20 ml) into a new 75 cm^2 tissue culture flask.
2. On the day before transfection, remove culture medium, wash twice with PBS, add a thin layer of trypsin/EDTA, and incubate 10 min at 37°C to allow cells to detach from the plate. Count cells using a hemacytometer, and plate 1.2×10^6 cells per 60-mm diameter tissue culture dish in 3 ml DMEM + 10% FBS.
3. For each 60-mm dish, place 100 μl Opti-MEM I reduced serum medium into each of two 15 ml conical tubes. To one tube, add 2 μg of HSV-1 BAC DNA and 0.2 μg of plasmid p116, which expresses Cre recombinase with an NLS (kindly

provided by K. Tobler, University of Zurich, Switzerland) (see Note 21). To the other tube, add 12 μl LipofectAMINE.

4. Combine the contents of the two tubes, mix well, and incubate 45 min at room temperature.
5. Wash the cultures prepared the day before (step 2) once with 2 ml Opti-MEM I. Add 1.1 ml Opti-MEM I to the tube from step 4 containing the DNA-LipofectAMINE transfection mixture (1.3 ml total volume). Aspirate medium from the culture, add the transfection mixture, and incubate for 5.5 h.
6. Aspirate the transfection mixture and wash the cells three times with 2 ml Opti-MEM I. After aspirating the last wash, add 3.5 ml DMEM + 6% FBS and incubate 2–3 days.
7. Scrape cells into the medium using a rubber policeman. Transfer the suspension to a 15 ml conical centrifuge tube and place the tube containing the cells into a beaker of ice water. Submerge the tip of the sonicator probe ~0.5 cm into the cell suspension and sonicate 20 s with 20% output energy. This disrupts cell membranes and liberates cell-associated virus particles.
8. Remove cell debris by centrifugation for 10 min at 960 × *g*, 4°C and inoculate fresh Vero cells or appropriate complementing cells for plaque purification.
9. Characterize plaque-purified virus as follows: confirm the absence of the BAC sequences (e.g., PCR), determine growth properties and titers, analyze the genotype of the recombinant virus.

3.3. Packaging of HSV-1 Amplicon Vectors Using a Replication-Competent, Packaging-Defective HSV-1 Genome Cloned as a BAC

3.3.1. Preparation of HSV-1 BAC DNA

1. Prepare a 17 × 100 mm sterile snap-cap tube containing 5 ml LB/chloramphenicol medium. Inoculate with frozen long-term culture of the HSV-1 BAC clone (fHSVΔpacΔ27ΔKn). Incubate 8 h at 37°C in a shaker.
2. Transfer 1 ml of the culture into each of four 2 l flasks containing 1,000 ml sterile LB/chloramphenicol medium, and incubate 16 h at 37°C, with shaking.
3. Distribute the 4 l overnight culture into six 250 ml polypropylene centrifuge tubes and pellet by centrifugation for 10 min at 4,000 × *g*, 4°C. Decant medium, fill polypropylene centrifuge tubes again with bacterial culture, and repeat centrifugation.
4. After the last centrifugation, invert each tube on a paper towel for 2 min to drain all liquid. Resuspend each of the pellets in 5 ml buffer P1 and combine the six aliquots. Add 130 ml buffer P1 and distribute to four fresh 250 ml polypropylene centrifuge tubes (40 ml per tube).
5. Add 40 ml buffer 2 to each centrifuge tube, mix by inverting the tubes four to six times, and incubate 5 min at room temperature.

6. Add 40 ml buffer P3 and mix immediately by inverting the tubes six times. Incubate the tubes for 20 min on ice. Invert the tube once more and centrifuge 30 min at 16,000 × *g*, 4°C.
7. Filter the supernatants through a folded filter (120-mm diameter) into four fresh 250 ml polypropylene centrifuge tubes.
8. Precipitate the DNA with 0.7 volumes (84 ml per tube) isopropanol, mix gently, and centrifuge immediately for 30 min at 17,000 × *g*, 4°C.
9. Remove the supernatants and mark the locations of the pellet. Wash the DNA pellet by adding 20 ml cold 70% ethanol to each and centrifuge for 30 min at 16,000 × *g*, 4°C.
10. Carefully remove the supernatants and resuspend each of the four pellets in 2 ml TE buffer, pH 7.4. Pool the four solutions (total volume 8 ml) and add 52 ml QGT buffer (final volume 60 ml).
11. Equilibrate two Qiagen tip 500 columns with 10 ml buffer QBT, and allow the columns to empty by gravity flow.
12. Transfer the solution through a folded filter (120-mm diameter) into Qiagen tip 500 columns (30 ml per column), and allow the liquid to enter the resin by gravity flow.
13. Wash each column twice with 30 ml buffer QC, and then elute DNA from each column with 15 ml prewarmed (65°C) buffer QF into a 30 ml centrifuge tube.
14. Precipitate the DNA with 0.7 volumes (10.5 ml) isopropanol, mix, and immediately centrifuge 30 min at 20,000 × *g*, 4°C.
15. Carefully remove the supernatants from step 14 and mark the locations of the pellets on the outside of the tubes. Wash the pellets with chilled 70% ethanol and, if necessary, re-pellet at the same settings as in step 14.
16. Aspirate the supernatants completely. Resuspend each pellet in 3 ml TE buffer (pH 7.4) for several hours at 37°C.
17. Prepare two Beckman Ultra-Clear Centrifuge tubes (13 × 51 mm) with 3 g CsCl and add the DNA solution from step 16 (3 ml per tube). Mix the solution gently until salt is dissolved. Add 300 μl ethidium bromide (10 mg/ml in H_2O) to the DNA/CsCl solution. Then overlay the solution with 300 μl paraffin oil and seal the tubes.
18. Centrifuge 17 h at 234,600 × *g*, 20°C.
19. Two bands of DNA, located in the center of the gradient, should be visible in normal light. The upper band consists of linear and nicked circular HSV-1 BAC DNA. The lower band consists of closed circular HSV-1 BAC DNA.
20. Harvest the lower band using a disposable 1 ml syringe fitted with a 21-gauge hypodermic needle under UV-light and transfer it into a microfuge tube.

21. Remove ethidium bromide from the DNA solution by adding an equal volume of *n*-butanol in TE/CsCl (3 g CsCl dissolved in 3 ml TE, pH 7.4).
22. Mix the two phases by vortexing and centrifuge at 210×*g* for 3 min at room temperature in a bench centrifuge.
23. Carefully transfer the lower, aqueous phase to a fresh microfuge tube. Repeat steps 21–23 four to six times until the pink color disappears from both the aqueous phase and the organic phase.
24. Add an equal volume of isopropanol, mix and centrifuge at 210×*g* for 3 min at room temperature. Transfer the aqueous phase to a fresh microfuge tube.
25. To remove the CsCl from the DNA solution, dialyze 6 h against TE, pH 7.4 at 4°C. Then, change the TE buffer and dialyse overnight. For dialysis, the DNA solution is injected into a dialysis cassette, Slide-A-Layzer 10K (10,000 MWCO) using a 1 ml disposable syringe fitted with a 36-gauge hypodermic needle. After dialysis, the solution is recovered from the dialysis cassette by using a fresh 1 ml disposable syringe fitted with a 36-gauge hypodermic needle. The DNA solution is then transferred to a clean microfuge tube and stored at 4°C. After characterization of the DNA (concentration and restriction enzyme analysis), store DNA at 4°C.
26. Determine the absorbance of the DNA solutions from step 25 at 260 nm (A_{260}) and 280 nm (A_{280}) using an UV spectrophotometer. From 4 l of bacterial cultures, HSV-1 BAC DNA yields are typically in the range of 200–300 µg.
27. Verify the HSV-1 BAC DNA by restriction endonuclease analysis (e.g., *Hin*dIII, *Kpn*I). Separate the fragments overnight by electrophoresis on a 0.4% agarose gel at 40 V in 1× TAE electrophoresis buffer (see Note 22), using high-molecular-weight DNA and 1 kb DNA ladder as size standards. Stain with ethidium bromide (1 mg/ml in H_2O) and compare restriction fragment patterns with the published HSV-1 sequence (35).

3.3.2. Preparation of Plasmid DNA (Maxiprep Protocol)

1. Prepare a 17×100 mm sterile snap-cap tube containing 5 ml LB/chloramphenicol medium. Inoculate with frozen long-term culture of the *E. coli* harboring the plasmid. Incubate 8 h at 37°C in a shaker.
2. Transfer 1 ml of the culture into a 1 l flask containing 200 ml sterile LB medium containing the appropriate antibiotic, and incubate 16 h at 37°C, with shaking.
3. Transfer the overnight culture into a 250 ml polypropylene centrifuge tube and pellet by centrifugation for 10 min at 4,000×*g*, 4°C. Decant medium and invert the tube on a

paper towel for 2 min to drain all liquid. Resuspend the pellet in 10 ml buffer P1.

4. Add 10 ml buffer 2, mix by inverting the tube four to six times, and incubate 5 min at room temperature.
5. Add 10 ml chilled buffer P3 and mix immediately by inverting the tube six times. Incubate the tube for 20 min on ice. Invert the tube once more and centrifuge 30 min at 16,000 × *g*, 4°C.
6. Filter the supernatants through a folded filter (120-mm diameter) into a 30 ml centrifuge tube.
7. Equilibrate a Qiagen tip 500 column with 10 ml buffer QBT, and allow the column to empty by gravity flow.
8. Transfer the solution from step 6 into the Qiagen tip 500 column, and allow the liquid to enter the resin by gravity flow.
9. Wash the column twice with 30 ml buffer QC, and then elute DNA from the column with 15 ml prewarmed (65°C) buffer QF into a 30 ml centrifuge tube.
10. Precipitate the DNA with 0.7 volumes (10.5 ml) isopropanol, mix, and immediately centrifuge 30 min at 20,000 × *g*, 4°C.
11. Carefully remove the supernatant from step 10 and mark the location of the pellet on the outside of the tube. Wash the pellet with chilled 70% ethanol and, if necessary, re-pellet at the same settings as in step 10.
12. Aspirate the supernatant completely. Resuspend the pellet in 200 μl TE buffer (pH 7.4), and determine the DNA concentration using a UV spectrophotometer.

3.3.3. Transfect Vero 2-2 Cells and Harvest, Concentrate, and Purify Packaged Amplicon Vectors

1. Maintain 2-2 cells in DMEM/10% FBS containing 500 μg/ml G418. Propagate the culture twice a week by splitting 1/5 in fresh medium (20 ml) into a new 75 cm^2 tissue culture flask (see Note 23).
2. On the day before transfection, remove culture medium, wash twice with PBS, add a thin layer of trypsin/EDTA, and incubate 10 min at 37°C to allow cells to detach from plate. Count cells using a hemacytometer and plate 1.2×10^6 cells per 60 mm diameter tissue culture dish in 3 ml DMEM/10% FBS.
3. For each 60-mm dish, place 100 μl Opti-MEM I reduced serum medium into each of two 15 ml conical tubes. To one tube, add 0.6 μg amplicon DNA and 2 μg of the HSV-1 BAC DNA and 0.2 μg pEBHICP27 DNA. To the other tube, add 12 μl LipofectAMINE.
4. Combine the contents of the two tubes, mix well, and incubate 45 min at room temperature.

5. Wash the cultures prepared the day before (step 2) once with 2 ml Opti-MEM I. Add 1.1 ml Opti-MEM I to the tube from step 4 containing the DNA-LipofectAMINE transfection mixture (1.3 ml total volume). Aspirate medium from the culture, add the transfection mixture, and incubate for 5.5 h.
6. Aspirate the transfection mixture and wash the cells three times with 2 ml Opti-MEM I. After aspirating the last wash, add 3.5 ml DMEM/6% FBS and incubate 2–3 days.
7. Scrape cells into the medium using a rubber policeman. Transfer the suspension to a 15 ml conical centrifuge tube and place the tube containing the cells into a beaker of ice water. Submerge the tip of the sonicator probe ~0.5 cm into the cell suspension and sonicate 20 s with 20% output energy. This disrupts cell membranes and liberates cell-associated vector particles.
8. Remove cell debris by centrifugation for 10 min at 1,400 × *g*, 4°C and filter the supernatant through a 0.45 μm syringe-tip filter attached to a 20 ml disposable syringe into a new 15 ml conical tube. Remove a sample for titration, then divide the remaining stock into 1 ml aliquots, freeze them in a dry ice/ethanol bath, and store at −80°C. Alternatively, concentrate (steps 9a and 10a) or purify and concentrate (steps 9b and 10b) the stock before storage.

Protocol a (pelleting):

9a. Transfer the vector solution from step 8 to a 30 ml centrifuge tube and spin 2 h at 20,000 × *g*, 4°C.

10a. Resuspend the pellet in a small volume (e.g., 300 μl) of 10% sucrose. Remove a sample of the stock for titration, then divide into aliquots (e.g., 30 μl) and freeze in a dry ice/ethanol bath. Store at −80°C.

Protocol b (gradient centrifugation):

9b. Prepare a sucrose gradient in a Beckman Ultra-Clear 25 × 89 mm centrifuge tube by layering the following solutions in the tube: 7 ml 60% sucrose; 7 ml 30% sucrose; 3 ml 10% sucrose. Carefully add the vector stock from step 8 (up to 20 ml) on top of the gradient and centrifuge 2 h at 100,000 × *g*, 4°C, using a Beckman SW28 rotor.

10b. The interface between the 30 and 60% sucrose layers appears as a cloudy band when viewed with a fiber-optic illuminator. Aspirate the 10 and 30% sucrose layers from the top and collect the virus band at the interface between the 30 and 60% layers. Transfer to a Beckman Ultra-Clear 14 × 95 mm centrifuge tube, add ~15 ml PBS, and pellet virus particles for 1 h at 100,000 × *g*, 4°C, using a Beckman SW40 rotor. Resuspend the pellet in a small volume (e.g., 300 μl) of 10%

sucrose. Divide into aliquots (e.g., 30 μl) and freeze in a dry ice/ethanol bath. Store at −80°C. Before freezing, retain a sample of the stock for titration.

3.3.4. Titration of HSV-1 Amplicon Vector Stocks

1. Plate cells (e.g., Vero-7b, BHK 21, or 293 cells) at a density of 1.0×10^5 per well of a 24-well tissue culture plate in 0.5 ml DMEM/10% FBS. Incubate overnight.
2. Aspirate the medium and wash each well once with PBS. Remove PBS and add 0.1, 1, or 5 μl samples collected from vector stocks, diluted to 250 μl each in DMEM/2% FBS.
3. Incubate 1–2 days. Remove the inoculums and fix cells for 20 min at room temperature with 250 μl of 4% paraformaldehyde, pH 7.0. Wash the fixed cells three times with PBS, then proceed (depending on the transgene) with a detection protocol such as green fluorescence (step 4a), X-gal staining (steps 4b and 5b), or immunocytochemical staining (steps 4c–6c).

Protocol a (fluorescence detection):

4a. Detect cells expressing the gene for EGFP (see Note 24): Examine the culture from step 3 (before or after fixation) using an inverted fluorescence microscope. Count green fluorescent cells and determine the vector titer in transducing units (t.u.)/ml by multiplying the number of transgene-positive cells by the dilution factor (see Notes 25 and 26)

Protocol b (X-gal detection):

4b. Detect cells expressing the *E. coli lacZ* gene: Add 250 μl X-gal staining solution per well of the 24-well tissue culture plate from step 3, and incubate 4–12 h (depending on the cell type and the promoter regulating expression of the transgene) at 37°C.

5b. Stop the staining reaction by washing the cells three times with PBS. Count blue cells using an inverted light microscope, and determine the vector titer in t.u./ml by multiplying the number of transgene-positive cells by the dilution factor.

Protocol c (immunocytochemical staining):

4c. Detect transgene expressing cells by immunocytochemical staining: Add 250 μl GST solution per well of the 24-well tissue culture plate from step 3 (to block nonspecific binding sites and to permeabilize cell membranes) and let stand 30 min at room temperature. Replace the blocking solution with the primary antibody (diluted in GST) and incubate overnight at 4°C.

5c. Wash the cells three times with PBS, leaving the solution in the well for 10 min each time. Add secondary antibody (diluted in GST) and incubate at least 4 h at room temperature.

6c. Wash the cells twice with PBS and develop according to the appropriate visualization protocol. Count transgene-positive cells using an inverted light microscope and determine the vector titer as t.u./ml by multiplying the number of the transgene-positive cells by the dilution factor.

3.4. Packaging of Amplicon Vectors Using a Replication-Incompetent Cre/loxP-Sensitive Helper Virus

3.4.1. Production, Purification, and Titration of HSV-1 LaLΔJ

1. HSV-1-LaLΔJ is a defective recombinant virus. Therefore, to prepare stocks of this virus, follow the instructions described in protocol Subheading 3.1.4. Since HSV-1-LaLΔJ is an ICP4 minus virus, it should be grown in ICP4 expressing cells, such as the 7b Vero-derived cell line (19). These cells grow in DMEM medium supplemented with 10% FBS, 2 mM L-glutamine, 100 units/ml penicillin and 100 μg/ml streptomycin. Geneticin (G418) should be added every four passages (1 mg/ml), to avoid losing the complementing ICP4 gene.
2. To titrate the helper virus stock, follow the instructions described in protocol Subheading 3.1.5. The virus should be titrated simultaneously in complementing cells, such as Vero-7b, and in noncomplementing Vero cells, to allow detection of undesired replication-competent particles that can sometimes be generated by recombination between the virus genome and the ICP4 gene located in the cellular chromosomes; such particles should produce lysis plaques in Vero cells. If this is the case, start the production again, infecting the complementing cells at very low MOI (lower than 0.05 PFU/cell), using plaque-purified defective virus.
3. To purify virus stock if required, follow the instruction described in protocol Subheading 3.1.6.

3.4.2. Production of Amplicon Vectors Using Cre/loxP Site-Specific Recombination

The protocol for producing amplicons using this system is a two-step protocol, described in detail in reference (31). In the first step, stocks of amplicons contaminated with large amounts of helper particles are produced in ICP4 complementing cells (Vero-7b cells). In the second step, stocks of vectors only barely or not contaminated with helper particles are prepared in cells expressing both ICP4 and Cre recombinase (TE-Cre-Grina cells). To this end, we usually proceed as follows:

3.4.2.1. Generation of P0 Stock (Helper Contaminated)

1. The day before transfection, plate 5.10^6 Vero-7b cells in a 75 cm^2 tissue culture flask with growth medium (see Subheading 2.1).
2. Transfect the amplicon plasmid. For one 75 cm^2 cell culture flask: Mix 6 μg amplicon plasmid DNA + 750 μl Opti-MEM + 30 μl plus reagent. Wait 15′ at room temperature (RT), and add a mix of 45 μl LipofectAmine + 750 μl Opti-MEM. After 15′ at RT, add the transfection mix to the

Vero-7b cells in 10 ml Opti-MEM medium and incubate at 37°C in 5% CO_2.

3. After 3 h add 10 ml Opti-MEM medium on the cells. Let until the following day.
4. Infect the transfected cells with helper virus as follows:
5. One day after transfection, discard medium from the flask: (see Note 27)
6. Rinse one time with maintenance medium (see Subheading 2.1), discard medium.
7. Add 3 ml maintenance medium containing the helper virus dilution at a MOI of 0.5 PFU/cell.
8. Let the flask on a shaker for 1 h 30 min, if possible under CO_2 atmosphere.
9. Discard medium, rinse two times with maintenance medium.
10. Add 20 ml maintenance medium.
11. Incubate 48 h at 37°C in 5% CO_2 incubator.
12. Collect helper-contaminated amplicon vectors. At 48 h post infection, most cells should be rounded and show open cytopathic effect (CPE) typical of HSV-1. Scrape the cells and put cells plus medium in a 50 ml Falcon tubes.
13. Spin down at $771 \times g$ for 10′ at 4°C.
14. Transfer the supernatant (SN) to a 35 ml oak ridge tube (this is SN1).
15. Resuspend the cell pellet in 1 ml PBS without serum and disrupt the cells by three cycles of freeze–thaw or using a water sonicator (three times 30″ in cold water). Then, spin down at $771 \times g$ for 10′ at 4°C.
16. Discard the pellet containing cell debris and keep the PBS supernatant containing viral particles (this is SN2). Keep SN2 at −80°C.
17. Spin down SN1 at $18{,}000 \times g$ at 4°C for 1 h 30 min. Discard the supernatant and resuspend the pellet containing virus particles in 1 ml PBS.
18. Add to SN2 and keep this final P0 vector stock at −80°C until titration.

3.4.2.2. Titration of Amplicons and Helper Virus in P0 Stocks

Follow the instructions described in protocol Subheading 3.1.5.

1. One day prior the titration of the P0 virus stock prepare six-well tissue culture plates with 1×10^6 cells per well of either Gli36 cells, Vero-7b cells, or Vero cells. These three cell lines are propagated in growth medium described in Subheading 2.1 (see Note 28).

2. Prepare a series of tenfold dilutions (10^{-2} to 10^{-10}) of the vector stock in Eppendorf tubes with 1 ml growth medium without serum.
3. Infect Gli36, Vero-7b, and Vero cell lines as described in Subheading 3.1.5.

3a. Determination of the titer of amplicon particles. One day following infection, if the amplicon vectors express GFP, count green fluorescent Gli36 cells and determine the average for each dilution (it should always be done in duplicate). Then multiply by a factor of 10 to get the number of transduction units (TU)/ml for each dilution. Multiply this number by 10 to the power of the dilution to achieve the titer of the stock in TU/ml. If the amplicon expresses LacZ, count Gli36 cells expressing the *E. coli* LacZ gene using the X-gal staining solution, as described in protocol Subheading 3.3.4.

3b. Determination of the titer of helper virus particles. Three days after infection, fix, stain, and count the number of plaques per well in the Vero 7b monolayers. Determine the average for each dilution (this should also be done in duplicate), and multiply by a factor of 10 to get the number of plaque-forming units/ml (PFU/ml) for each dilution. Multiply this number by 10 to the power of the dilution to achieve the titer in PFU/ml. See Table 1.

3c. Determination of the titer of replication-competent revertant virus. Proceed exactly as in 3b but infecting nontranscomplementing Vero cells (see Note 29).

3.4.2.3. Amplification from P0 to P1 and Titration of P1 Stocks (Helper Contaminated)

The stock of helper-contaminated amplicon vectors should be expanded by amplifying the P0 vector stock in Vero-7b cells.

1. The day before infection, plate 1.3×10^7 Vero-7b cells in one 175 cm^2 tissue culture flask.
2. Infect cells using the P0 vector stock:
3. Add 5 ml maintenance medium containing the vector stock dilution at a MOI of 0.3 PFU/cell.

Table 1
Titres, ratios, and amounts of amplicon vectors and helper particles (see Note 34)

	P0 (HC)	P1 (HC)	P2 (HC)	P3 (HF)
Titre amplicon (t.u./ml)	10^7	10^8	10^9	10^8
Titre helper (PFU/ml)	3×10^7	5×10^7	10^8	5×10^5
Ratio A/H	1:3	2:1	10:1	200:1
Amount (ml)	0.5	1	5–10	5–10

4. Let the flask on a shaker for 1 h 30 min, if possible under CO_2 atmosphere.
5. Discard medium, rinse two times with maintenance medium.
6. Add 30 ml maintenance medium.
7. Then proceed as in Subheading 3.4.2.1 (from point 11). You will obtain the P1 vector stock.
8. Titrate P1 vector stock as in Subheading 3.4.2.2 (see Note 30). See Table 1.

3.4.2.4. Amplification from P1 to P2 and Titration of P2 Stocks (Helper Contaminated)

1. It is often convenient to further amplify the vector stock once more. To this end, you can expand the amount of Vero-7b cells to be infected as much as required. If you prefer to use 175 cm^2 tissue culture flasks, proceed as in Subheading 3.4.2.3 but expanding the number of tissue culture flasks and scaling up the procedure. However, instead of using 175 cm^2 flasks, at this step we prefer to infect cells in roller bottles, which are easier to manipulate and allow to spend less medium than using several 175 cm^2 tissue culture flasks, but you need for this to have a gyratory system for roller bottles in your laboratory. If this is the case, the protocol is as follows:
2. Seed 2×10^7 Vero-7b cells/roller bottle in 100 ml of growth medium. Since cells in roller bottles are not incubated in a CO_2 atmosphere, you should add CO_2 to the growth medium using a pipette connected to a CO_2 tube, until CO_2 bubbles fill up the bottle.
3. Turn the roller bottles at a speed of 0.4 rounds per minute. Cells generally arrive to confluence (10^8 cells/bottle) in 4–5 days at 37°C.
4. Infect cells with the P1 vector stock. Discard medium from each roller bottle and add 20 ml of maintenance medium containing the vector stock dilution at a MOI of 0.3 PFU/cell.
5. Two hours later add maintenance medium up to 100 ml per roller bottle.
6. Incubate 48 h at 37°C always turning the bottles at a speed of 0.4 rounds per minute.
7. When CPE is maximum, which generally occurs at 48 h post infection, collect the particles as in Subheading 3.4.2.1 (from point 12, but scaling up the number of tubes). This is the P2 vector stock.
8. Titrate P2 vector stock as in Subheading 3.4.2.2 (see Note 31). See Table 1.

3.4.2.5. Production and Titration of P3 Amplicon Stocks (Helper Free)

1. Plate 1.3×10^7 TE-Cre-Grina cells in 175 cm^2 tissue culture flasks.
2. The following day, infect cells with the P2 stock at an MOI of 3 TU/cell of amplicons, which generally should correspond to about 0.3–0.5 PFU/cell of helper virus. If the amount of helper virus in the stock is too low, add more helper virus (see Note 32). For this, add 5 ml maintenance medium containing the P2 vector stock dilution per tissue culture flask.
3. Let the flask on a shaker for 1 h 30 min, if possible under CO_2 atmosphere.
4. Discard medium, rinse two times with maintenance medium.
5. Add 30 ml maintenance medium per flask.
6. Incubate 48 h at 37°C in 5% CO_2 incubator.
7. To collect the virus proceed as in protocol Subheading 3.4.2.1 (from point 12). This will be the P3 vector stock ("helper free" (HF) vector stock).
8. Then titrate the vectors and the helper particles as in Subheading 3.4.2.2 (see Note 33). See Table 1.
9. If you wish to purify your amplicon stocks, either helper contaminated or helper-free stocks, for in vivo inoculation, follow the protocol described in Subheading 3.1.6.

4. Notes

1. The *E. coli* SW102 strain is derived from *E. coli* DH10B and contains the λ prophage recombination system and a deletion in the galactokinase gene (*galK*). The *galK* function can be added in trans, which restores the ability of the bacteria to grow on galactose as carbon source.
2. Any plasmid that expresses Cre recombinase with a nuclear localization signal can be used.
3. An amplicon plasmid is any *E. coli* plasmid containing one origin of DNA replication and one cleavage/packaging signal *pac* from HSV-1. It usually carries also a reporter gene expressing GFP, LacZ, or luciferase, which allows to easily titrate the vector stock and to identify the infected cells. It contains, in addition, a multiple cloning site where the desired transgenic sequences can be introduced. It is produced and purified like any standard bacterial plasmid.
4. The 7b Vero-derived cell line (19) expresses the HSV-1 immediate early genes ICP4 and ICP27 required for replication of a recombinant virus deleted in both IE genes. Vero-7b

cells are subjected after several passages to 2-week long selection with 1 mg/ml G418 (Sigma).

5. Viral DNA is very long and fragile. It is thus critical to take extreme care when handling viral DNA by using wide-end pipette tips (tips for genomic DNA). Store the viral DNA at 4°C, do not freeze to avoid breaking the large DNA.
6. Example of a plasmid used to insert new sequences into the viral genome. pTZgJHE plasmid: The 2,036 bp *Sal*I–*Hin*dIII fragment from the HSV-1 genome (nucleotides 136308 to 138345) containing gene *Us5* (corresponding to HSV-1 glycoprotein J promoter and coding sequence) is cloned into *Sal*I–*Hin*dIII of pTZ18U plasmid (InVitrogen) using T4 DNA ligase (NEB). The resulting plasmid, pTZgJ, is used to generate pTZgJHE plasmid by insertion of GFP coding sequence driven by the cytomegalovirus (HCMV) promoter. *GFP* cassette is inserted as an *Nru*I–*Sph*I fragment isolated from pcDNA3.1HygroGFP into the *Sph*I (137626) and *Nru*I (137729) sites of the virus genome, making a deletion in *Us5* between the TATA box and the gJ coding sequence of pTZgJ. This plasmid can be used as a shuttle plasmid, where GFP can be replaced with a desired transgene. The pTZgJH*X* plasmid thus obtained can be used to insert the wanted gene into the chosen recombinant virus by cotransfection method.
7. Examples of recombinant HSV-1 viruses. (a) Replication defective viruses that require complementing cell lines: S0ZgJHE is a recombinant virus deleted in the ICP4 immediate early gene, with the GFP reporter gene driven by HCMV promoter placed in the *Us5* locus (glycoprotein J) and the lacZ reporter gene, under ICP0 promoter, placed in the *UL41* locus (vhs, virion host shutoff). TOZ-GFP is a recombinant virus deleted in the ICP4, ICP27, and ICP22 immediate-early genes, with the GFP reporter gene, under HCMV promoter, placed in the *Us1* locus (ICP22) and the lacZ reporter gene, under ICP0 promoter, in the *UL41* locus. In these cases, the viral DNAs and the recombinant plasmids containing the desired transgenes can be cotransfected into a complementing cell line (Vero-7b), which provides, in trans, the essential viral genes ICP4 and ICP27. (b) Replication-competent viruses that can be grow in Vero cells or other permissive cell lines are deleted in nonessential genes, such as UL41, γ34.5, or TK.
8. The use of wide boar Pipetman tips will help to prevent shearing of the viral DNA, increasing on this way the infectivity of the viral DNA preparation. The quality of the plasmid is crucial for the recombination efficiency. The size of the HSV-1 flanks is crucial and longer sequences are better. The size of the insert can affect stability if it is too large; part of the insert can be lost over time and it will not be possible to obtain a

purified isolate of the desired recombinant. It is important to linearize the plasmid to increase the recombination event compared with the supercoiled plasmid. The pH of the HBS is crucial for the transfection efficiency.

9. Prior to do the limiting dilution it is better to sonicate the virus stock for a few seconds before infection in order to separate out virus particles that might have clumped up together. This might cause a single plaque arising out of two viruses, which is not desirable after the limiting dilution.
10. Once the plaques have started opening up at this time it is better to not let the single virus plaque to replicate too much since this is the primary stock of it.
11. The specific HSV-1 gene locus targeted for deletion/insertion can affect the recombination event or affect the stability of the desired recombinant and sometimes it will not be possible to obtain a purified isolate of the desired recombinant.
12. Recombination into the repeat sequences can yield a mixture of viruses containing insertion into one copy of the gene, leading to rescue by the not deleted copy, and also in this case it will not be possible to obtain a purified isolate of the desired recombinant.
13. To prepare attenuated viral vectors, the amount of cells that should be infected can be lower since usually these recombinants grow well in vitro. These recombinants can infect attached cells. The cells can be prepared the day before and on the next day decant medium from the flasks and add the virus in an amount of medium, without serum, sufficient to cover the monolayer. The infection can be performed at an MOI of 0.01 PFU/cell.
14. The titer of a large viral stock should be done in duplicate.
15. Following purification of the virus, it is necessary to add glycerol to a final concentration of 10% to virus stock in order to cryo-preserve the virus. If the virus is prepared to be used in animal experiments it should be aliquoted without glycerol but in small volumes to avoid thawing the vials twice.
16. Virus stocks should be maintained at a low passage. Use one vial of a newly prepared stock as a stock for preparing all future stocks. In order to reduce the chance of rescuing wild-type virus during the propagation of viruses carrying deletions of essential gene(s), stocks should be routinely prepared from single plaque isolates.
17. *Dpn*I does not cut nonmethylated DNA amplified by PCR but cut the methylated template plasmid DNA isolated from *E. coli*.
18. A value of 1.0 for A_{260} is equivalent to 50 μg/ml of double-stranded DNA. Additionally, the ratio between A_{260} and A_{280}

provides information about DNA purity. Typically, pure DNA preparations have an A_{260}/A_{280} value of 1.8; do not use DNA preparations with a ratio below this value.

19. Washing with M9 salts is important to remove all residual rich medium from the bacteria before plating on minimal medium plates.
20. The efficiency of the *galK*-negative selection step is low, and the majority of the colonies that form under DOG selection are not correct (but may have point mutations in the *galK* gene or large deletions). To overcome the problem, longer homology arms could be designed to increase the frequency of recombination.
21. The HSV-1 sequences in BACs are normally flanked by loxP sites. This allows removing the bacterial sequences by Cre recombinase during reconstitution of virus following transfection of BAC DNA in mammalian cells.
22. Treat gel with care, 0.4% gels are very delicate.
23. Cells are incubated in a humidified 37°C, 5% CO_2 incubator throughout the protocol. All solutions and equipment coming into contact with cells must be sterile.
24. Use of an amplicon that expresses an easily detectable reporter gene (e.g., EGFP) is strongly recommended when establishing the packaging protocol in the laboratory. Although the quality of the DNA and condition of the cells are of prime concern, other components, e.g., lipids, DNA concentration, and incubation times may also influence transfection and packaging efficiency.
25. The titers expressed as transducing units per milliliter (t.u./ml) are relative. Factors influencing relative transduction efficiencies include: the cells used for titration, the promoter regulating the expression of the transgene, the transgene, and the sensitivity of the detection method.
26. The vector titers realized with amplicons that contain the standard ~1-kb *ori* should be in the range of 10^6 to 10^7 t.u./ml before concentration. The recovery of transducing units after concentration/purification is around ~50%.
27. Before infecting the transfected cells, confirm that transfection was efficient, resulting in at least 30% of cells expressing the reporter transgene (generally EGFP). If this is not the case, it is better to start transfection again, using fresh cells and/or optimizing the transfection procedure.
28. While the number of physical particles is an intrinsic property of the virus stock, independent of the cell types to be infected, the number of infectious particles, hence the titer of a virus or of a vector stock, strongly depends on the susceptibility of the cells. In the case of helper-free amplicon vectors, some cell

types, such as Gli36 cells (a human glioblastoma cell line), give very high vector titers, while Vero-derived cell lines give relatively lower vector titers. In contrast, Vero or Vero-derived cells give very good titers of the helper virus.

29. In a typical P0 situation we obtain an amplicon to helper ratio of about 1:3. We usually observe no generation of replication-competent virus particles.
30. At this step the ratio of amplicon to helper particles generally inverts in favor of amplicon particles (from 2:1 to 5:1). The titers of P1 are generally one order of magnitude higher than in P0.
31. At this step, the ratio of amplicon to helper particles increases in favor of amplicon particles (from 5:1 to 10:1) while the titers of the stock can be substantially increased, depending on the number of tissue culture flasks infected.
32. The critical point here is that each cell should receive at least one amplicon particle. The infected cells will become round but without displaying an open CPE, as the helper particles cannot spread in these cells.
33. We usually observe less than 1% contamination of the vector stock with defective helper particles (ratio of amplicon to helper particles ranges between 100:1 and 500:1). However, the titer of the amplicon vectors is generally one order of magnitude lower than that of the P2 stock used to infect TE-Cre-Grina cells.
34. Table 1 presents results obtained in a typical vector preparation. Values can be somewhat different depending on the nature and size of the amplicon plasmid, on the passage number of cell lines, and on the efficacy of transfection in P0. HC: helper-contaminated stocks, HF: helper-free stocks. Note that "helper-free" stocks obtained using this strategy can be slightly contaminated with defective helper particles.

References

1. Roizman, B., and Knipe, D.M. Herpes simplex viruses and their replication. In: Knipe DM, Howley PM (eds). *Fields Virology*. Lippincot, Williams and Wilkins: Philadelphia, PA, 2001, pp. 2399–2460.
2. Marozin, S., Prank, U., and Sodeik, B. (2004) Herpes simplex virus type 1 infection of polarized epithelial cells requires microtubules and access to receptors present at cell-cell contact sites. *J. Gen. Virol.* **85**, 775–786.
3. Honess, R.W., and Roizman, B. (1975) Regulation of herpesvirus macromolecular synthesis: sequential transition of polypeptide synthesis requires functional viral polypeptides. *Proc. Natl. Acad. Sci. USA.* **72**, 1276–1280.
4. Batterson, W., and Roizman, B. (1983) Characterization of the herpes simplex virion-associated factor responsible for the induction of alpha genes. *J. Virol.* **46**, 371–377.
5. Skepper, J.N., Whiteley, A., Browne, H., and Minson, A. (2001) Herpes simplex virus nucleocapsids mature to progeny virions by an envelopment > deenvelopment > reenvelopment pathway. *J. Virol.* **75**, 5697–5702.
6. Deshmane, S.L., and Fraser, N.W. (1989) During latency, herpes simplex virus type 1

DNA is associated with nucleosomes in a chromatin structure. *J. Virol.* **63**, 943–947.

7. Farrell, M.J., Dobson, A.T., Feldman, L.T. (1991) Herpes simplex virus latency-associated transcript is a stable intron. *Proc. Natl. Acad. Sci. U S A.* **88**, 790–794.
8. Umbach, J.L., Kramer, M.F., Jurak, I., Karnowski, H.W., Coen, D.M., and Cullen, B.R. (2008) MicroRNAs expressed by herpes simplex virus 1 during latent infection regulate viral mRNAs. *Nature* **454**, 780–783.
9. Preston, C.M. (2000) Repression of viral transcription during herpes simplex virus latency. *J. Gen. Virol.* **8**, 1–19.
10. Todo, T. (2008) Oncolytic virus therapy using genetically engineered herpes simplex viruses. *Front. Biosci.* **13**, 2060–2064.
11. Burton, E.A., Fink, D.J., and Glorioso, J.C. (2005) Replication-defective genomic HSV gene therapy vectors: design, production and CNS applications. *Curr. Opin. Mol. Ther.* 7, 326–336.
12. Oehmig, A., Fraefel, C., and Breakefield, X.O. (2004) Update on herpesvirus amplicon vectors. *Mol. Ther.* **10**, 630–643.
13. Cuchet, D., Potel, C., Thomas, J. and Epstein, A.L. (2007) HSV-1 amplicon vectors: a promising and versatile tool for gene delivery. *Expert Opin. Biol. Ther.* 7, 975–995.
14. Manservigi, R., Argnani, R., Marconi, P. and Epstein, A.L. (2007). Herpesvirus-based vectors for gene transfer, gene therapy, and the development of novel vaccines. In Virus Expression Vectors, pp205–246. Ed. Kathleen L. Hefferon. Transworld Research Network.
15. Krisky, D.M., Marconi, P.C., Oligino, T.J., Rouse, R.J., Fink, D.J., Cohen, J.B., Watkins, S.C., and Glorioso, J.C. (1998) Development of herpes simplex virus replication-defective multigene vectors for combination gene therapy applications. *Gene Ther.* **5**, 1517–1530.
16. Wu, N., Watkins, S.C., Schaffer, P.A., and DeLuca, N.A. (1996) Prolonged gene expression and cell survival after infection by a herpes simplex virus mutant defective in the immediate-early genes encoding ICP4, ICP27, and ICP22. *J. Virol.* **70**, 6358–6369.
17. Samaniego, L.A., Neiderhiser, L., and DeLuca N.A. (1998) Persistence and expression of the herpes simplex virus genome in the absence of immediate-early proteins. *J. Virol.* **72**, 3307–3320.
18. Berto, E., Bozac, A., and Marconi, P. (2005) Development and application of replication-incompetent HSV-1-based vectors. *Gene Ther.* **12** Suppl 1:S98–S102.
19. Krisky, D.M., Wolfe, D., Goins, W.F., Marconi, P.C., Ramakrishnan, R., Mata, M., Rouse, R.J., Fink, D.J., and Glorioso, J.C. (1998) Deletion of multiple immediate-early genes from herpes simplex virus reduces cytotoxicity and permits long-term gene expression in neurons. *Gene Ther.* **5**,1593–1603.
20. Advani, S.J., Weischelbaum, R.R., Whitley, R.J., and Roizman, B. (2002) Friendly fire: redirecting herpes simplex virus-1 for therapeutic applications. *Clin. Microbiol. Infect.* **8**, 551–563.
21. Argnani, R., Lufino, M., Manservigi, M., and Manservigi, R. (2005) Replication-competent herpes simplex vectors: design and applications. *Gene Ther.* **12** Suppl 1:S170–177.
22. Nawa, A., Luo, C., Zhang, L., Ushjima, Y., Ishida, D., Kamakura, M., Fujimoto, Y., Goshima, F., Kikkawa, F., and Nishiyama, Y. (2008) Non-engineered, naturally oncolytic herpes simplex virus HSV1 HF-10: applications for cancer gene therapy. *Curr. Gene Ther.* **8**, 208–221.
23. Gage, P.J., Sauer, B., Levine, M., and Glorioso, J.C. (1992) A cell-free recombination system for site-specific integration of multigenic shuttle plasmids into the herpes simplex virus type 1 genome. *J. Virol.* **66**, 5509–5515.
24. Rinaldi, A., Marshall, K.R., Preston, C.M. (1999) A non-cytotoxic herpes simplex virus vector which expresses Cre recombinase directs efficient site specific recombination. *Virus Res.* **65**, 11–20.
25. Stricklett, P.K., Nelson, R.D., and Kohan, D.E. (1998) Site-specific recombination using an epitope tagged bacteriophage P1 Cre recombinase. *Gene* **215**, 415–423.
26. Krisky, D.M., Marconi, P.C., Oligino, T., Rouse, R.J., Fink, D.J., and Glorioso, J.C. (1997) Rapid method for construction of recombinant HSV gene transfer vectors. *Gene Ther.* **4**, 1120–1125.
27. Saeki, Y., Ichikawa, T., Saeki, A., Chiocca, E.A., Tobler, K., Ackermann, M., Breakefield, X.O., and Fraefel, C. (1998) Herpes simplex virus type 1 DNA amplified as bacterial artificial chromosome in *Escherichia coli*: rescue of replication-competent virus progeny and packaging of amplicon vectors. *Human Gene Ther.* **9**, 2787–2794.
28. Tanaka, M., Kagawa, H., Yamanashi, Y., Sata, T., and Kawaguchi, Y. (2003) Construction of an excisable bacterial artificial chromosome containing a full-length infectious clone of herpes simplex virus type 1: viruses reconstituted from the clone exhibit wild-type properties in vitro and in vivo. *J. Virol.* 77, 1382–91.
29. Fraefel, C., Song, S., Lim, F., Lang, P., Yu, L., Wang, Y., Wild, P., and Geller, A.I. (1996).

Helper virus-free transfer of herpes simplex virus type 1 plasmid vectors into neural cells. *J. Virol.* **70**, 7190–7197.

30. Saeki, Y., Fraefel, C., Ichikawa, T., Breakefield, X.O., Chiocca, E.A. (2001) Improved helper virus-free packaging system for HSV amplicon vectors using an ICP27-deleted, oversized HSV-1 DNA in a bacterial artificial chromosome. *Mol. Ther.* **3**, 591–601.
31. Zaupa, C., Revol-Guyot, V. and Epstein, A.L. (2003) Improved packaging system for generation of high levels non-cytotoxic HSV-1 amplicon vectors using Cre-loxP site-specific recombination to delete the packaging signals of defective helper genomes. *Hum. Gene Ther.* **14**, 1049–1063.
32. Warming, S., Costantino, N., Court, D.L., Jenkins, N.A., and Copeland, N.G. (2005) Simple and highly efficient BAC recombineering using *galK* selection. *Nucleic Acids Res.* **33**:e36.
33. Smith, I.L., Hardwicke, M.A., and Sandri-Goldin, R.M. (1992) Evidence that the herpes simplex virus immediate early protein ICP27 acts post-transcriptionally during infection to regulate gene expression. *Virology* **186**, 74–86.
34. Kashima, T., Vinters, H.V., and Campagnoni, A.T. (1995) Unexpected expression of intermediate filament protein genes in human oligodendroglioma cell lines. *J. Neuropathol. Exp. Neurol.* **54**, 23–31.
35. McGeoch, D.J., Dalrymple, M.A., Davison, A.J., Dolan, A., Frame, M.C., McNab, D., Perry, L.J., Scott, J.E., and Taylor, P. (1988) The complete DNA sequence of the long unique region in the genome of herpes simplex virus type 1. *J. Gen. Virol.* **69**, 1531–15374.

Chapter 14

Manufacture of Measles Viruses

Kirsten K. Langfield, Henry J. Walker, Linda C. Gregory, and Mark J. Federspiel

Abstract

Measles viruses have shown potent oncolytic activity as a therapeutic against a variety of human cancers in animal models and are currently being tested in clinical trials in patients. In contrast to using measles virus as a vaccine, oncolytic activity depends on high concentrations of infectious virus. For use in humans, the high-titer measles virus preparations must also be purified to remove significant levels of cellular proteins and nucleic acid resulting from the cytolytic products of measles virus replication and release. Pleomorphic measles virus must be treated as >1-µm particles that are extremely shear sensitive to maximize recoveries and retain infectivity. Therefore, to maximize the recovery of sterile, high titer infectious measles viruses, the entire production and purification process must be done using gentle conditions and aseptic processing.

Here we describe a procedure applicable to the production of small (a few liters) to large (50–60 L) batches of measles virus amplified in Vero cells adapted to serum-free growth. Cell culture supernatant containing the measles virus is clarified by filtration to remove intact Vero cells and other debris, and then treated with Benzonase® in the presence of magnesium chloride to digest contaminating nucleic acid. The measles virus in the treated cell culture supernatant is then concentrated and purified using tangential flow filtration (TFF) and diafiltration. The concentrated and diafiltered measles virus is passed through a final clarifying filter prior to final vialing and storage at <–65°C. An infectivity assay to quantify infectious measles virus concentration based on the $TCID_{50}$ method is also described. This procedure can be readily adapted to the production and purification of measles viruses using good manufacturing practices (GMP).

Key words: Measles virus, Oncolytic virotherapy, Tangential flow filtration, Vero cells, Large-scale production, $TCID_{50}$ infectious virus assay

1. Introduction

Measles viruses (MV) have shown potent oncolytic activity as a therapeutic against a variety of human cancers in animal models and are currently being tested in clinical trials in patients (see recent reviews (1–3)). A reverse genetics system has enabled the construction and rescue of recombinant measles viruses containing

Otto-Wilhelm Merten and Mohamed Al-Rubeai (eds.), *Viral Vectors for Gene Therapy: Methods and Protocols*, Methods in Molecular Biology, vol. 737, DOI 10.1007/978-1-61779-095-9_14, © Springer Science+Business Media, LLC 2011

additional transcription units that add a gene encoding a detectable protein to monitor virus replication and/or a protein to increase therapeutic potency, or both (4). The location of the additional genes inserted in the measles virus genome can significantly alter virus replication kinetics and virus yields. In contrast to using measles virus as a vaccine, oncolytic activity depends on high concentrations of infectious virus. For use in humans, the high-titer measles virus preparations must also be purified to remove significant levels of cellular proteins and nucleic acid resulting from the cytolytic products of measles virus replication and release. While the size range of the pleomorphic measles virus is often quoted as 100–300 nm (5), in practice, to maximize recoveries and retain infectivity, measles viruses must be treated as >1-μm particles that are extremely shear sensitive. Therefore, to maximize the recovery of sterile, high titer infectious measles viruses, the entire production and purification process must be done using gentle conditions and aseptic processing.

Here we describe a procedure applicable to the production of small (a few liters) to large (50–60 L) batches of measles virus amplified in Vero cells adapted to serum-free growth (Fig. 1). Using Vero cells and serum-free growth conditions results in 50% of the measles virus in the supernatant and 50% staying associated with the cells. The cell culture supernatant containing the measles virus is clarified by filtration to remove intact Vero cells and other debris, and then treated with Benzonase® in the presence of magnesium chloride to digest contaminating nucleic acid. The measles virus in the treated cell culture supernatant is then concentrated and purified using tangential flow filtration (TFF) and diafiltration. The concentrated and diafiltered measles virus is passed through a final clarifying filter prior to final vialing and storage at <–65°C. This procedure can be readily adapted to the production and purification of measles viruses using good manufacturing practices (GMP).

2. Materials

2.1. Thawing and Expansion of Vero Cells

1. Vero cell bank adapted to grow in serum-free medium (see Note 1).
2. Serum-free medium. This process was developed using VP-SFM (Gibco/Invitrogen) supplemented with 4 mM L-Glutamine, without antibiotics (see Note 1).
3. TrypLE Select (Gibco/Invitrogen) or other cell dissociation reagent (see Note 2).
4. Dulbecco's Phosphate Buffered Saline (D-PBS) without calcium and magnesium.

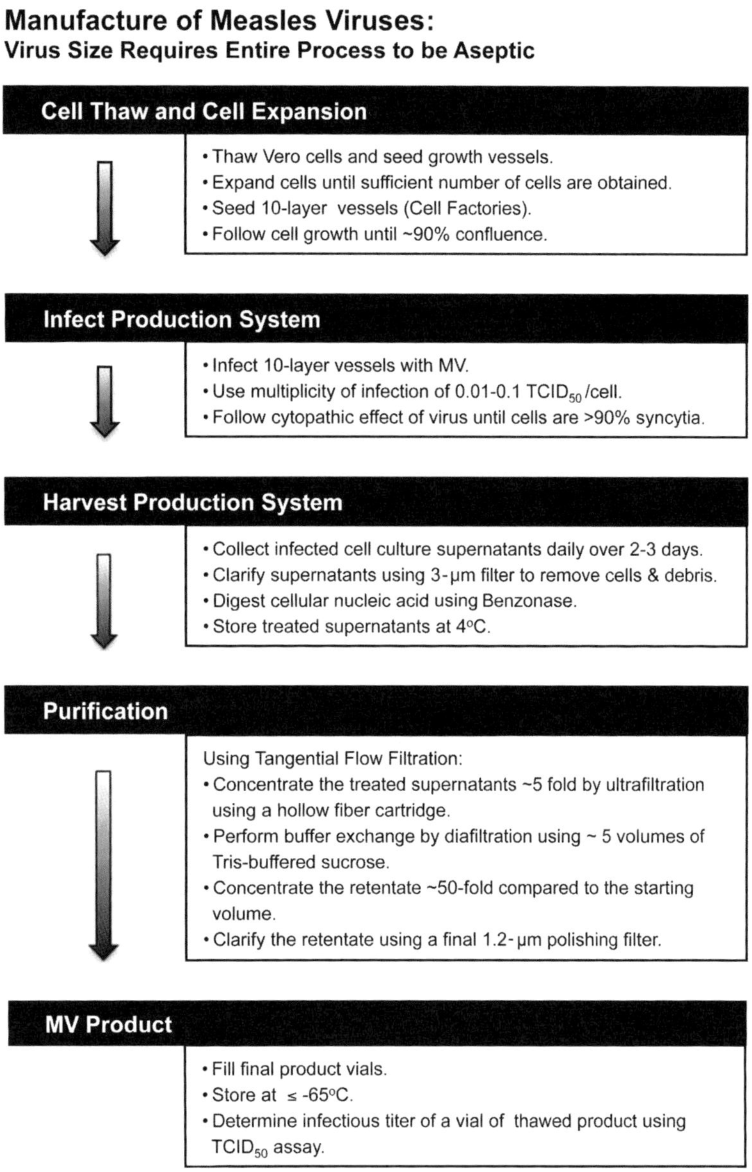

Manufacture of Measles Viruses:
Virus Size Requires Entire Process to be Aseptic
Cell Thaw and Cell Expansion
• Thaw Vero cells and seed growth vessels.
• Expand cells until sufficient number of cells are obtained.
• Seed 10-layer vessels (Cell Factories).
• Follow cell growth until ~90% confluence.
Infect Production System
• Infect 10-layer vessels with MV.
• Use multiplicity of infection of 0.01-0.1 $TCID_{50}$/cell.
• Follow cytopathic effect of virus until cells are >90% syncytia.
Harvest Production System
• Collect infected cell culture supernatants daily over 2-3 days.
• Clarify supernatants using 3-μm filter to remove cells & debris.
• Digest cellular nucleic acid using Benzonase.
• Store treated supernatants at 4°C.
Purification
Using Tangential Flow Filtration:
• Concentrate the treated supernatants ~5 fold by ultrafiltration using a hollow fiber cartridge.
• Perform buffer exchange by diafiltration using ~ 5 volumes of Tris-buffered sucrose.
• Concentrate the retentate ~50-fold compared to the starting volume.
• Clarify the retentate using a final 1.2-μm polishing filter.
MV Product
• Fill final product vials.
• Store at ≤ -65°C.
• Determine infectious titer of a vial of thawed product using $TCID_{50}$ assay.

Fig. 1. Overview of the measles virus manufacturing process.

5. Tissue culture flasks: T75, T175 and T500.
6. Sterile, disposable bottles, 250 and 1,000 mL bottle, sterile (Nunc Nalgene, Corning, or equivalent).
7. Cell Factory multilayer growth vessels, for example, 10-layer (CF10) or 40-layer (CF40) (Nunc Nalgene) (see Note 3).
8. Sterile fittings to enable venting of, additions to and removals from the Cell Factories (see Note 4).

2.2. Infection of Vero Cells in Cell Factories

1. Measles virus (MV) infecting stock (see Note 5).
2. Sterile, disposable bottles, 250 and 1,000 mL bottle, sterile (Nunc Nalgene, Corning, or equivalent).
3. Sterile bioprocess containers (BPC), for example, 2, 5, 50, and 100 L (Hyclone, Sartorius Stedim, other reputable vendor) depending on the volume of medium from the Cell Factories.
4. Sterile tubing assemblies with luer style and/or QDC (quick disconnect) style fittings (see Note 6).
5. Sterile male and female plug/cap assemblies for luer and or QDC fittings. As a source of sterile male and female plugs and caps (see Note 6).
6. Sterile feed caps, 1 and 2 L sizes or sterile 2 L bottle/cap assemblies (see Note 7).

2.3. Harvest of Measles Virus Infected Cell Culture Supernatant from Cell Factories

1. Sterile bioprocess containers (BPC), various sizes.
2. Sterile tubing assemblies with luer style and/or QDC (quick disconnect) style fittings (see Note 6).
3. Plastic totes for transporting and storing filled bioprocess containers.

2.4. Initial Processing of Measles Virus Infected Cell Culture Supernatant

1. Sterile bioprocess containers (BPC), various sizes.
2. Sterile tubing assemblies with luer style and/or QDC (quick disconnect) style fittings (see Note 6).
3. Sterile, 3-μm filter for removal of intact cells. For example, a 1,500 cm^2 Pall 3-μm Versapor capsule filter (Pall #12116), custom gamma-irradiated with tubing connectors attached (see Note 8).
4. Magnesium chloride, 1.0 M (Sigma #M1028).
5. Benzonase® endonuclease: Purity Grade 1 (>99%) for biotechnology (EM Industries, associate of Merck KGaA).

2.5. Purification of Measles Virus Using TFF

1. Sucrose buffer: 5% Sucrose, 50 mM Tris–HCl pH 7.4, and 2 mM $MgCl_2$.
2. Spectrum 50 nm polysulfone hollow fiber TFF module.
3. Laboratory scale pump drives and heads; for example Masterflex Model 07523-80 drive and easy load heads.
4. For larger-scale production runs, industrial process pump drives and heads; for example, Masterflex Model 77420-00 drive and 77601-10 head (see Note 9).
5. TFF system, various sizes and custom configurations (e.g., a Spectrum TFF system, see Subheading 3.5.2).
6. Digital pressure monitor. For example, SpectrumLabs.com, PendoTECH.

7. Data monitoring software, for example, Spectrum Labs Data Monitoring software.
8. Feed and retentate tubing segments, sterile, with disposable pressure transducers (see Notes 10 and 11).
9. Sterile tubing assemblies with luer style and/or QDC (quick disconnect) style fittings (see Note 6).
10. Sterile TFF reservoir (either a customized bioprocess container or any other vented reservoir).
11. Sterile bioprocess containers (BPC), various sizes.
12. For large-scale production runs: Plastic drum on dolly for holding 50 or 100 L bioprocess containers (BPC).
13. Sterile, 1.2-μm filter as a final polishing filter. For example, a 500 cm^2 Pall 1.2 μm Glass fiber/Versapor serum capsule filter (Pall #12168), custom gamma-irradiated with tubing connectors attached (see Note 12).

2.6. Final Filling of Measles Virus Purified Product

1. For smaller fill volumes, internally threaded, sterile, cryogenic vials may be used as product vials, for example 1.0, 1.8, and 4.5 mL sizes.
2. For larger fill volumes, fully assembled, Nalgene PETG diagnostic bottles various sizes may be used as product vials.
3. Labels for cryogenic vials, suitable for storage at ultra-low temperatures (e.g., Laser Cryo-tags, USA Scientific).
4. Repeat pipettor with sterile pipette tips for example Rainin Distriman.
5. (Optional). Semi-automated Flexicon PF6 filling system with disposable flow path (see Note 13).

2.7. Determination of Measles Virus Infectious Titer: $TCID_{50}$ Method

1. 96-Well tissue culture plates.
2. Vero cells growing in DMEM supplemented with 5% FBS and penicillin and streptomycin.
3. Medium (5% FBS DMEM with penicillin and streptomycin).
4. Trypsin–EDTA.
5. Hemocytometer.
6. Inverted microscope.
7. Dulbecco's Phosphate Buffered Saline (D-PBS) without calcium and magnesium.
8. Sterile reagent reservoirs for use with multichannel pipettors.
9. Nunc deep well plates, 96-well format (Nunc Nalgene #278743).
10. Pipettors: a variety of multichannel and repeat pipettors. Pipette tips with filters are used for pipetting virus.

11. Measles virus specimen to be tested.
12. Negative control: medium (5% FBS DMEM with penicillin and streptomycin) is used as the negative control.
13. Positive control: a quantitative positive control material may be developed from a MV stock preparation similar to the types of specimens to be assayed (see Note 14).

3. Methods

The methods described for measles virus production in this section may be applied to various production scales. Production is described using ten-layer Cell Factories (CF10) (EasyFill™ format) for Vero cell expansion. Additions can be made to EasyFill™ CF10 by simple pouring, but tubing assemblies and pumping are preferred for removing liquid from these units. For larger production batches CF40 flasks can reduce the number of manipulations, but all additions and removals require tubing assemblies and pumping (see Note 3). Twenty four CF10 are easily used for production of 50–60 L of infected cell supernatant, collected as two consecutive daily harvests of ~25 L each. An example is given for purification of 25 L of treated supernatant containing measles virus. The measles virus is too large to filter through a final sterilizing filter therefore aseptic processing techniques using sterile parts and reagents must be used throughout this process. Standard procedures for safe handling of the virus are also followed (see Note 15).

3.1. Thawing and Expansion of Vero Cells

1. Vero cells are thawed and expanded in T flasks using serum-free medium. We use VP-SFM with 4 mM L-glutamine, without antibiotics throughout this protocol.
2. The Vero cells are expanded stepwise into multiple T500 flasks over the course of approximately 7–10 days.
3. Typically, a T500 flask of confluent Vero cells is expanded to a ten-layer Cell Factory (CF10). This results in one ten-layer culture vessel being seeded per flask of confluent Vero cells.
4. The ten-layer flasks are placed in 37°C incubators (at 5–8% CO_2) until cells are grown to approximately 80–100% confluency (typically takes 3–6 days).

3.2. Infection of Vero Cells in Cell Factories

1. The amount of infecting MV used for each CF-10 will depend on the desired MOI (multiplicity of infection) and the specific virus used (see Note 16).
2. The MV infecting virus suspension for each CF-10 is prepared by pouring 250 mL serum-free medium into a sterile, disposable, 250 mL bottle. The appropriate volume of infecting

Procedure For Daily Harvest From 24-Ten-Layer Cell Factories:

i. Connect sterile tubing assemblies to a 50 L bioprocess container (BPC) so that the tubing assembly reaches into the biosafety cabinet (BSC) when the BPC is placed outside of the BSC in a plastic tote.
ii. Transfer two Cell Factories to the biosafety cabinet. Attach the tubing assembly from the BPC to a luer or QDC connection on the medium fitting of the first CF10.
iii. Pump the contents of the Cell Factory into the BPC using a laboratory scale peristaltic pump at speeds not exceeding 300 mL/min.
iv. When the Cell Factory is empty, transfer the tubing assembly to the next Cell Factory to be processed. The plug/cap from the second Cell Factory can be used to close the medium removal fitting on the first Cell Factory.
v. While the contents of the second Cell Factory are being pumped out, pour 1 L of serum-free medium into the first (i.e., emptied) Cell Factory.
vi. Repeat this procedure until all the Cell Factories have been harvested and re-fed.
vii. The harvested cell culture supernatant is clarified and Benzonase® treated prior to further storage.

Fig. 2. Example of daily harvest procedure using Cell Factories.

measles virus is thawed, added to the bottle and mixed by gentle swirling. Do *not* vortex.

3. Prior to infection, spent culture medium is removed from the Cell Factory; then the infecting virus suspension is added (see Fig. 2 for description of removals from and additions to CF10, EasyFill™ format). For example, to infect one CF10 at an M.O.I. of 0.01 using a measles virus stock with a titer of 4.5×10^6 $TCID_{50}$/mL, add 2.2 mL of the virus stock to the 250 mL medium in the bottle. This is a critical step because it has to make sure that the cells stay covered by the medium and that requires the CF10s to be very flat (leveling if necessary can be done using paper towels).
4. The ten-layer flasks are placed in a 37°C CO_2 incubator (at 5–8% CO_2) for 2 h, manually rocking the flask to evenly distribute medium at least once to prevent the cell layer from drying out. Without removing the infecting solution, 1 L of fresh serum-free medium is added to each CF-10 resulting in a volume of 1.25 L per CF10.
5. The infected ten-layer culture vessels are placed in a 32°C CO_2 incubator (at 5–8% CO_2) for 3–6 days. The date of infection is designated as Day 0 (see Note 17).

3.3. Harvest of Measles Virus Product

The measles virus-containing cell culture supernatant is harvested from each multilayer flask daily, typically for two or three consecutive days. Replacing the harvested supernatant with fresh medium results in an increased final product yield. The post-infection days chosen to be harvest days should be determined from empirical observations during process development using that particular type of measles virus. For example, observations of different recombinant measles viruses indicated that Days 4, 5, and 6 were optimal for harvesting cell culture supernatant for some and Days 4 and 5 were optimal for others. Altering the MOI will also change the optimal harvest days (see Note 18).

1. The first harvest is typically recovered from the ten-layer flasks, approximately 72–96 h following infection. In general, harvests are taken on Days 4, 5, and sometimes 6 following infection on Day 0.
2. Except on the final day of harvest, the infected cell supernatant removed from each ten-layer vessel is replaced with 1 L of fresh medium per vessel.
3. Each daily harvest results in an aseptic pool of the harvests from all the culture vessels processed on that day (e.g., the Day 4 harvest pool, the Day 5 harvest pool, etc.).
4. An example of a harvest procedure from 24 CF10 vessels is provided in Fig. 2.

3.4. Clarification and Benzonase® Treatment of the MV Product

Following harvest, the infected cell culture supernatant is clarified by filtration through a 3-μm filter to remove intact Vero cells (see Note 19). Then the clarified supernatant is treated with Benzonase® (an endonuclease) to degrade any Vero cell DNA present. Following Benzonase® treatment, the treated supernatant is stored at 4°C until further purification. (Data have shown that the treated supernatant may be stored at 4°C for at least 7–10 days without measurable loss of virus titer.)

3.4.1. Clarification

1. The infected cell culture supernatant is pumped through the 3-μm filter into an empty BPC using a lab-scale peristaltic pump at pumping speeds not exceeding 600 mL/min. If pressure is monitored, sustained feed pressures should not exceed 5 psig.
2. The total volume that can be filtered through a single 3-μm filter will depend on the surface area of the filter and the amount of whole cells and cell debris in the cell culture supernatant. In our process, a 1,500 cm^2 Pall 3-μm Versapor capsule filter is used for 10–12 L of infected cell culture supernatant. Therefore, for large harvests, it may be necessary to aseptically replace the filter with a fresh filter assembly part way through the filtration.

3.4.2. Benzonase® Treatment

1. Benzonase® is added to the clarified, infected cell culture supernatant so that the final concentration of Benzonase® is 10 U per mL (equivalent to 10,000 U Benzonase® per L).
2. To provide optimum conditions for Benzonase® activity, sterile magnesium chloride is added to the clarified cell culture supernatant to produce a final concentration of 2 mM magnesium chloride (see Note 20).
3. The BPC should be *gently* rocked by hand to mix the contents and then incubated at room temperature for 1 h.

4. Following incubation, the BPC containing the clarified, Benzonase®-treated MV bulk is refrigerated at 4°C for at least 24 h and up to 10 days prior to further purification.

3.5. Purification of Measles Virus Using a TFF System

Purification of infectious measles virus by TFF requires a design that will not disrupt the viral envelope. Hollow fiber tangential flow filtration provides conditions under which MV is gently separated from non-viral components while maintaining infectivity (Fig. 3) (see Note 21). Our TFF process for MV purification differs markedly from a TFF process designed for protein

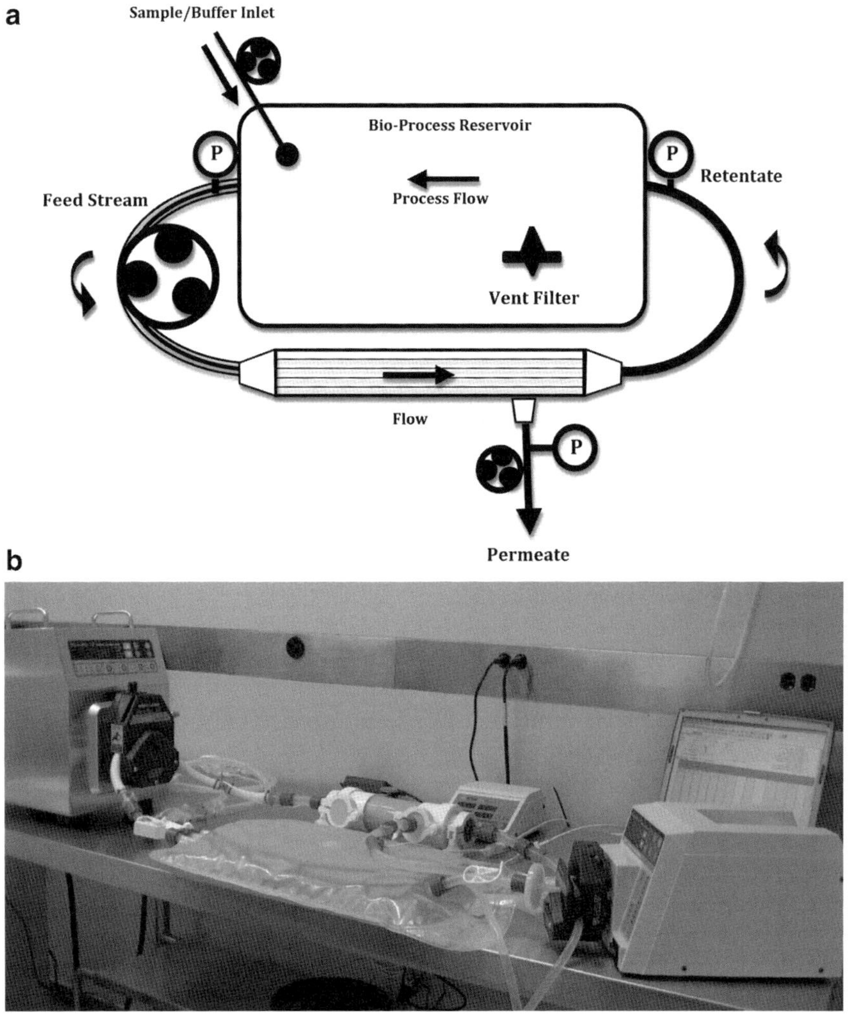

Fig. 3. Tangential flow filtration system. (**a**) Schematic overview of a TFF system. *Arrows* indicate direction of product flow. Pumps are represented by a *circle* with *three black discs*. Pressure transducers (P) are used to measure feed inlet pressure, permeate pressure and retentate pressure. The closed recirculation loop between the reservoir and the filter module is shown. Treated supernatant and/or diafiltration buffer is pumped in through the "sample/buffer inlet." Permeate leaves the system as shown under the control of the permeate pump. The vent filter is used to remove excess air from the TFF system. (**b**) Example of a TFF system. The horizontal orientation of the TFF module and reservoir are shown. A digital recirculation pump and lab scale permeate pump are shown; the feed pump is outside of the picture.

Examples of Spectrum TFF Modules with 20 cm Fiber Lengths.

Process Volume	Hold Up Volume	Spectrum Model	Membrane Matrix	Surface Area cm^2 / Pore Size	Fiber Bundles per Fiber ID	Pump Recirculation Rate for Shear Rate of 2000-4000 sec^{-1}	Permeate Flux
5 – 50 L	500 – 1000 mL	EZ-K30S-100-P01-1	Polysulfone	16000 / 50 nm	5000 / 0.5 mm	7250 – 14500 mL/min	280 mL/min/cm^2
< 2 L	50 – 100 mL	EZ-M10S-360-P02-1	Polysulfone	1050 / 50 nm	320 / 0.5 mm	464 – 928 mL/min	18 mL/min/cm^2

Fig. 4. Examples of operating characteristics for TFF systems using Spectrum modules.

purification. For example, our process uses low recirculation rates resulting in laminar feed flows, very low differential pressures, and low shear rates of 4,000 s^{-1} or less (Fig. 4). By definition, our process barely operates in a truly tangential mode. Use of digital pressure monitoring enables control of key parameters in our TFF process. Using our TFF system, clarified, Benzonase-treated supernatant is initially concentrated five- to tenfold to reduce the process volume. The partially concentrated product is then diafiltered against five volumes of Tris–HCl buffered sucrose to simultaneously purify the measles virus and perform buffer exchange. The final stage of TFF processing is to concentrate the treated supernatant ~50-fold compared to the starting volume before a final polishing step takes place.

3.5.1. Design of a TFF System for Purification of Measles Virus

Criteria to consider:

1. Short, high surface area TFF modules are preferred (see Fig. 4).
2. Horizontal orientation of the module and reservoir reduces pressure drops.
3. Flow pathway fittings are sized according to the TFF module surface area and to minimize turbulence. Fitting selections must be compatible with TFF module.
4. System requires a closed recirculation loop with capacity to eliminate air bubbles, for example through a vent filter (see Note 10; Fig. 3a).
5. Modular TFF system enables easy replacement of filtration modules.
6. Process control may be provided by digital pressure transducers and monitoring equipment (see Note 11).
7. The hollow fiber material and pore size are chosen according to manufacturing goals (see Note 22).

3.5.2. TFF Assembly and Set-Up

1. The TFF system is assembled from individual, modular TFF processing system components, or purchased as custom-manufactured, presterilized systems (Spectrum Labs, CA). Additional connections to the system including permeate,

sample, and/or diafiltration buffer containers are made in a biosafety cabinet before processing begins.

2. Pressure monitoring is accomplished by use of a pressure-monitoring unit compatible with digital pressure transducers placed in the system (see Note 11). Pressure data from sensors and digital peristaltic pumps may be collected and stored by TFF specific software for review and future reference.
3. Hold-up volume is the minimum volume a system requires in order to operate. Empirically determine the system hold-up gravimetrically or volumetrically prior to operation.
4. Prior to performing the initial concentration step, treated supernatant is pumped into the reservoir and cycled slowly until the permeate side of the membrane is filled. Zero the pressure monitor after filling the system with treated supernatant and before establishing operating conditions. With the permeate line fully restricted, the ultrafiltration membrane is wetted by gently recycling the treated supernatant through the system and removing all entrapped air from the system. Air bubbles trapped in the recirculation loop will damage the virus and should be avoided.
5. To set-up operating conditions, the recirculation pump rate is established with the permeate line closed. At the start of the concentration step, the permeate pump is used to control the permeate flux at the appropriate rate (see Fig. 4 and Subheading 3.5.3).

3.5.3. TFF: Initial Concentration by Ultrafiltration

In this step the treated supernatant is concentrated five- to tenfold compared to the starting volume. Ideal operating parameters include:

1. Sustained feed pressure of 4–5 psig or less.
2. Retentate pressure such that the transmembrane pressure drop is minimized and does not exceed 3 psig, assuming a permeate pressure of 0.
3. Maximum recirculation rate corresponding to a shear rate of 4,000 s^{-1} or less. Typical shear rates are on the order of 2,000 s^{-1} or less (refer to manufacturers recommendations).
4. Control of permeate flux by placing a pump on the permeate line *is required* when using TFF modules with membrane pore sizes of 50 nm. The empirically determined, maximum permeate flux for Spectrum modules with a 50 nm pore size and a 20 cm fiber length is approximately 0.02 mL/min/cm^2 of surface area. Control of permeate flux may be optional for membranes with pore sizes less than 50 nm.
5. During operation, factors such as temperature and viscosity of the process fluids will vary and it may be necessary to adjust

the retentate pressure via auxiliary clamp and/or the recirculation rate to maintain ideal operation conditions.

3.5.4. TFF: Purification by Diafiltration

1. Diafiltration may proceed directly from initial concentration without stopping the recirculation pump.
2. The partially concentrated, treated supernatant is diafiltered against at least five volumes of 5% sucrose, 50 mM Tris–HCl, pH 7.4, 2 mM $MgCl_2$ (referred to as sucrose buffer) to simultaneously purify the measles virus and perform buffer exchange.
3. The containers of sucrose buffer are connected aseptically to the reservoir.
4. Parameters for diafiltration are identical to those described for ultrafiltration above, except that the flow rate of the sucrose buffer is matched to the permeate flow by controlling buffer flow with a peristaltic pump to maintain constant reservoir volume.

3.5.5. TFF: Final Concentration

1. After diafiltration (buffer exchange) is completed, the partially concentrated and purified supernatant remaining in the reservoir is concentrated further. Final concentration may proceed directly from diafiltration without stopping the recirculation pump.
2. Concentration by ultrafiltration continues until the target retentate volume is reached. The minimum retentate volume is defined by the hold up volume of the TFF system (see Note 23). *Do not introduce bubbles into the feed stream* – this *rapidly* destroys virus infectivity, probably due to cavitation-like forces.
3. The retentate is recovered in a sterile container.

3.5.6. Final Polishing Step: 1.2-µm Filtration

1. After the tangential flow steps, the container of final retentate is aseptically connected to a filter assembly containing a 1.2-µm serum filter capsule (glass fiber/Versapor); the container may be left connected to the TFF system to provide venting.
2. The downstream side of the filter assembly is typically directly connected to an empty, sterile BPC or bottle fitted with a feed cap.
3. Using a peristaltic pump, the retentate is filtered through the 1.2-µm filter into the empty container at flow rates not exceeding 600 mL/min, and pressures less than 5 psig (if pressure is monitored).
4. The retentate that has been filtered through the 1.2-µm filter is the purified measles virus.
5. The purified measles virus may be stored at 4°C for up to 2 weeks prior to final vialing. A specific example of a purification process is included in Fig. 5.

Example Purification Scheme For 25 L Supernatant:

i. 25 L of treated cell culture supernatant in a 50 L BPC is aseptically connected to the reservoir of the Spectrum TFF system containing a Spectrum module with a polysulfone (PS) membrane, pore size of 50 nm, surface area of 1.6 m^2 and fiber ID of 0.5 mm.
ii. Prior to the initial (partial) concentration step, approximately 10 L of treated cell culture supernatant is pumped into the 10 L TFF reservoir.
iii. During the initial concentration step, treated cell culture supernatant is pumped into the TFF reservoir at a rate matching permeate withdrawal (approximately 240 mL/min). Permeate flux is controlled using a peristaltic pump on the permeate line.
iv. The recirculation rate corresponding to a shear of 2000 sec^{-1} is 7.25 L/min on the size 82 (1/2 inch ID) Pharmapure tubing of the feed inlet tubing. This is the maximum recirculation rate planned for this process.
v. The initial (partial) concentration step results in a retentate of approximately 4-6 L that is fully contained in the TFF reservoir. (A load cell may be used to estimate this volume.)
vi. The initial retentate is diafiltered against at least 25 L of sucrose buffer.
vii. The diafiltered retentate is then concentrated to a final volume of approximately 0.5-1.0 L.
viii. The final retentate is collected in a 5 L BPC connected to a port on the TFF reservoir via a tubing assembly.
ix. The BPC of final retentate is typically left connected to the TFF system so that the vent filter on the TFF system may be used during the polishing filtration step.
x. The final retentate is filtered through a 500 cm^2, 1.2-micron serum filter capsule (glass fiber/Versapor).

Fig. 5. Example of purification scheme for 25 L of treated supernatant.

3.6. Final Filling of Purified Measles Virus

1. The purified measles virus is vialed in a biosafety cabinet and stored at <−65°C.
2. Vialing may be performed manually or using a semi-automated Flexicon filling system (see Note 13).

3.7. Determination of Infectious Titer: $TCID_{50}$ Method

A critical parameter of a measles virus (MV) stock is the titer of infectious virus, i.e., the number of infectious units per unit volume. An infectious unit is defined as the smallest amount of virus capable of producing a *detectable* biological effect in an assay. Replication of laboratory-adapted and vaccine MV strains in Vero cells (African Green Monkey Kidney) produce a cytopathic effect, syncytia formation, which can be used to quantify the MV titer. The infectivity assay for determination of MV titers described here is a quantitative assay based on detecting the presence or absence of MV-induced syncytia in cell cultures infected with serial dilutions of the MV test sample. Syncytia in the cell cultures are detected by light microscopy. A 1/5 serial dilution series of each MV sample and a positive quantitative control is assayed in triplicate. The dilution of virus required to infect 50% of a given batch of cell cultures is defined as the *tissue culture infective dose* 50 ($TCID_{50}$). In this procedure, the $TCID_{50}$ is calculated using the Spearman–Karber method (6). The MV infectious titer is reported as $TCID_{50}$ U/mL (see Note 24).

In theory, the minimum virus concentration that can be measured using this assay is 45 $TCID_{50}$ U/mL (see Note 25); however, in practice this assay has not accurately measured the titer of virus suspensions that were expected to be 200 $TCID_{50}$ U/mL. In our laboratory, this assay appears to give accurate determinations down to about 2×10^3 $TCID_{50}$ U/mL for measles virus. In this assay, Nunc deep-well plates are used to create a dilution array for the TCID50 assay. Each sample is diluted serially from left to right across the plate in a single row. Then the contents of that single row are transferred to multiple 96-well plates containing Vero cells. Although called 2-mL plates, the listed maximum volume of each well is only 1.9 mL and care must be taken when diluting samples so that the wells do not overflow.

3.7.1. Set-Up of Vero Cells (24 h Prior to Infection)

1. Exponentially growing Vero cells are trypsinized, resuspended in 5% FBS DMEM with Penicillin and Streptomycin and counted using a hemocytometer.
2. Dilute the cell suspension to 1.4×10^5 cells/mL with medium and plate 50 μL/well to 96-well plates (effectively 7,000 cells/well).
3. The seeded plates are incubated overnight at 37°C, 5% CO_2.
4. The cells are typically 60–80% confluent the following day and ready for infection.

3.7.2. Dilution of Virus and Infection of Cells (Day 0 of Infection)

1. The samples and control for $TCID_{50}$ measurement are kept on ice or refrigerated at 4°C until use. Culture medium, 5% FBS DMEM with Penicillin and Streptomycin, is used as a negative control for each assay.
2. A dilution series is prepared for each sample in the Nunc deep-well plates. Typically, the initial dilution is a 1 in 100 dilution followed by nine, serial fivefold dilutions. This creates a series of dilutions, ranging from 10^{-2} to $10^{-8.3}$. An example dilution plate for six samples and two controls using this scheme is shown in Fig. 6. This scheme may be extended or reduced according to the expected titer of the sample. Follow the instructions below to generate the dilution series:
   (a) The volumes of 5% FBS DMEM with Penicillin and Streptomycin, as given in Fig. 6, are added to the appropriate wells in a deep-well plate using an appropriate multichannel or repeat pipettor.
   (b) The initial dilution (10^{-2}) for each sample is made by adding 10 μL of the given sample or control into 990 μL of 5% FBS DMEM using a P10 or P20 pipettor.

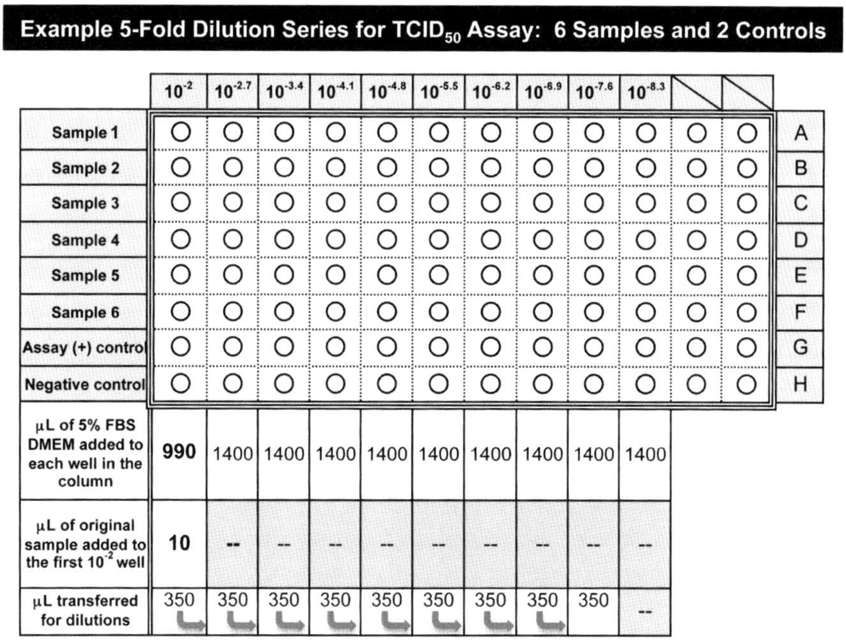
Example 5-Fold Dilution Series for TCID50 Assay: 6 Samples and 2 Controls
10-2 10-2.7 10-3.4 10-4.1 10-4.8 10-5.5 10-6.2 10-6.9 10-7.6 10-8.3
Sample 1
Sample 2
Sample 3
Sample 4
Sample 5
Sample 6
Assay (+) control
Negative control
A B C D E F G H
µL of 5% FBS DMEM added to each well in the column
990 1400 1400 1400 1400 1400 1400 1400 1400 1400
µL of original sample added to the first 10-2 well
10 -- -- -- -- -- -- -- -- --
µL transferred for dilutions
350 350 350 350 350 350 350 350 350 --

Fig. 6. $TCID_{50}$ sample dilution.

(c) A multi-channel pipette is used to complete the dilution scheme for all samples in the deep-well plate simultaneously as follows:

i. The Rainin Pipet-Lite manual multi-channel pipette (100–1,200 μL dispensing range) is used to mix the samples thoroughly and to transfer 350 μL to the next set of wells for the dilution series; then the tips are discarded.

ii. New tips are installed on the pipette and are used to mix the newly made dilutions and then make the transfer of 350 μL to the next set of wells for the dilution series.

iii. This dilution procedure is repeated until the dilution scheme is complete.

3. Label three (3) 96-well plates containing cultured Vero cells for each sample and the positive control.

4. Label one (1) 96-well plate containing cultured Vero cells for the negative control.

5. In this transfer step, the operator repeatedly pipettes from the dilution series, transferring the dilution series to the plated cells.

6. Using an electronic digital pipette with multi-channel end (capable of repetitive dispensing of 50 μL) and appropriate

a Example Plating Scheme For TCID$_{50}$ Assay:

	$10^{-3.4}$	$10^{-4.1}$	$10^{-4.8}$	$10^{-5.5}$	$10^{-6.2}$	$10^{-6.9}$	$10^{-7.6}$	$10^{-8.3}$				
A	O	O	O	O	O	O	O	O	O	O	O	O
B	O	O	O	O	O	O	O	O	O	O	O	O
C	O	O	O	O	O	O	O	O	O	O	O	O
D	O	O	O	O	O	O	O	O	O	O	O	O
E	O	O	O	O	O	O	O	O	O	O	O	O
F	O	O	O	O	O	O	O	O	O	O	O	O
G	O	O	O	O	O	O	O	O	O	O	O	O
H	O	O	O	O	O	O	O	O	O	O	O	O

b Example of a Scored TCID$_{50}$ Assay Plate:

	$10^{-3.4}$	$10^{-4.1}$	$10^{-4.8}$	$10^{-5.5}$	$10^{-6.2}$	$10^{-6.9}$	$10^{-7.6}$	$10^{-8.3}$				
A	+	+	+	+	–	–	–	–				
B	+	+	+	+	–	–	–	–				
C	+	+	+	+	–	–	–	–				
D	+	+	+	+	–	–	–	–				
E	+	+	+	–	–	–	–	–				
F	+	+	+	–	+	–	–	–				
G	+	+	+	+	+	–	–	–				
H	+	+	+	+	–	–	–	–				

+ = CPE observed **– = No CPE observed**

c Expression of pi for the Example Scored TCID$_{50}$ Assay Plate:

Dilution	$10^{-3.4}$	$10^{-4.1}$	$10^{-4.8}$	$10^{-5.5}$	$10^{-6.2}$	$10^{-6.9}$	$10^{-7.6}$	$10^{-8.3}$
pi: +/total wells	8/8	8/8	8/8	6/8	2/8	0/8	0/8	0/8

NOTE: Values of **pi** to be used to calculate **s** in this example are indicated by bold outline.

Fig. 7. TCID$_{50}$ assay example.

(300 or 1,000 μL) tips, transfer 50 μL of the $10^{-3.4}$ through $10^{-8.3}$ dilutions to the corresponding wells in each row (A through H) of a labeled 96-well plate.

7. Transfer dilutions to all *three* plates labeled for each sample and the positive control; transfer dilutions for the negative control to the *one* plate labeled "negative control." See Fig. 7a for an example of the plating layout for the dilutions. Typically, the $10^{-3.4}$ to $10^{-8.3}$ dilutions are plated, leaving the $10^{-2.0}$ and $10^{-2.7}$ dilutions behind.
8. Repeat step 5 for each sample and control using a fresh set of pipette tips for each sample.
9. The infected plates are placed in the incubator and incubated at 37°C until Day 6 post-infection.

3.7.3. Reading the Assay

The plates are examined for syncytia on Day 6 post-infection (the day of infection is Day 0). Each well is examined for the presence

(+) or absence (−) of syncytia using light microscopy and scored on a paper grid representing a 96-well plate array. An example of a scored plate is shown in Fig. 7b.

3.7.4. Calculations

1. For any sample, each 96-well plate represents a single replicate.
2. Calculate the virus titer for each replicate in $TCID_{50}$ units. The equation for the $TCID_{50}$ calculation is based on the Spearman–Karber method (6):

$$\mathrm{Log}_{10}\left(\frac{\mathrm{TCID}_{50}}{\mathrm{mL}}\right) = L + d(s-0.5) + \log_{10}\left(\frac{1}{v}\right)$$

L = negative $\log_{10}$ of the highest dilution (least concentrated) tested in which all wells are positive for CPE (syncytia), $d = \log_{10}$ of dilution factor (e.g., for a 1/5 dilution series, $d = \log_{10} 5 = 0.7$ and for a 1/10 dilution series, $d = \log_{10} 10 = 1$), pi = calculated proportion of positive wells for a given dilution (i.e., number of positive wells/total number of wells for the given dilution), s = sum of the individual proportion (pi) of the highest dilution (least concentrated) for which all wells are positive (i.e., pi = 1) and the pi values for all higher dilutions that contain positive wells, v = volume of viral dilution (mL/well).

Example calculation from the scoring illustrated in Fig. 7b:

(a) First, express pi for each dilution on the plate (no. of positive wells/total no. of wells) see Fig. 7c for example.

(b) Then, determine the values for all variables in the equation:

$$L = 4.8,\ d = \log_{10} 5 = 0.7,\ s = 8/8 + 6/8 + 2/8 = 2,$$
$$v = 0.05\ \mathrm{mL},\ 1/v = 1\ \mathrm{mL}/0.05\ \mathrm{mL} = 20$$

(c) Substitute values into equation and solve for $TCID_{50}$ result.

$$\mathrm{Log}_{10}\left(\frac{\mathrm{TCID}_{50}}{\mathrm{mL}}\right) = L + d(s-0.5) + \log_{10}\left(\frac{1}{v}\right)$$

$$\mathrm{Log}_{10}\left(\frac{\mathrm{TCID}_{50}}{\mathrm{mL}}\right) = 4.8 + 0.7(2-0.5) + \log_{10}(20) = 7.15$$

$$\mathrm{TCID}_{50}/\mathrm{mL} = 10^{7.15}/\mathrm{mL}$$

$$\mathrm{Titer} = \mathrm{TCID}_{50}/\mathrm{mL} = 1.4 \times 10^{7}/\mathrm{mL}$$

3. Calculate final $TCID_{50}$ result for each sample and the positive control. Average the results of the three replicates to give the final result.

3.7.5. Evaluation of Results

1. *Negative control.* All wells in the assay plate for the negative control must be negative for syncytia formation. If any well in the negative control plate is positive for syncytia, the entire assay is invalid.

2. *Positive control.* From the final result for the positive control (the average of the three replicate plates), determine the $\log_{10}$ of the control value. The transformed value is expected to be within the ±2 SD range established for the current quantitative measles virus control material (see Note 14).
3. *Unknown sample(s).* Result is usually the average of at least three replicates in units of $TCID_{50}$ U/mL.

4. Notes

1. This protocol was developed using Magenta WHO Vero cells originally sourced from BioReliance. These cells adapt readily to growth in VP-SFM serum-free medium (Gibco/Invitrogen) supplemented with 4 mM l-Glutamine. Other commercially available or proprietary serum-free media may be more suitable for different sources of Vero cells. We avoid supplementing the serum-free media with antibiotics.
2. TrypLE Select (Gibco/Invitrogen) is used instead of trypsin to avoid the use of animal sourced reagents as we develop our process for clinical applications. For this application, dilution of the residual TrypLE Select with VP-SFM serum-free medium during culture expansion is sufficient to neutralize the reagent without using any other inhibitors.
3. Cell Factories are available from Nunc in a variety of sizes and formats. For production volumes of a few liters, ten-layer Cell Factories (CF10) are appropriate: they are easy to handle, can be accessed by simple pouring operations, fit in small incubators and cell growth on one layer can be observed using an inverted microscope with the top objective removed. For larger production volumes, using 40-layer Cell Factories (CF40) minimizes the number of operating steps and can facilitate creating a semi-closed production system. However, CF40 have some disadvantages: they are cumbersome to handle, require large incubators, access by pouring is not feasible, and they cannot be viewed using a standard lab microscope.
4. Nunc have increased the fittings available for accessing and venting the Cell Factories. Sterile fittings for venting and enabling additions to and removals from the Cell Factories can be purchased from Nunc or the user can customize and prepare their own fittings. Improvements have also been made in the strength of the Cell Factories. However, we still use caution when pumping volume out of the Cell Factories to avoid creating vacuums and stressing the welds.
5. The measles virus used for infection should be of the lowest possible passage number following three rounds of plaque

purification and amplification. In our hands, both crude measles virus preparations from unpurified cell supernatant or lysates and purified measles virus in the presence of 5% sucrose appear to be indefinitely stable when stored at <–65°C. However, typically 50% of the virus infectivity is lost on each freeze/thaw cycle.

6. Tubing assemblies may be prepared using platinum-cured silicone tubing (Masterflex, Saint Gobain and others) and luer and QDC style fittings secured with cable ties. These tubing assemblies may be prepared in house and sterilized by autoclaving. Alternately, many vendors will prepare custom tubing assemblies and sterilize them by gamma irradiation (Hyclone, Nunc Nalgene and many others). Care must be taken to ensure that all parts are compatible with the sterilization method chosen.
7. Feed caps are available from Biovest International as "Cap assemblies." Sometimes known as "Endo Caps" or "sipper caps," these fittings are designed for aseptic liquid transfer between bottles and other vessels by means of sterile tubing and luer connections. They can be fitted with a vent filter and sterilized by autoclaving. The 1 L sizes are compatible with sterile, disposable, 1 L bottles (Nunc Nalgene #455-1000). The 2 L sizes are compatible with the sterile, disposable 2-L Erlenmeyer flasks (Nunc Nalgene #4112-2000). The feed caps are compatible with any bottle with a 45-mm media bottle-style thread. Typically we access the feed caps using a 48″ length of Size 25 Masterflex platinum-cured silicone tubing with a male slip luer on one end and a male luer lock on the other. Using feed caps requires some experience and for critical applications we use custom prepared, gamma-irradiated bottle/cap assemblies from Nunc Nalgene, which consist of a 2-L bottle with a customized "Top Works" fitting that contains a 50-mm vent filter and either a female luer or QDC fitting.
8. In our Vero process, the 3-µm Versapor filters (Pall) remove intact Vero cells without removing measles virus from the infected cell culture supernatant. However, they are not easily scalable. For larger scale clarification, we prefer the 3-µm, Profile Star polypropylene prefilter capsules from Pall's Kleenpak Nova product line. These filters are readily scalable and can either be sterilized by autoclaving, purchased sterile, or custom gamma-irradiated with tubing connectors attached (Pall). This same filter material is available for small-scale evaluation in the 90 cm^2, Mini-Profile Star capsule series (Pall). Due to its pleomorphic nature, measles virus behaves somewhat unpredictably as a >1µm particle. We cannot overemphasize the importance of small-scale evaluation of any filter membrane intended for processing measles virus.

9. Caution: Some peristaltic, high-performance pump heads have too vigorous a pumping action and completely eliminate measles virus infectivity when used as a recirculation pump for a TFF system.
10. Tubing selection for pumping segments is based on material durability. Pharmapure (Saint Gobain) is a common choice for the segment used for the recirculation pump. We avoid using gamma-irradiated thermoplastics as we have observed high spallation (degradation of the internal tubing wall adding fragments to the process stream) of these materials during pumping. Non-pumping segments may be silicone or other high quality tubing.
11. Multiple digital pressure transducers, sources and types may be used as long as they measure low-pressure levels. They may be integrated into tube sets and gamma irradiated (e.g. PendoTECH and Spectrum Labs transducers) or sterile transducers (e.g., Hospira style blood pressure transducers) may be inserted into the assembled system.
12. We use the 500 cm^2, 1.2-μm Glass Fiber/Versapor filters as the final polishing filter in our Vero-based measles virus process. This filtration step has the advantage of binding residual DNA, but reduces the final titer by an average of ~50%. Based on data from measles production in a different host cell we find that use of the 1.5-μm, Profile Star polypropylene prefilter capsules (Pall; see Note 8) results in better viral recoveries in the final polishing step.
13. For vialing larger batches of purified measles virus, we use the Flexicon PF6 semi-automated filling system (Watson-Marlow Flexicon). This system is simple to operate, portable, fits in a biosafety cabinet and can be used with a completely disposable, sterile flow path.
14. To develop a quantitative MV control material for the $TCID_{50}$ assay, we calculate the mean result and control ranges from statistical analysis of at least 30 repeated assays of the control material.
15. All procedural steps manipulating the Vero cells and measles virus in an open container should be performed in a biological safety cabinet. All solid waste that has come into contact with cells or virus should be decontaminated by steam sterilization. Liquid waste that has come into contact with cells or virus should be decontaminated either by steam sterilization or by treatment with 10% bleach for at least 1 h.
16. The estimated number of cells in a confluent CF10 is $1–2 \times 10^9$, thus infecting with an MOI of 0.1 would use 1×10^8 $TCID_{50}$ U of MV per ten-layer vessel and infecting with an MOI of 0.01 would use 1×10^7 $TCID_{50}$ U of MV per ten-layer vessel.

17. Lowering the incubation temperature after infection has been shown to significantly increase the yield of measles virus (7). In our process, lowering the temperature slows the infection, enabling multiple harvests and increasing the final yield of measles virus.
18. Criteria to determine which harvests will be processed for production are based on empirical observations of the % of CPE (i.e., syncytia), the confluence and appearance of the infected cell monolayer, and the % of cells that have detached from the culture surface.
19. For very small batches, clarification may be performed using low speed centrifugation.
20. For example, the Benzonase® and $MgCl_2$ may be placed in a small volume of medium in a 1,000-mL bottle. Do not place Benzonase® directly into 1 M $MgCl_2$ (ratio of medium to $MgCl_2$ solution should be at least 10:1). The Benzonase®/ $MgCl_2$/medium solution may be added to a BPC of clarified cell culture supernatant by fitting the 1,000-mL bottle with a 1-L feed cap and pumping the contents into the BPC via a sterile tubing assembly. Other methods of addition include using a syringe or a 2 L bottle/cap assembly.
21. Several publications by manufacturers discuss the basic concepts of TFF that are beyond the scope of this chapter. The ABCs of Filtration and Bioprocessing For The Third Millennium. Spectrum Labs Publication Number (PN) 420-11345-000 Rev.00. Millipore Technical PN TB032. Pall Life Sciences: Introduction to Tangential Flow Filtration for Laboratory and Process Development Applications PN33213 PN33289. GE Healthcare Hollow Fiber Membrane Separations Operating Handbook PN 18-1165-30 AB.
22. Either polysulfone (PS) or polyethersulfone (PES) membrane fibers materials with 50 nm pore sizes are easily wetted by recirculating the treated supernatant through the system for 10–15 min prior to concentration. Membranes with smaller pore sizes may need wetting prior to use, consult the manufacturer for details.
23. Typically we concentrate the final retentate ~25–50-fold compared to the starting material. However, it is possible to transfer the final retentate to a second, smaller TFF system and concentrate further. In our hands, concentration of the retentate to approximately 1/100 the starting volume has been shown to be proportional to increased virus concentration in the final retentate.
24. In our laboratory we find the $TCID_{50}$ assay more convenient to perform than a traditional plaque assay. For measles virus we have compared the infectious titer measured by the

$TCID_{50}$ assay to that determined by a plaque assay and find that they correspond with a 1:1 ratio (TCID50 U/mL: plaque forming units/mL). Such a comparison should be determined empirically on a virus-by-virus basis.

25. Calculation of theoretical minimum virus concentration that can be measured using this assay is 45 $TCID_{50}$ U/mL. This should be achievable when using a dilution scheme beginning with the undiluted virus followed by fivefold serial dilutions to $10^{-4.9}$ as the least concentrated dilution. A result where all wells plated with undiluted virus show CPE and no wells for other dilutions show CPE would be calculated as follows:

$$\mathrm{Log}_{10}\left(\frac{\mathrm{TCID}_{50}}{\mathrm{mL}}\right) = L + d(s - 0.5) + \log_{10}\left(\frac{1}{v}\right)$$

$$\mathrm{Log}_{10}\left(\frac{\mathrm{TCID}_{50}}{\mathrm{mL}}\right) = 0 + 0.7(1 - 0.5) + 1.3 = 1.65$$

$$\left(\frac{\mathrm{TCID}_{50}}{\mathrm{mL}}\right) = \left(\frac{10^{1.65}}{\mathrm{mL}}\right) = \left(\frac{45\ \mathrm{TCID}_{50}\mathrm{units}}{\mathrm{mL}}\right)$$

Acknowledgments

The authors wish to acknowledge the invaluable contributions of the other current and former members of the Viral Vector Production Laboratory: Guy Griesmann, Deborah Melder, Julie Sauer, Sharon Stephan, Cindy Whitcomb, and Troy Wegman. This work was supported by: the Mayo Foundation, the George M. Eisenberg Foundation for Charities, the Richard M. Schulze Family Foundation, Al and Mary Agnes McQuinn, the NIH NCI RAID Program and NIH grant CA15083.

References

1. Blechacz, B., and Russell, S. J. (2008) Measles virus as an oncolytic vector platform, *Curr. Gene Ther.* **8**, 162–175.
2. Msaouel, P., Dispenzieri, A., and Galanis, E. (2009) Clinical testing of engineered oncolytic measles virus strains in the treatment of cancer: an overview, *Curr. Opin. Mol. Ther.* **11**, 43–53.
3. Russell, S. J., and Peng, K. W. (2009) Measles virus for cancer therapy, *Curr. Top. Microbiol. Immunol.* **330**, 213–241.
4. Radecke, F., Spielhofer, P., Schneider, H., Kaelin, K., Huber, M., Dotsch, C., Christiansen, G., and Billeter, M. A. (1995) Rescue of measles viruses from cloned DNA., *EMBO* **14**, 5773–5784.
5. Griffin, D. E. (2007) Mealses Virus, in *Fields Virology* (Knipe, D. M., and Howley, P. M., Eds.) 5th ed., pp 1551–1585, Liipincott Williams & Wilkins, Philidelphia.
6. Hierholzer, J. C., and Killington, R. A. (1996) Virus isolation and Quantitation, in *Virological Methods Manual* (Mahy, B. W. J., and Kangro, H. O., Eds.), pp 25–46, Academic Press, San Diego.
7. Udem, S. A. (1984) Measles Virus: Conditions for the propagation and purification of infectious virus in high yield, *J. Virol. Meth.* **8**, 123–136.

Chapter 15

In Vivo Gene Delivery into hCD34$^+$ Cells in a Humanized Mouse Model

Cecilia Frecha, Floriane Fusil, François-Loïc Cosset, and Els Verhoeyen

Abstract

In vivo targeted gene delivery to hematopoietic stem cells (HSCs) would mean a big step forward in the field of gene therapy. This would imply that the risk of cell differentiation and loss of homing/engraftment is reduced, as there is no need for purification of the target cell. In vivo gene delivery also bypasses the issue that no precise markers that permit the isolation of a primitive hHSC exist up to now. Indeed, in vivo gene transfer could target all HSCs in their stem-cell niche, including those cells that are "missed" by the purification criteria. Moreover, for the majority of diseases, there is a requirement of a minimal number of gene-corrected cells to be reinfused to allow an efficient long-term engraftment. This requisite might become a limiting factor when treating children with inherited disorders, due to the low number of bone marrow (BM) CD34$^+$ HSCs that can actually be isolated. These problems could be overcome by using efficient in vivo HSC-specific lentiviral vectors (LVs). Additionally, vectors for in vivo HSC transduction must be specific for the target cell, to avoid vector spreading while enhancing transduction efficiency. Of importance, a major barrier in LV transduction of HSCs is that 75% of HSCs are residing in the G0 phase of the cell cycle and are not very permissive for classical VSV-G-LV transduction. Therefore, we engineered "early-activating-cytokine (SCF or/and TPO)" displaying LVs that allowed a slight and transient stimulation of hCD34$^+$ cells resulting in efficient lentiviral gene transfer while preserving the "stemness" of the targeted HSCs. The selective transduction of HSCs by these vectors was demonstrated by their capacity to promote selective transduction of CD34$^+$ cells in in vitro-derived, long-term culture-initiating cell colonies and long-term NOD/SCID repopulating cells. A second generation of these "early-acting-cytokine"-displaying lentiviral vectors has now been developed that is fit for targeted in vivo gene delivery to hCD34$^+$ cells. In the method presented here, we describe the in vivo gene delivery into hCD34$^+$ cells by intramarrow injection of these new vectors into humanized BALB/c *Rag2*^{null}/IL2*rgc*^{null} (BALB/c RAGA) mice.

Key words: In vivo gene delivery, Hematopoietic stem cells, Lentiviral vectors, RD114, TPO, SCF, Humanized mice

Otto-Wilhelm Merten and Mohamed Al-Rubeai (eds.), *Viral Vectors for Gene Therapy: Methods and Protocols*, Methods in Molecular Biology, vol. 737, DOI 10.1007/978-1-61779-095-9_15,

1. Introduction

1.1. The Human Hematopoietic Stem Cell

The maintenance of the integrity of an organism and its functions through life can be attributed to the role of adult stem cells. This role is derived from the unique combination of two characteristics that stem cells are endowed with: (1) multilineage potential and (2) self-renewal. A hallmark feature of adult stem cells is also their relative proliferative quiescence. Adult stem cells are diverse and specialized in giving rise to a certain lineage of cells. Taking into account the previous principles, the following characteristics must be included as criteria to define a hematopoietic stem cell (HSC) (1): (a) multipotency and (asymmetric) division to give rise to a differentiated hematopoietic progeny, (b) quiescence and self-renewal ability: 75% of HSC are in G0 and must occasionally enter the cell cycle to undergo self-renewing cell divisions (2, 3), (c) capacity to maintain an undifferentiated state in the stem-cell niches, which provides specific cell-fate signals, and (d) long-term repopulation and differentiation, and the ability to engraft and reconstitute the hematopoietic tissue upon transplantation.

Human HSCs can be isolated either from the BM, where they reside, or from peripheral blood upon cytokine-induced mobilization (MPB). An alternative HSC source is the umbilical cord blood (UCB). However, up to date, no unique or combination of surface markers has been found to identify the primitive human HSC. Nevertheless, in practice, the HSC population is identified and purified from the above-mentioned tissues by the surface marker CD34 (4, 5). The $CD34^+$ cell population is heterogeneous and contains the following: (a) *primitive self-renewing stem cells* (long term – LT-HSCs), (b) *more committed cells* that retain self-renewing ability (short term – ST-HSCs), and (c) *multipotent progenitors* (MPPs) that have lost their self-renewing ability and acquired differentiation and proliferation capacity (6). Additionally, a more primitive HSC population can be identified as $CD34^+/CD38^-$ and has the ability to sustain long-term hematopoiesis, as compared to the $CD34^+/CD38^+$ population (7). However, the phenotypic characterization of HSCs using these surface markers, even if commonly applied in both clinical practice and research, has some limitations, as evidenced by the identification of a $CD34^-$ population retaining in vivo repopulation capacity and giving rise to $CD34^+$ cells (8–11). More recently, the surface marker CD133 (12) as well as novel HSC markers such as the angiotensin-converting enzyme (CD143) (13) have been studied. Nevertheless, the most accepted way for the identification of HSCs is in terms of their function, that is, their ability to undergo long-term self-renewal and differentiation in in vitro and in vivo assays. Primitive stem cells must retain the ability to engraft an immunodeficient murine recipient while providing differentiated long-term progeny (so-called *severe combined immunodeficiency (SCID) repopulating cells* (SRCs) (14, 15). These SRCs are mainly contained in the $CD34^+/CD38^-$ subfraction of HSCs. *Long-term culture-initiating cells* (LTC-IC) have poorer self-renewal ability than SRCs and are identified upon 5-week in vitro culture on a stromal layer under minimal cytokine stimulation (16). Finally, *colony-forming cells* (CFCs) are already committed progenitors that give rise to myeloid and erythroid lineages and extensively proliferate during a 2-week culture (17).

1.2. Ex Vivo Hematopoietic Stem Cell Transduction

HSCs can be modified by lentiviral vectors and can be considered the ideal target for achieving long-term correction of diseases affecting the hematopoietic system (18). However, ex vivo manipulation is often an essential step to achieve efficient gene transfer. Vectors able to integrate in the host genome such as those derived from lentivirus are desirable when long-term expression in the HSC progeny is required. Lentiviral vectors (LV) have been shown to transduce $CD34^+/CD38^-$ cells efficiently upon a short ex vivo incubation, in the absence of fibronectin (FN) or cytokine stimulation while conserving their engraftment and long-term differentiation potential as well as long-term transgene expression (19, 20). However, the HSC population contains many cells with different degrees of restriction to LVs. Indeed, a subpopulation of HSCs remains refractory even when LVs are added at high doses (21). To increase transduction rates, the addition of early-acting cytokine cocktails containing IL-6, stem cell factor (SCF), thrombopoietin (TPO), and Flt3 ligand (Flt3L) is needed. Importantly, transduction of HSCs under these conditions generates a polyclonal pattern of vector integration without loss of NOD/SCID engraftment ability of transduced cells (20). Mechanistically, culture of $hCD34^+$ cells in the presence of these cytokines promotes cell-cycle progression from G0 to G1, a prerequisite for LV transduction of T-cells and HSCs (22). Additionally, this causes the downregulation of the proteasome, which acts as a restrictive factor for LV transduction, resulting in enhanced gene-transfer efficiency (19). LVs allow to reach higher levels of gene marking than oncoretroviral vectors with a shorter ex vivo manipulation resulting in long-term persistence of transduced cells in vivo, as assessed with primary and secondary transplants in immunodeficient mice (23, 24). However, ex vivo manipulation of HSCs may have negative effects on this population. First, stemness and function of HSCs might be affected by cell cycle entry induced by cytokine cocktails. It is generally accepted that most primitive cells reside in a quiescent state. In the human BM, 75% of primitive long-term repopulating stem cells are resting, thus residing in G0 and only rarely transit through early G1 (2, 3). Second, homing and extravasation ability and/or fate of the cells may also be affected by the cytokines used during the ex vivo culture, since circulating CD34$^+$ cells need BM cytokine activation to extravasate and home (25, 26). On the contrary, it has been previously suggested that cytokine stimulation (SCF, IL-6) might induce HSCs to express CXCR4, which attracts cells to the BM, thus increasing the migration and engraftment of stimulated cells (27). With a few exceptions, cytokine stimulation is mostly inducing cell cycle entry and proliferation, promoting differentiation rather than expansion of the HSC pool (28). These exceptions include TPO, SCF, and flk2/flt3L, which allow the cells to retain comparable engraftment abilities with respect to unmanipulated HSCs and multilineage differentiation in xenotransplantation models (29–32). However, thus far, the proper mixture of growth factors that retain stemness of HSCs while allowing high LV transduction has not yet been determined.

1.3. Advantages of In Vivo Lentiviral Gene Transfer

The delivery of a therapeutic or relevant gene directly into the organism is called in vivo gene delivery or in vivo gene transfer. By these means, the risk of cell differentiation or loss of homing/engraftment is avoided, as there is no need of extraction and purification of the target cell. In vivo gene delivery also bypasses the issue concerning the markers that permit the isolation of the true primitive HSCs. Indeed, in vivo gene transfer could target all HSCs, including those that are "missed" by the purification criteria. Moreover, for the majority of diseases, there is a requirement of a minimal number of gene-corrected cells to be reinfused to allow an efficient long-term engraftment. This requisite might become a limiting factor when treating children with inherited disorders, due to the low number of BM CD34$^+$ HSCs that can actually be isolated. This problem could be overcome by using efficient in vivo HSC-specific lentiviral vectors. On the contrary, the known limitation of lentiviral vectors to transduce cells residing in G0 cannot be avoided by current in vivo delivery vectors systems that use VSVG as pseudotyping glycoprotein. In this sense, the use of HSC-specific ligands (such as SCF and/or TPO) on the surface of these lentiviral vectors that can be further injected into the BM might increase targeting by slight and transient activation of the HSCs, allowing high local lentiviral gene transfer into these target cells. Indeed, extensive vector optimization is needed for in vivo delivery. The development of LVs allowing in vivo targeted gene delivery to HSCs is detailed in Subheading 1.4.

As it is explained before, exogenous cytokine stimulation of HSCs together with the need for high vector doses to ensure good transduction efficiency has associated risks and problems. First, insertional mutagenesis due to the presence of multiples copies of the vectors integrated into the genome of the target cell poses a risk (reviewed in ref. 33). In this respect, if the targeting capacity of the vector is precise, the vector dose can be markedly reduced. Second, vector spreading in the blood stream could cause loss of effectiveness by dilution of the vector and/or unknown side effects due to infection of nontarget cells. Thus, vectors for in vivo HSC transduction must be specific for the target cell, to avoid vector spreading while enhancing transduction efficiency. Another major challenge is the exposure to the host innate and specific immune system. Current VSVG-pseudotyped LVs are not suitable for a direct in vivo injection, since VSVG is rapidly inactivated by the human complement (34). In this respect, several pseudotyping partners are emerging in the field, such as the complement-resistant *gibbon ape leukemia virus* (GaLV) or the *endogenous feline leukemia virus* (RD114) (34) glycoproteins (gp). The development of targeted vectors suitable for in vivo gene delivery (reviewed in ref. 35) that preserves the undifferentiated/pluripotent character of the targeted HSCs is explained in detail in Subheading 1.4.

1.4. SCF/TPO Displaying Lentiviral Vectors to Target HSC In Vivo

1.4.1. Engineering of "Early-Acting-Cytokine"-Displaying Lentivectors for Gene Transfer into hCD34+ Cells (Fig. 1)

Two HSC-specific cytokines that have shown to promote HSC survival without inducing loss of homing and engraftment are SCF and TPO. Their cellular receptors are c-Kit and Mpl, respectively. Moreover, the relevance of SCF has been recently reassigned, based on emerging evidences of the role of c-Kit in regulating the maintenance of quiescent HSCs (36). To achieve an efficient functional presentation on the vector surface, the cytokines were fused to the N-terminus of the hemagglutinin influenza glycoprotein HA (TPOHA and SCFHA). However, these gps demonstrated next to a functional ligand–receptor binding a strongly reduced vector–cell fusion capacity. Thus, both cytokines needed to be displayed at the surface of LVs, together with a fusion partner gp, VSV-G, to allow for fusion of the vector with the target cell (Fig. 1).

Display of TPO, or both TPO and SCF, on the LV surface dramatically improved gene transfer into quiescent cord blood CD34+ cells, a population highly enriched in HSCs, by 55-fold or 100-fold, respectively, as compared to conventional lentiviral vectors (37).

Our data showed that after in vitro myeloid and lymphoid differentiation of the transduced hCD34+ cells, the level of transduced cells was consistently much higher for the TPO-, SCF-, and TPO/SCF-displaying LVs as compared to unmodified LVs in the presence of recombinant TPO and/or SCF. Most importantly, the selective transduction of HSCs by these vectors was demonstrated by their capacity to promote selective transduction of CD34+ cell in vitro-derived LTC-ICs and of long-term NOD/SCID repopulating cells (SRCs) (Fig. 2; (37)). Thus, these novel LVs allowed superior gene transfer of HSCs ex vivo as compared to conventional LVs in the presence of high concentration of recombinant cytokines used up to now. It is speculated that the superior performance of TPO-, SCF-, and TPO/SCF-displaying LVs might be due to increased specific activity of the cytokines

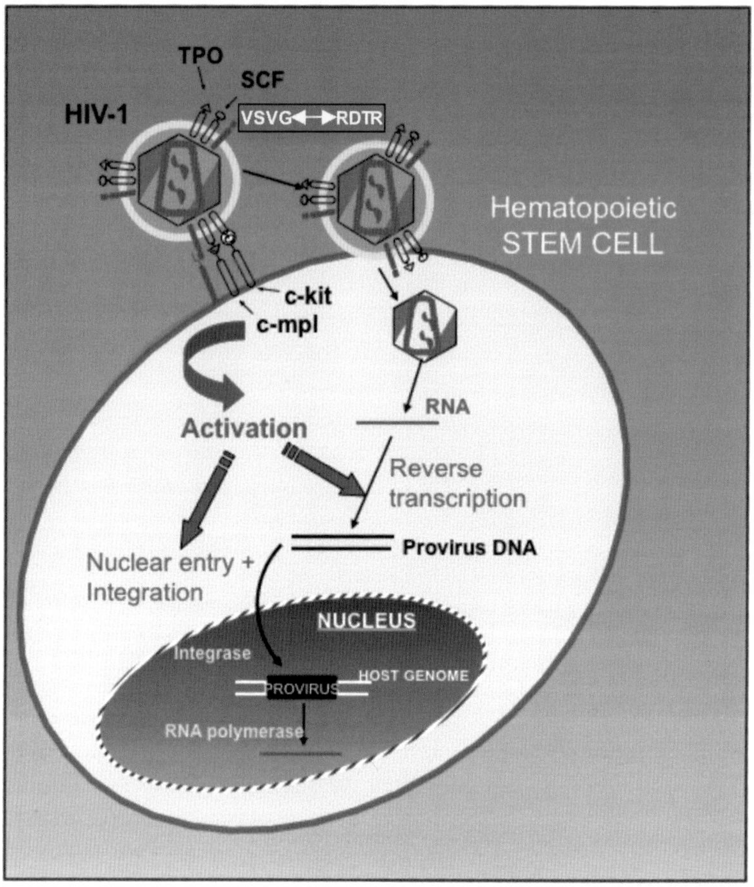

Fig. 1. ***Lentiviral vectors displaying stem cell factor (SCF) and thrombopoietin (TPO) by fusion to the envelope gp influenza HA.*** This allows for specific targeting of the vector particles to HSCs expressing c-Kit and c-Mpl, the receptors for SCF and TPO, respectively. VSV-G gp or RDTR gp is coexpressed on the vector surface to allow for efficient vector–cell fusion, since the chimeric cytokine displaying vectors showed a reduced vector–cell fusion capacity. After binding to the c-Kit or/and Mpl receptors, the cells get slightly activated, allowing all the steps of lentiviral transduction (reverse transcription of viral RNA into DNA, nuclear proviral DNA entry, and provirus integration into the host genome) to occur and resulting in productive transduction.

when presented on the viral surface as multivalent trimers or due to an increased targeting of HSCs. Moreover, the improved specificity of transduction of HSCs in the $CD34^+$ cell population allowed us to decrease the vector doses and thus reducing the genomic insertion site per cell without decreasing gene transfer efficacy, making these LVs safer vectors for gene therapy.

1.4.2. Upgraded LVs for In Vivo *Gene Delivery to HSCs*

In vivo targeted gene delivery to HSCs would mean a big step forward in the field of gene therapy as mentioned under Subheading 1.3. However, since the fusion glycoprotein VSV-G in VSV-G/TPO- and VSV-G/SCF- codisplaying lentiviral vectors is complement sensitive and its receptor is present on all tissues,

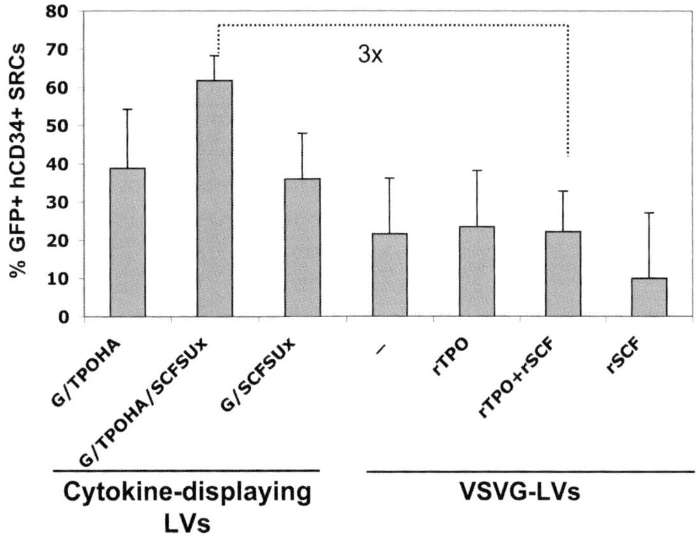

Fig. 2. ***Preferential transduction of NOD/SCID repopulating cells by TPO- and SCF- or TPO/SCF-displaying lentiviral vectors.*** NOD/SCID mice that received 2×10^5 CB CD34+ cells transduced with TPO- (G/TPOHA), SCF- (G/SCFSUx), TPO/SCF- (G/TPOHA/SCFSUx) or VSV-G pseudotyped vectors were analyzed for human engraftment in femur BM at 7 weeks posttransplantation. CB hCD34+ cells were transduced for 24 h at an MOI of 4 with TPO- (G/TPOHA), SCF- (G/SCFSUx), TPO/SCF- (G/TPOHA/SCFSUx), or VSV-G pseudotyped vectors in the absence (–) or presence of the counterpart cytokines in their soluble form. After repopulation into NOD/SCID mice, transduction levels of the subpopulation of human progenitor cells (hCD34+) present in the bone marrow (BM) were compared. SRCs = SCID repopulating cells.

it is unsuited for in vivo targeted gene delivery. Therefore, we exchanged it for another fusion partner, a mutant feline endogenous glycoprotein, RDTR (Fig. 1). RD114 is an attractive candidate as fusion partner for in vivo use because of multiple reasons: First, RD114 gp is resistant to the degradation by the human complement. Second, RD114-pseudotyped oncoretroviral vectors (MLV) are known to transduce CD34+ HSCs efficiently. Finally, high glycoprotein incorporation onto lentiviral vectors has been achieved by the exchange of the cytoplasmic tail of RD114 for the one of the MLV-gp (34). Of importance, RDTR single pseudotyped LVs allow only high CD34+ cell transduction in the presence of retronectin, which allows on one hand the binding of the cells via VLA-4 and VLA-5 receptors and on the other hand the envelope gp to be displayed on the vector. In this way, both vectors and cells are approached to allow efficient virus–cell binding and fusion. The resulting RDTR/SCF- and RDTR/SCF/TPO-displaying lentivectors were far more efficient in transducing hCD34+ cells (up to 40%) than RDTR vectors in the presence of rTPO and rSCF and in the absence of retronectin (<0.5%). Thus, of importance, these novel cytokine-displaying LVs are completely independent of retronectin. In addition, vector doses can be decreased with the concomitant decrease of the risk of insertional mutagenesis while maintaining high transduction efficiency. A selective transduction of $hCD34^+$ cells was obtained in an in vivo-like setting, namely, a complete cord blood sample, and these vectors transduced $hCD34^+$ cells with 100-fold selectivity as compared to T-cells. Together, these cytokine-displaying LVs are suited for in vivo use. In the method presented here, we describe the in vivo gene delivery into $hCD34^+$ cells by intramarrow injection of these new vectors into a humanized mouse model (see Subheading 1.5).

1.5. The Human Immune System (HIS) Mouse: A Model for In Vivo Testing of Lentiviral Vector Gene Delivery into HSCs

1.5.1. The Humanized Mice as Indispensable Model for In Vivo Studies

Complex biological mechanisms often require in vivo studies, and using mice as a model have already permitted many important research advances. However, studies on human immunology, immunopathology, or gene therapy and the direct translation of these results from rodents to humans often fails; indeed, "mice are not humans" (38). So it was critical to develop animal models that would allow in vivo studies on human cells, tissues, or organs to investigate fundamental and complex biological processes such as human immunity or hematopoiesis, which cannot be modeled in vitro or ex vivo.

Since the discovery of immunodeficient CB17-Scid mice in 1983 (39), humanized mice have been developed to try to overcome these constraints, and mouse–human chimeras are now an important research tool for the in vivo study of human cells, tissues, and biological systems.

Humanized mice can be defined by immunodeficient mice engrafted with functional human hematopoietic cells or tissues, but also include mice (either immunodeficient or immunocompetent) that transgenically express human genes (40–44). The humanized mice serve as in vivo human models for both physiological and pathological processes.

The development of mice humanized by xenotransplantation of HSCs or peripheral-blood mononuclear cells (PBMCs) provides a unique and efficient experimental system to study differentiation, function, and interaction of human blood cells or immune components in vivo that would not be possible otherwise. These HIS mice are able to develop and maintain functional hematopoietic cells.

1.5.2. Improving the HIS Mice

The discovery of the CB17-*scid* mice in 1983 (39) was the first major breakthrough in the ability of mice to be engrafted with human hematolymphoid cells. The $Prkdc^{scid}$ (DNA-dependent protein kinase) mutation results in the lack of T and B cells in mice. This discovery was soon followed by the first proof of principle that mice can be engrafted with PBMCs (45), fetal hematopoietic tissues (46), or HSCs (47) to study in vivo the human hematopoietic system. However, these first HIS mice models revealed rapidly their limitations: First, CB17-*Scid* mice maintain high levels of host natural killer (NK)-cell and other innate immune activity (48, 49), which limit the engraftment of the human hematopoietic compartment. Second, spontaneous generation of mouse T and B cells during aging (known as leakiness) (50) occurs and results de novo in their ability to support only very low level of human hematopoietic reconstitution. Third, *scid* mutation also results in defective DNA repair, and consequently, these mice are sensitive to mutagenesis (51).

Targeted deletions at the recombination-activating genes (*Rag*) 1 and 2 prevent adaptive immunity because of complete prevention of mature B- and T-cell development (52, 53) and do not cause leakiness or radiosensitivity, which offered incremental improvement in human cell xenotransplantation. However, residual NK cell activity remained (49).

The SCID mutation was then backcrossed in the NOD (non-obese diabetic) background, and this new model was called NOD/LtSz-*Prkdc*^{scid} (abbreviated as NOD-*scid*) (54). NOD-*scid* mice revealed reduced levels of NK cell activity, which is one of the principal hurdles for human cell engraftment and additional deficiencies in the innate immune system that allow higher levels of human PBMC (55) and HSC (56, 57) engraftment as compared to CB17-*scid* mice. These studies were the first to highlight the importance of the strain background (54, 55, 58–64).

For 10 years, the NOD-*scid* model has served as the standard reference for the studies of the human hematopoietic system, and many improvements have been introduced to decrease the host innate immune system (65, 66). However, the residual NK cells activity, short lifespan (37 weeks), and the development of thymic lymphomas (54, 67) are still insolvable limitations of this model. The last main breakthrough in the field of HIS mice was the generation of mice carrying homozygous targeted mutation at the interleukin 2-receptor common gamma chain (IL2*rgc*) locus (68–71). This molecule is an essential component of a number of cytokine receptors and is crucial for IL-2, IL-4, IL7, IL-9, IL-15, and IL-21 high-affinity ligand binding and signaling through these receptors. Absence of the IL2*rgc* leads to severe impairments in innate and adaptive immunity, resulting in severe defects in T- and B-cell development and function, and complete abrogation of NK-cell development (72, 73). IL2*rgc* deficiency causes X-linked SCID in humans (74). Several IL2*rgc*^{null} mouse strains have been developed and present some differences in terms of both the *IL2rgc*-targeted mutation (leading to a complete absence of IL2*rgc* or to a partial truncation at the intracytoplasmic domain) and the inbred strain background (69, 70, 72, 73, 75–77).

1.5.3. The Humanized Mice Model Today

Recently, NOD-*scid*/IL2*rgc*^{null} (68–70, 78, 79) and BALB/c *Rag2*^{null}/IL2*rgc*^{null} (71) (BALB/c RAGA) mice models have emerged, and engraftment of human HSCs and PBMCs in these immunodeficient strains of mice bearing the *Il2rg*-targeted mutations is consistently higher than in all previously described humanized mouse strains. In these strains, xenografts of human HSCs could repopulate a complete human immune system, including human thymocytes, peripheral mature B and T cells, myeloid cells, platelets, and RBCs (77).

Actually, results differ approximately in terms of repopulation efficiency, but until now, no study has directly compared these models (NOD-*scid*, NOD-*scid*/IL2*rgc*^{null}, and BALB/c RAGA) in the same conditions of xenograft, routes of injection, sources of human donor cells, age of recipient, leaving the optimal strain and conditions undetermined.

Indeed, depending on the recipient, several routes of injection have been reported. Conditioned newborn immune-deficient mice have also been repopulated after intraperitoneal (80, 81), intravenous (IV) (68), and, more recently, intrahepatic injections (71, 82). In the adult HIS model, the most commonly used route of human cell injection is the IV route via tail vein (70). In an attempt to overcome the BM homing requirement of human HSCs, direct injection of human HSCs was performed in a supportive microenvironmental stem cells "niche," which is essential for the development, maintenance, and differentiation of HSCs (adult BM (83, 84), newborn liver (71), or in utero (85)).

An equally important component for the development of HIS mice is the source of engrafting cells. Several subpopulations of HSCs (human CD34$^+$ (29, 71, 78), CD133$^+$ (86), CD34$^+$/CD133$^+$ (87), CD34$^+$/CD38$^-$ (68, 88), and CD34$^+$/CD38$^-$/Lin$^-$ (89) cells) have been used to repopulate immunodeficient mice. It is now well established that fetal HSCs have more potential than adult HSCs (29) and that UCB is a better source than BM to obtain HSCs with high engraftment levels (90). It is estimated that injection of at least 5×10^4 UCB CD34$^+$ cells leads to the detection of differentiated human cells in various organs (91). The hematopoietic potential of PBMCs has also been studied in HIS mice (55, 57); however, the best reconstitution has resulted from the introduction of human HSCs.

1.5.4. Focusing on the BALB/c Rag2^{null}/IL2rg^{null} (BALB/c RAGA) Mice

The BALB/c RAGA mice is yet one of the most promising models. Goldman and colleagues were the first to use the *Rag2*^{null} and IL2*rg*^{null} mutation combination in a mixed background, and already, engraftment level of human peripheral blood leukocytes and proliferation of B-lymphoblastoid cells were higher than those observed in NOD-*scid* mice (92). Afterward, H2^d-*Rag2*^{null}/IL2*rgc*^{null} (93), C57BL/6 J-*Rag2*^{null}/IL2*rgc*^{null}, NOD/*shi-Rag2*^{null}/IL2*rgc*^{null}, and the BALB/c-RAGA mice were reported (91). In 2004, fully functional HIS mice were generated by injection of UCB CD34$^+$ cells in newborns via the intrahepatic route, considering that liver acts as a major hematopoietic organ during fetal and neonatal periods (71). The BALB/c RAGA HIS model is efficiently engrafted with other sources of cells or routes of injection, while similar level of engraftment has been obtained with human fetal liver CD34+ after IP or intrahepatic injection (94). Contrary to NOD-*scid IL2rgc*^{null} mice (NOG (69) and NSG (70)), BALB/c-RAGA mice lack the SCID mutation and support higher doses of irradiation because of their X-ray resistance (generally 1 Gy vs. 3.5 Gy) (71, 80, 91). One other advantage of BALB/c-RAGA mice is their low incidence of thymoma as compared to NOD-*scid-IL2rgc*^{null} mice (71, 82).

Recently, several teams have used this functional model to study human disease and particularly infection by HIV (82, 95, 96).

BALB/c-RAGA and NOD-*scid*/IL2*rgc*^{null} mice are actually highly efficient xenograft recipients because of their high levels of immunosuppression, both for adaptive and innate immune systems. These mice are considered as the best model for HIS mice. As remaining limitations and improvements in these two models are obtained, the use of these HIS mice will become a powerful tool for fundamental and clinical research.

The method described here shows the evaluation of the new SCF/TPO-displaying LVs for in vivo targeted gene transfer into highly immature hematopoietic progenitors including HSCs in the humanized BALB/c-RAGA model. Humanization of BALB/c-RAGA mice gives an efficient BM repopulation by human CD34+ cells, which are the targets for in vivo gene transfer (Figs. 3 and 4).

2. Materials

2.1. Buffers and Solutions

1. Phosphate-buffered saline (PBS) without calcium and magnesium, without sodium bicarbonate, sterile (Invitrogen, France).
2. Trypsin–ethylenediaminetetraacetic acid (EDTA) 1× Hank's balanced salt solution without calcium and magnesium, sterile (Invitrogen, France).
3. Lymphoprep, sterile (Fresenius Kabi Norge, Norway).

2.2. Media

1. Fetal calf serum (FCS), sterile (Lonza, BioWhittaker, Belgium).
2. DMEM (Dulbecco's modified Eagle medium; Lonza, BioWhittaker, Belgium) with 0.11 g/l sodium pyridoxine and pyruvate. DMEM is supplemented with 10% FCS, 100 μg/l streptomycin, 100 U/ml penicillin (stored at 4°C).
3. RPMI medium (Invitrogen, France) is supplemented with 10% FCS, 100 μg/l streptomycin, 100 U/ml penicillin, rIL-3 (stored at 4°C; for rIL-3 supplementation see Note 5).
4. Serum-free CellGro medium (CellGenix, Germany).

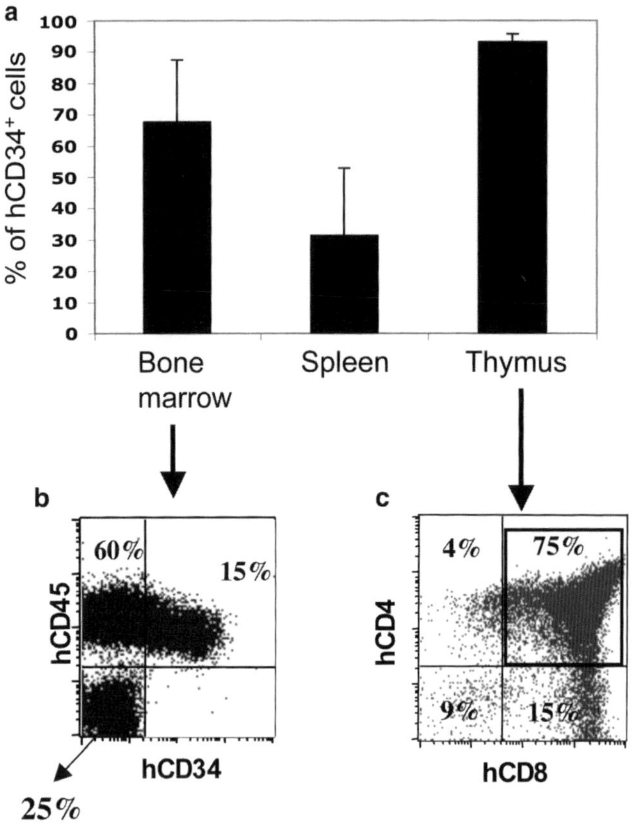

Fig. 3. ***Efficient reconstitution of Balb/c-RAGA mice with human CD34⁺ cells***. In (**a**) the total number of reconstituting human cells (hCD45⁺ cells) is given for the BM, spleen, and thymus. In (**b**), the percentage of early progenitors including HSCs (hCD34⁺ cells, *upper right quadrant*) in the total human cell population (hCD45⁺) in the BM is indicated. In (**c**), the different subpopulations (single positive cells: CD4⁺CD8⁻ and CD4⁻CD8⁺ cells, double positive cells: CD4⁺CD8⁺, and double negative cells: CD4⁻CD8⁻) of the thymus are shown.

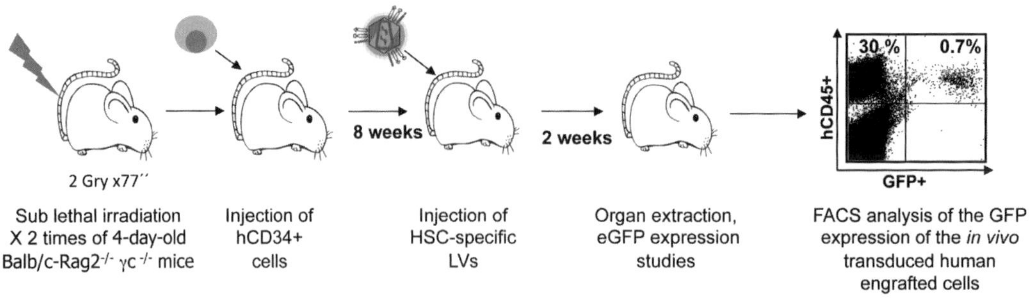

Fig. 4. Scheme of the general protocol for in vivo transduction of hCD34⁺ engrafted cells in Balb/c-RAGA mice. 2- to 4-day-old Balb/c-RAGA mice are submitted to two rounds of sublethal irradiation to allow a high engraftment. Human CD34⁺ cells are injected intrahepatically, and after 8–10 weeks the percentage of human engraftment is assessed by FACS analysis of hCD45⁺ cells in the mouse peripheral blood. HSC-targeted LVs expressing SCF and/or TPO in combination with RDTR as a fusion glycoprotein are injected intrafemurally. Two weeks later, transduction efficiency in the human compartment is measured by FACS. The plot shows the percentage of GFP⁺ cells vs. hCD45⁺ cells in the BM of a humanized mouse injected with the vectors.

2.3. Nucleic Acids

1. pHIVSIN-SFFV-GFP-WPRE (97): lentiviral vector DNA encoding for an HIV-1-derived self-inactivating vector (SIN) with the internal SFFV (spleen focus forming virus) promoter driving the expression of the reporter gene GFP.
2. Envelope glycoprotein expressing plasmids: fusion glycoprotein: RD114-TR glycoprotein (34) (see Note 1).
3. Activating and targeting glycoproteins for HSCs: (1) TPO-HA (TPO fused N-terminally to the influenza hemagglutinin transmembrane domain) and (2) SCF-HA (37) (SCF fused N-terminally to the influenza hemagglutinin transmembrane domain).
4. Virus structural protein (gag-pol) expressing plasmid (pCMV8.91) (37).

2.4. Cells and Tissue

1. 293T cells (ATCC CRL-11268).
2. Baf-3-c-Kit and Baf-3-Mpl cells (37).
3. Source of HSCs: cord blood CD34+ cells (see Note 2).

2.5. Animals

BALB/c-Rag2$^{-/-}$γc$^{-/-}$ immunodeficient mice (Dr. Mamoro Ito, CIEA, Kawasaki, Japan; Taconic, Japan). Animals are kept under specific pathogen-free conditions. Animal handling must be carried out under sterile conditions.

2.6. Special Equipment

1. VIVASPIN concentration columns (cut-off: 100,000 MW; Sartorius Stedim Biotech, Germany).
2. Magnetic separation device (Miltenyi Biotech, Germany).
3. Gamma irradiator.
4. Sterile dissection material: dissecting scissors, insulin syringe, 25 G and insulin needles (Dutcher, France), suture thread, surgical needle (World Precision Instruments, FL, USA).
5. Fluorescence-activated cell sorter (FACS; FACSCanto II, BD, France).

2.7. Additional Reagents

1. MACS CD34+ cell isolation kit (Miltenyi Biotech, Germany) (see Note 3).
2. 24-well cell tissue culture plates (Corning Inc., NY, USA).
3. 0.45-μm filter to filter the vector supernatants (Millipore, France).
4. Anesthetic solution: ketamine/xylazine; stock concentrations: Ketamine 50 (PANPHARMA, France) 50 mg/ml, Xylazine hydrochloride (Rompun®, Germany) 2%.
5. Analgesic drug: Doliprane® 2.4%.
6. Mouse monoclonal antibodies for identification of the degree of reconstitution with human differentiated cells: anti-hCD45-PERCP, anti-hCD34-APC, anti-hCD3-APC,

anti-hCD4-APC, anti-hCD8-PE-Cy7, anti-hCD19-APC, anti-hCD14-PE, anti-hCD13-APC, anti-hCD20-PE, anti-hCD41a-PE, anti-hCD56-APC, anti-hCD3-PE and corresponding PE, APC, and PERCP-conjugated mouse IgG controls (BD, Pharmingen Biosciences, France).

3. Methods

3.1. Production HSC-Targeted Lentiviral Vectors and Titration

1. Day 0: 2.5×10^6 293T cells are seeded the day before transfection in 10-cm plates in a final volume of 10 ml DMEM.
2. Day 1: Cotransfection of HIV packaging construct (8.6 μg) with the lentiviral gene transfer vector (8.6 μg) and two or three glycoproteins: (a) RD114-TR (7 μg) and (b) TPOHA or/and (c) SCFHA (2 μg/each) using the Clontech calcium-phosphate transfection system.
3. Day 2: 15 h after transfection, the medium is replaced with 6 ml fresh CellGro medium.
4. Day 3: 36 h after transfection, the vectors are harvested, filtered through a 0.45-μm pore-sized membrane and concentrated using a Vivaspin filter concentration system by overnight centrifugation at $3{,}000 \times g$, 4°C. Concentrated vectors are aliquoted and stored at –80°C for 2–3 months (see Note 4).

3.2. Titer Determination of GFP-Encoding LVs

1. Day 1: 293T cells are seeded in DMEM at a density of 2×10^5 cells per well in 6-well plates in a final volume of 2 ml.
2. Day 0: Serial dilutions of concentrated vector preparations are added to 293T cells and incubated O/N.
3. Day 1: The medium on the cells is replaced with 2 ml fresh DMEM and the cells are incubated for 72 h.
4. Day 4: The cells are trypsinized and transferred to FACS tubes. The percentage of green fluorescent protein (GFP)-positive cells is determined by FACS analysis.

3.3. Determination of Cytokine-Displaying LV Activity

1. Day 0: Baf-3 cells engineered to express c-Kit (SCF receptor) or Mpl (TPO receptor) are seeded in complete RPMI at a density of 1×10^5 cells per well in 24-well plates in a final volume of 1 ml (see Note 5).
2. Serial dilutions of concentrated vector preparations are added to the cells and incubated O/N on the day of cell seeding (Day 0).
3. Day 4: The cells are transferred to FACS tubes. The level of survival of Baf-3-cKit or Baf-3-Mpl cells is estimated by the percentage of living cells determined by FACS analysis as a

result of the activity of the cytokine-displaying lentiviral vector. The cells are briefly incubated with propidium iodide (PI) for 20 min, which colors dead cells, leaving living cells unstained. The dead cells are analyzed by FACS and revealed by excitation with 488 nm laser and emission in the red channel.

3.4. Analysis of Transduction and Titer

1. Infectious titers: They are provided as transducing units (TU)/ml and can be calculated by using the following formula: Titer = %inf × (N/100) × d/ml; where N is the number of cells at Day 0, d is the dilution factor of the viral supernatant, and %inf is the percentage of GFP-positive cells as determined by FACS analysis using dilutions of the viral supernatant that results in 5–10% of GFP-positive cells.
2. For FACS analysis, the cells are detached by incubation with Trypsin for 3–5 min at 37°C and resuspended in PBS/2%FCS in FACs tubes.
3. The cells are analyzed by FACS by excitation with 488 nm laser and detected in the GFP channel in a dot plot showing GFP on the y-axis and the forward size scatter (FSC) size marker on the x-axis.
4. Multiplicities of infection (MOI): the ratio between infectious particles and target cells that are required

3.5. hCD34+ Cell Isolation from Human Cord Blood

1. Dilute cord blood at 1:1 with PBS and gently overlay 35 ml diluted product on 15 ml Lymphoprep in a 50-ml tube.
2. Centrifuge the cells at 850 g for 30 min, 20°C without using brake and collect the layer containing the mononuclear cells.
3. Wash the collected mononuclear cell interface in PBS/2% FCS at 850 g, 20°C for 10 min and determine the cell number. Proceed to magnetic purification as in the steps below.
4. Resuspend the cells at $1–2 \times 10^8$/ml and add anti-hCD34+ microbeads according to the manufacturer's instructions, and incubate for 30 min while rocking.
5. Wash cells to remove the unbound antibody and resuspend in PBS/2% FCS.
6. Hydrate the MAC separation column with PBS/2% FCS and then pass labeled cells through a first column put on the MAC magnetic device; wash once with PBS/2% FCS and then remove the column to flush out the CD34+ cells with 1 ml PBS and repeat this procedure once more. The purity of the CD34+ cells is routinely 90–95% and is verified by FACS analysis.
7. Keep the purified CD34+ cells in CellGro medium at a density of 5×10^6 cells/ml, at 37°C overnight.

3.6. Conditioning and Reconstitution of Balb-c Rag2$^{-/-}$, γc$^{-/-}$ Mice

1. 2- to 4-day-old newborn BALB/c Rag2$^{-/-}$, γc$^{-/-}$ mice are subjected to a sublethal irradiation of 2 × 1.5 Gy with a minimum 2-h interval (see Note 6).
2. 2 × 10^5 hCB-CD34$^+$ cells are injected intrahepatically into the newborns. The animals are kept for 8–10 weeks under sterile conditions (see Note 7).
3. Peripheral blood is taken from the facial vein to check the reconstitution efficiency by analyzing the percentage of hCD45$^+$ cells (human common leukocyte marker) by FACS.
4. For FACS analysis, white blood cells are incubated in 100 μl PBS with 3 μl anti-hCD45-PERCP antibody combined with 3 μl antibody for blood cells lineage markers (e.g., anti-hCD3-APC) for 30 min at 4°C. The cells are washed with 4 ml PBS/2%FCS, centrifuged at 600 × *g* in FACS tubes and then passed through FACS.
5. The white cell populations, gated by granularity and size (FSC vs. SSC (side size scatter)), are evaluated by surface staining with anti-hCD45-PERCP combined with anti-hCD3-APC (total T-cells), anti-hCD19-APC (total B-cells), or anti-hCD14-APC (monocytes) (see Note 8) by FACS.

3.7. Intrafemural Vector Injection of BALB-c Rag2$^{-/-}$, γc$^{-/-}$ Mice (Fig. 4)

1. The mice are anesthetized by intraperitoneal injection of a 70:30 mixture of ketamine and xylazine diluted in 100 μl PBS at a dose of 100 ml/25 g (see Note 9).
2. A small incision is made at the top of the kneecap using a pair of small scissors. The animals are injected with a maximum of 20 μl vectors with titers ranging from 1 × 10^7 to 3 × 10^7 TU/ml in the femoral lower epiphysis. The incision is stitched up (one to two stitches using suture thread and surgical needle). The mice have to be monitored from the day after injection until 1 week. An analgesic drug is added to the drinking water (Doliprane 0.6 g/l) (see Note 10).

3.8. FACS Analysis of In Vivo Transduction

1. The mice are sacrificed by cervical dislocation in agreement with bioethical procedures, and BM, spleen, thymus, and peripheral blood are extracted. Cells are separated from the tissues using a mesh, and the mononuclear cell fraction is obtained from a Lymphoprep density gradient.
2. For the detection of LV transduction of the engrafted cells, flow cytometry analysis is performed using PERCP-conjugated anti-hCD45 antibody for the detection of total human cell engraftment in each hematopoietic tissue (BM, thymus, peripheral blood, and spleen). In combination, APC-coupled antibodies are used for the detection of hCD3 (total T cells), hCD56 (natural killer), hCD19 (B cells), hCD34 (progenitor cells), hCD13 (more mature progenitors), and hCD4 (monocytes/CD4 T cells). PE-coupled antibodies are used for the

detection of hCD20 (mature B cells), hCD14 (myeloid cells), hCD41 (megakaryocytes), and hCD8 (T cells). Simultaneous marking of thymocytes with anti-hCD3-PE, anti-hCD8-PECy7, and anti-hCD4-APC is performed to screen thymic subpopulations.

3. For FACS analysis, $5 \times 10^5 - 1 \times 10^6$ total white blood cells are incubated in 100 μl PBS with 3 μl anti-hCD45PERCP antibody combined with 3 μl of antibody for blood cell lineage markers (e.g., anti-CD3APC) for 30 min at 4°C.
4. The cells are washed with 4 ml PBS/2%FCS, centrifuged at 600 × *g* in FACS tubes and then passed through FACS. During acquisition, a gate according to size and granularity of the cells (FSC vs. SSC) is created. A mean of 10,000–100,000 events is acquired in this gate depending on the percentage of human cell reconstitution in the different tissues.
5. To carry out the analysis, a first gate in a dot plot showing the FSC vs. SSC is designed corresponding to the mononuclear cells. The gated population is analyzed in a 2D dot plot where the PERCP + population is allowed to appear on the *y*-axis and the APC + population on the *x*-axis. The double-positive cell population will carry the human marker (CD45-PERCP) plus the specific marker of the population of interest (CD34-APC for immature progenitors, CD3-APC for T cells, CD19-APC for B cells, or CD14-APC for monocytes). We can evaluate the degree of reconstitution from the percentage of CD45-PERCP+ cells. As an example, one can also estimate the level of reconstitution of human T-cells from the percentage of CD45/CD3 double positive cells.
6. For the determination of the lentiviral vector transduction efficiency in the in vivo human transduced cells, a dot plot is created where the gated population is analyzed for its CD45-PERCP expression on the *y*-axis and for GFP expression on the *x*-axis. The efficiency of transduction is given by the percentage of hCD45/GFP double positive cells as compared to the percentage of total hCD45+ cells. Double marking including a lineage marker (e.g., anti-CD34) is additionally useful to investigate if a different expression pattern in the different blood cell types exists.

4. Notes

1. RD114-TR is a fusion protein in which the cytoplasmic tail of RD114 gp was exchanged with the one of MLV gp to increase the efficiency of glycoprotein incorporation onto lentiviral cores (34).

2. $CD34^+$ cells represent approximately 1% of the PBMC fraction in the umbilical cord blood.
3. The $CD34^+$ magnetic bead purification method consists in positive selection of target cells by the presence of the $CD34^+$ antigen on their surface. Purity of eluted cells is 90–95% on a regular basis after two consecutive elution processes.
4. The Vivaspin device can be filled up to a maximum volume of 20 ml of vector supernatant. Centrifugation time may vary between 30 min and 2 h, until the whole volume of vector supernatant has passed through the membrane. The final volume of Vivaspin concentrators is approximately 0.5 ml. Depending on the initial vector volume, several rounds of centrifugation in the same device can be done to increase the final vector concentration.
5. Baf-3-c-Kit and Baf-3-Mpl cells are derived from the murine Baf-3 cell line engineered to stably express the hSCF receptor c-Kit or the hTPO receptor Mpl. These cells grow in suspension in complete RPMI medium with the addition of IL-3. IL-3 is obtained from the supernatant of subconfluent semi-adherent WEHI cells. The supernatant is filtered and diluted 1/15 in the Baf-3 growth medium. When starved from IL-3, Baf-3 cells are exclusively dependent on hSCF or hTPO to proliferate. The Baf-3-c-Kit or Baf-3-Mpl cell lines are excellent tools to test the functional activity of the SCF/TPO-displaying LVs. Cell expansion and survival, upon addition of serial dilutions of the vectors, are measured by FACS.
6. In order to achieve an efficient high-level engraftment, it is important to apply two rounds of irradiation to the newborn mice with an interval of at least 2 h between each irradiation.
7. The liver of 2- to 4-day-old mice is visible through the skin just above the stomach. A 1–2 mm injection of a maximum of 20–30 μm of cell suspension is enough to deliver the $hCD34^+$ cells intrahepatically.
8. Humanization of the mice is achieved when the percentage of peripheral blood $hCD45^+$ cells is equal to or above 5% at week 8–10. However, occasionally, this time frame is too short to observe reconstitution in the peripheral blood, and it is necessary to wait until week 12 or later.
9. This mouse model reacts to the anesthesia within 15 min, and the effect may last for 2 h. However, extreme caution must be taken, and mice must be followed during the whole intervention and during wake-up period. Indeed, these animals are very sensitive to the anesthesia, even if the dose is adapted to the animal weight.
10. First, the femur is pierced with a 25 G needle. Subsequently, the vectors are injected using an insulin needle to have a very low dead volume.

Acknowledgements

The first two authors, Cecilia Frecha and Floriane Fusil, have equally contributed to the writing of this manuscript. Further, we would like to thank Caroline Costa for her excellent technical assistance and Dr. Mamoro Ito (CIEA, Kawasaki, Japan), and Taconic (Japan) for sharing the BALB/c-Rag2$^{-/-}$γc$^{-/-}$ immunodeficient mice with us and the animal facility at the ENS de Lyon (PBES). Further, we would like to acknowledge the support by the following grants: the "Agence Nationale pour la Recherche contre le SIDA et les Hépatites Virales" (ANRS), the "Agence Nationale de la Recherche" (ANR), the "Association française pour la myopathie," AFM and the European Community (FP7-HEALTH-2007-B/222878 "PERSIST"). C.F. is supported by an ANRS postdoctoral fellowship.

References

1. Martinez-Agosto, J. A., Mikkola, H. K., Hartenstein, V., and Banerjee, U. (2007) The hematopoietic stem cell and its niche: a comparative view, *Genes Dev* **21**, 3044–3060.
2. Orford, K. W., and Scadden, D. T. (2008) Deconstructing stem cell self-renewal: genetic insights into cell-cycle regulation, *Nat Rev Genet* **9**, 115–128.
3. Cheshier, S. H., Morrison, S. J., Liao, X., and Weissman, I. L. (1999) In vivo proliferation and cell cycle kinetics of long-term self-renewing hematopoietic stem cells, *Proc Natl Acad Sci U S A* **96**, 3120–3125.
4. Strauss, L. C., Rowley, S. D., La Russa, V. F., Sharkis, S. J., Stuart, R. K., and Civin, C. I. (1986) Antigenic analysis of hematopoiesis. V. Characterization of My-10 antigen expression by normal lymphohematopoietic progenitor cells, *Exp Hematol* **14**, 878–886.
5. Baum, C. M., Weissman, I. L., Tsukamoto, A. S., Buckle, A. M., and Peault, B. (1992) Isolation of a candidate human hematopoietic stem-cell population, *Proc Natl Acad Sci U S A* **89**, 2804–2808.
6. Morrison, S. J., and Weissman, I. L. (1994) The long-term repopulating subset of hematopoietic stem cells is deterministic and isolatable by phenotype, *Immunity* **1**, 661–673.
7. Terstappen, L. W., Huang, S., Safford, M., Lansdorp, P. M., and Loken, M. R. (1991) Sequential generations of hematopoietic colonies derived from single nonlineage-committed CD34+CD38- progenitor cells, *Blood* **77**, 1218–1227.
8. McKenzie, J. L., Gan, O. I., Doedens, M., and Dick, J. E. (2007) Reversible cell surface expression of CD38 on CD34-positive human hematopoietic repopulating cells, *Exp Hematol* **35**, 1429–1436.
9. Kimura, T., Asada, R., Wang, J., Morioka, M., Matsui, K., Kobayashi, K., Henmi, K., Imai, S., Kita, M., Tsuji, T., Sasaki, Y., Ikehara, S., and Sonoda, Y. (2007) Identification of long-term repopulating potential of human cord blood-derived CD34-flt3- severe combined immunodeficiency-repopulating cells by intra-bone marrow injection, *Stem Cells* **25**, 1348–1355.
10. Goodell, M. A., Rosenzweig, M., Kim, H., Marks, D. F., DeMaria, M., Paradis, G., Grupp, S. A., Sieff, C. A., Mulligan, R. C., and Johnson, R. P. (1997) Dye efflux studies suggest that hematopoietic stem cells expressing low or undetectable levels of CD34 antigen exist in multiple species, *Nat Med* **3**, 1337–1345.
11. Dao, M. A., Arevalo, J., and Nolta, J. A. (2003) Reversibility of CD34 expression on human hematopoietic stem cells that retain the capacity for secondary reconstitution, *Blood* **101**, 112–118.
12. Mizrak, D., Brittan, M., and Alison, M. R. (2008) CD133: molecule of the moment, *J Pathol* **214**, 3–9.
13. Jokubaitis, V. J., Sinka, L., Driessen, R., Whitty, G., Haylock, D. N., Bertoncello, I., Smith, I., Peault, B., Tavian, M., and Simmons, P. J. (2008) Angiotensin-converting enzyme (CD143) marks hematopoietic stem cells in human embryonic, fetal, and adult hematopoietic tissues, *Blood* **111**, 4055–4063.

14. Meyerrose, T. E., Herrbrich, P., Hess, D. A., and Nolta, J. A. (2003) Immune-deficient mouse models for analysis of human stem cells, *Biotechniques* **35**, 1262–1272.
15. Larochelle, A., Vormoor, J., Hanenberg, H., Wang, J. C., Bhatia, M., Lapidot, T., Moritz, T., Murdoch, B., Xiao, X. L., Kato, I., Williams, D. A., and Dick, J. E. (1996) Identification of primitive human hematopoietic cells capable of repopulating NOD/SCID mouse bone marrow: implications for gene therapy, *Nat Med* **2**, 1329–1337.
16. Sutherland, H. J., Lansdorp, P. M., Henkelman, D. H., Eaves, A. C., and Eaves, C. J. (1990) Functional characterization of individual human hematopoietic stem cells cultured at limiting dilution on supportive marrow stromal layers, *Proc Natl Acad Sci U S A* **87**, 3584–3588.
17. Dick, J. E., Kamel-Reid, S., Murdoch, B., and Doedens, M. (1991) Gene transfer into normal human hematopoietic cells using in vitro and in vivo assays, *Blood* **78**, 624–634.
18. Eckfeldt, C. E., Mendenhall, E. M., and Verfaillie, C. M. (2005) The molecular repertoire of the 'almighty' stem cell, *Nat Rev Mol Cell Biol* **6**, 726–737.
19. Santoni de Sio, F. R., Cascio, P., Zingale, A., Gasparini, M., and Naldini, L. (2006) Proteasome activity restricts lentiviral gene transfer into hematopoietic stem cells and is down-regulated by cytokines that enhance transduction, *Blood* **107**, 4257–4265.
20. Ailles, L., Schmidt, M., Santoni de Sio, F. R., Glimm, H., Cavalieri, S., Bruno, S., Piacibello, W., Von Kalle, C., and Naldini, L. (2002) Molecular evidence of lentiviral vector-mediated gene transfer into human self-renewing, multi-potent, long-term NOD/SCID repopulating hematopoietic cells, *Mol Ther* **6**, 615–626.
21. Guenechea, G., Gan, O. I., Inamitsu, T., Dorrell, C., Pereira, D. S., Kelly, M., Naldini, L., and Dick, J. E. (2000) Transduction of human CD34+ CD38- bone marrow and cord blood-derived SCID-repopulating cells with third-generation lentiviral vectors, *Mol Ther* **1**, 566–573.
22. Sutton, R. E., Reitsma, M. J., Uchida, N., and Brown, P. O. (1999) Transduction of human progenitor hematopoietic stem cells by human immunodeficiency virus type 1-based vectors is cell cycle dependent, *J Virol* **73**, 3649–3660.
23. Woods, N. B., Bottero, V., Schmidt, M., von Kalle, C., and Verma, I. M. (2006) Gene therapy: therapeutic gene causing lymphoma, *Nature* **440**, 1123.
24. Capotondo, A., Cesani, M., Pepe, S., Fasano, S., Gregori, S., Tononi, L., Venneri, M. A., Brambilla, R., Quattrini, A., Ballabio, A., Cosma, M. P., Naldini, L., and Biffi, A. (2007) Safety of arylsulfatase A overexpression for gene therapy of metachromatic leukodystrophy, *Hum Gene Ther* **18**, 821–836.
25. Voermans, C., Gerritsen, W. R., von dem Borne, A. E., and van der Schoot, C. E. (1999) Increased migration of cord blood-derived CD34+ cells, as compared to bone marrow and mobilized peripheral blood CD34+ cells across uncoated or fibronectin-coated filters, *Exp Hematol* **27**, 1806–1814.
26. Ahmed, F., Ings, S. J., Pizzey, A. R., Blundell, M. P., Thrasher, A. J., Ye, H. T., Fahey, A., Linch, D. C., and Yong, K. L. (2004) Impaired bone marrow homing of cytokine-activated CD34+ cells in the NOD/SCID model, *Blood* **103**, 2079–2087.
27. Peled, A., Petit, I., Kollet, O., Magid, M., Ponomaryov, T., Byk, T., Nagler, A., Ben-Hur, H., Many, A., Shultz, L., Lider, O., Alon, R., Zipori, D., and Lapidot, T. (1999) Dependence of human stem cell engraftment and repopulation of NOD/SCID mice on CXCR4, *Science* **283**, 845–848.
28. Sorrentino, B. P. (2004) Clinical strategies for expansion of haematopoietic stem cells, *Nat Rev Immunol* **4**, 878–888.
29. Ueda, T., Yoshida, M., Yoshino, H., Kobayashi, K., Kawahata, M., Ebihara, Y., Ito, M., Asano, S., Nakahata, T., and Tsuji, K. (2001) Hematopoietic capability of CD34+ cord blood cells: a comparison with CD34+ adult bone marrow cells, *Int J Hematol* **73**, 457–462.
30. Piacibello, W., Gammaitoni, L., Bruno, S., Gunetti, M., Fagioli, F., Cavalloni, G., and Aglietta, M. (2000) Negative influence of IL3 on the expansion of human cord blood in vivo long-term repopulating stem cells, *J Hematother Stem Cell Res* **9**, 945–956.
31. Luens, K. M., Travis, M. A., Chen, B. P., Hill, B. L., Scollay, R., and Murray, L. J. (1998) Thrombopoietin, kit ligand, and flk2/flt3 ligand together induce increased numbers of primitive hematopoietic progenitors from human CD34+Thy-1+Lin- cells with preserved ability to engraft SCID-hu bone, *Blood* **91**, 1206–1215.
32. Chen, B. P., Galy, A., Kyoizumi, S., Namikawa, R., Scarborough, J., Webb, S., Ford, B., Cen, D. Z., and Chen, S. C. (1994) Engraftment of human hematopoietic precursor cells with secondary transfer potential in SCID-hu mice, *Blood* **84**, 2497–2505.
33. Baum, C., Dullmann, J., Li, Z., Fehse, B., Meyer, J., Williams, D. A., and von Kalle, C.

(2003) Side effects of retroviral gene transfer into hematopoietic stem cells, *Blood* **101**, 2099–2114.
34. Sandrin, V., Boson, B., Salmon, P., Gay, W., Negre, D., Le Grand, R., Trono, D., and Cosset, F. L. (2002) Lentiviral vectors pseudotyped with a modified RD114 envelope glycoprotein show increased stability in sera and augmented transduction of primary lymphocytes and CD34+ cells derived from human and nonhuman primates, *Blood* **100**, 823–832.
35. Verhoeyen, E., and Cosset, F. L. (2004) Surface-engineering of lentiviral vectors, *J Gene Med* **6 Suppl 1**, S83–94.
36. Thoren, L. A., Liuba, K., Bryder, D., Nygren, J. M., Jensen, C. T., Qian, H., Antonchuk, J., and Jacobsen, S. E. (2008) Kit regulates maintenance of quiescent hematopoietic stem cells, *J Immunol* **180**, 2045–2053.
37. Verhoeyen, E., Wiznerowicz, M., Olivier, D., Izac, B., Trono, D., Dubart-Kupperschmitt, A., and Cosset, F. L. (2005) Novel lentiviral vectors displaying "early-acting cytokines" selectively promote survival and transduction of NOD/SCID repopulating human hematopoietic stem cells, *Blood* **106**, 3386–3395.
38. Roep, B. O., Atkinson, M., and von Herrath, M. (2004) Satisfaction (not) guaranteed: re-evaluating the use of animal models of type 1 diabetes, *Nat Rev Immunol* **4**, 989–997.
39. Bosma, G. C., Custer, R. P., and Bosma, M. J. (1983) A severe combined immunodeficiency mutation in the mouse, *Nature* **301**, 527–530.
40. Gotoh, M., Takasu, H., Harada, K., and Yamaoka, T. (2002) Development of HLA-A2402/K(b) transgenic mice, *Int J Cancer* **100**, 565–570.
41. Krimpenfort, P., Rudenko, G., Hochstenbach, F., Guessow, D., Berns, A., and Ploegh, H. (1987) Crosses of two independently derived transgenic mice demonstrate functional complementation of the genes encoding heavy (HLA-B27) and light (beta 2-microglobulin) chains of HLA class I antigens, *EMBO J* **6**, 1673–1676.
42. Banuelos, S. J., Shultz, L. D., Greiner, D. L., Burzenski, L. M., Gott, B., Lyons, B. L., Rossini, A. A., and Appel, M. C. (2004) Rejection of human islets and human HLA-A2.1 transgenic mouse islets by alloreactive human lymphocytes in immunodeficient NOD-scid and NOD-Rag1(null)Prf1(null) mice, *Clin Immunol* **112**, 273–283.
43. Bock, T. A., Orlic, D., Dunbar, C. E., Broxmeyer, H. E., and Bodine, D. M. (1995) Improved engraftment of human hematopoietic cells in severe combined immunodeficient (SCID) mice carrying human cytokine transgenes, *J Exp Med* **182**, 2037–2043.
44. Takaki, T., Marron, M. P., Mathews, C. E., Guttmann, S. T., Bottino, R., Trucco, M., DiLorenzo, T. P., and Serreze, D. V. (2006) HLA-A*0201-restricted T cells from humanized NOD mice recognize autoantigens of potential clinical relevance to type 1 diabetes, *J Immunol* **176**, 3257–3265.
45. Mosier, D. E., Gulizia, R. J., Baird, S. M., and Wilson, D. B. (1988) Transfer of a functional human immune system to mice with severe combined immunodeficiency, *Nature* **335**, 256–259.
46. McCune, J. M., Namikawa, R., Kaneshima, H., Shultz, L. D., Lieberman, M., and Weissman, I. L. (1988) The SCID-hu mouse: murine model for the analysis of human hematolymphoid differentiation and function, *Science* **241**, 1632–1639.
47. Lapidot, T., Pflumio, F., Doedens, M., Murdoch, B., Williams, D. E., and Dick, J. E. (1992) Cytokine stimulation of multilineage hematopoiesis from immature human cells engrafted in SCID mice, *Science* **255**, 1137–1141.
48. Christianson, S. W., Greiner, D. L., Schweitzer, I. B., Gott, B., Beamer, G. L., Schweitzer, P. A., Hesselton, R. M., and Shultz, L. D. (1996) Role of natural killer cells on engraftment of human lymphoid cells and on metastasis of human T-lymphoblastoid leukemia cells in C57BL/6 J-scid mice and in C57BL/6 J-scid bg mice, *Cell Immunol* **171**, 186–199.
49. Greiner, D. L., Hesselton, R. A., and Shultz, L. D. (1998) SCID mouse models of human stem cell engraftment, *Stem Cells* **16**, 166–177.
50. Bosma, G. C., Fried, M., Custer, R. P., Carroll, A., Gibson, D. M., and Bosma, M. J. (1988) Evidence of functional lymphocytes in some (leaky) scid mice, *J Exp Med* **167**, 1016–1033.
51. Fulop, G. M., and Phillips, R. A. (1990) The scid mutation in mice causes a general defect in DNA repair, *Nature* **347**, 479–482.
52. Mombaerts, P., Iacomini, J., Johnson, R. S., Herrup, K., Tonegawa, S., and Papaioannou, V. E. (1992) RAG-1-deficient mice have no mature B and T lymphocytes, *Cell* **68**, 869–877.
53. Shinkai, Y., Rathbun, G., Lam, K. P., Oltz, E. M., Stewart, V., Mendelsohn, M., Charron, J., Datta, M., Young, F., Stall, A. M., and et al. (1992) RAG-2-deficient mice lack mature lymphocytes owing to inability to initiate V(D)J rearrangement, *Cell* **68**, 855–867.
54. Shultz, L. D., Schweitzer, P. A., Christianson, S. W., Gott, B., Schweitzer, I. B., Tennent, B., McKenna, S., Mobraaten, L., Rajan, T. V.,

Greiner, D. L., and et al. (1995) Multiple defects in innate and adaptive immunologic function in NOD/LtSz-scid mice, *J Immunol* **154**, 180–191.

55. Hesselton, R. M., Greiner, D. L., Mordes, J. P., Rajan, T. V., Sullivan, J. L., and Shultz, L. D. (1995) High levels of human peripheral blood mononuclear cell engraftment and enhanced susceptibility to human immunodeficiency virus type 1 infection in NOD/LtSz-scid/scid mice, *J Infect Dis* **172**, 974–982.
56. Lowry, P. A., Shultz, L. D., Greiner, D. L., Hesselton, R. M., Kittler, E. L., Tiarks, C. Y., Rao, S. S., Reilly, J., Leif, J. H., Ramshaw, H., Stewart, F. M., and Quesenberry, P. J. (1996) Improved engraftment of human cord blood stem cells in NOD/LtSz-scid/scid mice after irradiation or multiple-day injections into unirradiated recipients, *Biol Blood Marrow Transplant* **2**, 15–23.
57. Pflumio, F., Izac, B., Katz, A., Shultz, L. D., Vainchenker, W., and Coulombel, L. (1996) Phenotype and function of human hematopoietic cells engrafting immune-deficient CB17-severe combined immunodeficiency mice and nonobese diabetic-severe combined immunodeficiency mice after transplantation of human cord blood mononuclear cells, *Blood* **88**, 3731–3740.
58. Takizawa, H., and Manz, M. G. (2007) Macrophage tolerance: CD47-SIRP-alpha-mediated signals matter, *Nat Immunol* **8**, 1287–1289.
59. Takenaka, K., Prasolava, T. K., Wang, J. C., Mortin-Toth, S. M., Khalouei, S., Gan, O. I., Dick, J. E., and Danska, J. S. (2007) Polymorphism in Sirpa modulates engraftment of human hematopoietic stem cells, *Nat Immunol* **8**, 1313–1323.
60. Piganelli, J. D., Martin, T., and Haskins, K. (1998) Splenic macrophages from the NOD mouse are defective in the ability to present antigen, *Diabetes* **47**, 1212–1218.
61. Ogasawara, K., Hamerman, J. A., Hsin, H., Chikuma, S., Bour-Jordan, H., Chen, T., Pertel, T., Carnaud, C., Bluestone, J. A., and Lanier, L. L. (2003) Impairment of NK cell function by NKG2D modulation in NOD mice, *Immunity* **18**, 41–51.
62. O'Brien, B. A., Huang, Y., Geng, X., Dutz, J. P., and Finegood, D. T. (2002) Phagocytosis of apoptotic cells by macrophages from NOD mice is reduced, *Diabetes* **51**, 2481–2488.
63. Greiner, D. L., Shultz, L. D., Yates, J., Appel, M. C., Perdrizet, G., Hesselton, R. M., Schweitzer, I., Beamer, W. G., Shultz, K. L., Pelsue, S. C., and et al. (1995) Improved engraftment of human spleen cells in NOD/LtSz-scid/scid mice as compared with C.B-17-scid/scid mice, *Am J Pathol* **146**, 888–902.
64. Dorshkind, K., Pollack, S. B., Bosma, M. J., and Phillips, R. A. (1985) Natural killer (NK) cells are present in mice with severe combined immunodeficiency (scid), *J Immunol* **134**, 3798–3801.
65. Christianson, S. W., Greiner, D. L., Hesselton, R. A., Leif, J. H., Wagar, E. J., Schweitzer, I. B., Rajan, T. V., Gott, B., Roopenian, D. C., and Shultz, L. D. (1997) Enhanced human CD4+ T cell engraftment in beta2-microglobulin-deficient NOD-scid mice, *J Immunol* **158**, 3578–3586.
66. Shultz, L. D., Banuelos, S., Lyons, B., Samuels, R., Burzenski, L., Gott, B., Lang, P., Leif, J., Appel, M., Rossini, A., and Greiner, D. L. (2003) NOD/LtSz-Rag1nullPfpnull mice: a new model system with increased levels of human peripheral leukocyte and hematopoietic stem-cell engraftment, *Transplantation* **76**, 1036–1042.
67. Prochazka, M., Gaskins, H. R., Shultz, L. D., and Leiter, E. H. (1992) The nonobese diabetic scid mouse: model for spontaneous thymomagenesis associated with immunodeficiency, *Proc Natl Acad Sci U S A* **89**, 3290–3294.
68. Ishikawa, F., Yasukawa, M., Lyons, B., Yoshida, S., Miyamoto, T., Yoshimoto, G., Watanabe, T., Akashi, K., Shultz, L. D., and Harada, M. (2005) Development of functional human blood and immune systems in NOD/SCID/IL2 receptor {gamma} chain(null) mice, *Blood* **106**, 1565–1573.
69. Ito, M., Hiramatsu, H., Kobayashi, K., Suzue, K., Kawahata, M., Hioki, K., Ueyama, Y., Koyanagi, Y., Sugamura, K., Tsuji, K., Heike, T., and Nakahata, T. (2002) NOD/SCID/gamma(c)(null) mouse: an excellent recipient mouse model for engraftment of human cells, *Blood* **100**, 3175–3182.
70. Shultz, L. D., Lyons, B. L., Burzenski, L. M., Gott, B., Chen, X., Chaleff, S., Kotb, M., Gillies, S. D., King, M., Mangada, J., Greiner, D. L., and Handgretinger, R. (2005) Human lymphoid and myeloid cell development in NOD/LtSz-scid IL2R gamma null mice engrafted with mobilized human hemopoietic stem cells, *J Immunol* **174**, 6477–6489.
71. Traggiai, E., Chicha, L., Mazzucchelli, L., Bronz, L., Piffaretti, J. C., Lanzavecchia, A., and Manz, M. G. (2004) Development of a human adaptive immune system in cord blood cell-transplanted mice, *Science* **304**, 104–107.
72. Cao, X., Shores, E. W., Hu-Li, J., Anver, M. R., Kelsall, B. L., Russell, S. M., Drago, J.,

Noguchi, M., Grinberg, A., Bloom, E. T., and et al. (1995) Defective lymphoid development in mice lacking expression of the common cytokine receptor gamma chain, *Immunity* **2**, 223–238.

73. DiSanto, J. P., Muller, W., Guy-Grand, D., Fischer, A., and Rajewsky, K. (1995) Lymphoid development in mice with a targeted deletion of the interleukin 2 receptor gamma chain, *Proc Natl Acad Sci U S A* **92**, 377–381.
74. Leonard, W. J. (1996) Dysfunctional cytokine receptor signaling in severe combined immunodeficiency, *J Investig Med* **44**, 304–311.
75. Jacobs, H., Krimpenfort, P., Haks, M., Allen, J., Blom, B., Demolliere, C., Kruisbeek, A., Spits, H., and Berns, A. (1999) PIM1 reconstitutes thymus cellularity in interleukin 7- and common gamma chain-mutant mice and permits thymocyte maturation in Rag- but not CD3gamma-deficient mice, *J Exp Med* **190**, 1059–1068.
76. Ohbo, K., Suda, T., Hashiyama, M., Mantani, A., Ikebe, M., Miyakawa, K., Moriyama, M., Nakamura, M., Katsuki, M., Takahashi, K., Yamamura, K., and Sugamura, K. (1996) Modulation of hematopoiesis in mice with a truncated mutant of the interleukin-2 receptor gamma chain, *Blood* **87**, 956–967.
77. Shultz, L. D., Ishikawa, F., and Greiner, D. L. (2007) Humanized mice in translational biomedical research, *Nat Rev Immunol* **7**, 118–130.
78. Hiramatsu, H., Nishikomori, R., Heike, T., Ito, M., Kobayashi, K., Katamura, K., and Nakahata, T. (2003) Complete reconstitution of human lymphocytes from cord blood CD34+ cells using the NOD/SCID/gammacnull mice model, *Blood* **102**, 873–880.
79. Yahata, T., Ando, K., Nakamura, Y., Ueyama, Y., Shimamura, K., Tamaoki, N., Kato, S., and Hotta, T. (2002) Functional human T lymphocyte development from cord blood CD34+ cells in nonobese diabetic/Shi-scid, IL-2 receptor gamma null mice, *J Immunol* **169**, 204–209.
80. Gimeno, R., Weijer, K., Voordouw, A., Uittenbogaart, C. H., Legrand, N., Alves, N. L., Wijnands, E., Blom, B., and Spits, H. (2004) Monitoring the effect of gene silencing by RNA interference in human CD34+ cells injected into newborn RAG2-/- gammac-/- mice: functional inactivation of p53 in developing T cells, *Blood* **104**, 3886–3893.
81. Legrand, N., Cupedo, T., van Lent, A. U., Ebeli, M. J., Weijer, K., Hanke, T., and Spits, H. (2006) Transient accumulation of human mature thymocytes and regulatory T cells with CD28 superagonist in "human immune system" Rag2(-/-)gammac(-/-) mice, *Blood* **108**, 238–245.
82. Berges, B. K., Wheat, W. H., Palmer, B. E., Connick, E., and Akkina, R. (2006) HIV-1 infection and CD4 T cell depletion in the humanized Rag2-/-gamma c-/- (RAG-hu) mouse model, *Retrovirology* **3**, 76.
83. Mazurier, F., Doedens, M., Gan, O. I., and Dick, J. E. (2003) Rapid myeloerythroid repopulation after intrafemoral transplantation of NOD-SCID mice reveals a new class of human stem cells, *Nat Med* **9**, 959–963.
84. Wang, J., Kimura, T., Asada, R., Harada, S., Yokota, S., Kawamoto, Y., Fujimura, Y., Tsuji, T., Ikehara, S., and Sonoda, Y. (2003) SCID-repopulating cell activity of human cord blood-derived CD34- cells assured by intrabone marrow injection, *Blood* **101**, 2924–2931.
85. Schoeberlein, A., Schatt, S., Troeger, C., Surbek, D., Holzgreve, W., and Hahn, S. (2004) Engraftment kinetics of human cord blood and murine fetal liver stem cells following in utero transplantation into immunodeficient mice, *Stem Cells Dev* **13**, 677–684.
86. Kuci, S., Wessels, J. T., Buhring, H. J., Schilbach, K., Schumm, M., Seitz, G., Loffler, J., Bader, P., Schlegel, P. G., Niethammer, D., and Handgretinger, R. (2003) Identification of a novel class of human adherent CD34- stem cells that give rise to SCID-repopulating cells, *Blood* **101**, 869–876.
87. de Wynter, E. A., Buck, D., Hart, C., Heywood, R., Coutinho, L. H., Clayton, A., Rafferty, J. A., Burt, D., Guenechea, G., Bueren, J. A., Gagen, D., Fairbairn, L. J., Lord, B. I., and Testa, N. G. (1998) CD34+AC133+ cells isolated from cord blood are highly enriched in long-term culture-initiating cells, NOD/SCID-repopulating cells and dendritic cell progenitors, *Stem Cells* **16**, 387–396.
88. Bhatia, M., Wang, J. C., Kapp, U., Bonnet, D., and Dick, J. E. (1997) Purification of primitive human hematopoietic cells capable of repopulating immune-deficient mice, *Proc Natl Acad Sci U S A* **94**, 5320–5325.
89. Bonnet, D., Bhatia, M., Wang, J. C., Kapp, U., and Dick, J. E. (1999) Cytokine treatment or accessory cells are required to initiate engraftment of purified primitive human hematopoietic cells transplanted at limiting doses into NOD/SCID mice, *Bone Marrow Transplant* **23**, 203–209.
90. Kim, D. K., Fujiki, Y., Fukushima, T., Ema, H., Shibuya, A., and Nakauchi, H. (1999) Comparison of hematopoietic activities of human bone marrow and umbilical cord blood

CD34 positive and negative cells, *Stem Cells* **17**, 286–294.

91. Ito, M., Kobayashi, K., and Nakahata, T. (2008) NOD/Shi-scid IL2rgamma(null) (NOG) mice more appropriate for humanized mouse models, *Curr Top Microbiol Immunol* **324**, 53–76.
92. Goldman, J. P., Blundell, M. P., Lopes, L., Kinnon, C., Di Santo, J. P., and Thrasher, A. J. (1998) Enhanced human cell engraftment in mice deficient in RAG2 and the common cytokine receptor gamma chain, *Br J Haematol* **103**, 335–342.
93. Kirberg, J., Berns, A., and von Boehmer, H. (1997) Peripheral T cell survival requires continual ligation of the T cell receptor to major histocompatibility complex-encoded molecules, *J Exp Med* **186**, 1269–1275.
94. Huntington, N. D., and Di Santo, J. P. (2008) Humanized immune system (HIS) mice as a tool to study human NK cell development, *Curr Top Microbiol Immunol* **324**, 109–124.
95. Baenziger, S., Tussiwand, R., Schlaepfer, E., Mazzucchelli, L., Heikenwalder, M., Kurrer, M. O., Behnke, S., Frey, J., Oxenius, A., Joller, H., Aguzzi, A., Manz, M. G., and Speck, R. F. (2006) Disseminated and sustained HIV infection in CD34+ cord blood cell-transplanted Rag2-/-gamma c-/- mice, *Proc Natl Acad Sci U S A* **103**, 15951–15956.
96. Gorantla, S., Sneller, H., Walters, L., Sharp, J. G., Pirruccello, S. J., West, J. T., Wood, C., Dewhurst, S., Gendelman, H. E., and Poluektova, L. (2007) Human immunodeficiency virus type 1 pathobiology studied in humanized BALB/c-Rag2-/-gammac-/- mice, *J Virol* **81**, 2700–2712.
97. Frecha, C., Costa, C., Negre, D., Gauthier, E., Russell, S. J., Cosset, F. L., and Verhoeyen, E. (2008) Stable transduction of quiescent T cells without induction of cycle progression by a novel lentiviral vector pseudotyped with measles virus glycoproteins, *Blood* **112**, 4843–4852.

Chapter 16

In Vivo Evaluation of Gene Transfer into Mesenchymal Cells (In View of Cartilage Repair)

Kolja Gelse and Holm Schneider

Abstract

Gene transfer of specific growth factors is suitable for inducing chondrogenic differentiation of mesenchymal cells to be used for cartilage regeneration. However, extent and quality of repair tissue formation also depend on biomechanical and metabolic influences that can only be studied in vivo. We describe three methods to evaluate viral gene transfer into mesenchymal cells in animal models of articular cartilage defects, e.g., mouse, rat and miniature pig models, focussing on the repair of hyaline cartilage tissue.

Key words: Gene transfer, Animal model, Mesenchymal cells, Cartilage repair, Cell transplantation

1. Introduction

Periosteum and perichondrium are abundant tissues that contain mesenchymal precursor cells suitable for cartilage repair. However, the differentiation of these cells into mature chondrocytes does not occur spontaneously (1, 2) but requires certain stimuli such as mechanical load, low oxygen tension, or specific growth factors (3, 4). Among the latter, certain members of the TGF-β superfamily (e.g., TGF-β, BMP-2, BMP-7) are very potent inducers of cartilage-specific gene expression and chondrogenesis (5, 6). Application of these factors in form of recombinant proteins, however, is expensive, and the desired effects are limited by their short half-lives. Chondrogenic differentiation of growth-factor-stimulated mesenchymal progenitor cells may actually be fragile in vivo, resulting in dedifferentiation of stimulated precursor cells to a fibroblastic phenotype subsequent to cell transplantation. Therefore, gene transfer has been considered a promising strategy

Otto-Wilhelm Merten and Mohamed Al-Rubeai (eds.), *Viral Vectors for Gene Therapy: Methods and Protocols*,
Methods in Molecular Biology, vol. 737, DOI 10.1007/978-1-61779-095-9_16, © Springer Science+Business Media, LLC 2011

to combine the supply of chondrogenic cells with the production of therapeutic proteins directly at the implantation site. This strategy is suitable for achieving a more sustained protein delivery at therapeutic levels (7, 8). Cells that have been infected ex vivo by the respective viral or nonviral vectors can either be injected at random into the joint cavity, where they reach predominantly the recessus (9), or can be applied directly to the region of interest, e.g., to circumscribed cartilage defects (2, 10). The transgene-stimulated cells can be transplanted in the form of a cell suspension or bound to a biodegradable scaffold. In some cases, the fate of repair cells needs to be monitored over a longer period of time, or the transplanted cells are to be detected within the repair tissue. For that purpose, the cells can be labelled efficiently prior to transplantation, e.g., by transfer of a β-galactosidase or green fluorescent protein (GFP) transgene.

We describe the application of cell-mediated gene transfer in three animal models that are characterized by different types of cartilage lesions: spontaneous osteoarthritis in knee joints of STR/ORT mice (11), surgically induced osteoarthritis-like cartilage fissures in a rat model (Fig. 1) (10), and standardized circumscribed or large cartilage defects in the knee joints of miniature pigs (Fig. 2) (2).

2. Materials

2.1. Cell Isolation and Culture

1. Phosphate-buffered saline (PBS): Prepare 10× stock with 1.37 M NaCl, 27 mM KCl, 100 mM Na_2HPO_4, 18 mM KH_2PO_4 (adjust to pH 7.4 with HCl if necessary).
2. Trypsin (Gibco Life Technologies, Paisley, UK) is dissolved at 0.2% in PBS.
3. Clostridial collagenase (Roche, Mannheim, Germany) is dissolved at 0.02% in PBS.
4. Dulbecco's modified Eagle's Medium (DMEM) (Gibco Life Technologies, Paisley, UK) supplemented with 10% fetal calf serum (FCS; Gibco), and 1% Penicillin/Streptomycin (Gibco).
5. OptiMEM (Gibco Life Technologies, Paisley, UK).
6. Solution of trypsin (0.1%) and ethylenediamine tetraacetic acid (EDTA, 0.025%) (Gibco).
7. Sterile cell strainer with 100-μm nylon mesh (BD Biosciences, Heidelberg, Germany).
8. Culture dishes with diameters of 6 and 15 cm (TPP, Trasadingen, Switzerland).

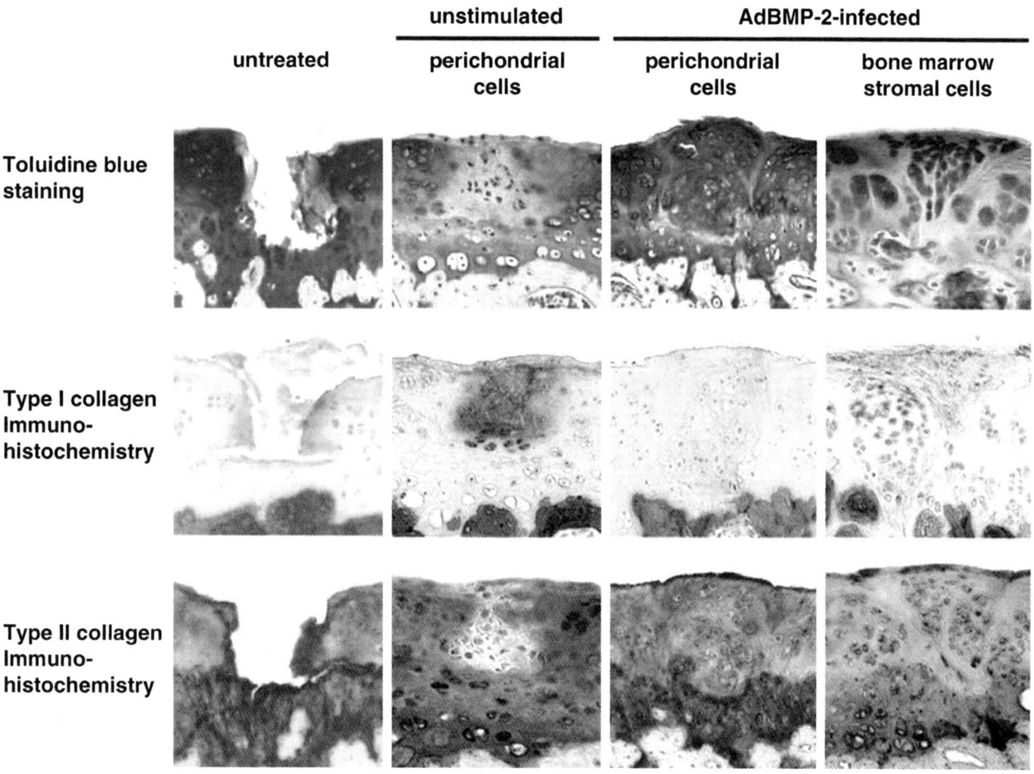

Fig. 1. Articular cartilage repair in the rat model described above using transgene-activated mesenchymal cells of different origin. Partial-thickness cartilage lesions in the patellar groove of the femur were treated by transplantation of native perichondrial cells or by applying perichondrial or bone marrow stromal cells that had been infected ex vivo with an adenoviral vector carrying BMP-2 cDNA. Eight weeks after transplantation, the joints were analyzed by toluidine blue staining. Representative serial sections were also investigated by immunohistochemistry for type I collagen, a marker for fibrous tissue, and for type II collagen, a major and specific component of hyaline cartilage. Untreated lesions did not heal spontaneously. The transplantation of unstimulated cells gave only rise to fibrous tissue. AdBMP-2-stimulated perichondrial cells and AdBMP-2-stimulated bone marrow stromal cells produced a proteoglycan-rich, type II collagen-positive matrix with only faint staining for type I collagen.

2.2. Manufacturing of Adenoviral and AAV Vectors

See Chapters 5, 6, and 9.

2.3. Animal Care and Anesthesia

1. STR/ORT mice, female (Harlan Winkelmann, Borchen, Germany).
2. Wistar rats, female (Charles River, Sulzfeld, Germany).
3. Göttingen miniature pigs, female (Ellegaard, Dalmose, Denmark).
4. Ketamine (Ketavet; Pfizer, New York, USA).
5. Midazolam (Dormicum, Roche, Mannheim, Germany).
6. Pentobarbitone (Sagatal, Rhone Merieux, Harlow, UK).
7. Isoflurane (Baxter, McGaw Parl, IL, USA).

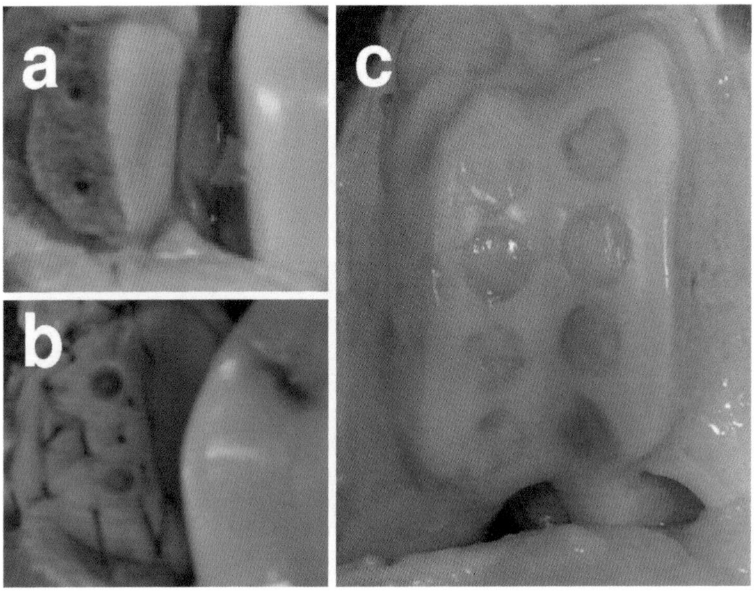

Fig. 2. Two models for investigating cartilage repair in the knee joint of miniature pigs: (**a**) Large articular cartilage defect comprising the entire medial half of the patella. This defect was treated by a cell-loaded PGA matrix fixed with resorbable pins, (**b**) which was additionally covered with a collagen matrix (Chondrogide). (**c**) Multiple circumscribed articular cartilage lesions (diameter of 5 mm) in the femoral trochlea of the knee joint of miniature pigs. Macroscopic analysis demonstrates variable degrees of healing 12 weeks after application of different repair approaches including transplantation of cell-loaded collagen matrices or microfracturing.

8. Xylazine (Rompun, Bayer, Leverkusen, Germany).
9. Fibrin glue (Beriplast; Aventis-Behring, Marburg, Germany).
10. Type I/III bilayer Collagen-Matrix (Chondrogide; Geistlich Biomaterials, Wolhusen, Switzerland).
11. PGA matrix (Soft PGA Felt, Alpha Research, Berlin, Germany).
12. PGA pins (Resor-Pin, Geistlich Biomaterials, Wolhusen, Switzerland).
13. 5-0 Polydioxanone suture (Ethicon, Norderstedt, Germany).
14. Resorbable vicryl suture (Ethicon, Norderstedt, Germany).

2.4. Sample Processing

1. Fixation solution: 4% paraformaldehyde (PFA).
2. Decalcification solution: 0.5 M EDTA.
3. Paraffin (Roth, Karlsruhe, Germany).
4. Xylol (Roth, Karlsruhe, Germany).

2.5. Histology

1. Toluidine blue staining solution: Dissolve 0.1 g Toluidine blue in 100 ml 2.5% Na_2CO_3.
2. Mayer's Hemalaun staining solution (Merck, Darmstadt, Germany).

3. 0.3% Eosin staining solution (Roth, Karlsruhe, Germany).
4. Safranin-O staining solution (Roth, Karlsruhe, Germany): dissolve 1 g Safranin-O in 100 ml pure ethanol and add 50 ml H_2O to create a stock solution; dilute this stock solution 1:10 with distilled water to obtain fresh staining solution.
5. Fast Green staining solution (Merck, Darmstadt, Germany): create a 0.5% Fast Green staining solution by dissolving 1 g Fast Green in 200 ml 0.5% ethanol.
6. Fast Red (Sigma-Aldrich, Taufkirchen, Germany): dissolve one tablet of Fast Red in 1 ml Tris-HCL.
7. X-gal staining solution: 1 mg of X-Gal/ml (Promega, Madison, WI, USA), 5 mM $K_3Fe(II)_6$, 5 mM $K_4Fe(II)_6$, 2 mM $MgCl_2$, 0.2% (v/v) Nonidet P40 (Sigma-Aldrich, Taufkirchen, Germany).
8. Entellan embedding agent (Electron Microscopy Sciences, Hatfield, PA).
9. Aquatex embedding agent (Merck, Darmstadt, Germany).

2.6. Immunohistochemistry

1. Hyaluronidase (Roche, Mannheim, Germany) is dissolved at 0.2% in PBS.
2. Pronase (Sigma-Aldrich, Taufkirchen, Germany) is dissolved at 0.2% in PBS.
3. Tris-buffered saline (TBS; washing buffer): Tris is dissolved at 5 mM in 0.9% NaCl and adjusted to pH 7.35.
4. Antibodies: monoclonal mouse anti-human type I collagen antibody (MP Biomedicals, Irvine, CA, USA); mouse anti-human type II collagen antibody (MP Biomedicals, Irvine, CA, USA); biotinylated donkey anti-mouse secondary antibody (Dianova, Hamburg, Germany).

3. Methods

3.1. Cell Isolation and Culture

3.1.1. Murine Dermal Fibroblasts to Be Injected into Knee Joints of Mice (see Note 1)

1. Isolate murine dermal fibroblasts from the skin of newborn mice.
2. Cut skin tissue into small pieces using a scalpel and digest the tissue with 0.1% trypsin for 20 min, followed by treatment with 0.02% clostridial collagenase for 16 h at 37°C.
3. Purify the cells using a sterile cell strainer with a 100-μm nylon mesh.
4. The cells are to be cultivated in monolayer culture (e.g., on 6-cm dishes) in DMEM supplemented with 10% FCS and 1% Penicillin/Streptomycin.

3.1.2. Rat Perichondrial Cells to Be Administered to the Knee Joints of Rats (see Note 1)

1. To isolate perichondrial cells from the chondral part of the ribs of adult Wistar rats, anesthetize the animals by a single intraperitoneal injection of 20 mg ketamine and 2 mg xylazine prior to sacrificing them by cervical dislocation.
2. Expose the thorax by a skin incision and excise the anterior part of the thorax using surgical scissors.
3. Dissect the chondral parts carefully from the surrounding muscles and expose to 0.2% trypsin for 20 min, followed by treatment with 0.02% clostridial collagenase in DMEM supplemented with 10% FCS. To isolate predominantly mesenchymal perichondrial cells and to minimize the release of centrally located chondrocytes, the digestion period should be limited to 12 h at 37°C.
4. Suspended perichondrial cells are separated from the cartilage cores and purified using a sterile cell strainer with a 100-μm nylon mesh.
5. Then, cells are cultivated in monolayer culture (e.g., on 6-cm dishes) in DMEM supplemented with 10% FCS and 1% Penicillin/Streptomycin.

3.1.3. Rat Bone Marrow Stromal Cells as an Alternative to the Periosteal Cells

1. To isolate rat bone marrow stromal cells (BMSC) from the femur and tibia of Wistar rats, anesthetize the animals by a single intraperitoneal injection of 20 mg ketamine and 2 mg xylazine prior to sacrificing them by cervical dislocation.
2. Disconnect tibia or femur and cut off the proximal and distal meta-epiphyses.
3. Flush the bone marrow cavity with a syringe under sterile conditions with 10 ml DMEM supplemented with 10% FCS and 1% Penicillin/Streptomycin.
4. Filter the resulting cell suspension with a 100-μm nylon mesh and transfer the suspension to culture vessels.
5. The cells are to be cultivated in monolayer culture (e.g., on 6-cm dishes) in DMEM supplemented with 10% FCS and 1% Penicillin/Streptomycin.
6. After 4 days, nonadherent cells are removed by careful washing with PBS.

3.1.4. Miniature Pig Periosteal Cells for Autologous Transplantation (see Note 1)

1. Anesthetize miniature pigs by an intramuscular injection of 1 mg/kg midazolam and 30 mg/kg ketamine followed by ventilation with isoflurane at 2 L/min.
2. Expose the medial aspect of the proximal tibia by a skin incision. Outline the periosteal flap with a scalpel and carefully scrape it off from the bone. It is important to harvest the lower cambium layer of the periosteal flap completely, since this layer contains most mesenchymal precursor cells.

3. Mince the flap with a scalpel. Digest the tissue fragments with 0.2% trypsin for 20 min, followed by exposure to 0.02% clostridial collagenase dissolved in DMEM with 10% FCS at 37°C for 10 h.
4. Purify the suspended periosteal cells from debris with a sterile cell strainer with a 100-μm nylon mesh and resuspend the cells in DMEM.
5. The cells are cultivated in monolayer culture (e.g., on 15-cm dishes) in DMEM supplemented with 10% FCS and 1% Penicillin/Streptomycin.

3.2. Viral Gene Transfer Ex Vivo

Grow the cells in monolayer culture using DMEM supplemented with 10% FCS and 1% Penicillin/Streptomycin to obtain the desired total cell number (e.g., $0.5–1 \times 10^6$ cells/cm^2 of transplanted matrix (miniature pig model); 1×10^6 cells per joint (rat or mouse)). 24 h prior to infection, the cells should be detached using 0.1% trypsin/0.025% EDTA and plated at a density of 5×10^5 cells per 6-cm dish or 1×10^7 cells per 15-cm dish to reach 70–90% confluence at the time of infection.

3.2.1. Adenoviral Gene Transfer

1. For adenoviral infection, the cells are preincubated in 6-cm dishes with 500 μl OptiMEM without FCS for 30 min.
2. Adenoviral vectors diluted in PBS to a total volume of 50 μl are applied at suitable multiplicities of infection (MOI) between 10 and 500 infectious particles (i.p.)/cell (i.p. values corresponding to plaque-forming units). The virus solution should be applied dropwise all over the dish to ensure equal distribution (see Note 2).
3. At 45 min after infection, 4 ml complete culture medium is added (DMEM, 10% FCS, 1% Penicillin/Streptomycin).
4. Following incubation for another 4 h, the cells are washed with PBS, mobilized using 0.1% trypsin/0.025% EDTA, resuspended in DMEM containing 10% FCS, and washed twice with serum-free DMEM to eliminate noninternalized viral particles.
5. Aliquots of 1×10^6 cells are suspended in 500 μl serum-free DMEM and may be kept for 1–4 h in suspension at 37°C until transplantation into the joint or seeding onto a scaffold.

3.2.2. AAV-Mediated Gene Transfer

1. For infection with AAV vectors, the cells (e.g., 1×10^7 cells per 15-cm dish) should be preincubated with OptiMEM without FCS for 15–30 min.
2. AAV vectors are applied at doses up to 1,000 infectious particles/cell in a total of 10 ml serum-free DMEM. Optimal vector dose depends on the cell type and must be determined

in advance by titration experiments evaluating the efficiency of gene transfer.

3. Two hours after infection, 10 ml DMEM with 20% FCS is added to the dish.
4. Following incubation for another 24 h, the cells are washed with PBS, mobilized using 0.1% trypsin/0.025% EDTA, resuspended in DMEM containing 10% FCS, and washed twice with DMEM to remove noninternalized viral particles.
5. Cells may be kept in serum-free DMEM at 37°C for up to 4 h in suspension prior to transplantation into the joint or seeding onto a scaffold.

3.3. Preparation of Cell-Loaded Scaffolds (Miniature Pig Model)

1. Suspended cells ($0.5\text{–}1 \times 10^7$ cells/cm² scaffold) are centrifuged at 550 × *g* for 5 min.
2. (a) For seeding onto a PGA scaffold (soft PGA Felt), the cell pellet and a remainder of approx. 30 μl medium are resuspended in 30 μl fibrinogen and seeded onto a PGA matrix with an area of 2 cm² (shaped correspondingly to the cartilage defect area).
   (b) Gel formation is achieved by adding 30 μl thrombin solution to both sides of the cell-loaded scaffold, which is then incubated for another 24–48 h in DMEM with 10% FCS to allow cell attachment.
3. (a) For seeding onto a type I/III collagen matrix (Chondrogide® Membrane), $0.5\text{–}1 \times 10^7$ cells are resuspended in 100 μl DMEM containing 10% FCS and applied to the rough side of the bilayer collagen matrix.
   (b) The cell-loaded matrices are incubated at 37°C for 30 min without any further medium to enable the cells to settle and attach to the matrix.
   (c) Afterward, the cell-loaded matrix should be transferred to a 6-cm dish containing a larger volume of medium (5 ml DMEM with 10% FCS), followed by incubation for 48 h to allow firm cell attachment.
4. Prior to transplantation into the joint, the cell-loaded scaffolds are washed in serum-free DMEM to eliminate FCS.

3.4. Surgical Procedures

3.4.1. Mouse

1. STR/ORT mice are fed a standard laboratory diet ad libitum. The animal house should be air-conditioned and should provide a constant temperature of 20–25°C with a relative humidity of 40–55% and a light–dark cycle of 12 h.
2. Adult STR/ORT mice are anesthetized with a single intramuscular application of 2.5 mg ketamine and 0.25 mg xylazine.
3. The hair around the knee joint is shaved and the skin is treated with 70% ethanol.

4. 1×10^6 of the cells expressing the transgene are suspended in 15 μl serum-free DMEM. The suspension is injected into the knee joint through the patellar ligament using a 1-ml syringe with a 29-gauge needle. As a control, noninfected cells may be injected into the other knee joint of the same animal. After the injection, the animals are warmed under an infrared lamp and allowed to move freely in the cage.

3.4.2. Rat

1. Female Wistar rats should have reached skeletal maturity (age at least 18 weeks; weight at least 250 g) (see Note 3). The rats are fed a standard laboratory diet ad libitum. The animal house should be air-conditioned and should provide a constant temperature of 20–25°C with a relative humidity of 40–55% and a light–dark cycle of 12 h.
2. The rats (240–270 g) are anesthetized with a single intraperitoneal application of 20 mg ketamine and 2 mg xylazine.
3. The hair around the knee joint is shaved, the skin is disinfected, and the leg is put through a small hole of a sterile surgical drape. After a median skin incision, the knee joint is exposed by a medial parapatellar incision of the joint capsule and the patella is displaced laterally.
4. Eight partial-thickness cartilage defects in the femoral trochlea are created by transverse movement of a custom-made device consisting of four parallel 21-gauge needles clamped between two metal spacers with the needle tips exceeding the edge of the spacers by 0.3 mm (Fig. 3). Since the thickness of rat cartilage at that site is approximately 0.5 mm, this device

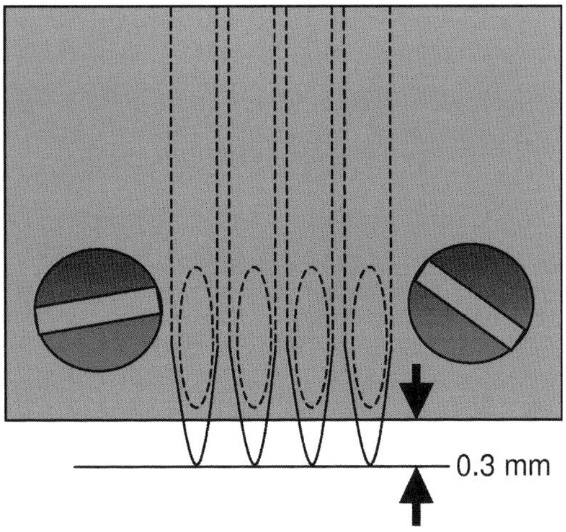

Fig. 3. Custom-made device creating partial-thickness cartilage lesions on the articular surface of the rat femoral trochlea. The device consists of four parallel 21-gauge needles clamped between two metal spacers with the needle tips exceeding the edge of the spacers by 0.3 mm.

prevents subchondral bone plate injuries, which is confirmed by a complete absence of bleeding. It is important that the procedure is carried out carefully to prevent the ingrowth of stem cells from bone marrow spaces, which may interfere with repair cartilage formation by the transplanted cells.

5. Optionally, the exposed cartilage surface may be treated with a hyaluronidase solution (hyaluronidase dissolved at 0.1% in PBS) to allow better cell attachment.
6. The cartilage surface should then be washed thoroughly with PBS using a syringe and dried completely with sterile cotton-tipped applicators.
7. The prepared cell suspension (see above) is centrifuged at 970 × g for 1 min. The supernatant is discarded completely, resulting in a cell pellet with a total volume of approximately 2 μl. To allow for better cell attachment to the articular cartilage, the cell pellet is mixed with 0.5 μl of a fibrinogen/fibronectin solution and 0.5 μl of a thrombin solution. Immediately after mixing the suspension, aliquots (1×10^6 cells per joint) are applied to the area of the lesions using a pipette. To facilitate complete filling of the lesions, the cell suspension is spread carefully over the area of the defects by applying minimal pressure with the lateral side of the pipette tip.
8. The cell suspension is then left for 5 min to adhere and form a gel.
9. Joint capsule and skin are closed with resorbable 5-0 sutures.

3.4.3. Miniature Pig

1. Skeletally mature miniature pigs (18 months, 35–40 kg) are anesthetized by an intramuscular injection of 30 mg midazolam and 300 mg ketamine, followed by ventilation with isoflurane at 2 L/min. The skin around the knee joint is washed, shaved, disinfected, and draped. Following a median skin incision in the anterior aspect of the knee joint, the joint capsule is opened by a medial parapatellar incision and the patella is displaced laterally.
2. Two different types of cartilage defects are described in the following:

   (a) Large partial-thickness defect

   - A large partial-thickness cartilage defect comprising the entire medial half of the articular surface of the patella is created using a custom-made plane-like device in which the blade exceeds the basis by 0.6 mm (Fig. 4). Since the cartilage thickness at this site was determined to be 0.7–0.9 mm, this device does not injure the subchondral bone plate, as confirmed by a complete absence of bleeding.

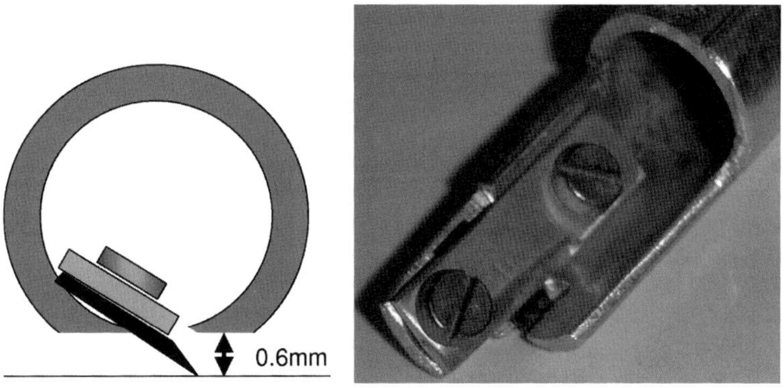

Fig. 4. Custom-made plane-like device creating large partial-thickness cartilage lesions in the knee joint of miniature pigs.

- The cell-loaded PGA scaffold is trimmed to the defect size and fixed with two PGA pins to the subchondral bone (Fig. 2a). To avoid damage to the opposing joint surface, the head of the pins has to be lowered beneath the surface level by milling the superficial part of the drill hole.
- The PGA matrix is mechanically stable but characterized by a relatively rough surface, which may damage the opposing joint surface. To provide a smooth surface, a collagen matrix (Chondrogide) was fixed onto the PGA scaffold by a noninterrupted suture with 5-0 Polydioxanon material (Fig. 2b).

(b) Circumscribed partial-thickness defects

- Smaller partial-thickness cartilage defects allow investigation of different treatment approaches within the same joint. A dermal punch with a diameter of 5 mm is used to cut the rim of cartilage defects. Cartilage tissue within this rim is removed by a curette (see Note 4). The defects may be located in the femoral trochlea or the femoral condyle.
- The same dermal punch is used to cut the cell-loaded collagen matrix scaffold-discs (Chondrogide) with a diameter corresponding to the size of the defect.
- The cell-loaded matrices are press-fitted into the defects with the cell-loaded rough aspect facing towards the subchondral bone plate. The height of the moistured bilayer collagen matrix is comparable to the height of the articular cartilage in the knee joints of miniature pigs. If the size of the cell-loaded matrices fits, they should be retained stably in the lesions. Optionally, fibrin glue may be applied to provide further fixation of the scaffolds (Fig. 2c).

3. The patella is reduced and the stability of the graft is tested by repeated flexion of the knee. The joint capsule and the skin are then closed by resorbable vicryl sutures. After this treatment, the animals are allowed to move freely in their cages.

3.5. Sample Processing

1. Mice are sacrificed by cervical dislocation. Rats are killed by intraperitoneal application of 20 mg pentobarbitone. Miniature pigs are sacrificed by an intramuscular injection of 400 mg ketamine and 30 mg midazolam, followed by intravenous bolus injection of 80 mval KCl.
2. The joints of the animals are carefully dissected and the articular surfaces are first assessed macroscopically.
3. For histological examination, the knee joints have to be fixed in 4% paraformaldehyde followed by decalcification. Mouse and rat knee joints can be fixed and decalcified without further dissection of the tissue. Fixation with 4% paraformaldehyde should be performed for 12–18 h followed by decalcification with 0.5 M EDTA for 3 weeks. To limit the time-consuming decalcification process of the larger minipig samples, it is recommended to resect the treated defect areas in form of osteochondral tissue blocks from the whole joint surface using a miniature rotating saw. Tissue blocks are fixed in 4% PFA for 24 h, followed by decalcification using 0.5 M EDTA. Dependent on the size of the samples, the decalcification process may take up to 12 weeks. EDTA should be replaced at least weekly.
4. The samples are washed with a 0.9% NaCl solution for 12–24 h, followed by exposure to 50% ethanol for 2 h, 80% ethanol for 2 h, 100% ethanol for 2 h, 100% isopropanol for 2 × 2 h, xylol for 2 × 2 h, and hot paraffin for 2 × 2 h.
5. The samples are embedded in paraffin and serial 5-μm sections are cut with a microtome.

3.6. Histology

1. Slides must be deparaffinized by treatment with xylol (3°× 10 min), followed by hydration with 100% ethanol (3°× 2 min), 80% ethanol (2 min), 50% ethanol (2 min) and final rinsing (twice) with distilled water.
2. Staining:
   (a) Safranin-O/Fast Green staining: The slides are covered with Safranin-O staining solution for 5–6 min, followed by rinsing with running H_2O until bone tissue is faintly stained. They may be counterstained with 0.5% Fast Green for 5 min.
   (b) Hematoxylin/eosin staining: The slides are covered with Mayer's Hemalaun for 6 min, rinsed with distilled water,

placed in 1% HCl/99% ethanol, and rinsed for 2 min with running H_2O, followed by staining with 0.3% Eosin for 5 min.

(c) Toluidine blue staining (Fig. 1): The slides are covered with Toluidine blue staining solution for 4 min.

3. The stained slides are rinsed with distilled water for 2 min, followed by dehydration using 60% ethanol for 2 min, 80% ethanol for 2 min, 100% ethanol (2×2 min), and xylol for 3–5 min.
4. Entellan embedding agent is used for final mounting.

3.7. Immunohistochemistry

1. To assess the quality of cartilage repair tissue, it is recommended to detect both type I and type II collagen by immunohistochemistry (Fig. 1) (see Note 5).
2. Sections must be deparaffinized and hydrated as described under Subheading 3.6, step 1.
3. For immunohistochemical staining of type I and type II collagen, deparaffinized sections are pretreated with 0.2% hyaluronidase in PBS (pH 5.0) at 37°C for 60 min and with Pronase (2 mg/ml in PBS, pH 7.3) for 60 min at the same temperature.
4. Wash the sections with 5 mM Tris-buffered saline (TBS).
5. Incubate sections with a monoclonal mouse anti-human type I collagen antibody (diluted in PBS at a ratio of 1:200) or with a mouse anti-human type II collagen antibody (diluted in PBS at a ratio of 1:500), followed by careful washing with TBS.
6. Incubate sections with a biotinylated donkey anti-mouse secondary antibody for 30 min.
7. After careful washing with TBS, a complex of streptavidin and biotinylated alkaline phosphatase is added.
8. The sections are developed with Fast red for 25 min, rinsed with H_2O and then counterstained with Hematoxylin for 20 s, followed by short rinsing with running water.
9. Aquatex embedding agent is used for final mounting.

3.8. Localization of Transplanted Cells (e.g., Following β-Galactosidase Reporter Gene Transfer (see Note 6))

1. Knee joints or tissues treated with cells that had been infected with an adenoviral or AAV vector encoding the β-galactosidase reporter gene are fixed in 2% PFA for 1 h.
2. Wash thoroughly with PBS containing 2 mM $MgCl_2$ and 0.2% Nonidet P40 (NP40).
3. Incubate the specimen in X-Gal staining solution containing 1 mg X-Gal/ml, 5 mM $K_3Fe(II)_6$, 5 mM $K_4Fe(II)_6$, 2 mM $MgCl_2$, and 0.2% (v/v) NP40 for 12 h.

4. Stained tissue should be assessed macroscopically prior to postfixation in 4% PFA for at least 8 h and decalcification for 3–8 weeks.
5. After standard processing, sections should be stained with Safranin-O to differentiate green-stained AdLacZ-positive cells from the background staining of the matrix.
6. The slides are investigated by conventional light microscopy.

4. Notes

1. In small animals (e.g., mice, rats), syngeneic cells may be used. No critical immune responses/rejections have been observed in our previous studies. In large animals (miniature pig), the isolation of autologous cells should be preferred.
2. An optimal MOI should be determined for each cell type in advance, which often lies close to the threshold for toxic effects on the cells. Adenoviral doses above 500 MOI are not recommended.
3. Physes of small animals (mice, rats) do not close throughout their life.
4. If the study aims at investigating the impact of transplanted cells, the subchondral bone plate must be left intact to avoid ingrowth of cells from the bone marrow, which may influence repair tissue formation significantly. However, if combined effects of transplanted cells and ingrowing bone marrow stromal cells are of interest, the defect may be supplemented by additional microfractures with a depth of 2 mm using a 1-mm K-wire.
5. Type I collagen is primarily detected in fibrous repair tissue and indicates insufficient chondrogenic differentiation of the cells. Type II collagen is the major collagen of hyaline articular cartilage and can be considered as a marker of hyaline-like repair cartilage of higher quality (Fig. 1).
6. To localize transgene-expressing cells within knee joints of the animal models described, a LacZ marker gene appears to be superior to green fluorescent protein (GFP), since the GFP fluorescence signal may deteriorate during the long period of sample decalcification.

Acknowledgements

This work was supported by the Interdisciplinary Centre for Clinical Research of the University of Erlangen-Nuernberg.

References

1. Luyten F. P. DBC, Dell`Accio F. Identification and characterization of human cell populations capable of forming stable hyaline cartilage in vivo. In: Hascall V.C. KKE, editor. The many faces of osteoarthritis. Basel: Birkhaeuser; 2002. p. 67–76.
2. Gelse K, Muhle C, Franke O, Park J, Jehle M, Durst K, et al. Cell-based resurfacing of large cartilage defects: Long-term evaluation of grafts from autologous transgene-activated periosteal cells in a porcine model of osteoarthritis. Arthritis Rheum 2008;58:475–488.
3. Robins JC, Akeno N, Mukherjee A, Dalal RR, Aronow BJ, Koopman P, et al. Hypoxia induces chondrocyte-specific gene expression in mesenchymal cells in association with transcriptional activation of Sox9. Bone 2005;37:313–22.
4. Gelse K, Muhle C, Knaup K, Swoboda B, Wiesener M, Hennig F, et al. Chondrogenic differentiation of growth factor-stimulated precursor cells in cartilage repair tissue is associated with increased HIF-1alpha activity. Osteoarthritis Cartilage 2008;16:1457–65.
5. Klein-Nulend J, Semeins CM, Mulder JW, Winters HA, Goei SW, Ooms ME, et al. Stimulation of cartilage differentiation by osteogenic protein-1 in cultures of human perichondrium. Tissue Eng 1998;4:305–13.
6. Wozney JM, Rosen V, Celeste AJ, Mitsock LM, Whitters MJ, Kriz RW, et al. Novel regulators of bone formation: molecular clones and activities. Science 1988;242:1528–34.
7. Gelse K, Schneider H. Ex vivo gene therapy approaches to cartilage repair. Adv Drug Deliv Rev 2006;58:259–84.
8. Evans CH, Ghivizzani SC, Robbins PD. Gene therapy for arthritis: what next? Arthritis Rheum 2006;54:1714–29.
9. Gelse K, Jiang QJ, Aigner T, Ritter T, Wagner K, Poschl E, et al. Fibroblast-mediated delivery of growth factor complementary DNA into mouse joints induces chondrogenesis but avoids the disadvantages of direct viral gene transfer. Arthritis Rheum 2001;44: 1943–53.
10. Gelse K, von der Mark K, Aigner T, Park J, Schneider H. Articular cartilage repair by gene therapy using growth factor-producing mesenchymal cells. Arthritis Rheum 2003;48: 430–441.
11. Gelse K, Pfander D, Obier S, Knaup KX, Wiesener M, Hennig FF, et al. Role of hypoxia-inducible factor 1 alpha in the integrity of articular cartilage in murine knee joints. Arthritis Res Ther 2008;10:R111.

Chapter 17

Ethical Consideration

Michael Fuchs

Abstract

The twofold distinction between interventions into the germ line and interventions into somatic cells on the one hand and between the treatment of diseases and enhancement on the other hand resulted in the concept of somatic gene therapy. There is a nearly unanimous agreement that somatic gene therapy has a high-ranking moral objective and uses methods that extend current techniques for treating diseases in a morally acceptable way. In its experimental phase principles of research ethics as the autonomy and the informed consent of the patient or the test person, a fair selection of test persons and a careful weighing of risks and benefits have to be taken into account and several specific points have to be considered. Experimental somatic gene therapy requires a positive vote of a competent and independent ethics committee.

Key words: Somatic gene therapy, Enhancement, Informed consent, Effects on the germ line, Gene doping, Risk management

1. Introduction

The procedure of preparing, accompanying, and following clinical trials in the field of gene therapy is not only the consequence of an intensive debate concerning technical, biological, and medical requirements, but also owes to an ethical judgment formation. All prerequisites for both the risk management and the risk research result from the moral conviction that risks for the affected patient or test person as well as risks for third parties are to be avoided, minimized, and balanced with the respective benefit. In the case of somatic gene therapy, it has to be stressed that the concept itself is the outcome of an intensive ethical discussion.

The discussion concerning corrective genetic interventions into the human being has come into being at a very early point in time. More strongly than in other fields of innovative research,

Otto-Wilhelm Merten and Mohamed Al-Rubeai (eds.), *Viral Vectors for Gene Therapy: Methods and Protocols*, Methods in Molecular Biology, vol. 737, DOI 10.1007/978-1-61779-095-9_17, © Springer Science+Business Media, LLC 2011

it was the researchers involved who promoted this discussion. They were guided by the thought that such interventions would be ethically acceptable only if there was a clear-cut distinction between the legitimate goals of medicine and the intentions of the breeding of humans. They were the precursors of gene therapy, in particular, who clearly distanced themselves from all kinds of utopia regarding the improvement of the human species (1) like it had been presented at the CIBA Foundation symposium held in London in 1962 on the theme "Man and His Future" (2).

As an instrument for the differentiation, a twofold distinction was established: the one between interventions into the germ line and interventions into somatic cells as well as the one between, on the one hand, the treatment and therapy of diseases and, on the other hand, the improvement and enhancement of other bodily features (3). This twofold distinction led to one major result: the therapeutic intervention into somatic cells – the somatic gene therapy – had been established as a concept, which was distinguished from all genetic interventions of an ethically doubtful nature.

2. Somatic Gene Therapy in Public Perception

Attempts to express a transgene in human somatic cells for therapeutic reasons have, in general, been regarded a high-ranking moral objective. Poll data provided by the Louis Harris organization in 1986 and 1992 showed the openness of the American public to the use of genetic technology in the war against disease. The more urgent the medical need, the more likely respondents were to approve the use of gene therapy. The data also reflected a strong support of research in that area. On this topic, Walters and Palmer found 28 policy statements by governmental authorities, groups of experts consulted by state authorities, churches and medical associations published from 1980 through 1993. All 28 policy statements agreed that somatic cell gene therapy for the treatment of serious diseases is, in principle, ethically acceptable. "There are few issues in all of biomedical ethics on which one would be able to discover such unanimous agreement" (4). The main reason for this strong agreement is that both in the scientific community and in public perception "somatic cell gene therapy has increasingly come to be viewed as a natural and logical extension of current techniques for treating disease" (5, cf. also 4).

In medical ethics, the argument that a new technology is morally acceptable because there is an analogy between this new technology and an old one that is ethically non-problematic is always an important argument. If such an analogy can be found, it only has to be asked whether the old technology is really not linked to any difficulties and whether there are no morally significant differences between the old and the new technology. The analogy which can be found most frequently in literature is the one between somatic cell gene therapy and organ or tissue transplantation (6–9). Comparisons to other medical procedures have also been proposed. One argument, for instance, assumed that the new therapy closely resembles medications or enzyme therapies that were currently being in use. Since some applications of gene therapy function as vaccinations, this has been regarded as another similarity between gene therapy and common types of vaccines. All these analogies have been used in favor of the "gene-therapy-as-extension view" (4). I shall come back to this question when commenting on the appropriate regulation procedure for gene therapy. Given the gene-therapy-as-extension view, the necessary ethical analyses of somatic gene therapy do, however, not seem to differ categorically from conventional experimental therapy. Some researchers even argue that somatic cell gene therapy is less invasive than the allotransplantation of a major organ or the allotransplantation of tissues, such as bone marrow, inasmuch as the probability that the cells will be rejected by the immune system of the patient is much lower.

3. Major Ethical Issues to be Considered in Somatic Gene Therapy

The long period of ethical reflections prior to the first authorized clinical trials made it possible to find at least a consensus on what the major ethical issues connected to somatic gene therapy are. They can be formulated as follows:

1. What is the disease that is to be treated?
2. What alternative interventions are available for the treatment of this particular disease?
3. What is the anticipated or potential harm of the experimental gene therapy procedure?
4. What is the anticipated or potential benefit of the experimental gene therapy procedure?
5. What procedure will be implemented to ensure fairness in the selection of patient-subjects?
6. What steps will be taken to ensure that patients, or their parents or guardians, give informed and voluntary consent to their participation in the research?
7. How will the privacy of patients and the confidentiality of their medical information be protected? (cf. 4, 10).

1. What is the disease that is to be treated?

The first aspect of the first question posed is, again, the distinction between therapy and enhancement as well as the problem of drawing a line between these two. When comparing eye diseases and deafness as possible candidates for gene therapy, it becomes obvious that it is neither convincing to understand health as species-typical functioning nor to regard health, illness, disease, or malady only as conventionally defined by a given society. It seems more appropriate to begin with the concept of disease understood as an action-related concept, i.e., a treatment-related concept that is ascribed to a specific situation as a result of a dialogue between patient and physician. Although deafness might be seen as a fault of the species-typical functioning, many deaf people do not ask for a cure even if it is available.

It might be concluded that it is wise to start clinical trials with diseases where there is no doubt about their status as a disease. Moreover, in view of the high risk, it becomes apparent that it would be best to begin with serious diseases. The question which serious disease is chosen as a target disease is usually being discussed in terms of its feasibility and potential effectiveness. Taking a closer look at this question reveals that it is also an ethical question, a question of utility and justice. What is the right balance when taking into account the burden of disease, the number of patients affected and a genuine research strategy? It seems as if neither literature nor expert discourses provide a clear answer concerning this matter. If we accept that there are limits of planning efficient research strategies we have to ask what degree of freedom for the researchers is acceptable from justice's point of view.

The genuine research opportunity seems to coincide with the priority for targeting orphan diseases. On the other hand, we have to admit that research, and biomedical and clinical research in particular, is not independent from the market. Ever since its beginnings, gene therapy has been caught in the tension between, on the one hand, physicians' aspirations to help their own patients and, on the other hand, long-term research strategies for developing a significant alternative therapy. The poles of these two conflicting fields can become more salient during various phases of research caused by the involvement of research funding institutions and sponsors; they can thus turn out to be of direct relevance for the decision-making in the preclinical and clinical procedures. Acting in such a conflicting environment does not only pose a moral challenge but requires thorough ethical reflection. This is due to the fact that with each decision taken to select a certain disease as a model case, another serious decision is made regarding the risks considered to be reasonable and the prospects for curing this genetic ailment. At the same time, the responsibility for today's generation is weighed against that for future generations; collective groups of patients of different sizes are also balanced against one another. Different stakeholders look after their own interests and their moral intuitions that serve as a foundation of their individual position. The interplay of interests and moral intuitions is very complex, which is the precise reason for its need to undergo an ethical analysis. In conjunction with acknowledging groups of patients and research strategies, another difficulty emerges, namely that therapies are often developed for curing serious and rare diseases, hoping that the costs of development will be covered by applying and marketing the therapy for related and more frequently found diseases. With a focus on justice, the question of appropriate selection criteria arises; when focusing on dignity, a more fundamental question needs to be addressed concerning the ethical limits prohibiting to regard patients solely as means and to use them as such.

2. What alternative interventions are available for the treatment of this particular disease?

The second question is linked both with the first and with the fifth question. Most policy statements argued that gene therapy should start with diseases for which no alternative therapy was available. Where an alternative treatment was, in fact, available, trials should only involve patients for whom no treatment was available. This was, for instance, the case with patients affected with ADA-SCID who had no matching bone marrow donor. ADA-SCID is a hereditary, severe combined immune disease due to a deficiency of the enzyme adenosine deaminase (ADA). With the progress of gene therapy techniques as well as risk assessment strategies and evidences of therapeutic success, the question arises as to when the requirement of having no effective alternative can be or needs to be loosened. The question should be posed whether there is a need for controlled studies that compare somatic cell gene therapy with the alternative approach before patient recruitment policies can be changed.

3. What is the anticipated or potential harm of the experimental gene therapy procedure? and 4. What is the anticipated or potential benefit of the experimental gene therapy procedure?

A clear connection can also be seen between question 3 and question 4: Researchers are asked to provide the results of studies performed in their own laboratories or in those of others. The common view is that "if a gene therapy protocol provides a satisfactory answer to the question about harm, it must also offer at least a low probability of benefit to the patients who are invited to enter the protocol, and it must have an excellent scientific design, so that the information gathered from studying the early patients will be useful to later patients and to the entire field of gene therapy research" (4). The circumstances of the death of Jesse Gelsinger might indicate the problem of pure toxicology studies in the field of gene therapy. In September 1999, 18-year-old Jesse Gelsinger died a few days after having been injected with a high dose of adenoviruses in an experimental trial. The adenoviruses carried a gene, which, it was hoped, may cure a severe deficiency that impairs the urea cycle and can cause death even in early childhood. Gelsinger participated voluntarily in the trial which – as is standard practice in Phase I trials – was intended to determine toxicity. He himself suffered from the deficiency of the ornithine transcarbamylase (OTC) enzyme in the liver that impairs the urea cycle and which subsequent phases of the trial were intended to remedy. Although he was not entirely free of the symptoms of his condition, they were largely under control and were not life-threatening. Jesse Gelsinger died of the immune reaction to the injected adenoviruses.

Unclear risks without potential benefit should not be permitted even if a formal consent is given. Weighing harms and benefits is always difficult especially when both are potential. After the ADA and SCID-X1 trials in Paris, Milano and London we know more about the positive outcomes as well as about the risks. At the Necker Hospital in Paris, Alain Fischer had treated children suffering from the severe combined immune deficiency SCID-X1 – the X-linked severe combined immunodeficiency due to a mutation in the gene encoding the common γ (γ c) chain – which is usually fatal in early childhood. He had removed haematopoietic stem cells from his patients and corrected the genetic defect in these cells using gene transfer. Trials have since also been conducted in Milan and London that also reveal the effectiveness of somatic gene therapy for significant disorders of the immune system on the basis of adult stem cells. Yet the hopes of the patients and their relatives, especially in the case of SCID-X1, were soon dampened by the diagnosis of unusual leukemia that occurred in a number of the Paris patients and later on in one London patient. Based on the various cases, conclusions can be drawn regarding dangerous doses and perhaps concerning age groups that are particularly at risk. Yet, the patients in question are not only at risk due to leukemia caused by gene therapy, but also – and in the first place – due to their severe immune deficiency. Without effective therapy, the children treated by Alain Fischer would probably no longer be alive.

Further security studies in mice fuelled ideas about how to prevent the risks; nevertheless, a situation of uncertainty still prevails. Is it ethical to start trials now only as a compassionate response to requests of dying patients or their parents (or legal representatives) following the principle of rescue? Is there any evidence that the term "hopelessly incurable" is used to justify dangerous experimentation, as Henry Beecher feared at the very outset of gene therapy? Would it be better to wait for further results from security research seen from the patients' point of view? Would it be better to wait for further results following the logic of research? There is evidence showing that stem cell research has opened the door for efficient gene therapy. One might argue that new vectors will probably provide a higher degree of security. Apart from the efforts to construct customized vectors for particular therapeutic genes on a viral basis, the ability of zink finger proteins to bind specifically to DNA-sequences in order to develop new gene vectors is of high importance for security research and its clinical application. Would it be ethical to wait?

5. What procedure will be implemented to ensure fairness in the selection of patient-subjects?

Investigators ought to assure a fair procedure in the selection of patients. For many cancer patients being included in a clinical trial represents a last hope even if the chance for a cure is, indeed, very small. In target diseases such as SCID-X1 or Wiskott–Aldrich Syndrome (WAS), the situation presents itself as being entirely different. Here, the investigators have to hope for getting enough patients to eventually obtain a convincing protocol. Another vital question to be considered is about whether children ought to be included in clinical trials. Since gene therapy still involves more than minimal risk, they are only allowed to participate when there are clear evidences for a potential benefit.

These questions concerning minors in research acquire special relevance because the intervention at the earliest possible stage is closely linked to the paradigm of gene therapy as a causal therapy; particularly relevant problems are further constituted by risks and uncertainties. Investigators should be able to make sure that the potential benefit for each individual patient is higher than that of the standard treatment.

6. What steps will be taken to ensure that patients, or their parents or guardians, give informed and voluntary consent to their participation in the research?

There is no doubt that gene therapy requires researchers to convey to potential patient-subjects the facts about their condition, the major alternative treatments, and the procedures and strategies applied in the research. Since inherited diseases are sometimes very rare and patients thus come from all over the world to see specialized researchers in clinical centers working with gene therapy, it may be difficult to give oral presentations and to discuss the relevant issues in the native language of the patient. Patients should understand the need of follow-up studies and comply with it.

7. How will the privacy of patients and the confidentiality of their medical information be protected?

Compared to other developments in modern medicine, gene therapy has proven to treat its patients and their families very respectfully; the media have acted alike, at least there seem to not have been any media circuses in that field endangering the patients' privacy. The question how to deal with the press and the media is not so much a concern for the confidentiality of medical information on the side of the patients but much rather a question of creating hopes and assuring transparency on the side of the researchers.

Researchers should be aware that many people and especially scientists from fields other than gene therapy still think that gene therapy has been promising too much and yet provided too little. They have to explain why scientific and therapeutic success needs time. The question of transparency was also addressed by the American review system. All meetings of NIH Recombinant DNA Advisory Committee are to be open to the public. Nevertheless, the review system was accused of being partly responsible for the death of Jesse Gelsinger. That is why the review system deserves to be looked at more closely.

4. Defining an Adequate Ethical Review System (see Box 1)

A principal goal of public policy in the field of medical research and clinical testing is the protection of human subjects involved in clinical trials. Since the General Assembly of the World Medical Association (WMA) took place in Tokyo in 1975, the establishment and the work of research ethics committees are seen as the essential means of ensuring protection of human subjects. Meanwhile, these committees are also considered guarantors of high scientific quality and protectors of researchers and their respective institutions. In 1978, the National Commission for the Protection of Biomedical and Behavioral Research in the USA recommended that research ethics committees should be "located in institutions where research involving human subjects is conducted. Compared to the possible alternatives of regional or national review process, local committees have the advantage of greater familiarity with the actual conditions surrounding the conduct of research" (11). Since human gene therapy as a result of discussions was evaluated as an appropriate goal as well as an acceptable means and as an extension of customary medical means, protocols for gene therapy have been prepared and were hence subjected to an ethical review process. Bearing in mind the potentially high risks and the scientific and ethical difficulties of their evaluation, the regulation procedure that was established in

Box 1
Important Moral Concerns About Somatic Gene Therapy

1. What is the disease that is to be treated? Is it a severe disease?
2. What alternative interventions are available for the treatment of this particular disease?
3. What is the anticipated or potential harm of the experimental gene therapy procedure? Does the insurance cover all costs that could result from the procedure?
4. What is the anticipated or potential benefit of the experimental gene therapy procedure? Do the potential benefits for the individual justify the potential harms?
5. What procedure will be implemented to ensure fairness in the selection of patient-subjects? Do the inclusion criteria take into account an individualized risk-benefit analysis?
6. What steps will be taken to ensure that patients, or their parents or guardians, give informed and voluntary consent to their participation in the research?
7. What measures have been taken to avoid a therapeutic misconception?
8. How will the privacy of patients and the confidentiality of their medical information be protected?
9. How can the social context of the experimental gene therapy be described? Are the decisions of the researchers and physicians independent from commercial interests?
10. Is the public adequately informed about the procedure and its positive and negative outcomes?

the USA presents itself as being of a very complicated nature. Control of government-funded trials occurs at both the local and the national level. At the local level, facilities where experiments would potentially take place are required to have Institutional Review Boards to ensure the protection of human subjects and Institutional Biosafety Committees to approve the gene insertion in advance. At the national level, the Director of the NIH must approve each human gene therapy proposal. He seeks advice from the Recombinant DNA Advisory Committee (RAC). All meetings to the RAC are to be open to the public. Many private companies submit proposals to NIH for review although they do not receive federal funding. In addition, the FDA must also review and approve clinical trials involving gene therapy. For a long time, this complex system was held up as a model for transparency, expertise and synthesis of the local principle as well as the requirement of harmony. After the death of Jesse Gelsinger, the same system was criticized because of a lack of communication between the committees involved and the inadequate control for conflicts of interest.

Compared to the American situation, there is no harmony in Europe as far as the procedures for approving gene therapies are concerned. In some countries, the national level is even stronger than in the USA. In the Netherlands, for instance, the central committee (Centrale Commissie Mensgebonden Onderzoek) has the direct authority to supervise research areas such as gene therapy. Hence, each protocol in gene therapy needs the approval of the committee. In France or Germany, on the other hand, the regional and the local level are more important than the national one. When the European Directive for Good Clinical Practice had to be implemented into German law, discussions arose about whether the central committee for the evaluation of gene therapy trials at the federal chamber of physicians should be mentioned in the new law; the legislator eventually decided to keep up the local principle (12). One might argue that the division of responsibility has shown to be problematic in the USA.

Another issue that is to be addressed is whether special regulations for gene therapy experiments are really needed. Are the risks really higher than it is the case with regular drug trials? Are local committees able to evaluate research procedures for which a distinction between phase 1 and phase 2 is difficult or even impossible to make? Do all trials in the area of somatic gene therapy deserve the same answer to these questions or do gene therapy protocols follow different strategies deserving different requirements?

The consensus about the gene-therapy-as-extension view mentioned above left open the question concerning what could be defined as the exact starting point of the extension and in which direction this extension is headed. Looking back on the theoretical debate and the history of clinical trials, different answers are possible. As regards the ideal of gene repair as a concept of causal therapy, the deviation from classical therapies is remarkable. Thinking along this line, we have to be aware of the fact that germ line therapy could be seen as the logical next step. Similarities to common types of therapy can be observed regarding different conceptions of gene addition. The best analogy is not easy to define. The use of stem cells, bone marrow, and other tissues renders some studies on gene therapy comparable to classical tissue transplantation. It is a question of the state of the art in security research whether or not the remaining differences justify a specific regulation procedure for gene therapy trials. Other therapies are seen in analogy to medications. Finally, the analogy of vaccines is important for some kinds of studies. The specific ethical question at hand here is the distinction between prevention and enhancement. Talking about an adequate regulation system and the remaining ethical questions, we should keep these different approaches in mind.

5. Germ Line Therapy and Inadverted Effects on the Germ Line

Although several attempts have been made to fuel a debate on germ line therapy, the discussion on this issue has remained theoretical because a scientific venture with consequences in policy-making has failed to appear.

When looking at the early documents of various ad-hoc committees in several European countries, we find that the majorities of these committees were in favor of somatic-cell gene therapy while being opposed to germ line therapy and genetic enhancement (6, 7). If we look a little bit closer here, we can recognize that, while their arguments are quite similar, some propose a moratorium for germ line therapy while others demand a formal prohibition of all attempts to deliberately modify the genome of germinal cells and even of any gene therapy involving the risk of such a modification (7). But even this consensus concerning a critical attitude is incomplete and the debate on germ line gene therapy is being led continuously all over the world. Some scientists as well as ethicists argue that it is medically necessary to prevent certain classes of diseases and that it should not only be allowed but even advisable since it fits with the duty to remove harm. The European Society of Gene and Cell Therapy (ESGCT) shares the ethical and social concerns against germ line cell modification and consequently does not support it (9). The American Association for the Advancement of Science (AAAS) called for a moratorium on inheritable genetic modifications technology (13). The documents of the academic associations do not reflect the whole spectrum of positions though. An opinion poll conducted at the beginning of our millennium showed that a clear majority of responding scientists (64%) from the American Society of Human Genetics view germ line intervention to prevent serious diseases as ethically acceptable, if and when gene repair or replacement were to become a safe and validated technique (5).

The debate is very much focused on germ line therapy. Nevertheless, some of the guidelines and legal documents also address the question of inadverted effects on the germ line. As far as the normative assessment of intentional and inadverted germ line modification is concerned, there are three possible positions to be held.

A first position holds that both intentional and unintentional inheritable genetic modifications have to be legally excluded. CCNE, the French National Ethics Advisory Council, may serve as an example: "Gene therapy should be restricted in its scope to somatic cells, and there should be a formal prohibition of all attempts to deliberately modify the genome of germinal cells, and of any gene therapy involving the risk of such a modification. For the same reasons any transfer of genes by viral vectors into the human embryo should be prohibited, because of the risk of damaging germinal cells" (7).

For a second position, germ line interventions that are directly wanted are less problematic than inadverted interventions. A report that was prepared for the AAAS by Mark S. Frankel and Audrey R. Chapman argued in that sense: "The possibility of genetic problems occurring as a result of the unintended germ line side-effects of somatic cell therapy seems at least as great or greater than those that might arise from intentional IGM. Presumably, if researchers were conducting intentional IGM they would be using methods designed to cause the least possible genetic disruption in germ cells. Further, if they were using in vitro embryos, they would attempt to monitor the effects of the genetic manipulation before they implanted an embryo. With intentional IGM, there would be at least some safeguards for minimizing the possibility that a person would be born with iatrogenic genetic damage. The same cannot be said of an inadvertent germ line modification" (13, cf. also, e.g., 14).

Finally, international law and a European Directive prohibit intentional germ line modifications but not procedures that can have such an effect without this intention: "An intervention seeking to modify the human genome may only be undertaken for preventive, diagnostic, or therapeutic purposes and only if its aim is not to introduce any modification in the genome of any descendants." (15); *see* also EU Directive on good Clinical Practice, Guidelines of the 6th Framework Program.

What could be the rationale underlying this third position? In moral philosophy, the so-called doctrine of double effects maintains that it may be permissible to perform a good act with the knowledge that bad consequences will ensue, but that it is always wrong to do a bad act intentionally for the sake of good consequences that will ensue. In an ethical framework where germ line intervention is categorically excluded, it opens the space for somatic gene therapy in which the risk of germ line modifications cannot entirely be excluded. The underlying distinction between intention and side-effect may operate within a non-absolutist framework provided that it functions as a deontic constraint over a certain range of action.

But why is intentional germ line modification prohibited even if the intention is to cure a disease? There are some arguments against germ line therapy. They correspond to the procedure of developing a germ line intervention, to the goal setting and the demand as well as to societal consequences. As regards the procedure of developing a germ line intervention, especially the scenario of a consumptive embryo research is to be discussed. Both advocates of an absolute as well as those for an alleviated claim for protection of the human embryo have to articulate concerns here. In many cases, malformations would arise in the experimental stadium in a development phase of the human being, in which its dignity and claim for protection are entirely beyond doubt. Concerning the goals, it has to be asked whether an intervention at a later point in time, which is targeted at the individual alone, would yield more success. Is there a clear and undisputed demand of such an intervention into the germ line, which would make the consent of future generations seem to not be essential? Looking at the societal consequences, the need arises to pick out the delimitation from enhancement and eugenics, apart from questions concerning the equality of allocation that arise anytime when it comes to new options for acting. Oftentimes, reasons are being given for such a delimitation to be even more problematic than it is the case with an intervention into somatic cells. Undoubtedly, these reasons are of high importance and plausible to some degree. However, a further discussion is not rendered unnecessary because these reasons have debatable presuppositions and they do not seem to be coercive for everybody.

6. Enhancement and Gene Doping

In the field of bioethics, the term "enhancement" generally describes a correcting intervention into the human body, which does not treat an illness, or rather, which is not medically indicated. Oftentimes, the term "enhancement" is being used interchangeably with the term "enhancement genetic engineering" for the targeted manipulation of human characteristics that are genetically determined. A distinction is being made between an enhancement genetic engineering, which supports or originates a specific characteristic, which an individual person wishes to have for him- or herself (somatic cell genetic engineering), and such an intervention, by which the desired feature is being aspired for the offspring too (germ cell genetic engineering) (16). Apart from "enhancement" and "enhancement genetic engineering," the term "improvement" (Fletcher) has also been made use of. Anderson distinguishes enhancement genetic engineering from eugenic genetic engineering; in doing so, he relates the former to the changes of single known features, while relating the latter to the attempt to correct complex human features, meaning features that are being coded through a large number of genes, such as personhood, intelligence, or character (1). This differentiation has, however, not established itself (4).

The question concerning the qualification of distinguishing between treating a disease and enhancement as well as the application of this differentiation for ethically drawing boundaries is controversially being discussed. Starting points for this controversy were not, initially, clinical projects but thought experiments regarding the qualification of experimental interventions into the genetic fundamentals of intelligence and aggression (17). Anderson, in particular, advocates the rejection of genetic enhancement. He thus ties up to earlier considerations according to which the tasks of a physician do, indeed, include the treatment of diseases as well as their prevention, but do not encompass an enhancement of health and the improvement of the human organism by means of psychosocial and biological methods (18). By making use of the example of the implementation of an additional growth hormone gene in the case of a normal child (1), Anderson lists the medical risks emerging due to the introduction of alien genes into the human genome, which are not justified by the battling of a severe disease; he also names the problems concerning equality as to the decision who will be given the opportunity of genetic enhancement as well as the risks of discriminating people carrying features that are unwanted by society (16). Although this drawing of boundaries is subject to the difficulty to operate with the criterion of disease, which is, in fact, not selective, Anderson deems genetic engineering's orientation toward the goal of battling severe diseases as unproblematic. Based on more expansive experiences, the procedure could then be expanded to other groups of diseases. This proposal concerning the drawing of boundaries has met a lot of approval (Royal Commission, United Kingdom, Federal Ministry of Health, Germany, CCNE, France) and has found its way into the European Council's Convention of Human Rights on Biomedicine, which permits interventions into the human genome only for preventive, diagnostic, or therapeutic purposes (15). At the same time, the proposal is subject to criticism from various sides. Critics of genetic engineering advocate that the differentiation is not sufficiently clear as to be able to prevent the impending slippery slope from occurring (17). Others criticize the distinction of therapy and enhancement and the moral ban of enhancement through genetic engineering for its undue elimination of options for acting (19, 20). In particular, it is being asserted that opportunities for intervention belonging into the scope of traditional medical action would be declared illegitimate.

Torres refers to the fact that there are, indeed, medical interventions that are not linked to a treatment of disease, such as a cosmetic surgical correction of nose and breasts, but which are nevertheless being considered parts of regular and accepted medical practice (21). He advocates a differentiation between forms of genetic engineering that constitute an end in themselves and forms of genetic engineering by means of which no diseases are treated either, but which do improve the prerequisites for a possible therapy. He thus deems a genetic engineering intervention as legitimate, which, for instance, renders bone marrow cells resistant against the undesired side effects of chemotherapy.

Against the backdrop of similar examples as well as clinical protocols, the threshold between medically indicated action and enhancement is being called into question, be it in order to criticize the project of genetic engineering in its totality and to demand a narrower drawing of boundaries (22), or be it to further the goal of avoiding, in advance, premature general bans concerning a differentiated assessment of the social consequences (4).

If the exclusion of a non-medically indicated use of genetic engineering on the human being is to be retained, a clear demarcation between prevention and enhancement is to be strived at. Concerning this matter, Juengst suggested to recur to a notion of disease, which is epistemically robust. He asserts that our efforts to prevent diseases can be considered as medically legitimate requirements only if a settlement regarding the definition of disease is successful (23). In later works, he expresses doubts about the possible success of such a demarcation between legitimate preventions and a clear enhancement. Each and every reasons in favor of the permissiveness of a treatment of disease by means of genetic engineering and against the ban of enhancement – no matter where the exact boundary is being drawn – is based on the insight that an ethical judgment is in need not only concerning the choice of means but also as regards the choice of goals. Undoubtedly, battling diseases is a goal of high value. In the course of assessing risks and expected benefit, it is being assigned more weight than the desired correction of a particular other feature. In cases of interventions into the germ line, in particular, the difficulty arises as to which correction actually entails an enhancement for the person concerned. Not without further ado can a clear criterion for exclusion for a gene-enhancing intervention into somatic cells be gained from this balancing of goals, insofar as this intervention is carried out after a consultation has been provided and the consent of the person concerned has been given. The criterion for exclusion is obtained only by means of additionally recurring to the fundamentals of the medical professional ethics or to assumptions of naturalness, which also include the assumption of the natural imperfection of the human being.

Although somatic gene transfer in the fight against diseases registers successes only slowly and experiences setbacks again and again, the same procedure seems to open up possibilities by genetic intervention into the body of sportsmen.

The opportunity for enhancement through genetic engineering has hence assumed a concrete form, which could even become reality in the not too distant future. Enhancements such as those that have already been achieved in athletes through medical-technical interventions can now be aimed at by means of genetic engineering: increasing the oxygen saturation of the blood or affecting genes, which restrain or promote muscle growth.

Among experts of different relevant disciplines it is currently contentious how effective such gene transfer methods can be in the sporty range, when they will be applicable to athletes and whether valid and practicable proof procedures for the Anti-Doping Agencies will be available in time. Such attempts would undoubtedly be connected with substantial health risks for the athletes. However, experience shows that the respective readiness to assume a risk is very high, at least with a part of the athletes in competitive and professional sports; it is increasingly so in the leisure range.

It has to further be discussed whether, as a result of new procedures and beyond the difficulty of proof, additional problems arise concerning the depth of the interference and the extent of the manipulation or also the effects that those procedures might have on third parties. Does the danger of the instrumentalization and self-instrumentalization of sportsmen intensify, one must ask, by gene doping? Does the face of sport change, and does the sportsman's part of the achievement diminish in favor of the sport physician and the molecular biologist? Or does gene transfer make available only an additional means to the long time well-known goal of doping?

The risk analysis of gene doping proceeds from the well-known risks of somatic gene therapy. In this range, an extensive safety research is under way, which is not by any means able to quantify the risk potential of innovative gene therapies today though; this is due to the difference of the treated diseases and the partially small numbers of patients included in studies as well as due to the novelty of the subject.

Moreover, with risk analysis of gene doping it is assumed that as a result of the illegal framework additional problems will arise since requirements of control and safety conditions would only be pronounced very little if at all. In a comparative ethical analysis one would have to examine, if necessary, how these arguments could be deployed if gene doping were no longer illegal.

At present, gene transfer procedures hold a high danger potential. It is connected with substantial risks and uncertainties. Only in individual cases can these endangerments be justified, because only that way does a healing chance exist or does an advantage result in consideration with the risks and dangers of alternative, conventional therapy procedures. This individual case consideration cannot be foregone even if a consent of the concerning individual is present.

From an ethical point of view, the use of gene doping cannot be endorsed. Especially health endangerments, injuries of rules of fairness and the risk of instrumentalization represent crucial obstacles for the permission of gene doping (24). The ethical evaluation of hazard potentials depends on the empirical validity of their analysis. Even if danger assumptions represent only one column of the argumentation against doping, from a normative perspective an additional validating is nevertheless advised when regarding both the dangers of single procedures concerning the individual sportsman and the probability of imitation effects in the leisure range, which ought to be clarified sociologically.

As a consequence, the safety standards generally necessary for gene transfer must be examined and supervised to the effect that possibilities of illegal application with the goal of the enhancement of performance in sports remain impossible or, at least, limited. Generally, genetic procedures do, however, not experience completely different evaluation than is the case with non-genetic procedures. In consequence of the prohibition of gene doping, a renewed discussion of procedures so far considered as legal or not clearly as doping is hence required under the criterion of normative consistency, e.g., concerning procedures for pain management and for mental fitness in competitions.

7. Conclusion

It is to be noted that in the future, increasing experiences will make gene therapy appear no longer just an ultima ratio, but an equal, if not preferable option of treatment. However, the use outside of very severe, life-threatening or the quality of life substantially impairing diseases does not seem to become an ethically legitimate option in foreseeable time. There will nevertheless be concrete scopes of action within which the option of a non-therapeutic usage arises for individuals or particular groups of people. These options demand a societal discussion not only of the technical risks of intervention and those that concern a person's health but of the goals as well. Researchers shall contribute to this discussion as experts and as citizens. This also applies to the intended intervention into the germ line for which the reflection of reasons for rejecting an intervention has not yet yielded a lasting result.

In the perspective of ethics, the concept of somatic gene therapy has nonetheless proven to be of success. It excludes applications that still have to be considered cautiously and skeptically and thus opens up to medical science an interesting instrument for curing diseases. Due to this instrument's technical complexity as well as because of the targeted diseases' heterogeneity, there are numerous issues that have to be considered and clarified critically from an ethical point of view. The consensus concerning the exact nature of these issues has remained stable over the years of handling the experimental therapy. It is the answers given respectively that differ. Which vectors are more risky, which are less risky, which risk can be handled in regards of a potential risk and which cannot be born? Which experiences from gene therapy concerning a specific disease can be applied to another disease?

Responsible answers to these questions can be found, respectively, when different experts of the fields of virology, security research and clinical research work cooperatively, discuss their strategies with groups of patients and present them publically. Dialogue and transparency are ethical claims enabling a lasting success of a new therapeutic strategy.

References

1. Anderson, W. F. (1985) Human gene therapy: Scientific and ethical considerations *J Med Phil* **10**, 275–91.
2. Wolstenholme, G. E. W. (ed.) (1963) Man and his future: A CIBA foundation volume (Boston: Little, Brown).
3. Walters, L. (1991) Ethical issues in human gene therapy *J Clin Eth* **2**, 267–74.
4. Walters, L., Palmer, J. G. (1997) The ethics of human gene therapy (New York, Oxford: Oxford University Press).
5. Rabino, I. (2003) Ethical issues *Theor Med Bioeth* **24**, 31–58.
6. Der Bundesminister für Forschung und Technologie (ed.) (1985) In-vitro-Fertilisation, Genomanalyse und Gentherapie. Bericht der gemeinsamen Arbeitsgruppe des Bundesministers für Forschung und Technologie und des Bundesministers der Justiz (München: Schweitzer).
7. Comité Consultatif National d'Éthique pour les Sciences de la Vie et de la Santé (CCNE) (1990) Opinion on gene therapy. N° 22 – December 13, 1990.
8. Montgolfier, S. (2000) La thérapie génique humain, http://www.ethique.inserm.fr/inserm/ethique.nsf/937238520af658aec125704b002bded2/fd217f8102758fefc12570a500515225? OpenDocument (2009-11-05).
9. European Society of Gene Therapy (ESTG) (2002) Position paper on social, ethical and public awareness issues in gene therapy.
10. De Wachter, M. A. M. (1993) Experimental (somatic) gene therapy. Ethical concerns and control (Maastricht: Instituut voor Gezondheidsethiek).
11. National Commission for the Protection of Biomedical and Behavioral Research (1978) Report and recommendations institutional review boards (Washington, DC: US Government Printing Office).
12. Fuchs, M., Heyer, M., Fischer, N. (2006) Provision of support for producing a European directory of local ethics committees (LECs): Draft final report.
13. Frankel, M., Chapman, A. (2000) Human inheritable genetic modifications. Assessing scientific, ethical, religious and policy issues (Washington, DC: American Association for the Advancement of Science).
14. Rehmann-Sutter, C. (2003) Keimbahnveränderungen in Nebenfolge? Ethische Überlegungen zur Abgrenzbarkeit der somatischen Gentherapie, in: Rehmann-Sutter, C., Müller, H. (ed.) Ethik und Gentherapie. Zum praktischen Diskurs um die molekulare Medizin (2nd ed. Tübingen: Francke), 187–205.
15. Council of Europe (COE) Convention for the Protection of Human Rights and Dignity of the Human Being with regard to the Application of Biology and Medicine: Convention on Human Rights and Biomedicine.
16. Anderson, W. F. (1989) Human gene therapy: Why draw a line? *J Med Phil* **14**, 681–93.
17. Glover, J. (1984) What sort of people should there be? (Harmondsworth: Penguin).
18. Redlich, F. C. (1976) Editorial reflections on the concepts of health and disease *J Med Phil* **1**, 269–80.
19. Silvers, A. (1994) "Defective" agents. Equality, difference and tyranny of the normal *J Soc Phil* **25**, 154–75.
20. Sinsheimer, R. L. (1987) The prospect of designed genetic change, in: Chadwick, R. F. (ed.) Ethics, reproduction and genetic control (London, New York, Sydney: Routledge), 136–46.
21. Torres, J. M. (1997) On the limits of enhancement in human gene transfer: Drawing the line *J Med Phil* **22**, 43–53.
22. Krimsky, S. (1990) Human gene therapy: Must we know to stop before we start? *Hum Gene Ther* **1**, 171–3.
23. Juengst, E. T. (1997) Can enhancement be distinguished from prevention in genetic medicine? *J Med Phil* **22**, 125–42.
24. Fuchs, M., Lanzerath, D., Sturma, D. (2008) Natürlichkeit und Enhancement: Zur ethischen Beurteilung des Gendopings; Gutachten des Instituts für Wissenschaft und Ethik e.V. (IWE) im Auftrag des Deutschen Bundestages *Jahrbuch für Wissenschaft und Ethik* **13**, 263–302.

Chapter 18

Clinical Trials of GMP Products in the Gene Therapy Field

Kathleen B. Bamford

Abstract

Advances in gene therapy are increasingly leading to clinical assessment in many fields of medicine with diverse approaches. The basic science stems from approaches aimed at different functions such as correcting a missing/abnormal gene, altering the proportion or expression of normal genes to augment a physiological process or using this principle to destroy malignant or infected cells. As the technology advances, it is increasingly important to ensure that clinical trials answer the questions that need to be asked. In this chapter we review examples of published clinical trials, resources for accessing information about registered trials, the process of regulating trials, good clinical practice, and good manufacturing practice as well as summarising the approach taken by regulatory authorities in reviewing applications for the introduction of products for use in the clinic.

Key words: Clinical trial, Good clinical practice, GMP, Gene therapy, Gene therapy regulation, Clinical trials directive

1. Introduction

Advances in gene therapy are increasingly reaching the point where clinical trials and further clinical development are needed. Embarking on a clinical trial is a significant undertaking for both participants and investigators. Trials involving viral gene therapy products raise all the same issues, but have further layers of complexity than many other investigative medicinal products that must be addressed. In bringing a product to the clinic, investigators must navigate regulatory processes involving legislation governing the manufacture and potential release or containment of the product and registration of the investigational product as well as designing and conducting a trial that conforms to the principles of good clinical practice. These processes are designed to protect both research participants and the researcher, ensure that clinical

Otto-Wilhelm Merten and Mohamed Al-Rubeai (eds.), *Viral Vectors for Gene Therapy: Methods and Protocols*,
Methods in Molecular Biology, vol. 737, DOI 10.1007/978-1-61779-095-9_18, © Springer Science+Business Media, LLC 2011

research is of the highest possible quality, and that it effectively addresses appropriate clinical questions and minimises risk to the participants, others, and the environment. None of these processes are intrinsically difficult, but in combination they are exacting requiring determination and attention to detail. The relevant legislation varies between countries and continents or economic areas.

The route for an investigative or manufacturing team to bring a viral gene therapy product to clinical trial will be easier if they actively engage with their relevant regulatory bodies and competent authorities as early as possible. It is important that they familiarise themselves with the processes and requirements of the authorities in all the countries where they envisage conducting a trial or ultimately bringing a product to the clinic as there are geographic differences in regulatory systems that apply. These can most easily be divided into those pertinent to researchers in Europe, the United States of America, and the rest of the world. As the regulatory processes are most evolved in Europe and in the USA, this review of the processes involved will concentrate on these regions.

Effective gene therapy is dependent firstly on the principle that altering the presence and or expression of a gene in the human body will result in a therapeutic benefit. It is dependent then on being able to deliver a therapeutic gene or genes to a relevant part of the body, results in appropriate expression, and an effective therapeutic effect without unacceptable unwanted effects. This has been harder to deliver in practice than first anticipated. Although there are an increasing number of gene therapy products entering clinical trials; as yet there are no licensed products on the market or in routine use.

However the results of a number of phase I and II trails have been published. Examples of these are discussed in the next section.

There have been a number of well publicised and investigated episodes where adverse events have broadened the debate around gene therapy. What emerges from these is the need for scrupulous scrutiny emphasising the need for high quality and adherence to the principles of good clinical practice (1–4). There are a number of areas where the results of phase I and II trials have been published.

2. Clinical Trials of Gene Therapy Using Viral Vectors

2.1. Published Clinical Trials

Clinical trials for gene therapy can be grouped either by the vector used to deliver the therapeutic gene or by the clinical condition being addressed. Here we consider examples of trials published by early 2010 based on the underlying nature of the disease – whether it is a heritable single gene disorder, a multifactorial condition, malignancy, or infection.

2.1.1. Single Gene Disorders

Work began initially on single gene disorders such as cystic fibrosis. Early difficulties were encountered because therapeutic genes were only expressed for a limited time. This led to repeated administration followed by development of immunity to the vectors such as adenoviruses, which in turn led to the elimination of the vector by the hosts' immune response. This is a particular issue for all genes that require more than one delivery to tissue such as respiratory epithelium where there is a high turn over. Therefore while there has been some promise, e.g., with AAV, the CF community has directed effort towards nonviral vector and gene delivery systems as well as engineered integrating vectors (1–9).

Issues have also arisen in the use of viruses as vectors for the delivery of factor VIII gene therapy in severe hemophilia, again through development of immune responses that limit the usefulness of both the vector and transgene. The development of immune responses to vector components has been noted in other clinical trials. For example capsid specific CD4 cells and IgG responses have been activated in a trial that used IM injection of adenoassociated vector coding lipoprotein lipase. It is not yet known how this will affect the successful expression of the transgene and associated lowering of plasma triglyceride levels (10).

In other single-gene defect disorders there has been more success in using viral vectors to deliver a therapeutic gene that has had a clinical benefit.

One of the most dramatic approaches using gene therapy has been the successful use of retroviral vectors to correct X-linked severe combined immunodeficiency. None the less this has not been without serious unwanted effects. The theoretical possibility of insertional mutagenesis associated with integration of the vector has been borne out. Only further time will tell whether this will extend to other participants in this and other trials. Until vector science develops further addressing random insertion associated oncogenesis or to overcome random insertion, this highlights the appropriateness of reserving treatment with integrating vectors to individuals who have no other therapeutic options (11–15).

The oxidase activity in peripheral blood neutrophils in 2/3 patients with X-linked chronic granulomatous disease has been successfully corrected to a degree that results in full or partial resolution of infection using an MFGS retroviral vector encoding gp91(phox) in combination with busulfan conditioning (16). Despite success in relation to the intended outcome, this has not been wholly successful as subsequently one of these patients died and for both patients there was silencing of the transgene expression; in addition, the appearance of clones was observed).

Hemophilia is a major gene therapy target with the main aim being to ameliorate bleeding tendencies. Adeno associated viruses have been used with success in gene therapy for hemophilia, however it appears that of the viruses explored to date retroviruses may show most promise for sustained effect (17, 18).

Leber's congenital amaurosis (LCA) is a group of retinal dystrophies with onset in childhood. They cause progressive visual deterioration in children and adults. A number of trials have shown that this group of conditions is amenable to gene therapy using subretinal injection. Phase I trial of three doses of a single subretinal injection of an adeno-associated virus containing a gene for isomerohydrolase activity of the retinal pigment epithelium (65 kD protein (RPE65): AAV2-hRPE65v2 (19); AAV2-hRPE65 (20)) has provided evidence where the safety, extent, and stability of improvement support further use of this approach particularly in early disease (19). Measurable response (20) and acceptable safety features (21) in young adults have been reported. The trials in LCA have been reviewed. These underlie the need for analysis of the patients' genotype, but do demonstrate improvement in vision for some patients (22, 23). As follow-up periods extend, evidence is now available that suggests sustained immunological responses to the vector do not cause a problem in the eye and that functional improvement or amelioration of visual loss is sustained for 1–1.5 years in some patients (24, 25). Other causes of inherited retinal degeneration may also be suitable for similar approaches (26, 27).

Duchenne muscular dystrophy is an X-linked disorder that results in progressive loss of muscle with eventual death due to respiratory failure. Lentiviruses are showing promise as vectors for integrating the therapeutic gene in muscle cells (28). Adeno-associated viruses have also been reviewed and explored in clinical trials to develop a therapeutic approach to Duchene muscular dystrophy (29). Finding a systemic approach for integration using both viral and nonviral vectors remains a goal for this disorder (30).

2.1.2. Multifactorial Disorders

A number of nonmalignant multifactorial disorders are the subject of efforts to ameliorate symptoms or pathogenic processes by means of gene therapy. Two examples are Parkinson's disease and disorders associated with vascular insufficiency most notably coronary and peripheral artery disease.

In Parkinson's disease, symptoms progress as insufficient dopamine synthesis gets worse due to loss of dopaminergic cells in key areas of the central nervous system. In a clinical trial of patients with moderately advanced Parkinson's disease intraputamenal injection of an adeno-associated virus with a human aromatic 1-amino acid decarboxylase (AADC) has been carried out. AADC converts levodopa (a drug used in Parkinson's disease) to dopamine. This enzyme reduces in the putamen as the disease progresses therefore therapy becomes less effective. While there were surgery related risks, a significant dose related therapeutic effect has been reported at 6 months from therapy (31). Adeno-associated viruses have also been used to deliver the glutamic acid decarboxylase gene to the subthalamus on one side in an open label safety and tolerability study. This demonstrated an effect that was sustained for 12 months (32). A very promising approach has been developed using the equine anemia virus (EIAV) expressing three key dopamine biosynthetic enzymes. Preclinical studies in a rat and then nonhuman primate model have recently been extended to a phase I/II clinical trial with bilateral administration to the sensorimotor putamen (33, 34).

Revascularization of either the coronary or peripheral arteries is the target of gene therapy. Various studies have used Adenovirus-mediated gene transfer of vascular endothelial growth factor (VEGF) on the principle that this will generate new or collateral vessel development. The approach has been shown to improve lower limb ischemia (35). Favorable anti-ischemic effects have also been found with an adenovirus that delivers fibroblast growth factor 4 compared with placebo in some (36, 37) but not other (38) studies. In a Phase II study, a significant increase in myocardial perfusion was detected in patients who had intracoronary delivery of VEGF-adenovirus compared with plasmid (39) although plasmid based gene therapy has given encouraging results following direct intramyocardial injection (40). A concern about this approach is that stimulating vascularization could promote growth in undetected malignancies.

2.1.3. Gene Therapy Trials in Cancer

Different approaches have been used in different cancers. One mechanism is to introduce a suicide gene either one such as thymidine kinase gene that allows a nontoxic precursor drug to be converted into a toxic radical within or close to the malignant cell, or alternatively by introducing the human p53 gene, often with an additional upregulatory gene to induce apoptosis of the malignant target cells.

Ovarian cancer cells modified to express thymidine kinase as a suicide gene have been shown to home to mesothelioma in the human pleural space thus potentially allowing them to be used to target the toxic effect to bystander mesothelioma cells (41). Thymidine kinase genes have also been used with promise as a way of controlling graft versus host disease following donor lymphocyte infusion to treat relapse of hematological malignancies (42).

In other studies, adenovirus vectors have been used to deliver p53 to a variety of cancers by varied routes, e.g., intratumoral injection of non-small cell lung cancers via a trans-bronchial route (43) and intratumoral injection of metastatic breast cancers (44).

A recombinant adenovirus coding p53 has also been used via intraepithelial injection to treat dysplastic oral leukoplakia, again with some promise (45). Intraoperative injection of p53 coding adenovirus for resectable squamous cell carcinoma of the oral cavity in a phase II study while inconclusive has encouraged further work, but highlighted regulatory issues as a potential barrier to accrual in trial settings (46).

Cancers are also a target of different forms of immunotherapy using a gene therapeutic approach. In some, direct injection of a virus (virotherapy) such as vaccinia has been effective in increasing the inflammatory and therefore antitumor effect (47, 48). In others, injection of a virus coding a cytokine gene, e.g., adenovirus coding interferon-gamma, into skin deposits has been found to be safe and has modified the local immune response (49). By using a replication defective adenovirus to transduce, the gene for CD 40 ligand into CLL cells in chronic lymphoblastic leukemia, a population of antitumor CTLs were generated *in vivo* associated with drop in leukemia cell counts (50).

In patients with malignant melanoma, genetically modified highly reactive lymphocytes have been produced by immunizing transgenic mice with melanoma antigens, cloning the genes encoding the T-cell receptors and using a retroviral vector to transduce the melanoma specific T-cell receptor into autologous peripheral lymphocytes which are then administered to the patients. These had antitumor activity, but also affected normal melanocytes in skin, hair, and eye (51). A retroviral vector has also been used to transduce peripheral blood lymphocytes from patients with late stage melanoma so that they recognize MAGE-A3, a cancer germ line gene. The approach was targeted to those with tumors expressing MAGE. In this trial, it was demonstrated that the reintroduced genetically modified lymphocytes localized to the tumors and contributed to an inflammatory response. Further studies are needed to confirm efficacy (52).

2.1.4. Infection

There are relatively few published reports of gene therapy approaches to HIV infection, although there have been lessons that demonstrate the importance of T-helper support for the immune response to HIV (53). Infusion of CMV-specific cytotoxic T lymphocytes stimulated with an adenoviral vector encoding CMV protein modified dendritic cells has been used with success as prophylaxis for CMV disease following allogeneic hematopoietic stem cell transplant (54).

In summary, there is a great diversity of approaches to gene therapy using different viral vectors in a wide range of clinical settings. While there are as yet relatively few published trials, there are many more planned and in the pipeline. There are a range of resources through which investigators can keep abreast.

2.2. Registers of Ongoing Clinical Trials

There are a variety of sources for information about ongoing or unreported clinical trials. In Europe and the USA and most other industrialized nations worldwide there are open access registers of

Table 1
Registers of gene therapy trials

Registry/country	Website
Australia New Zealand Clinical Trials Registry	http://www.anzctr.org.au/default.aspx
NIH Clinical Center	http://clinicalstudies.info.nih.gov/
University Hospital Medical Information network, Japan	http://www.umin.ac.jp/english/
ClinicalTrials.gov	http://clinicaltrials.gov/
Gene Therapy Review Database	http://www.genetherapyreview.com/gene-therapy-clinical-trials/clinical-trials-database.html
Wiley Interscience	http://www.wiley.co.uk/genetherapy/clinical/
EudraCT Database	http://eudract.emea.eu.int/
Current Controlled Trials	http://www.controlled-trials.com/
Cancer Clinical Trials	http://www.cancertrials.org.uk/
International Clinical Trials Registry Platform (ICTRP)	http://www.who.int/ictrp/en/
US Government Database	http://www.clinicaltrials.gov/

all clinical trials involving investigational medical products. There are also sources of information specific to gene therapy through professional and scientific societies. Examples of these are given in Table 1.

3. Good Clinical Practice for Clinical Trials

The principles of good clinical practice were developed by a tripartite steering group of the regulatory authorities of Japan, the US and EU, the International conference on Harmonisation (ICH) to provide international assurance that both the confidentiality, rights, safety and well being of individuals participating in clinical trials are maintained, and, that the trials are conducted in such a way that the data generated is accurate and the results credible. This approach was initially promoted as best practice, but is now enforced by legislation in the relevant areas, e.g., in the EU the Medicines for Human Use (Clinical Trials) Regulations 2004, EU Directive on Good Clinical Practice, are now legally required for all trials of investigational medicinal products in the EU. Similar relevant legislation applies in the US and other parts of the world. The principles set out in this legislation are summarised below

1. Clinical trials should be conducted in accordance with the ethical principles that have their origin in the Declaration of

Helsinki, and that are consistent with GCP and the applicable regulatory requirement(s).

2. Before a trial is initiated, foreseeable risks and inconveniences should be weighed against the anticipated benefit for the individual trial subject and society. A trial should be initiated and continued only if the anticipated benefits justify the risks.
3. The rights, safety, and well-being of the trial subjects are the most important considerations and should prevail over interests of science and society.
4. The available nonclinical and clinical information on an investigational product should be adequate to support the proposed clinical trial.
5. Clinical trials should be scientifically sound, and described in a clear, detailed protocol.
6. A trial should be conducted in compliance with the protocol that has received prior institutional review board (IRB)/ independent ethics committee (IEC) approval/favorable opinion.
7. The medical care given to, and medical decisions made on behalf of subjects should always be the responsibility of a qualified physician or, when appropriate, of a qualified scientist.
8. Each individual involved in conducting a trial should be qualified by education, training, and experience to perform his or her respective task(s).
9. Freely given informed consent should be obtained from every subject prior to clinical trial participation.
10. All clinical trial information should be recorded, handled, and stored in a way that allows its accurate reporting, interpretation, and verification.
11. The confidentiality of records that could identify subjects should be protected, respecting the privacy and confidentiality rules in accordance with the applicable regulatory requirement(s).
12. Investigational products should be manufactured, handled, and stored in accordance with applicable good manufacturing practice (GMP) (see paragraph 5). They should be used in accordance with the approved protocol.
13. Systems with procedures that assure the quality of every aspect of the trial should be implemented.

An important tenant and core to application of these principles is the maintenance of records that provide evidence of compliance. The burden of responsibility for this lies with the researcher, clinicians, the institutions, and the regulators involved.

Developments in gene therapy are considered by a specialist Gene therapy discussion group involving experts and representatives of the regulatory bodies from the three ICH regions (EU, Japan, and USA), the European Free Trade Association (EFTA), Health Canada, the World Health Organization (WHO), and industry

4. The Regulatory Environment

The regulation of clinical trials involves aspects relating to the investigational product and ethical considerations relating to the study of human participants. Institutional review board and/or ethical review are a vital and intrinsic part of this process. The issues relating to this are addressed in an earlier chapter, but in practice is an inextricable part of bringing a trial to fruition.

In the United States, the Department of Health and Human Services (DHHS) has oversight of clinical trials. Two organizations within DHHS, the Office for Human Research Protections (OHRP) and the US Food and Drug Administration (FDA), have specific authority described in the Code of Federal Regulations (CFR). All investigators must comply with these regulations when conducting clinical gene therapy trials. The OHRP mandates that all research involving human subjects undergo Institutional Review Boards (IRB) review and approval. An IRB is charged with evaluating research risk to subjects and must approve research protocols and informed consent documents prior to beginning a study.

Another DHHS agency, the National Institutes of Health (NIH), oversees the conduct of federally funded clinical trials through a series of guidelines that add additional requirements to those specified in the CFR.

Before a trial can be conducted, the FDA will oversee a review of the proposed investigational product. There are extensive guidelines to this process, which aims to ensure that safety and quality are assured. Investigators need to be prepared to keep themselves up to date through interaction with the resources most appropriate to their regulatory area.

The legal framework for the conduct of clinical trials in the UK and EU are embodied in "The Medicines for Human Use (Clinical Trials) Regulations 2004". The Regulations implement the EU clinical trials directive (2001/20/EC) and will soon be amended to reflect the GCP directive (2005/28/EC). In the UK Integrated Research Application System (IRAS) provides an integrated web-based approach to the regulatory bodies' application systems.

As well as Product Manufacturing Licenses and appropriate agreements that ensure the production and use of GMP product is authorized (see below), a legal framework is also likely to be needed to:

(a) Transfer samples between study sites (Material Transfer Agreement)

Table 2
Online resources

US Guidance on the conduct of clinical trials	http://www.fda.gov/ScienceResearch/SpecialTopics/RunningClinicalTrials/default.htm
US Guidance on the conduct of gene therapy clinical trials	http://www.fda.gov/BiologicsBloodVaccines/GuidanceComplianceRegulatoryInformation/Guidances/CellularandGeneTherapy/default.htm
UK Clinical trials legislation	http://www.opsi.gov.uk/si/si2004/20041031.htm
European clinical trials legislation	http://europa.eu.int/eur-lex/lex/LexUriServ/LexUriServ.do?uri=CELEX:32001L0020:EN:HTML
UK integrated research application site	https://www.myresearchproject.org.uk/Signin.aspx
Routemap in UK for regulatory system involving genetics and cellular applications	http://www.advisorybodies.doh.gov.uk/genetics/gtac/interimukscroutemap120309.pdf
Listings of GMP facilities relevant to viral gene therapy vector manufacture	www.esgct.eu/information-and-resources/gmp-facilities/
European medicine agencies advanced therapies site	http://www.ema.europa.eu/htms/human/advanced_therapies/intro.htm
European union common website for medicines authorities	http://www.hma.eu/
Links to national medicines agencies websites	http://www.hma.eu/list
Orphan Drug designation	http://www.orpha.net/consor/cgi-bin/index.php http://www.ema.europa.eu/pdfs/human/comp/29007207en.pdf
Guidance for Industry CGMP for Phase 1 investigational Drugs	http://www.fda.gov/downloads/Drugs/GuidanceComplianceRegulatoryInformation/Guidances/ucm070273.pdf
International conference on Harmonisation	http://www.ich.org/cache/compo/276-254-1.html
Guideline for good clinical practice	http://www.ema.europa.eu/pdfs/human/ich/013595en.pdf

(b) Define the consistency and ways of working between centers (Clinical trials agreement)

(c) Service contracts necessary for the conduct of work by third parties

In the UK many of the regulatory hurdles that apply to gene therapy also apply to bringing stem cell research to the clinic. It is therefore likely to be helpful to those embarking on the design of clinical trials to consider the joint advisory and regulatory bodies' route map as an aid to developing an overview of the permissions and approvals that are required.

Products for gene therapy for rare conditions may be developed under the orphan drug designation by the European Commission. This provides a series of incentives such as protocol assistance with scientific advice during the product-development phase, marketing authorization providing up to 10 years marketing exclusivity and well as other financial incentives through fee reductions or exemptions. Further resources and criteria are available through the EMEA and orphan drug websites (Table 2)

5. Principles of GMP for Investigational Viral Vector Gene Therapy Agents

To bring an investigational product to the clinic it must be produced by a process that complies with GMP standards.

GMP is that part of quality assurance which ensures that medicinal products are consistently produced and controlled throughout the production process to the quality standards appropriate to their intended use and as required by the marketing authorization (MA) or product specification.

The term GMP in relation to a therapeutic agent, in this context a viral gene therapy product, is a standard of production or manufacture of the agent that assures its consistent quality and suitability for purpose. This standard is set and monitored by competent authorities. The competent authorities are the regulatory agencies in each country or jurisdiction that licenses the facility, process, and product through an inspection process. This measures production including the facilities and practices as well as the final product against predetermined standards. For a therapeutic agent to be used in a particular country, the GMP standard must comply with that of the country in which the agent is to be used in the clinic.

Each manufacturer appoints a responsible qualified person (QP)(s) who is(are) signatory(ies) on each license. They undertake their duties in accordance with a professional Code of Practice. An example of this is set out in the UK in Article 56 of Council Directive 2001/82/EC and/or Article 52 of Council Directive 2001/83/EC [http://www.mhra.gov.uk/Howweregulate/Medicines/Inspectionandstandards/GoodManufacturingPractice/Guidanceandlegislation/index.htm].

A key aspect of producing a medicinal product including an investigational viral gene therapy agent is the licensing of the process and premises.

Many will contract this process out to registered GMP facilities where there is already appropriate expertise.

Oversight of the production of a product or therapeutic agent can be divided into a number of areas that correspond to different sections in the regulations. Many of these will contribute to the investigators brochure of the agent and will justify the move to a first in man or developing portfolio of clinical trials leading towards licensing. The approach to this is divided into a number of areas.

1. The manufacture of the product

   The components involved – This will include details of the vector, cells used, reagents, excipients, and other materials used.

   The procedures – Vector production and purification, storage/interim steps, final formulation, and ex vivo gene modification of cells.

2. Validation and quality assurance

   This includes the testing associated with consistent quality assurance of the agent at each step in the pathway as well as validated determinations of standard of each batch released for use.

   It will also include sterility testing, assurance that adventitious or other infective agents including mycoplasmas are not present in the product at each stage of the manufacture process (an example is presented in Chapter 11, Fig. 1)

   It will include a confirmation of the identity – this is increasingly achieved by sequencing the vector and assessment of the presence of residual contaminants, endotoxin or pyrogens as well as a final assessment of the potency/viability of the product.

   Batch release/rejection criteria will be required and evidence of its stability under recommended storage/transport conditions and the stability of the agent under conditions of use.

   A full description of how the product will be presented (Labeling, packaging transport, and storage) and tracked.

3. Issues related to containment and release, the environmental impact, and product tracking

   This will relate to the jurisdiction and guidance on contained use or deliberate release of a genetically modified organism as well as any other environmental considerations relating to use and manufacture. Evidence for any claims made that support a thorough risk assessment will be required.

4. Preclinical studies

   These should summarize the proof of principle and provide evidence in a suitable model to supports use in humans. Toxicity studies are an important component and should be carried out in an appropriate model.

This is especially relevant to viral vectors as the cellular receptors for different vectors may have species variation or specificity.

In addition, distribution studies are considered to be important – these must include gonadal distribution and shedding (Urine, stool, semen, and respiratory secretions)

5. Clinical studies

If this is a first use of the product in humans, there will be no previous clinical studies to support safety; however, there could be evidence from studies that support the clinical principle or the use of related or similar products. This might be the use of the transgene in a different vector, or use of a related vector. If this is not a first in man use, the agent in its current form will have been previously administered to humans. In this case, data relating to toxicity, safety, distribution, and shedding in the clinical setting as well as any evidence supporting efficacy is required. The extent of this part of the portfolio will relate to the extent to which the product has already been used in a clinical setting.

Details of the protocol should include the study population with inclusion and exclusion criteria, the dose, frequency, and route of administration. Details of any genetic testing as well as biochemical and immunological testing should include pre-exposure and development of immunity to the vector. Informed consent details should also be included here as well as forming a key part of the ethical review.

6. Genetic Modification Regulatory Oversight

Oversight of a recombinant DNA advisory committee (US) or its national equivalent is also appropriate.

In the UK this is currently served by a genetic modification safety committee according to jurisdiction and guidance administered by the Health and Safety Executive (HSE) (contained use regulations) or in the case of deliberate release, the department of the environment food and agriculture (DEFRA); however, the regulations governing the use of genetically modified organisms are currently under review.

European Medicines Agency (EMA), currently seated in London, was established by Council Regulation (EEC) No 2309/93 of 22 July 1993. The agency coordinates scientific resources in Member States to oversee medicinal products for human and veterinary use. Based on an EMA opinion, the European Commission authorizes marketing of new medicinal products. Information about regulation of gene and cell therapy is available on the EMA Web Site. Currently, under "Human

Medicines" the section "Advanced Therapies" has information relevant for gene therapy.

There is a common Website for the medicines authorities in the European Union (Heads of Medicines Agencies). In addition, Clinigene is a European Network for the Advancement of Clinical Gene Transfer and Therapy: EC funded network of excellence fostering interaction of all stakeholders in the field in order to facilitate and help harmonize Ethical, Quality, Safety, Efficacy, and Regulatory issues. This NoE prolongs and extends the action of the former Euregenethy 1 and 2 networks supported by EC-DG research FP4 and FP5 programmes.

The National medical agencies have links to the EU Member State Medicines Agencies under "Choose your country."

Similar information is available via the FDA web site and there are links via the gene therapy communities scientific web pages.

All these agencies review the product characterization. It is therefore useful to outline a systematic approach to this as described in Subheading 6.2.

6.1. GMP and Research-Grade Vector Production Resources

Manufacturing biotherapeutics to GMP standard for clinical trials can be one of the most difficult hurdles to overcome. There are a number of facilities worldwide capable of manufacturing clinical and preclinical grade therapeutics including plasmid, oligonucleotides, and viral vectors. There is also a selection of facilities capable of conducting the range of tests necessary to meet the safety standards. Many of these are listed on the European society of gene and cell therapies website.

6.2. Characterization of the Product

To bring a gene therapy product to the point of clinical trial, a clear description, including sequence data for the vector, the steps involved in its modification and the therapeutic gene is required. With regard to bringing a product to the point where it can be produced under GMP conditions, records about use of animal derived tissue culture cells and products (e.g., fetal calf serum) throughout the development and manufacture process are required. Where possible, the use of any animal product should be eliminated.

These data will be used in the risk assessments carried out by both the licensing authorities and the relevant ethics committees in coming to an opinion. Again, these are best addressed in dialogue with the relevant regulatory bodies. Information presented as part of this portfolio should consider the following.

(a) *The vector*: The origin and nature of the species and strain of vector backbone should be described. The history, including natural host and disease profile of the parent virus is an important part of the description of an attenuated strain that is used as a therapeutic delivery agent. This will also influence whether or not the vector and or gene delivered is expected to persist in the target cell and whether there will be integration and if so at what site(s) with attendant risks of insertional mutagenesis (4). Information about potential shedding or recombination with wild type/other viruses and the possibility of persistence in the environment or potential transmission are important considerations that must all be addressed, preferably with experimental evidence.

(b) *Therapeutic gene/insert*: A description of the therapeutic gene should include inserts to modify its expression and or regulation as well as the expected consequences of its action. Consideration should be given to the intended physiological effects of over and under expression not only in the target tissue, but also at sites other than the target tissue in the event of unplanned dissemination. As part of this evaluation, theoretical possibilities must also be considered and the potential implications assessed.

(c) *Construction of the therapeutic agent*: A clear detailed description of the steps, verification, and controls assurance process that were taken throughout the initial construction as well as any differences that will occur with each subsequent batch as the agent is developed for clinical use need to be given.

(d) *The Gene therapy product*: Any additional considerations relevant to the assembled therapeutic agent should be addressed. This may relate for example to regulatory elements or potential effects of attenuation and again the likelihood of adventitious recombination in vivo. In addition, the issue of how the therapeutic agent and gene will be tracked and the limits of detection should be addressed.

The GMP and quality control (QC) process facilities and laboratories are inspected usually when the laboratory undertakes the following testing of either a marketed, or in this case, investigational medicinal product, which is therefore governed by a "Manufacturer investigational medicinal products" licence.

The QC process will include

- Microbiological testing at all stages of the manufacture process including sterility (and nonsterility) testing of starting materials, intermediate steps, and finished product batches.
- Chemical and physical analysis of components from starting materials, to finished products with stability studies on finished products under storage conditions.
- Environmental monitoring of manufacturing areas including microbiological testing.

Further details of these requirements are available via the relevant competent authorities in national jurisdictions, e.g., FDA and MHRA. Further information on these resources and the websites used are given in Table 2.

References

1. Thrasher, A. J. (2008) Gene therapy for primary immunodeficiencies. *Immunol. Allergy Clin. North Am.* **28**, 457–471.
2. Wilson, J. M. (2009) Lessons learned from the gene therapy trial for ornithine transcarbamylase deficiency. *Mol. Genet. Metab.* **96**, 151–157.
3. Stolberg, S. G. (1999) F.D.A. officials fault Penn team in gene therapy death. *N.Y. Times (Print)*, A22.
4. Rans, T. S. and England, R. (2009) The evolution of gene therapy in X-linked severe combined immunodeficiency. *Ann. Allergy Asthma. Immunol.* **102**, 357–362.
5. Ferrari, S., Geddes, D. M., and Alton, E. W. (2002) Barriers to and new approaches for gene therapy and gene delivery in cystic fibrosis. *Adv. Drug Deliv. Rev.* **54**, 1373–1393.
6. Alton, E. W. (2004) Use of nonviral vectors for cystic fibrosis gene therapy. *Proc. Am. Thorac. Soc.* **1**, 296–301.
7. Griesenbach, U. and Alton, E. W. (2009) Gene transfer to the lung: lessons learned from more than 2 decades of CF gene therapy. *Adv. Drug Deliv. Rev.* **61**, 128–139.
8. Mitomo, K., Griesenbach, U., Inoue, M., Somerton, L., Meng, C., Akiba, E. et al. (2010) Toward gene therapy for cystic fibrosis using a lentivirus pseudotyped with Sendai virus Envelopes. *Mol. Ther.* in press, doi:10.1038/mt.2010.13.
9. Moss, R. B., Rodman, D., Spencer, L. T., Aitken, M. L., Zeitlin, P. L., Waltz, D. et al. (2004) Repeated adeno-associated virus serotype 2 aerosol-mediated cystic fibrosis transmembrane regulator gene transfer to the lungs of patients with cystic fibrosis: a multicenter, double-blind, placebo-controlled trial. *Chest* **125**, 509–521.
10. Mingozzi, F., Meulenberg, J. J., Hui, D. J., Basner-Tschakarjan, E., Hasbrouck, N. C., Edmonson, S. A. et al. (2009) AAV-1-mediated gene transfer to skeletal muscle in humans results in dose-dependent activation of capsid-specific T cells. *Blood* **114**, 2077–2086.
11. Muul, L. M., Tuschong, L. M., Soenen, S. L., Jagadeesh, G. J., Ramsey, W. J., Long, Z. et al. (2003) Persistence and expression of the adenosine deaminase gene for 12 years and immune reaction to gene transfer components: long-term results of the first clinical gene therapy trial. *Blood* **101**, 2563–2569.
12. Cassani, B., Montini, E., Maruggi, G., Ambrosi, A., Mirolo, M., Selleri, S. et al. (2009) Integration of retroviral vectors induces minor changes in the transcriptional activity of T cells from ADA-SCID patients treated with gene therapy. *Blood* **114**, 3546–3556.
13. Aiuti, A., Cassani, B., Andolfi, G., Mirolo, M., Biasco, L., Recchia, A. et al. (2007) Multilineage hematopoietic reconstitution without clonal selection in ADA-SCID patients treated with stem cell gene therapy. *J. Clin. Invest.* **117**, 2233–2240.
14. Schwarzwaelder, K., Howe, S. J., Schmidt, M., Brugman, M. H., Deichmann, A., Glimm, H. et al. (2007) Gammaretrovirus-mediated correction of SCID-X1 is associated with skewed vector integration site distribution in vivo. *J. Clin. Invest.* **117**, 2241–2249.
15. Hacein-Bey-Abina, S., Le, D. F., Carlier, F., Bouneaud, C., Hue, C., De Villartay, J. P. et al. (2002) Sustained correction of X-linked severe combined immunodeficiency by ex vivo gene therapy. *N. Engl. J. Med.* **346**, 1185–1193.
16. Kang, E. M., Choi, U., Theobald, N., Linton, G., Long Priel, D. A., Kuhns, D. et al. (2010) Retrovirus gene therapy for X-linked chronic granulomatous disease can achieve stable long-term correction of oxidase activity in peripheral blood neutrophils. *Blood* **115**, 783–791.
17. Manno, C. S., Pierce, G. F., Arruda, V. R., Glader, B., Ragni, M., Rasko, J. J. et al. (2006) Successful transduction of liver in hemophilia by AAV-Factor IX and limitations imposed by the host immune response. *Nat. Med.* **12**, 342–347.
18. Powell, J. S., Ragni, M. V., White, G. C., Lusher, J. M., Hillman-Wiseman, C., Moon, T. E. et al. (2003) Phase 1 trial of FVIII gene transfer for severe hemophilia A using a retroviral construct administered by peripheral intravenous infusion. *Blood* **102**, 2038–2045.
19. Maguire, A. M., High, K. A., Auricchio, A., Wright, J. F., Pierce, E. A., Testa, F. et al. (2009) Age-dependent effects of RPE65 gene therapy for Leber's congenital amaurosis: a phase 1 dose-escalation trial. *Lancet* **374**, 1597–1605.

20. Bainbridge, J. W., Smith, A. J., Barker, S. S., Robbie, S., Henderson, R., Balaggan, K. et al. (2008) Effect of gene therapy on visual function in Leber's congenital amaurosis. *N. Engl. J. Med.* **358**, 2231–2239.
21. Maguire, A. M., Simonelli, F., Pierce, E. A., Pugh, E. N., Jr., Mingozzi, F., Bennicelli, J. et al. (2008) Safety and efficacy of gene transfer for Leber's congenital amaurosis. *N. Engl. J. Med.* **358**, 2240–2248.
22. Chung, D. C. and Traboulsi, E. I. (2009) Leber congenital amaurosis: clinical correlations with genotypes, gene therapy trials update, and future directions. *J. AAPOS* **13**, 587–592.
23. MacLaren, R. E. (2009) An analysis of retinal gene therapy clinical trials. *Curr. Opin. Mol. Ther.* **11**, 540–546.
24. Simonelli, F., Maguire, A. M., Testa, F., Pierce, E. A., Mingozzi, F., Bennicelli, J. L. et al. (2010) Gene therapy for Leber's congenital amaurosis is safe and effective through 1.5 years after vector administration. *Mol. Ther.* **18**, 643–650.
25. Cideciyan, A. V., Hauswirth, W. W., Aleman, T. S., Kaushal, S., Schwartz, S. B., Boye, S. L. et al. (2009) Human RPE65 gene therapy for Leber congenital amaurosis: persistence of early visual improvements and safety at 1 year. *Hum. Gene Ther.* **20**, 999–1004.
26. Smith, A. J., Bainbridge, J. W., and Ali, R. R. (2009) Prospects for retinal gene replacement therapy. *Trends Genet.* **25**, 156–165.
27. Cideciyan, A. V., Swider, M., Aleman, T. S., Tsybovsky, Y., Schwartz, S. B., Windsor, E. A. et al. (2009) ABCA4 disease progression and a proposed strategy for gene therapy. *Hum. Mol. Genet.* **18**, 931–941.
28. Quenneville, S. P., Chapdelaine, P., Skuk, D., Paradis, M., Goulet, M., Rousseau, J. et al. (2007) Autologous transplantation of muscle precursor cells modified with a lentivirus for muscular dystrophy: human cells and primate models. *Mol. Ther.* **15**, 431–438.
29. Rodino-Klapac, L. R., Chicoine, L. G., Kaspar, B. K., and Mendell, J. R. (2007) Gene therapy for duchenne muscular dystrophy: expectations and challenges. *Arch. Neurol.* **64**, 1236–1241.
30. Foster, K., Foster, H., and Dickson, J. G. (2006) Gene therapy progress and prospects: Duchenne muscular dystrophy. *Gene Ther.* **13**, 1677–1685.
31. Christine, C. W., Starr, P. A., Larson, P. S., Eberling, J. L., Jagust, W. J., Hawkins, R. A. et al. (2009) Safety and tolerability of putaminal AADC gene therapy for Parkinson disease. *Neurology* 73, 1662–1669.
32. Kaplitt, M. G., Feigin, A., Tang, C., Fitzsimons, H. L., Mattis, P., Lawlor, P. A. et al. (2007) Safety and tolerability of gene therapy with an adeno-associated virus (AAV) borne GAD gene for Parkinson's disease: an open label, phase I trial. *Lancet* **369**, 2097–2105.
33. Jarraya, B., Boulet, S., Ralph, G. S., Jan, C., Bonvento, G., Azzouz, M. et al. (2009) Dopamine gene therapy for Parkinson's disease in a nonhuman primate without associated dyskinesia. *Sci. Transl. Med.* 1, 2ra4.
34. Jarraya, B., Ralph, S., Lepetit, H., Stratful, H., Boulet, S., Jan, C. et al. (2009) A phase I/II trial for Parkinson's disease using a lentiviral vector (Prosavin). *Mol. Ther.* **17** Supplement 1, S197;514.
35. Rajagopalan, S., Shah, M., Luciano, A., Crystal, R., and Nabel, E. G. (2001) Adenovirus-mediated gene transfer of VEGF(121) improves lower-extremity endothelial function and flow reserve. *Circulation* **104**, 753–755.
36. Grines, C. L., Watkins, M. W., Mahmarian, J. J., Iskandrian, A. E., Rade, J. J., Marrott, P. et al. (2003) A randomized, double-blind, placebo-controlled trial of Ad5FGF-4 gene therapy and its effect on myocardial perfusion in patients with stable angina. *J. Am. Coll. Cardiol.* **42**, 1339–1347.
37. Grines, C. L., Watkins, M. W., Helmer, G., Penny, W., Brinker, J., Marmur, J. D. et al. (2002) Angiogenic Gene Therapy (AGENT) trial in patients with stable angina pectoris. *Circulation* **105**, 1291–1297.
38. Rajagopalan, S., Mohler, E. R., III, Lederman, R. J., Mendelsohn, F. O., Saucedo, J. F., Goldman, C. K. et al. (2003) Regional angiogenesis with vascular endothelial growth factor in peripheral arterial disease: a phase II randomized, double-blind, controlled study of adenoviral delivery of vascular endothelial growth factor 121 in patients with disabling intermittent claudication. *Circulation* **108**, 1933–1938.
39. Hedman, M., Hartikainen, J., Syvanne, M., Stjernvall, J., Hedman, A., Kivela, A. et al. (2003) Safety and feasibility of catheter-based local intracoronary vascular endothelial growth factor gene transfer in the prevention of post-angioplasty and in-stent restenosis and in the treatment of chronic myocardial ischemia: phase II results of the Kuopio Angiogenesis Trial (KAT). *Circulation* **107**, 2677–2683.
40. Kastrup, J., Jorgensen, E., Ruck, A., Tagil, K., Glogar, D., Ruzyllo, W. et al. (2005) Direct intramyocardial plasmid vascular endothelial growth factor-A165 gene therapy in patients with stable severe angina pectoris A randomized double-blind placebo-controlled study: the

Euroinject One trial. *J. Am. Coll. Cardiol.* **45**, 982–988.

41. Harrison, L. H., Jr., Schwarzenberger, P. O., Byrne, P. S., Marrogi, A. J., Kolls, J. K., and McCarthy, K. E. (2000) Gene-modified PA1-STK cells home to tumor sites in patients with malignant pleural mesothelioma. *Ann. Thorac. Surg.* **70**, 407–411.
42. Traversari, C., Marktel, S., Magnani, Z., Mangia, P., Russo, V., Ciceri, F. et al. (2007) The potential immunogenicity of the TK suicide gene does not prevent full clinical benefit associated with the use of TK-transduced donor lymphocytes in HSCT for hematologic malignancies. *Blood* **109**, 4708–4715.
43. Weill, D., Mack, M., Roth, J., Swisher, S., Proksch, S., Merritt, J. et al. (2000) Adenoviral-mediated p53 gene transfer to non-small cell lung cancer through endobronchial injection. *Chest* **118**, 966–970.
44. Cristofanilli, M., Krishnamurthy, S., Guerra, L., Broglio, K., Arun, B., Booser, D. J. et al. (2006) A nonreplicating adenoviral vector that contains the wild-type p53 transgene combined with chemotherapy for primary breast cancer: safety, efficacy, and biologic activity of a novel gene-therapy approach. *Cancer* **107**, 935–944.
45. Zhang, S., Li, Y., Li, L., Zhang, Y., Gao, N., Zhang, Z. et al. (2009) Phase I study of repeated intraepithelial delivery of adenoviral p53 in patients with dysplastic oral leukoplakia. *J. Oral Maxillofac. Surg.* **67**, 1074–1082.
46. Yoo, G. H., Moon, J., Leblanc, M., Lonardo, F., Urba, S., Kim, H. et al. (2009) A phase 2 trial of surgery with perioperative INGN 201 (Ad5CMV-p53) gene therapy followed by chemoradiotherapy for advanced, resectable squamous cell carcinoma of the oral cavity, oropharynx, hypopharynx, and larynx: report of the Southwest Oncology Group. *Arch. Otolaryngol. Head Neck Surg.* **135**, 869–874.
47. Antonia, S. J., Seigne, J., Diaz, J., Muro-Cacho, C., Extermann, M., Farmelo, M. J. et al. (2002) Phase I trial of a B7-1 (CD80) gene modified autologous tumor cell vaccine in combination with systemic interleukin-2 in patients with metastatic renal cell carcinoma. *J. Urol.* **167**, 1995–2000.
48. Gomella, L. G., Mastrangelo, M. J., McCue, P. A., Maguire, H. C., Jr, Mulholland, S. G., and Lattime, E. C. (2001) Phase i study of intravesical vaccinia virus as a vector for gene therapy of bladder cancer. *J. Urol.* **166**, 1291–1295.
49. Dummer, R., Hassel, J. C., Fellenberg, F., Eichmuller, S., Maier, T., Slos, P. et al. (2004) Adenovirus-mediated intralesional interferon-gamma gene transfer induces tumor regressions in cutaneous lymphomas. *Blood* **104**, 1631–1638.
50. Wierda, W. G., Cantwell, M. J., Woods, S. J., Rassenti, L. Z., Prussak, C. E., and Kipps, T. J. (2000) CD40-ligand (CD154) gene therapy for chronic lymphocytic leukemia. *Blood* **96**, 2917–2924.
51. Johnson, L. A., Morgan, R. A., Dudley, M. E., Cassard, L., Yang, J. C., Hughes, M. S. et al. (2009) Gene therapy with human and mouse T-cell receptors mediates cancer regression and targets normal tissues expressing cognate antigen. *Blood* **114**, 535–546.
52. Fontana, R., Bregni, M., Cipponi, A., Raccosta, L., Rainelli, C., Maggioni, D. et al. (2009) Peripheral blood lymphocytes genetically modified to express the self/tumor antigen MAGE-A3 induce antitumor immune responses in cancer patients. *Blood* **113**, 1651–1660.
53. Mitsuyasu, R. T., Anton, P. A., Deeks, S. G., Scadden, D. T., Connick, E., Downs, M. T. et al. (2000) Prolonged survival and tissue trafficking following adoptive transfer of CD4zeta gene-modified autologous CD4(+) and CD8(+) T cells in human immunodeficiency virus-infected subjects. *Blood* **96**, 785–793.
54. Micklethwaite, K. P., Clancy, L., Sandher, U., Hansen, A. M., Blyth, E., Antonenas, V. et al. (2008) Prophylactic infusion of cytomegalovirus-specific cytotoxic T lymphocytes stimulated with Ad5f35pp65 gene-modified dendritic cells after allogeneic hemopoietic stem cell transplantation. *Blood* **112**, 3974–3981.

Index

Otto-Wilhelm Merten and Mohamed Al-Rubeai (eds.), *Viral Vectors for Gene Therapy: Methods and Protocols*, Methods in Molecular Biology, vol. 737, DOI 10.1007/978-1-61779-095-9, © Springer Science+Business Media, LLC 2011

D

E

F

G

H

Y